Optical Signal Processing

Wiley Series in Pure and Applied Optics

The Wiley Series in Pure and Applied Optics publishes outstanding books in the field of optics. The nature of these books may be basic ("pure" optics) or practical ("applied" optics). The books are directed towards one or more of the following audiences: researchers in universities, government, or industrial laboratories; practitioners of optics in industry; or graduate-level courses in universities. The emphasis is on the quality of the book and its importance to the discipline of optics.

Optical Signal Processing

ANTHONY VANDERLUGT
North Carolina State University
Raleigh, North Carolina

A Wiley-Interscience Publication

John Wiley & Sons, Inc.

New York / Chichester / Brisbane / Toronto / Singapore

Library of Congress Cataloging in Publication Data:

VanderLugt, Anthony, 1937–
Optical signal processing/Anthony VanderLugt.
p. cm. — (Wiley series in pure and applied optics)
"A Wiley-Interscience publication."
Includes bibliographical references and index.
ISBN 0-471-54682-8
1. Optical data processing. 2. Signal processing. I. Title.
II. Series.
TA1632.V28 1992
621.36'7—dc20 91-23378
CIP

Printed and bound in the United States of America by Braun-Brumfield, Inc.

10 9 8 7 6 5 4 3 2 1

For Marilyn, Beth, and Rob;
and for all who would discover that optics is fun.

Preface

This book is an introduction to the theory of optical signal processing with descriptions of selected processing applications. The focus is on processing two-dimensional signals such as images or one-dimensional signals that are functions of time. The processing operations are most often linear integral operations such as spectrum analysis or correlation; some nonlinear adaptive operations are also discussed. This book is intended for use both as a senior–graduate-level textbook and as a reference book for workers in optical signal processing. Familiarity with linear systems and communication theory is helpful to understand the analogies between optics and these subjects. In the first six chapters we concentrate on the fundamentals of optical signal processing; this material, augmented by selected topics in holography or image processing, is suitable for a senior-level undergraduate or first-year graduate course. The remaining chapters are useful for a more advanced course dealing with real-time signal processing.

Chapter 1 provides a brief review of the fundamental characteristics of signals and linear systems. The signals we process, such as images or wideband time signals, are generally analog in nature and do not need to be sampled before processing. Nevertheless, many important characteristics of the signals, as well as the performance of the optical architectures required to process them, are best described in terms of sampling theory. The sample spacing d_0 is used extensively throughout the book to indicate spatial or frequency resolution, and the sample function $\text{sinc}(x/d_0)$ is also our model of a point source. Sampling theory leads to the concept of time bandwidth product, which is a convenient parameter that determines the number of resolvable frequencies in a spectrum analyzer and the correlation gain of a matched-filtering system.

In Chapter 2, we review concepts from geometrical optics useful for analyzing optical signal-processing systems. Although the content of this book deals mostly with physical optics, we often use geometrical optics to more easily determine the position of Fourier and image planes or to determine the scale of light distributions. Traditional topics from geometrical optics, such as the optical invariant, take on new meanings when studied in the light of modern signal processing because the optical invariant is closely related to the space bandwidth product. We show how to recognize the primary lens aberrations and how to arrange experimental systems to minimize them.

In Chapter 3, we discuss the basic elements of physical optics. We discuss interference produced by light waves from two sources and show how this interference phenomena can be extended to N sources. Although the interference produced by N sources is usually called a diffraction pattern, we simplify the concepts and provide new ways to handle diffraction theory by starting from an interference viewpoint. Furthermore, when the N sources are allowed to assume arbitrary magnitudes and phases, they become the sample functions that represent a signal. Bandlimited signals can be represented by a set of weighted sample functions, which we relate to Rayleigh's resolution criterion, or by a set of weighted exponential functions, which we relate to Abbe's resolution criterion.

We place diffraction on a more secure footing by developing the Fresnel-Kirchhoff integral solution to the scalar wave equation. The propagation of light through free space is characterized by a Fresnel transform, which we relate to linear FM or chirp signals. The Fresnel transform, along with the appropriate placement of lenses, leads to the more commonly used Fourier transform, which exists as a complex-valued light distribution at specific planes in an optical system. The Fourier transform is derived initially by using some simplifying assumptions and is illustrated by examples. A more detailed derivation of the Fourier transform covers the wide range of optical configurations under which it is produced. We also develop the conditions for which the capacity and the packing density of the optical system are maximized. We close this chapter with a short discussion of temporal and spatial coherence.

Because it exists as a physical light distribution, the Fourier transform can be measured directly to create a spectrum analyzer, or modified by a spatial filter to implement a generalized convolution operation. In Chapters 4–6 we discuss these major applications of the Fourier transform for processing two-dimensional signals. Spectrum analysis is implemented simply by placing a photodetector array at the plane where the Fourier transform exists. Feature analysis and statistical studies of images are made possible with a fairly simple optical processor consisting of a light source, a spatial light modulator, a photodetector array, and a few lenses. We show how an aperture weighting function controls sidelobe levels so that weak signals can be detected, and we determine the impact that the apperture weighting function has on frequency resolution. Signal-to-noise ratio establishes the single-tone dynamic range, and the nonlinearity of the spatial light modulator establishes the multiple-tone spur-free dynamic range.

In Chapter 5 we discuss methods for implementing spatial filters that modify the response of the system. Because the magnitude and phase can be chosen independently, the usual filter-design rules invoked by the principle of causality do not apply; we can therefore implement processing operations optically that have no direct counterparts in electronic signal processing. We introduce topics from communication and linear system theory as needed to support the development of optical matched filtering. Examples are given for several filter types, leading to the development of the most general type of filter in which amplitude and phase modulation of a spatial carrier frequency encodes the required complex-valued filter, commonly called the VanderLugt filter.

In Chapter 6 we consider some practical aspects of implementing spatial filtering systems, including comments on how to set up a laboratory system, how to establish the proper threshold level, and how to set the optimum reference-beam level. We describe searching methods for the orientation and scale of signals and how the performance of the system is affected by errors in the position of the filter. Because the noise spectral density for images is rarely uniform, we describe a dual frequency-plane processor and a transposed processor for optimum matched filtering. Finally, we illustrate signal-processing operations such as pattern recognition, motion analysis, and photogrammetry with experimental results.

Chapter 7 begins the second part of the book, in which we concentrate on processing wide-bandwidth signals in real time. Because the computational intensity of a real-time processing operation is proportional to the square of the signal bandwidth, we are concerned mostly with processing wideband time signals. An acousto-optic cell is a spatial light modulator that converts a time signal into a function of both space and time; these devices have been developed to a high degree of performance and can accept either analog or digital signals. The connection between the diffracted light angles and the frequency of the input signal is established early in this chapter, followed by the development of input/output relationships for both incoherent and coherent light. We discuss how acousto-optic devices scan a light beam in an information recording system, such as in the raster-scanned format discussed in Chapter 4. This material also provides the basis for understanding how chirp signals are used to provide the distributed local oscillator in the time-integrating spectrum analyzer described in Chapter 14.

In Chapter 8 we show how instantaneous power spectrum analysis is performed; the basic principles are similar to those covered in Chapter 4, but with the interesting difference that light diffracted by the acousto-optic cell retains the temporal frequency content of the input time signal. This feature has several consequences, chief among them is that systems using heterodyne detection can be easily implemented by separating the desired cross-product term from the bias terms.

We introduce the theory of heterodyne optical systems in Chapter 9 by returning to the simple instance of interference produced by two sources with different temporal frequencies and wavefront curvatures. We develop the key conditions for which a heterodyne output is obtained: the light rays must overlap and be parallel. These concepts are illustrated by describing an optical FM radio as a prelude to the development of a heterodyne spectrum analyzer.

The application of heterodyne detection to spectrum analysis is given in Chapter 10. The mixed transform is introduced to determine the relationships between the temporal and spatial frequencies of signal waveforms. Several methods for creating the distributed local oscillator needed for heterodyne detection are described and the required post-detection filtering is established. In Chapter 11 we discuss techniques for reducing the complexity of the postdetection circuitry and show how these techniques are modified to implement cross-spectrum analysis for determining the direction of arrival of an electronic signal. The general theory of heterodyne optical systems is given in Chapter 12 with application to signal excision in the frequency plane and to the implementation of an arbitrary filtering operation.

Correlation is another important signal-processing operation. In Chapter 13 we describe the basics of space-integrating correlators. When we apply correlation to images, implementing the filtering operation in the frequency domain has the advantage that no scanning of the image is needed to locate the signal. In a one-dimensional acousto-optic system the scanning action occurs naturally, because the signal moves through the system; implementing the filtering operation in the space domain with a reference function is therefore preferred. We show that correlation can be performed while simultaneously searching for Doppler if we append a spectrum analyzer to the basic correlator; the correlation function simply appears at a different position at the output plane. Both direct and heterodyne detection techniques are discussed.

In a space-integrating optical system, the frequency resolution and the correlation gain are determined by the time duration of the signal within the acousto-optic cell. When more correlation gain or better frequency resolution is needed, we turn to a class of time-integrating systems described in Chapter 14. We begin by showing how chirp signals are used to create the required distributed local oscillator for a spectrum analyzer whose frequency resolution can be made arbitrarily fine by increasing the integration time. The frequency resolution and frequency range are controlled by the chirp duration. Increased correlation gain is obtained in time-integrating correlators which multiply functions in space and integrate the result over a long period of time on a photodetector array; the correlation function therefore is produced as a function of space.

In Chapter 15 we extend basic two-product operations such as spectrum analysis and correlation to triple-product processors. These systems display the ambiguity function, Wigner-Ville distributions, or a two-dimensional real-time spectrum of a time signal. We also show how this flexible architecture implements adaptive processing algorithms and can be applied to processing signals collected by phased arrays, range/Doppler radars, and related systems.

The problems often require the student to extend his/her knowledge of the subject beyond that contained in the book. As a result, they frequently experience "ah ha's" when they solve the problems. The number of homework problems for each chapter generally reflects the amount of time spent discussing that material. Additional problems are easily generated by inverting the problems or by changing the conditions of the problem. I also frequently ask students to pose and solve a problem that is suitable for homework or an exam. Students report that developing their own problems significantly promotes their understanding of the material.

While writing this book, I recalled with great pleasure conversations with Andrew Bardos during our 15-year association at the Harris Corporation. I value his insights into the phenomena that I happened to be studying and his help in sharpening what were sometimes largely intuitive and occasionally fuzzy notions. I also envy his mentoring skills, ones that I have tried to emulate with my students when I joined the academic community in 1986. Thank you, Andy! In addition, Andy coordinated a review of a draft of this book by some of my former colleagues at Harris. My thanks go to Bill Beaudet, Mike Lange, Mark Koontz, Rob Montgomery, Linda Ralston, and Jerry Wood. I also thank students at North Carolina State University who critiqued drafts of this book. Their demands for more clarity and detail forced me to rethink and embellish several sections in this book.

Bobby Dean Guenther had significant influence in the development of this book. After our first meeting in the early 1970's, we met on an irregular basis for several years. In 1980 we began a closer working relationship and it is largely due to his efforts that I joined the faculty at North Carolina State University. While Bob was writing *Modern Optics*, he encouraged me to write this book. Since 1986 we have had breakfast weekly at Courtney's in Cary, where we discuss a wide range of topics about optics, woodworking, Macintoshes, and world affairs.

I acknowledge the influence that a group of people at the University of Michigan had on my early career. Professor L. J. Cutrona first introduced me to the wonderful world of coherently illuminated optical systems, Emmett Leith shared his knowledge of how optical systems could perform computational algorithms, and Bill Brown mentored me on the finer points of presenting and disseminating scientific results, as well as showing me how to secure research contracts. Many others at the University of Michigan contributed to my understanding of optical signal processing. Of these, Carmen Palermo deserves special thanks. His professional career and mine intertwined at diverse times and places for nearly three decades. His continued interest and commitment to optical signal processing, as well as his continuing friendship, are much appreciated. My thanks also go to Professor H. H. Hopkins for teaching me much of what I know about optics and for his hospitality during our two-year stay at the University of Reading in England.

On a personal note, I thank Taibi Kahler for sharing his keen insights into personality types and for showing me how to identify and meet my needs. His ideas have contributed significantly to improving my quality of life. Marilyn, Beth, and Rob have provided a family support system that I value greatly. I have learned much from Marilyn's personal quest for excellence; thank you for your encouragement, enthusiasm, and love.

ANTHONY VANDERLUGT

Cary, North Carolina
November 1991

Contents

Optical Signal Processing

1

Basic Signal Parameters

1.1. INTRODUCTION

The roots of optical signal processing date back to the work of Fresnel and Fraunhofer nearly 200 years ago. But the connection between optics and information theory did not take shape until the 1950's. In 1953, Norbert Wiener published a paper in the *Journal of the Optical Society of America* entitled "Optics and the Theory of Stochastic Processes" (1). That same issue contained articles by Elias on "Optics and Communication Theory" (2), and by Fellgett on "Concerning Photographic Grain, Signal-to-Noise Ratio, and Information" (3). Other interesting papers of that decade include those written by Linfoot on "Information Theory and Optical Imagery" (4), by Toraldo on "The Capacity of Optical Channels in the Presence of Noise" (5), and the seminal paper by O'Neill on "Spatial Filtering in Optics" (6). These papers represent the early infusion of information theory into classical optics.

A powerful feature of a coherently illuminated optical system is that the Fourier transform of a signal exists in space. As a result, we can implement filtering operations directly in the Fourier domain. This feature was anticipated in papers by Fresnel, Fraunhofer, and Kirchhoff, and had been demonstrated, before the turn of this century, by Abbe in connection with his work on images produced by microscopes.

It is one thing, of course, to recognize that images can be changed by modifying their spectral content; it is another matter to implement the change. In their image-processing work, Marechal (7) and O'Neill (6) used elementary spatial filters to illustrate the principles of optical spatial filtering and to perform mathematical operations such as differentiation and integration. Such was the status of optical signal processing in the early 1960's.

A major impetus to optical signal processing was the need to process data generated by synthetic aperture radar systems. These radar systems were a significant departure from conventional ones because they proved that a small antenna, when used appropriately, provides better resolution than that achieved by a large one. This result, at first glance, is surprising. No physical principles are violated, however, because the small antenna *samples and stores* the radar returns as a means to synthesize a large antenna. To display the radar maps, we need to process the two dimensionally formatted radar returns; because digital computers could not handle the computational load, powerful new signal-processing tools were required.

Photographic film stored the extensive information collected by the radar system. Range information was stored across the film and azimuth information was stored along the film. When the film was illuminated with coherent light, the desired radar map was created by the propagation of light through free space, coupled with the use of some special lenses (see Chapter 5, Section 5.6, for more details). Generating radar maps was the first routine use of optical processing and was the first application for which the matched spatial filter included complicated phase functions such as lenses. It is hard to overestimate the influence that radar processing had on optical signal processing and holography. The classic paper by Cutrona, Leith, Palermo, and Porcello on "Optical Data Processing and Filtering Systems" (8) is important because it presented the basic concepts in a remarkably complete way.

To expand the capabilities of optical filtering to more general operations, such as matched filtering for pattern recognition, we needed to construct filters for which amplitude and phase responses were arbitrary. A solution to the difficult problem of recording the phase information was developed in the early 1960's (9). Because every sample of an input object contributes light to every sample in the matched filter, these two planes are globally interconnected. The computational power of such systems is high because many complex multiplications and additions are performed in parallel. The performance of pattern-recognition systems from that decade has yet to be exceeded.

1.2. CHARACTERIZATION OF A GENERAL SIGNAL

Optical signal processing is based on the same fundamental principles used with other signal-processing technologies. In the remainder of this chapter, we briefly review these fundamentals, primarily to establish the linkage in terminology between spatial and temporal signal processing. We do so without rigor; the reader is invited to consult communication and signal-processing texts for more details. For many readers, the descriptions given here are reminders of what they already know, but possibly never connected with spatial signals. A signal is a signal is a signal. . . .

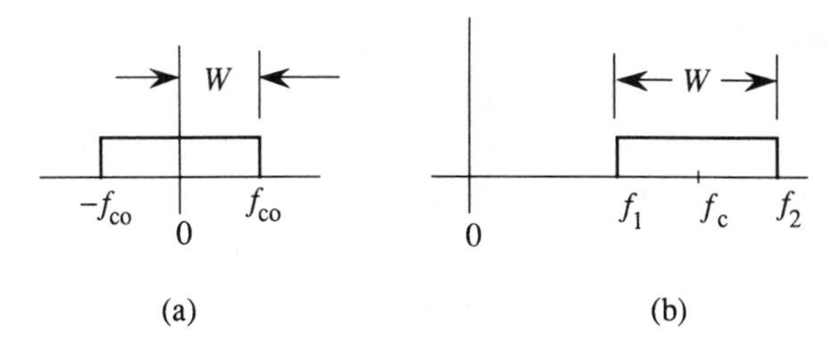

Figure 1.1. Signal spectra: (a) baseband spectrum and (b) bandpass spectrum.

1.2.1. By Bandwidth

One way to characterize a signal $f(t)$ is by its bandwidth. Suppose that all the signal energy is contained in a temporal frequency band W. That is, the signal contains no frequencies higher than $f_{co} = W$, where f_{co} is the *cutoff frequency* and W is the *bandwidth* of the signal. If the two-sided spectrum of a signal occupies the frequency range from $-W$ to W, as illustrated in Figure 1.1(a), it is a *baseband* signal. If the signal has energy only in a band of frequencies $W = f_2 - f_1$, it is a *bandpass* signal, as shown in Figure 1.1(b).

1.2.2. By Time

Signals are generally bounded in time, either because they are generated with a finite time duration, or because we restrict the time duration while processing the signal. For example, we sometimes segment a long-duration signal into shorter segments of duration T for spectrum analysis or correlation. This *time duration* T, as shown in Figure 1.2, is an important signal-processing parameter.

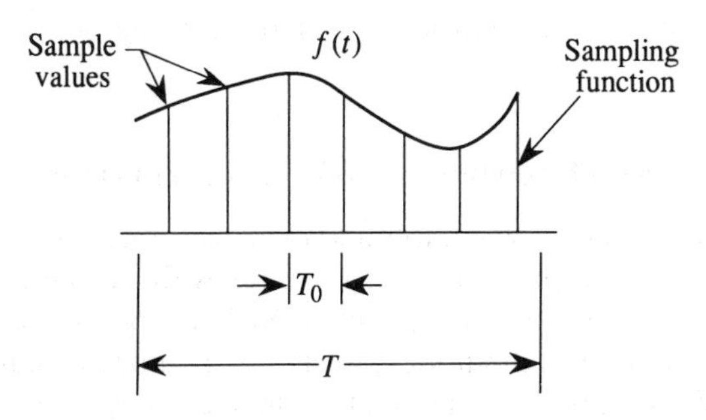

Figure 1.2. Sampling of an analog signal.

1.2.3. By Sample Interval

The sampling theorem states that a signal bandlimited to a frequency range $|f| \leq W$ can be accurately represented by its sample values if the signal is sampled at time intervals T_0, where

$$T_0 = \frac{1}{2f_{co}} = \frac{1}{2W}. \tag{1.1}$$

The signal must be sampled at the *Nyquist sampling rate* $R = 1/T_0$ so that each period of the highest frequency in the signal is sampled twice. Clearly, the sampling rate is just twice the signal bandwidth: $R = 2W$. In Figure 1.2 we show the signal $f(t)$ and a few of the sampling positions spaced at intervals of T_0.

1.2.4. By Number of Samples

If the signal is bounded in time T and in bandwidth W, the total *number of samples* needed to accurately represent the signal is

$$N = \frac{T}{T_0} = 2TW. \tag{1.2}$$

The product TW is generally called the *time bandwidth product* of a signal and is a standard measure of its complexity. For example, a signal with low frequencies and a short time duration contains less information than one with high frequencies and a long time duration. The time bandwidth product is therefore a strong indicator of the computational intensity of a processing operation.

It is not possible, in theory, for a signal to be bounded in both the time and frequency domains. Nevertheless, the assumption is reasonably accurate, in practice, for the purpose of characterizing signals by these two parameters.

1.2.5. By Number of Amplitude Levels or Signal Features

Binary signals have just two amplitude or phase states. We require $n = 2^r$ levels, however, to represent analog signals with an adequate degree of accuracy. For example, we may quantize each sample of an audio signal to at least $r = 16$ bits in amplitude to achieve a sufficient number of signal levels. Two-dimensional signals, such as images, may require n amplitude levels, generally referred to as the *gray scale*. Furthermore, we may require c colors, h hues, s saturation levels, and p polarizations to fully represent the image. We therefore require $g = nchsp$ values to characterize each sample.

1.2.6. By Degrees of Freedom

Because each sample may require g values to determine its state, a signal has gN *degrees of freedom*.

1.3. THE SAMPLE FUNCTION

Although optical signals are generally handled in analog form throughout the system, we introduce the sampling theorem for a variety of reasons. First, the sampling theorem provides a convenient way to characterize the complexity of an analog signal, such as an image, in terms of the required number of samples, generally called *pixels*. Optical images, in contrast to time signals, do not have a fixed underlying metric. For example, images can be magnified or demagnified, thereby changing their areas and their spatial frequency bandwidths, but the number of samples remains the same. Second, the Fourier transform of an optical signal exists in space and is often detected directly by a photodetector array. Because the elements of the array sample the spectrum, we need to understand how the sampling process affects the accuracy with which the spectrum is measured. Third, some images are originally sampled by collection devices, such as solid-state cameras, which contain two-dimensional photodetector arrays. In these cases, the input signals to the optical system have already been sampled and we must account for the impact of sampling on subsequent processing operations. Fourth, we frequently use point sources in optical systems. Throughout this book, we treat a point source as a bandlimited signal containing exactly one sample. Finally, the sampling theorem, when associated with bandlimited signals whose duration in time or space is finite, leads to the optical invariant that is important for the analysis and design of optical systems.

Consider a signal $f(t)$, whose spectrum is $F(f)$, bandlimited to the range $|f| \leq W$. This signal, when sampled by an infinite train of delta functions, is represented by

$$f_s(t) = f(t) \sum_{n=-\infty}^{\infty} \frac{1}{T_0} \delta(t - nT_0), \tag{1.3}$$

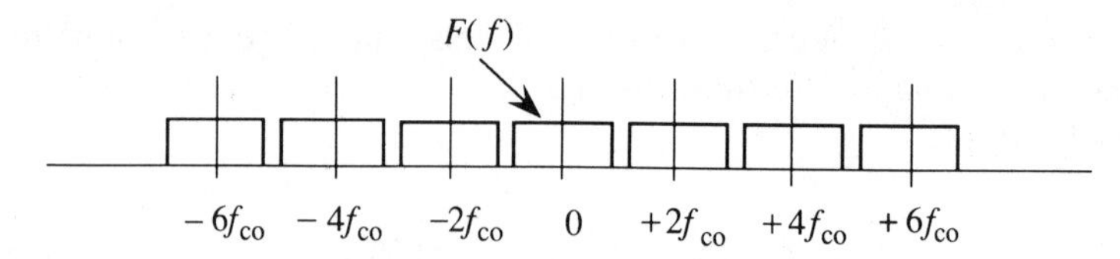

Figure 1.3. Spectrum of the sampled analog signal.

so that the spectrum of the sampled signal becomes

$$F_s(f) = F(f) * \sum_{n=-\infty}^{\infty} \delta(f - 2nf_{co}), \tag{1.4}$$

where $*$ indicates convolution, and the factor $1/T_0$ in Equation (1.3) ensures conservation of the integrated area between the function $F(f)$ and its sampled representation $F_s(f)$. The spectrum $F_s(f)$ is shown in Figure 1.3 with the baseband spectrum $F(f)$ centered at $f = 0$; replicas of $F(f)$ are centered at $f = \pm 2f_{co}$, $\pm\, 4f_{co}$ and other even multiples of the cutoff frequency.

The signal $f(t)$ can be recovered from the sampled signal $f_s(t)$ by passing $f_s(t)$ through a low-pass filter whose frequency response $\text{rect}(f/2W)$ is equal to one for $|f| \leq W$ and equal to zero elsewhere, as shown in Figure 1.4. The filter, whose impulse response is $h(t)$, admits only the central spectral components $F(f)$ from the sampled signal

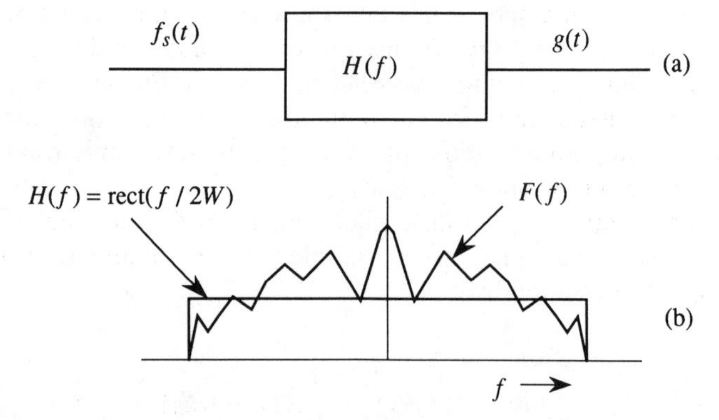

Figure 1.4. Filtering operation and spectrum of a bandlimited signal.

spectrum $F_s(f)$. The impulse response of the filter is

$$\begin{aligned} h(t) &= \int_{-\infty}^{\infty} H(f) e^{j2\pi ft}\, df \\ &= \int_{-\infty}^{\infty} \text{rect}(f/2W) e^{j2\pi ft}\, df \\ &= \int_{-W}^{W} e^{j2\pi ft}\, df \\ &= (2W)\text{sinc}(2Wt) = (1/T_0)\,\text{sinc}(t/T_0), \end{aligned} \tag{1.5}$$

where $2W = 1/T_0$ is a scaling factor similar to that used in Equation (1.3), and where

$$\text{sinc}(x) \equiv \frac{\sin(\pi x)}{\pi x}. \tag{1.6}$$

The sinc function, as defined by Equation (1.6) and shown in Figure 1.5, is generally called the *interpolation function* because it allows an interpolation between the sample intervals T_0 to recover $f(t)$ for all time.

As we noted before, concepts associated with the sampling theorem help in understanding optical signal processing, even if the signals are in an analog form. To cover the wide range of uses discussed at the beginning of this section, we refer to the interpolation function as the *sample function*. We therefore distinguish among the sampling function which is the train of delta functions shown in Figure 1.2, the sample function which is the sinc function wrapped around each delta function as shown in Figure 1.5, and the sampled values which are the values of the sampled signal $f_s(t)$.

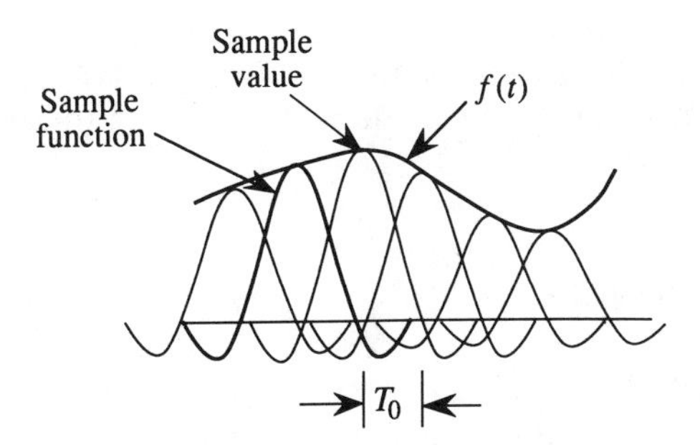

Figure 1.5. Sample function as an interpolator.

1.4. EXAMPLES OF SIGNALS

The bandwidths of commonly encountered signals vary over a considerable range:

• Underwater sound	4 Hz
• Speech (phone quality)	3,000 Hz
• Audio range	20,000 Hz
• Color television	6,000,000 Hz
• Wideband communication system	100,000,000 Hz
• Color motion picture	5,000,000,000 Hz
• Visible spectrum (blue to red)	10,000,000,000,000 Hz

The bandwidth of the visible spectrum is extremely wide relative to that of any of the listed signals, implying enormous signal-processing potential. This wide bandwidth is, in part, why optics is useful for performing communication and signal-processing functions.

The complexity of an analog signal-processing operation is typically proportional to $N = 2TW$. For example, if we compute the Fourier transform of each of the listed signals for a one-second interval, the complexity becomes significantly higher as we progress down the list. For example, the audio signal requires a 40,000-point FFT to determine the frequency content of the signal to a resolution of 1 Hz or a 4,000-point FFT to resolve frequencies to 10 Hz. For *real-time* operation the computational complexity is generally very high. Correlation, for example, requires that we perform $2TW$ multiplications and additions in the time interval T_0 so that the computational rate, in operations per second, is

$$R = \frac{2TW}{T_0} = 4TW^2, \tag{1.7}$$

which increases as the *square* of the bandwidth. Hence, it is difficult to process wideband signals in real time without using the power of optical processing.

1.5. SPATIAL SIGNALS

The signals cited above are all one-dimensional time signals. Pictorial signals, however, generally originate as three-dimensional signals; for example, a motion picture is a function of two space variables and a time

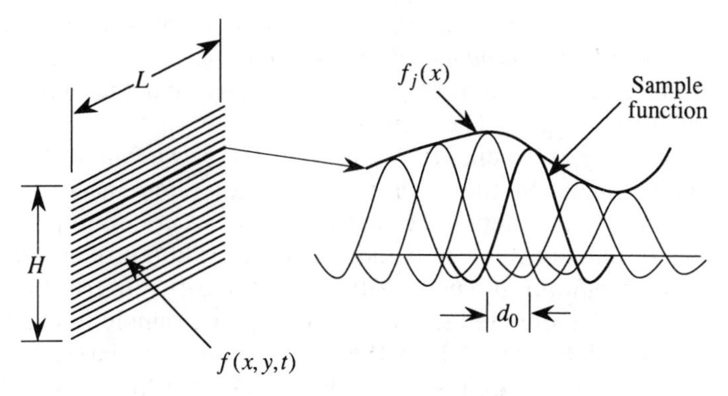

Figure 1.6. Raster-scanned image.

variable. For transmitting such images, we convert the three-dimensional signal $f(x, y, t)$ to a one-dimensional time signal $r(t)$ through a raster-scanning operation, which is easily visualized as a sampling operation in both spatial coordinates. A television receiver converts $r(t)$ to the same format as the original signal $f(x, y, t)$. A holographic or stereographic movie is an example of a four-dimensional signal $f(x, y, z, t)$.

We use spatial frequencies to describe spatial signals such as images, similar to our use of temporal frequencies for describing time signals. We encounter spatial frequencies every day, but may not recognize them as such. For example, the regular pattern of lines in a screen, a checkerboard, a piece of cloth, or a cornfield form spatial frequencies in different spatial directions. Irregular spatial frequencies are evident in water surface waves and most street patterns.

Figure 1.6 shows a raster-scanned signal $f(x, y, t)$ such as that produced by a single frame of a television signal. The signal has length L, height H, and exists for a time duration T_f. To illustrate the fundamental concepts, we select the jth line and display it as $f_j(x)$. This signal, whose highest *spatial frequency* is α_{co}, requires a *sampling interval* d_0, where

$$d_0 = \frac{1}{2\alpha_{co}}. \tag{1.8}$$

In general, we use α and β to indicate spatial frequencies, expressed in cycles/millimeter, in the x and y directions; they are the optical equivalents of a temporal frequency f. In turn, the sample interval d_0 is equivalent to the temporal sample interval T_0.

The number of samples in the x direction is $N_x = L/d_0 = 2\alpha_{co}L$. The *length bandwidth product* of the spatial signal in the x direction is the product of the length of the spatial signal and the cutoff spatial frequency: LBP $= \alpha_{co}L$. The number of samples in the y direction is $N_y = H/d_0 = 2\beta_{co}H$, where HBP $= \beta_{co}H$ is the *height bandwidth product* of the spatial signal in the y direction. We therefore need a total of $N = N_xN_y$ samples, sometimes called pixels, to accurately represent the image. The product of the length and height bandwidth products gives the overall *space bandwidth product*: SBP = (LBP)(HBP) $= \alpha_{co}\beta_{co}LH$.

If $f(x, y)$ is a bandlimited signal, the appropriate sample function is the product $\text{sinc}(x/d_0)\text{sinc}(y/d_0)$, in the same sense that $\text{sinc}(t/T_0)$ is the appropriate sample function for time signals. We can therefore represent the signal on the jth scan line as

$$f_j(x) = \sum_{n=-\infty}^{\infty} a_n \, \text{sinc}\left[\frac{x - nd_0}{d_0}\right] \text{rect}\left(\frac{x}{L}\right), \tag{1.9}$$

where the real-valued a_n are the sampled values, $\text{sinc}(x/d_0)$ is the sample function, and the rect function defines the region occupied by the signal. An alternative and completely equivalent way to represent $f_j(x)$ is by using exponential functions:

$$f_j(x) = \sum_{n=-\infty}^{\infty} b_n e^{j2\pi n\alpha_0 x} \, \text{rect}(n\alpha_0/2\alpha_{co}), \tag{1.10}$$

where the b_n are complex-valued weights for the exponential functions, $\alpha_0 = 1/L$ is the lowest, or fundamental spatial frequency contained in the signal, and the rect function shows that the signal is bandlimited by the cutoff frequency α_{co}. Either Equation (1.9) or Equation (1.10) is an appropriate way to represent the signal; which is most useful depends on the particular signal or signal-processing application.

PROBLEMS

1.1. A baseband time signal has a bandwidth of 6 MHz. What is the proper sampling interval T_0? If each sample is characterized by one of 32 voltage levels, calculate the degrees of freedom if the signal duration is 100 ms.

1.2. We record a 6-MHz bandwidth baseband video signal onto photographic film using a raster-scanned format. If the film resolution is $2\ \mu \times 2\ \mu$ (equivalent to the spatial sampling interval), calculate the area required to store 30 minutes of information. Assume a square recording format and that the film can support the required 32 information levels at each sample position.

1.3. For the parameters given in Problem 1.2, what is the highest spatial frequency required of the film? Calculate the two-dimensional space bandwidth product of the stored signal and the time bandwidth product of the time signal.

1.4. Find the required sample interval and the optimum sampling function in the frequency domain if the signal is limited to a time duration from $|t| \leq T$. The bandwidth of the signal is not constrained in any way. Hint: This problem is the dual of the sampling theorem in the space or time domains.

2

Geometrical Optics

2.1. INTRODUCTION

The main emphasis in this book is on physical optics, which describes how light interacts to produce diffraction effects useful in optical signal processing. Although most of the results in this chapter can be obtained from physical optics, we first provide a working knowledge of geometrical optics because it often provides the same results through more straightforward calculations. Geometrical optics is the characterization of optical systems based on an assumption that the wavelength of light is zero and that light travels only along ray paths.

As it turns out, we cannot isolate a single ray. If we attempt to do so, we find that the harder we try, the more difficult it becomes. In both Figure 2.1(a) and Figure 2.1(b) we successfully introduce apertures that reduce the spatial extent of the incident light beam. Figure 2.1(c), however, shows that a further reduction in the size of the aperture does not isolate a ray; in fact, light actually diverges after the aperture. The finite wavelength of light causes this spreading action or *diffraction*, as discussed extensively in Chapter 3, and is the foundation on which much of optical signal processing is built. In this chapter we proceed as though we can actually isolate a ray.

2.2. REFRACTIVE INDEX AND OPTICAL PATH

The velocity c of light in vacuum is approximately $3(10^8)$ m/sec. The velocity v of light in any other medium, however, is lower than c. The inverse relative velocity $n = c/v$ is the *refractive index* of the medium for monochromatic light of wavelength λ. Frequency, wavelength, and velocity are connected by the relationship $v = \lambda f$. Because the frequency of light remains unchanged in passing from a medium whose refractive index is n_1

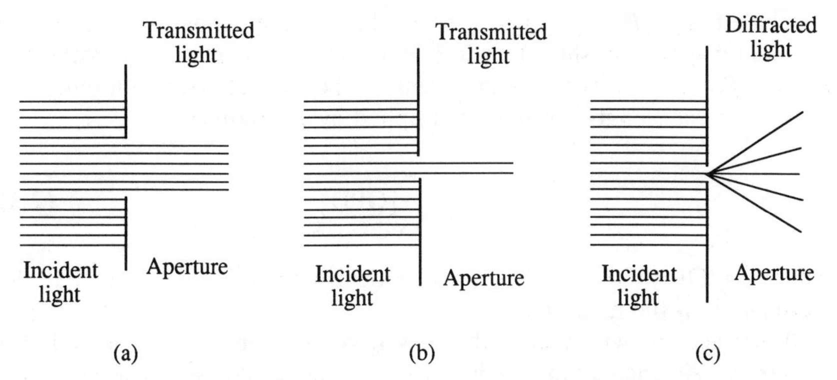

Figure 2.1. Effect of an aperture on a bundle of rays of light: (a) large aperture, (b) small aperture, (c) pinhole aperture.

into a medium of index n_2, we find that

$$v_1 = \lambda_1 f,$$
$$v_2 = \lambda_2 f, \tag{2.1}$$

from which we conclude that

$$n_1 \lambda_1 = n_2 \lambda_2, \tag{2.2}$$

so that the wavelength shortens when light passes into a medium with a higher refractive index, often called a more *dense* medium.

The index of refraction is generally a function of the wavelength of light. The *dispersive power* of a medium is defined as $(n_b - n_r)/(n_y - 1)$, where n_b is the refractive index for blue light, n_r is the refractive index for red light, and n_y is the refractive index for yellow light. This second-order property of the index of refraction is generally not important for us because we mostly study optical signal-processing systems for which light is monochromatic. We therefore use λ and n, without color-referenced subscripts, to indicate wavelength and refractive index.

A distance, multiplied by the appropriate refractive index, is defined as the *optical path*. By convention, we use brackets to indicate an optical path [OP] or an optical path difference [OPD]. In a time interval t_0, a light disturbance always traverses the same optical path length. For example, consider light entering media of refractive indices n_1 and n_2. The distances traveled in time t_0 are $D_1 = v_1 t_0$ and $D_2 = v_2 t_0$, from which we conclude that $n_1 D_1 = n_2 D_2 =$ [OP] so that the optical paths are equal. In a similar way we can show that if light arrives at a point via two paths, the *phase difference* between them is simply the optical path difference expressed in wavelengths of light, multiplied by 2π radians:

$$\Delta\phi = \frac{2\pi}{\lambda}[\mathrm{OPD}], \tag{2.3}$$

where $[\mathrm{OPD}] = [\mathrm{OP}_2 - \mathrm{OP}_1]$ is the optical path difference and λ is the wavelength in the medium.

A surface on which all light rays have the same phase is called a *wavefront*. At each point on an arbitrary wavefront we construct *wave normals* as suggested in Figure 2.2. As an example, if the medium is *isotropic*, which means that the index of refraction is the same in all directions, a point source emits light into expanding spherical wavefronts. In this case, the times of flight from the source to every point on the wavefront surface are equal. Furthermore, all the wavefront normals are rays that have the point source as their common origin. The use of rays is most convenient for the study of geometrical optics, whereas we use wavefronts to develop the theories of interference, diffraction, and lens aberrations.

Energy is transported along ray paths; the rays, in an isotropic medium, are normal to the wavefront. In an *anisotropic* medium, for which the indices of refraction are not the same in all directions, the rays still define the directions in which energy propagates; they are not, however, generally normal to the wavefront.

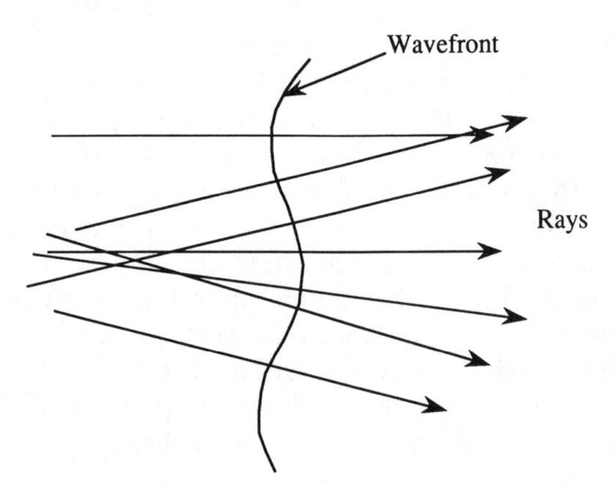

Figure 2.2. Rays are surface normals to wavefronts.

2.3. BASIC LAWS OF GEOMETRICAL OPTICS

The basic laws of geometrical optics are the laws of reflection and refraction, Snell's law, and Fermat's principle. These laws provide the basic tools for tracing rays through a system consisting of optical elements such as mirrors, prisms, and lenses.

2.3.1. Law of Reflection

We first examine how a wavefront changes when it is *reflected* at a surface. Consider a plane wave AB, incident on the reflective surface S shown in Figure 2.3. The refractive index of the medium is n_1. An incident ray, associated with the wavefront at A, arrives at the *angle of incidence* I_1 with respect to the surface normal. Note, too, that the incident wavefront AB forms an angle I_1 with respect to the surface S. At some time t_0, the wavefront has advanced so that it has just arrived at point C. The time of flight is

$$t_0 = \frac{BC}{v_1} = \frac{BC}{c/n_1} = \frac{n_1 BC}{c} = \frac{\text{optical path}}{\text{velocity of light in vacuum}}. \tag{2.4}$$

When t_0 seconds have elapsed, we know that the energy arriving at point A has been reflected, but we do not know the direction. To find the direction, we construct a circle of radius $n_2 AD = n_1 BC$, with center at A,

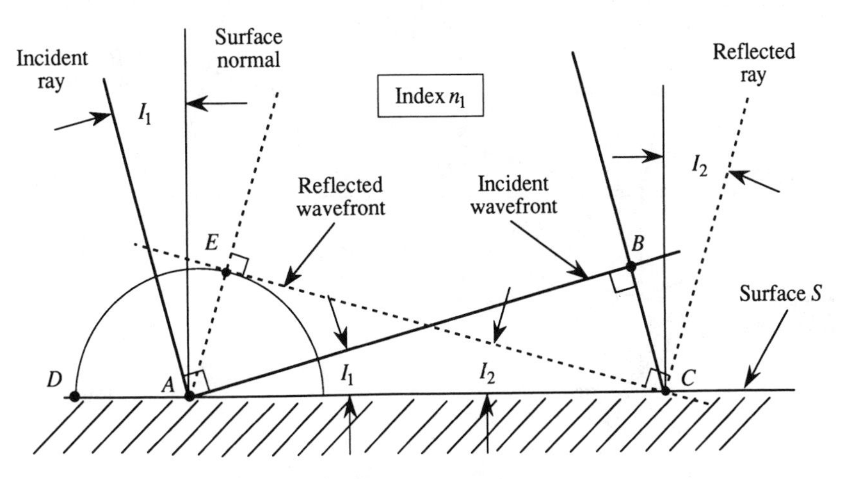

Figure 2.3. Law of Reflection.

where $n_2 = n_1$ refers to the index of refraction after the wavefront is reflected. Because the reflected wavefront must be tangent to this circle, the angle that EC makes with respect to the surface is the *angle of reflection* I_2, and the optical path $[AE]$ is equal to $n_2 AD = n_1 BC$. As the incident and reflected rays are in the same medium, we use the fact that $n_1 = n_2$ to obtain the result that

$$|I_1| = |I_2|. \tag{2.5}$$

The absolute value signs used in Equation (2.5) allow for the possibility that I_2 is equal to $-I_1$; we address the sign conventions for angles in Section 2.5.1. The *law of reflection* states that the magnitude of the angle of reflection is equal to that of the angle of incidence. The incident and reflected rays are therefore equally inclined relative to the normal of the reflecting surface. They are also in the same plane; that is, the reflected ray is in the plane defined by the incident ray and the surface normal.

2.3.2. Law of Refraction

When light passes through a surface separating media of different refractive indices, rays are *refracted* so that they change directions. The development of the law of refraction is similar to that for the law of reflection. A wavefront AB arrives at a surface S shown in Figure 2.4. The refractive

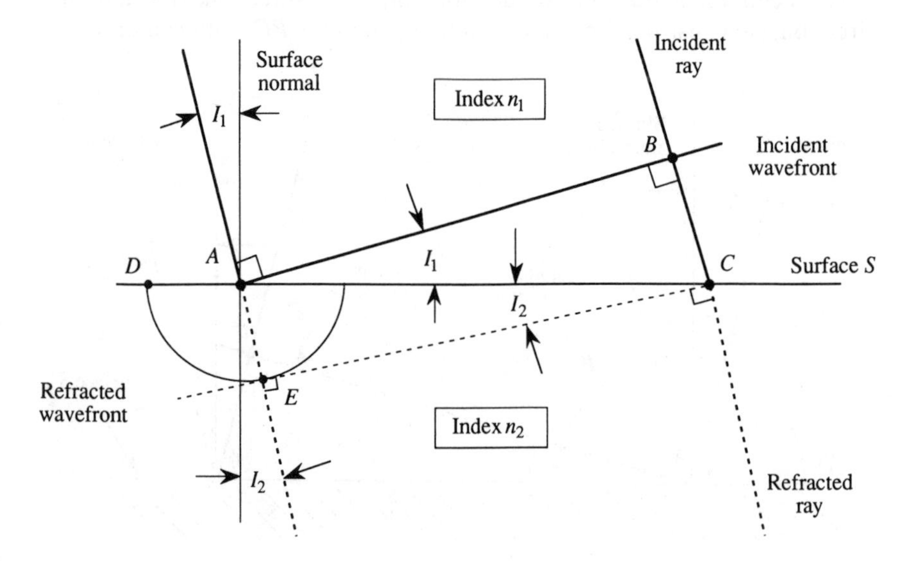

Figure 2.4. Law of Refraction.

indices of the two media are n_1 and n_2. At time $t_0 = BC/v_1$, the wavefront has just arrived at point C. During this same time interval, the wavefront element entering at A, with incidence angle I_1 to the surface normal, has traveled an optical path $n_2 AD = n_1 BC$. The new wavefront must be tangent to the circle of radius $n_2 AD$. From the diagram we find that

$$\sin I_1 = \frac{BC}{AC} \quad \text{and} \quad \sin I_2 = \frac{AD}{AC}, \tag{2.6}$$

or that

$$AD \sin I_1 = BC \sin I_2. \tag{2.7}$$

Because equal optical paths are traveled in equal time, we find that $n_1 BC = n_2 AD$ and that

$$\boxed{n_1 \sin I_1 = n_2 \sin I_2,} \tag{2.8}$$

which relates the angle of incidence to the angle of refraction. When $n_2 > n_1$ we find that $I_2 < I_1$. Thus, in passing into a denser medium, a ray is bent toward the surface normal. The *law of refraction*, as given in Equation (2.8), is also known as *Snell's law* and is the foundation on which geometrical optics is based.

2.3.3. Fermat's Principle

The laws governing the behavior of rays are combined in Fermat's principle. *Fermat's principle* states that the time of flight for a light packet traveling from one point to another along a ray is, to a first approximation, equal to the time of flight experienced by light packets on nearby rays; that is, *the time of flight has a stationary value*. Fermat originally stated that the time of flight, and therefore the optical path, is a *minimum*; however, the actual time of flight may be a maximum, a minimum, or neither, as we show shortly.

Consider the path of a ray in Figure 2.5 from point P_1 in a medium of index n_1 to the point P_2 in a medium of index n_2, where $n_1 < n_2$. The intersection point at the surface is found when the path length, or time of flight, has a *stationary value*. The total time to travel from P_1 to P_2 is

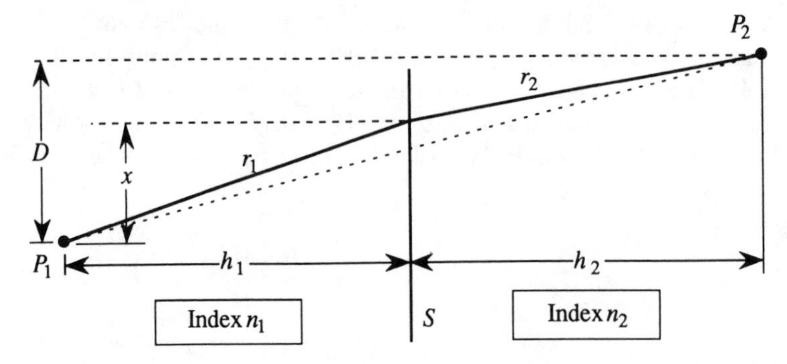

Figure 2.5. Fermat's principle.

$$t_{12} = t_1 + t_2 = \frac{r_1}{v_1} + \frac{r_2}{v_2}$$

$$= \frac{\sqrt{h_1^2 + x^2}}{v_1} + \frac{\sqrt{h_2^2 + (D-x)^2}}{v_2}. \tag{2.9}$$

When t_{12} is at a stationary value, the partial derivative $\partial t_{12}/\partial x$ is zero:

$$\frac{\partial t_{12}}{\partial x} = \frac{\frac{1}{2}(2x)}{v_1\sqrt{h_1^2 + x^2}} + \frac{\frac{1}{2}(-2)(D-x)}{v_2\sqrt{h_2^2 + (D-x)^2}} = 0, \tag{2.10}$$

from which we conclude that

$$\frac{n_1 x}{r_1} = \frac{n_2(D-x)}{r_2}, \tag{2.11}$$

which implies that

$$n_1 \sin I_1 = n_2 \sin I_2. \tag{2.12}$$

Snell's law, as shown by Equation (2.12), is therefore implicitly contained in Fermat's principle.

Figure 2.6 shows several examples in which the ray paths are at their stationary values, not necessarily at their minimum or maximum values. In each case, we consider light traveling from point P_1 to point P_2. The true

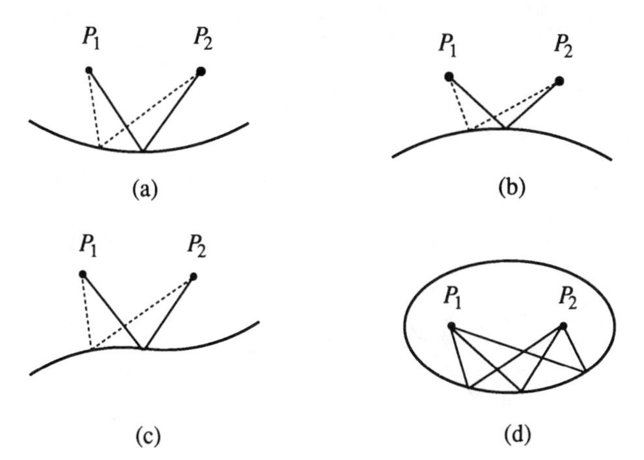

Figure 2.6. Ray Paths: (a) maximum length path, (b) minimum length path, (c) inflection length path, and (d) equal length path.

ray paths are shown as solid lines, and candidate ray paths are shown as dotted lines. Aside from the obvious fact that the direct path from P_1 to P_2 is a minimum in all cases, we are concerned with the paths of rays that reflect from the surfaces. The reflected path is a *maximum* in Figure 2.6(a); the path is a *minimum* in Figure 2.6(b); the path is at an *inflection point* in Figure 2.6(c); the paths are all *equal* in Figure 2.6(d), because P_1 and P_2 are the foci of an ellipsoid.

2.3.4. The Critical Angle

When a ray passes from a less dense medium into one that is more dense, the ray is bent *toward* the normal. When the ray propagates in the opposite direction, the ray bends *away* from the normal. Figure 2.7 shows that ray *AB*, in a medium of index n_1, has an angle of incidence I_1 with respect to the surface normal. When I_2 reaches 90°, Snell's law shows that ray *AB* cannot enter the less dense medium and is completely reflected at the surface. The angle at which *total internal reflection* occurs is called the *critical angle* I_c:

$$I_c = \sin^{-1}\left(\frac{n_2}{n_1}\right). \tag{2.13}$$

As an example, suppose that $n_1 = 1.5$ and $n_2 = 1.0$; the critical angle I_c for which total internal reflection occurs is then $I_c = \sin^{-1}(1/1.5) = 41.8°$.

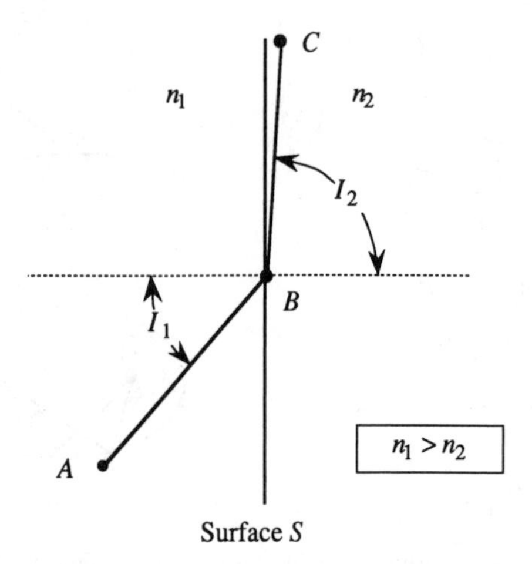

Figure 2.7. Critical angle.

Although Equation (2.13) shows that rays whose incidence angles are beyond the critical angle I_c cannot penetrate into the less dense medium, Snell's law shows that rays can always penetrate into a more dense material. Total internal reflection is found in many common optical systems, such as in the prisms used in cameras and binoculars.

2.4. REFRACTION BY PRISMS

Prisms are sometimes used in optical signal-processing systems to bend a set of parallel rays, perhaps for folding the system to make it more compact. Sometimes the magnification of a prism is used to change the size of a light beam, such as that produced by an injection laser diode, in only one dimension.

2.4.1. Minimum Deviation Angle

A symmetric prism, shown in Figure 2.8, is constructed from a material whose index of refraction is n_2; the medium on either side of the prism is air, with index $n_1 = n_3 = 1$. The entrance ray forms the angle I_1 with respect to the normal. We apply Snell's law to the incident and refracted rays at each surface of the prism:

$$\begin{aligned} n_1 \sin I_1 &= n_2 \sin I_2 \\ n_2 \sin I_3 &= n_3 \sin I_4. \end{aligned} \tag{2.14}$$

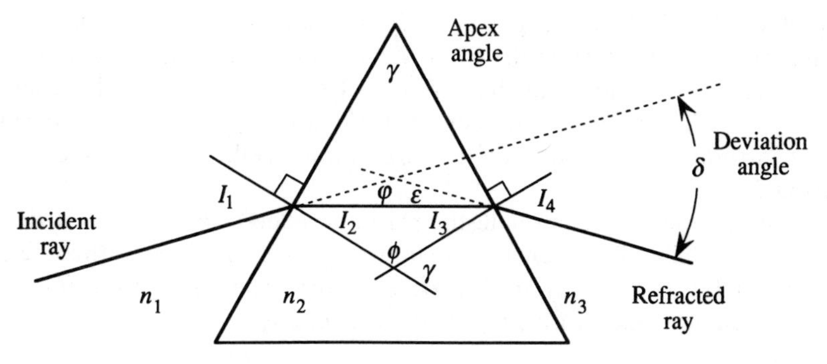

Figure 2.8. Refraction of a ray by a prism.

From Figure 2.8 we see that $I_2 + I_3 = \gamma$, where γ is the apex angle of the prism.

The *deviation angle* δ is the amount by which the entrance ray is bent in passing through the prism. From the diagram, we find that the deviation angle is

$$\begin{aligned}\delta &= \varphi + \varepsilon \\ &= I_1 - I_2 + I_4 - I_3 \\ &= I_1 - I_2 + I_4 - (\gamma - I_2) \\ &= I_1 + I_4 - \gamma. \end{aligned} \tag{2.15}$$

The deviation angle is therefore not a direct function of the internal angles I_2 and I_3. We use Equation (2.14) in Equation (2.15) to find that

$$\begin{aligned}\delta &= I_1 - \gamma + \sin^{-1}[n_2 \sin I_3] \\ &= I_1 - \gamma + \sin^{-1}[n_2 \sin(\gamma - I_2)] \\ &= I_1 - \gamma + \sin^{-1}\left[n_2 \sin\left\{\gamma - \sin^{-1}\left(\frac{\sin I_1}{n_2}\right)\right\}\right]. \end{aligned} \tag{2.16}$$

The result given by Equation (2.16) is valid for all angles of incidence, provided that the ray striking the second surface is not at, or beyond, the critical angle. The condition for avoiding total internal reflection is that

$$n_2 \sin\left\{\gamma - \sin^{-1}\left(\frac{\sin I_1}{n_2}\right)\right\} < 1. \tag{2.17}$$

A plot of the deviation angle δ, as a function of I_1, reveals that the *minimum deviation angle* occurs when $I_1 = I_4$ so that the entrance and exit rays make equal angles with the surfaces of the prism. Under this condition $\delta_{\min} = 2I_1 - \gamma$, as can be seen from Equation (2.15), and the aberrations introduced by the prism are minimized, as we show in Section 2.9.

When the apex angle is small, the prism is called a *thin prism*. Furthermore, if the angle of incidence is small, we replace the sines of the angles by the angles themselves in Equation (2.16). The expression for the deviation angle is therefore simplified considerably:

$$\boxed{\delta = (n_2 - 1)\gamma,} \tag{2.18}$$

and we see that the deviation angle is not a function of the angle of incidence.

We sometimes refer to the *power* of a prism defined as the deflection of a ray, in millimeters, at a distance of one meter; the unit of measurement is a *prism diopter*. A prism whose power is one prism diopter deflects a ray 10 mm at a distance of 1 m from the prism.

2.4.2. Dispersion by a Prism

The refractive index of an optically transparent material is dependent on the wavelength of light. As the index is typically higher for blue than for red wavelengths, blue wavelengths are bent more toward the normal within the prism and are bent even further away from the normal when they exit the prism, a phenomenon called *dispersion*. Dispersion explains why white light separates into its color spectrum, as reported by Newton. Also, since the dispersion is generally not linear ($\partial^2 n/\partial\lambda^2 \neq 0$), the spectrum is typically spread more in the blue region than in the red region.

2.4.3. Beam Magnification by a Prism

Prisms can be configured so that the magnification is different in orthogonal directions; this condition is called *anamorphic magnification*. Consider the prism shown in Figure 2.9, for which light is incident normal to the entrance face. As there is no ray deviation at the first surface, we apply Snell's law immediately to the second surface:

$$n_2 \sin I_2 = n_3 \sin I_3. \tag{2.19}$$

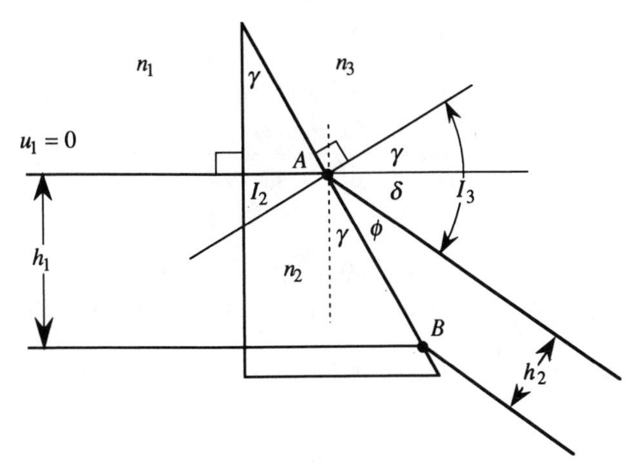

Figure 2.9. Beam magnification by a prism.

As $n_3 = 1$ and $I_2 = \gamma$, we find that $\sin I_3 = n_2 \sin \gamma$. Thus, we find that

$$\begin{aligned} I_3 &= \sin^{-1}[n_2 \sin \gamma], \\ &= \delta + \gamma, \end{aligned} \tag{2.20}$$

from which we conclude that the deviation angle is

$$\delta = \sin^{-1}[n_2 \sin \gamma] - \gamma. \tag{2.21}$$

The result given by Equation (2.21) is valid only when $n_2 \sin \gamma < 1$ so that internal reflection at the second surface is avoided.

The beam magnification of the prism is $M = h_2/h_1$. From Figure 2.9 we see that

$$\begin{aligned} h_1 &= AB \cos \gamma \\ h_2 &= AB \sin \phi, \end{aligned} \tag{2.22}$$

so that the beam magnification is $M = \sin \phi / \cos \gamma$. Because $\phi = 90° - \gamma - \delta$, we find that

$$M = \frac{\cos[\sin^{-1}(n_2 \sin \gamma)]}{\cos \gamma}. \tag{2.23}$$

After using some trigonometric substitutions and algebraic manipulations, we find an alternative form for the magnification:

$$M = \frac{\sqrt{1 - n_2^2 \sin^2 \gamma}}{\sqrt{1 - \sin^2 \gamma}}, \tag{2.24}$$

valid for $n_2 \sin \gamma < 1$ so that total internal reflection is avoided. The magnification is always less than 1 when light travels from left to right, as shown in Figure 2.9.

We solve Equation (2.24) to find the required apex angle γ in terms of M and n_2:

$$\sin \gamma = \sqrt{\frac{1 - M^2}{n_2^2 - M^2}}. \tag{2.25}$$

Large changes in the beam size are obtained with prisms of modest parameters, as shown in the following table. We calculate the exit angle I_3 with respect to the surface normal, as a function of increasing the apex angle of the prism; in all cases the refractive index is $n_2 = 1.70$:

γ	I_3	M
30	58.2	0.61
35	77.2	0.27
36	87.7	0.05

Because the magnification changes rapidly as $\gamma \to 36°$, it is a sensitive function of manufacturing errors in the apex angle or of errors in the refractive index of the prism. Because the exit angle is close to grazing at the exit face when the magnification is small, we risk total internal reflection; we reduce this risk by rotating the prism slightly toward the minimum deviation angle. In general it is better to use two or more prisms in series when a large change in the beam size is needed to avoid the most sensitive operating configuration for any one prism.

Magnifications greater than 1 are most easily handled by initially assuming that the magnification is less than 1. We use the relationships developed here and then reverse the prism configuration relative to the direction in which light travels. Although the relationships developed in this section are valid only when light enters normal to the entrance face of the prism, similar relationships can be developed for arbitrary incidence angles.

2.4.4. Counter-Rotating Prisms

An equivalent prism having a variable deviation angle is implemented by using two prisms, each with an apex angle γ, as shown in Figure 2.10. When the prisms are rotated so that their powers add, as shown in Figure 2.10(a), the equivalent deviation angle is 2δ. When the prisms are counter rotated so that their powers subtract, as shown in Figure 2.10(b), the equivalent deviation angle is zero. A pair of such prisms is therefore useful to bend light through an arbitrary angle ranging from zero to 2δ. The prisms are generally mounted so that the second prism is rotated relative to the first to obtain the desired deviation angle. The entire prism assembly is then rotated so that the deviation occurs in the desired plane.

2.4.5. The Wobble Plate

A parallel plate of glass can be used to slightly displace a beam of light. Figure 2.11 shows a glass plate of thickness z_{12} and index of refraction n_2. By straightforward calculations (see Problem 2.5), the displacement for small angles of incidence is

$$h = \frac{z_{12} I_1 (n_2 - 1)}{n_2}. \tag{2.26}$$

Fine control over ray displacements is obtained by such a *wobble plate*, without deviating the angles of the incident rays.

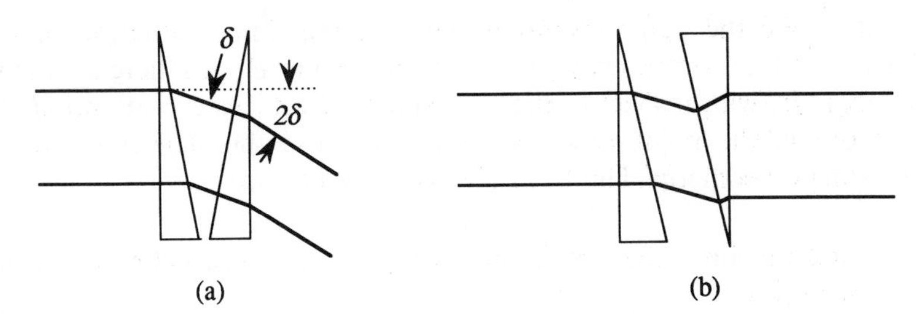

Figure 2.10. Counter-rotating prisms: (a) prism powers adding and (b) prism powers subtracting.

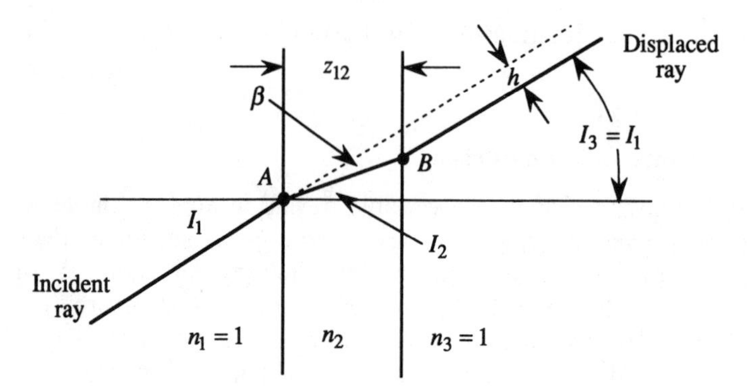

Figure 2.11. Displacement of a ray by a plane parallel plate.

2.5. THE LENS FORMULAS

Snell's law is the basic method for tracing rays through optical elements with curved surfaces. To obtain the key relationships we must first establish a sign convention. Unfortunately, no sign convention is used uniformly in all optics texts. The sign convention can be chosen freely, however, provided that it is used consistently throughout all calculations.

We want our sign convention to be consistent with our notion of positive and negative *temporal* frequencies, as well as with our notion of positive and negative *spatial* frequencies. The sign convention also influences the sign of the kernel function in the temporal and spatial Fourier-transform relationships. As a result of these requirements, we establish a sign convention that unifies the results from geometrical optics, physical optics, and Fourier-transform theory.

2.5.1. The Sign Convention

To illustrate the sign convention, consider the simple situation shown in Figure 2.12, in which a ray passes through plane P_0 at a height h_0 and at an angle u_0 with respect to the optical axis. We focus our attention at the first origin, the point O at plane P_0, and use the sign conventions of coordinate geometry. The basic sign conventions are

- Heights above the optical axis are positive; those below the axis are negative.
- Distances to the right of the *current origin* are positive; distances to the left of the current origin are negative.

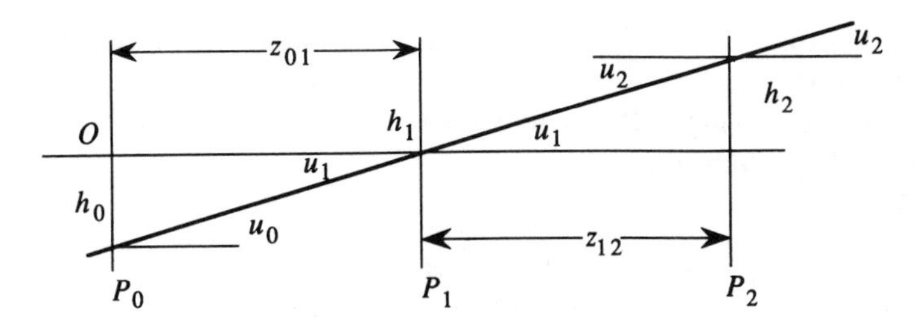

Figure 2.12. Sign convention for rays and distances.

- The *acute* angle that a ray makes with the axis, as measured *from the axis to the ray*, is positive if the rotation is counterclockwise; acute angles are negative if the rotation is clockwise. As shown, u_0, u_1, and u_2 are each positive. All angles are measured in radians.

Much of geometrical optics deals with rays that are nearly parallel to the optical axis; these rays are called *paraxial rays* and the approximations $\sin u \approx \tan u \approx u$ are valid. In the following development, we use the angles themselves as a substitute for the sines or the tangents of the angles; we use the trigonometric functions of finite angles, whenever necessary, for computational accuracy.

We illustrate how the sign convention works by starting with plane P_0 as our current origin and find that

$$h_1 = h_0 + u_0 z_{01}, \tag{2.27}$$

where u_0 is the paraxial angle of the ray between P_0 and P_1 and z_{01} is the distance from P_0 to P_1. From the diagram we note that $h_1 = 0$ so that

$$h_0 = -u_0 z_{01}. \tag{2.28}$$

Because z_{01} is positive and h_0 is negative, u_0 must be positive, as claimed, to satisfy our sign convention. Similarly, by shifting our current origin to plane P_1, we find that

$$h_2 = h_1 + u_1 z_{12}, \tag{2.29}$$

and, since $h_1 = 0$, we conclude that

$$h_2 = u_1 z_{12}. \tag{2.30}$$

Again, as both z_{12} and u_1 are positive, h_2 is positive as required by the sign convention.

The *transfer equation* that allows us to trace the ray heights through a system has the general form

$$\boxed{h_{n+1} = h_n + u_n z_{n,n+1},} \tag{2.31}$$

which is used to calculate the ray intersection heights at successive planes in an optical system. The current origin for which Equation (2.31) applies is at plane P_n.

2.5.2. Refraction at a Curved Surface

Consider the refraction of a ray at a curved surface of radius R_1 that delineates regions of refractive indices n_1 and n_2, as shown in Figure 2.13. The *curvature* of the surface is $c_1 = 1/R_1$. An auxiliary sign convention is that the curvature of a convex surface, as viewed from the direction that the ray travels, is positive; the curvature of a concave surface is negative. For the moment, we assume that the refractive index is n_2 everywhere to the right of the spherical surface intersecting the points O and P; the vertex of the sphere defines the position of plane P_1. Based on our sign

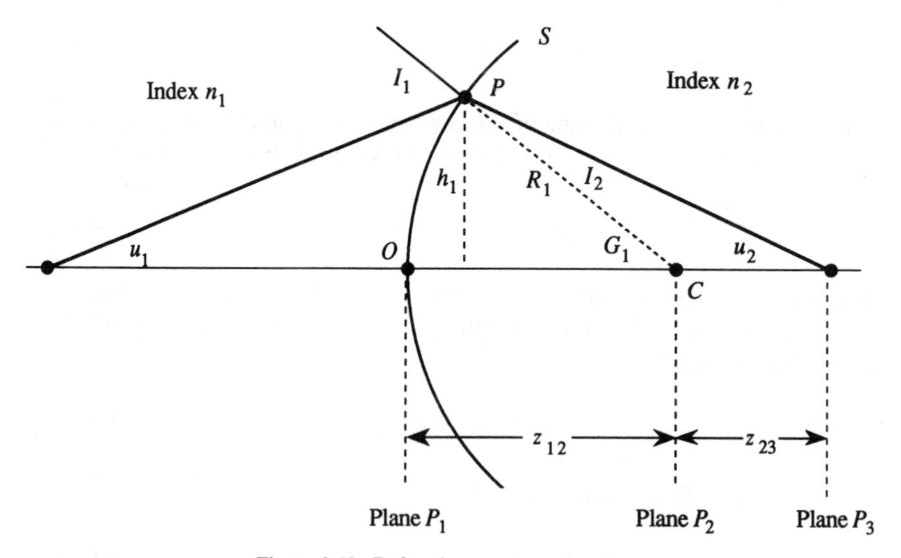

Figure 2.13. Refraction at a curved surface.

convention, we see that u_1 is positive, u_2 is negative, and h_1 is positive. The angle G_1 of the *normal to the surface* is given by the angle OCP. Because we use the same sign convention for wavefront normals as for rays, G_1 is negative.

We begin by setting our current origin at the vertex of the surface and applying Snell's law to the ray as it intersects the surface S:

$$n_1 I_1 = n_2 I_2, \tag{2.32}$$

where we use the paraxial ray approximation. Although all the angles are small in practice, we exaggerate them in our figures for clarity. From the sketch we see that the angle of incidence is equal to the sum of the angles u_1 and G_1. Because G_1 is negative, we find that

$$I_1 = u_1 - G_1. \tag{2.33}$$

The sign of the angle of incidence is dictated by the ray and normal angles so that relationship (2.33) is not overspecified. As shown, I_1 is positive because u_1 is positive and G_1 is negative. In a similar fashion, we find that

$$I_2 = u_2 - G_1. \tag{2.34}$$

The sign of I_2 is also positive because, although both u_2 and G_1 are negative, $|G_1| > |u_2|$. We use these values for the angles of incidence and refraction in Equation (2.32) to find that

$$n_1(u_1 - G_1) = n_2(u_2 - G_1). \tag{2.35}$$

As $G_1 = -h_1/R_1 = -h_1 c_1$, we find that

$$\boxed{n_2 u_2 = n_1 u_1 - h_1 c_1(n_2 - n_1),} \tag{2.36}$$

which is the *refraction equation* for a surface in its generalized form. The quantity $c_1(n_2 - n_1)$ is the *power of the surface* and is denoted by K_1. The power of a surface is expressed in *diopters* when the curvature is expressed in reciprocal meters. From Equation (2.36) we see that when $c_1 = 0$ the curved surface degenerates into a flat surface and Equation (2.36) reduces to Snell's law, as expected, because the power of the surface is then zero. Also, if $n_2 = n_1$, the power of the surface is zero for all values of c_1. This trivial situation simply shows that an optical surface cannot be defined when $n_2 = n_1$.

2.5.3. The Refraction Equation for Combined Surfaces

The refraction equation for a surface can be applied repeatedly to successive surfaces to develop the refraction equation for a lens. Suppose that the ray encounters a second surface of curvature c_2 separating a region of index n_2 from a region of index n_3, as shown in Figure 2.14. We apply the transfer equation (2.31) to find the height at which the ray penetrates the second surface of the lens:

$$h_2 = h_1 + u_1 z_{12}, \tag{2.37}$$

where z_{12} is the distance between the vertices of the two surfaces of the lens. In Section 2.5.6 we define the principal planes of a thin lens, whose parameters allow us to set h_2 equal to h_1. With $h_2 = h_1 = h$, we apply the refraction equation directly at the second surface of the lens to find that

$$n_3 u_3 = n_2 u_2 - h c_2 (n_3 - n_2). \tag{2.38}$$

We substitute the value of $n_2 u_2$ from Equation (2.36) into Equation (2.38) to find that

$$\boxed{n_3 u_3 = n_1 u_1 - h[c_2(n_3 - n_2) + c_1(n_2 - n_1)].} \tag{2.39}$$

The result given by Equation (2.39) is the most general statement of the refraction equation for lenses and can be used in all situations. It is valid for glass lenses in air or, as sometimes used in medical instruments, for air lenses in glass.

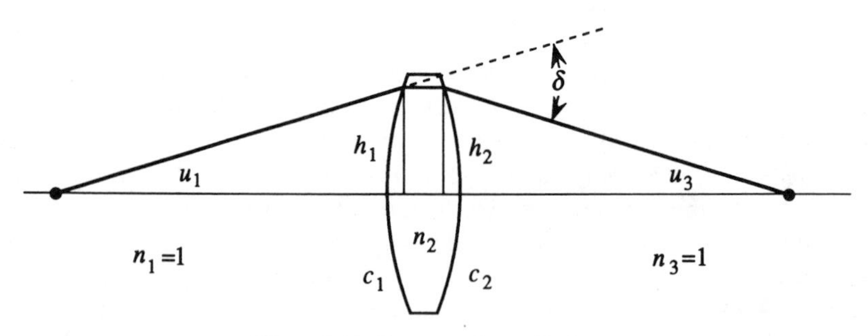

Figure 2.14. Ray trace for a thin lens.

The power of the lens is the sum of the powers of the two surfaces:

$$K = K_1 + K_2 = c_1(n_2 - n_1) + c_2(n_3 - n_2). \tag{2.40}$$

For the special situation of a glass lens in air, as sketched in Figure 2.14, $n_1 = n_3 = 1$ is the refractive index for the two air spaces and n_2 is the refractive index of the lens. We find that Equation (2.39) then yields

$$u_3 = u_1 - h(c_1 - c_2)(n_2 - 1), \tag{2.41}$$

which is the refraction equation for a thin lens in terms of the constructional parameters of the lens.

The *power* of a thin lens of index n_2, embedded in air, and with surface curvatures c_1 and c_2, is given by

$$\boxed{K = (c_1 - c_2)(n_2 - 1).} \tag{2.42}$$

As c_2 is negative and $n_2 > 1$, the power of the lens is positive. The *deviation* of the ray is given by $\delta = u_1 - u_3$, which is equal to hK, the product of the power of the lens and the ray height. The ray at the maximum height, called the *marginal ray*, is bent the most, whereas the axial ray is not deviated because $h = 0$.

2.5.4. The Condenser Lens Configuration

A condenser lens configuration occurs when all incident rays are parallel, as shown in Figure 2.15(a). The refraction equation, using the lens as our current origin, states that

$$u_3 = u_1 - hK, \tag{2.43}$$

so that, for $u_1 = 0$, we find that

$$u_3 = -hK. \tag{2.44}$$

As $u_3 = -h/z_{23}$ for small angles, we find that $z_{23} = 1/K = F$, which is the focal length of the lens. Thus, a thin lens *condenses* a bundle of parallel rays at a distance equal to its focal length. This plane is called the *back focal plane* of the lens.

Parallel rays entering the lens at an off-axis angle also focus at the back focal plane of the lens. Consider the upper entrance ray in Figure 2.16,

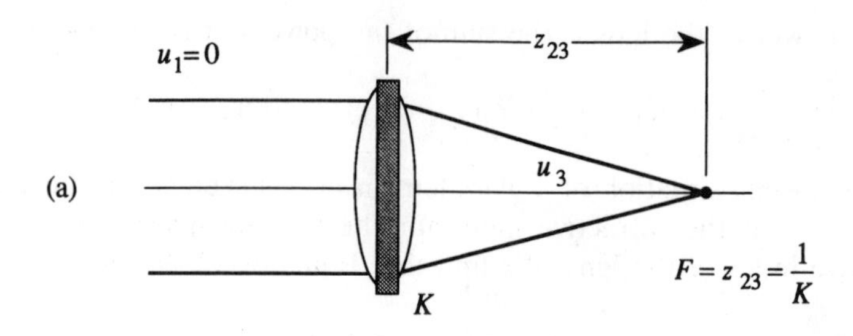

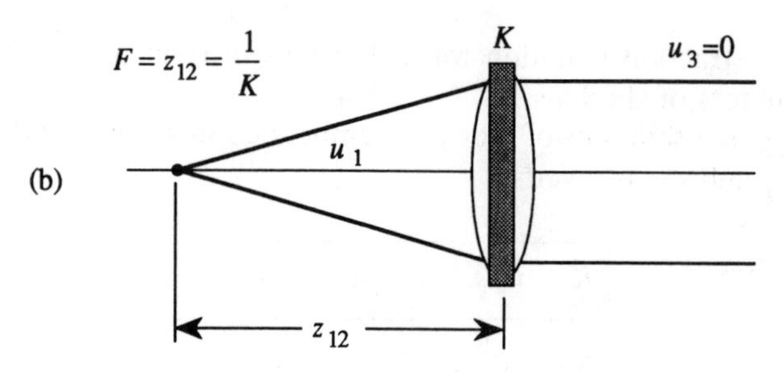

Figure 2.15. Thin lenses: (a) condenser and (b) collimator.

which makes an angle u_1 with respect to the optical axis. By the refraction equation we find that

$$u_3 = u_1 - h_2 K, \tag{2.45}$$

where h_2 is the height of the ray at the lens. By the transfer equation, we find that

$$h_3 = h_2 + u_3 z_{23}, \tag{2.46}$$

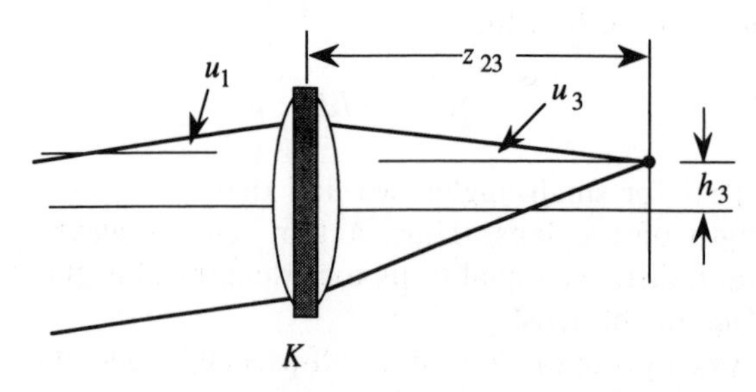

Figure 2.16. Imaging of an off-axis bundle of rays.

where the distance from the lens to the image plane is undetermined for the moment. The focal plane is found by tracing a second ray through the lens. The ray passing through the center of the lens is a convenient one to trace; for it, we find that

$$h_3 = u_1 z_{23}. \tag{2.47}$$

The parallel rays entering the lens are due, in effect, to an object sample at infinity. Because the image of a sample requires that all rays intersect, we require that the values of h_3 from Equation (2.46) and Equation (2.47) must be equal so that

$$\begin{aligned} u_1 z_{23} &= h_2 + u_3 z_{23} \\ &= h_2 + (u_1 - h_2 K) z_{23} \\ &= u_1 z_{23} + h_2 - h_2 K z_{23}, \end{aligned} \tag{2.48}$$

which, in turn, shows that z_{23} must be equal to F. This argument is easily extended to show that all parallel rays focus to a single point at the back focal plane of the lens.

2.5.5. The Collimating Lens Configuration

Suppose that we want to create parallel rays for which $u_3 = 0$, as shown in Figure 2.15(b). The refraction equation, as applied at the plane of the lens, becomes $u_1 = hK$. In a fashion similar to that shown in Section 2.5.4, we find that $z_{12} = 1/K = F$ so that the lens *collimates* a point source of light located at the *front focal plane* of the lens. Also, by using equations similar to those developed above, we find that light from any off-axis object sample, at the front focal plane, produces a parallel beam of light that propagates at an off-axis angle. A lens system is therefore *bilateral* so that its operation on rays traveling in one direction is the same as its operation on rays traveling in the opposite direction.

2.5.6. Principal Planes

In our derivation of the refraction formula, we assumed that $h_2 = h_1$ at the surfaces of the lens. In the case of thick lenses, where the assumption does not hold, we can still retain the formalism that $h_2 = h_1$ by introducing the concept of the principal planes of a lens. The first principal plane is found by tracing a ray from the front focal point O_1, as shown in

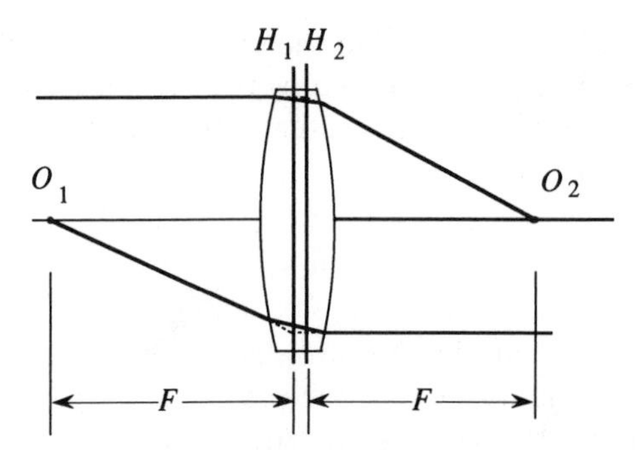

Figure 2.17. Principal planes for a thin lens.

Figure 2.17, to the lens surface. This ray exits the lens, parallel to the axis, at some point on the second surface of the lens, which establishes its height from the axis. We extend these two rays to their intersection point to define the position of plane H_1, called the *first principal plane* of the lens.

We follow a similar procedure for an entrance ray parallel to the axis, which passes through the back focal point O_2 after refraction. The intersection of these two rays defines the positions of plane H_2 and establishes the *second principal plane* of the lens. The principal planes provide the following properties for simplifying the representation of a lens:

- The front and back focal lengths of the lens are measured from the principal planes H_1 and H_2, respectively. For some lenses, principal plane H_2 may occur before principal plane H_1, generally referred to as a *crossover* of the principal planes.
- The principal planes are *unit magnification planes* because the magnification between them is $M = +1$. Thus, any ray that intersects plane H_1 at some height h_1 is transferred to plane H_2 at the *same* height h_1, where the entire bending action of the refraction equation is applied.
- Based on unit magnification, we collapse the space between H_1 and H_2 to represent the lens as a true thin lens with a single plane where all of the power of the lens is concentrated.

Principal planes may become curved surfaces for lenses with high relative apertures or for special lenses such as wide-angle lenses. The principal planes are generally flat for the types of lenses encountered in optical signal processing.

2.5.7. Thin-Lens Systems

We now generalize the refraction equation for a two-lens system, as shown in Figure 2.18(a), to find the equivalent power of a lens pair. Suppose that a parallel ray, for which $u_1 = 0$, enters the first lens at height h_1. This ray is bent by the first lens of power K_1 and intercepts the second lens at height h_2. The second lens bends the ray to its final value u_3 and the ray intercepts the axis at plane P_3. Suppose that we want to replace these two lenses with one having the equivalent power necessary to bend the incoming ray to the *same final angle* u_3, as shown in Figure 2.18(b). The first principal plane of the equivalent lens lies in the same plane as the first principal plane of the first lens of the pair. Although the equivalent lens *must* produce the same angle u_3 as the two-lens system, we do not require that the distances from the first principal planes to the focal planes be the same.

We begin by noting, from Figure 2.18(a), that

$$u_2 = u_1 - h_1 K_1 \tag{2.49}$$

by virtue of the refraction equation, that

$$h_2 = h_1 + u_2 z_{12} \tag{2.50}$$

by virtue of the transfer equation, and that

$$u_3 = u_2 - h_2 K_2 \tag{2.51}$$

by virtue of a second application of the refraction equation. We substitute

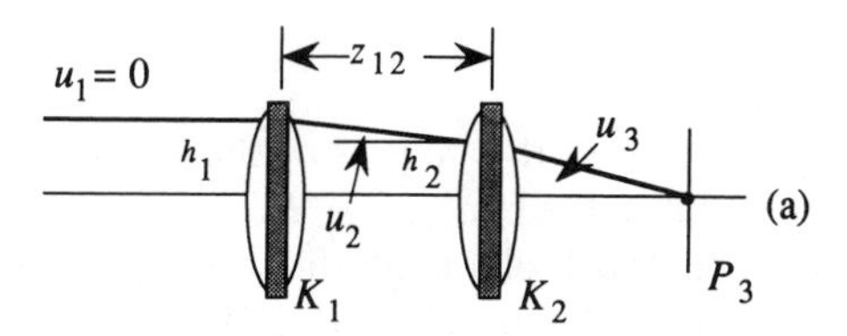

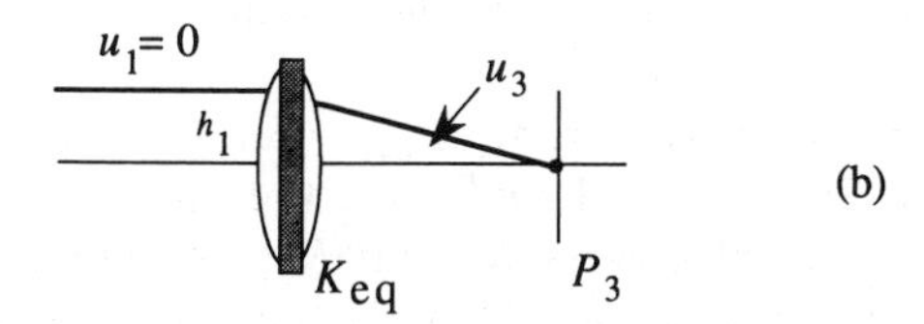

Figure 2.18. Ray traces: (a) two-lens system and (b) single-lens equivalent.

Equations (2.50) and (2.49) into Equation (2.51) to obtain

$$u_3 = (u_1 - h_1K_1) - [h_1 + (u_1 - h_1K_1)z_{12}]K_2. \tag{2.52}$$

Because $u_1 = 0$, we find that

$$u_3 = -h_1[K_1 + K_2 - z_{12}K_1K_2]. \tag{2.53}$$

For the equivalent lens shown in Figure 2.18(b), we see that

$$u_3 = -h_1K_{eq}. \tag{2.54}$$

By comparing Equation (2.53) with Equation (2.54) we see that the *equivalent power* of the two-lens system is

$$\boxed{K_{eq} = K_1 + K_2 - z_{12}K_1K_2,} \tag{2.55}$$

where K_1 and K_2 are the powers of the individual lenses and z_{12} is the distance between the two lenses. In this development, the sign of z_{12} is positive. The result given in Equation (2.55) is extremely useful for finding the proper geometric arrangement for two lenses to obtain a lens whose equivalent power is different from that of either lens.

To illustrate the use of the equivalent lens concept, consider the situation for two positive lenses. The maximum power is obtained when z_{12} is zero so that the two thin lenses are in contact and the equivalent focal power is

$$K_{eq} = K_1 + K_2, \tag{2.56}$$

in a fashion analogous to how the powers of the surfaces of the lenses add. A special case is obtained when $z_{12} = F_1$ so that the second lens is positioned at the back focal plane of the first lens. We then find that $K_{eq} = K_1$, which means that the lens with power K_2 has no contribution to the overall power; this lens is in a "no power plane." Such lenses are sometimes used as field lenses to help contain ray bundles, without contributing to image quality.

Because the power varies from $K_{max} = K_1 + K_2$ to $K_{min} = K_1$, each lens must have a focal length longer than the one we wish to synthesize. Also, given two lenses of powers K_1 and K_2, we obtain the largest range of powers when the lower-power lens is the first lens of the pair. This arrangement also tends to minimize the aberrations of the combination because the *relative apertures*, defined as the ratio of the lens aperture to its focal length, are more nearly equal for the two lenses.

There are no restrictions on the value of z_{12} or on the signs of the powers of the lenses *provided that u_3 has the proper numerical value and sign at the output of the equivalent lens*. For example, if the separation between the lenses is so large that the equivalent power is negative, we find that the ray angle u_3, for a parallel entrance ray a distance h_1 above the axis, must be negative, violating the constraint. Note that the equivalent power for a pair of negative lenses is always negative, independent of the value of z_{12}. If one lens is positive and one is negative, the constrained value of z_{12} is a function of the values of K_1 and K_2. Finally, we note that the total power of the two-lens system is zero when $z_{12} = F_1 + F_2$, a case that we now consider in more detail.

2.5.8. Afocal or Telescopic Configurations

In optical signal-processing systems we often consider signals that are in either the front or the back focal plane of the lens. Using these planes considerably simplifies the mathematical analysis of the system, with little loss of generality, and offers opportunities to reduce aberrations in a laboratory system (see Section 2.9). Furthermore, the plane under consideration may be simultaneously the back focal plane of one lens and the front focal plane of the next lens in cascaded systems.

For example, plane P_1, as shown in Figure 2.19(a), is the back focal plane of the lens whose power is K_1 and is simultaneously the front focal plane of the lens whose power is K_2. When arranged as shown, these two lenses are in an *afocal* or *telescopic configuration*. Such a configuration causes no net bending of the incident and exit rays, as shown by the fact that, for an *arbitrary* entrance ray at an angle u_1, we have

$$u_3 = u_1 - h_{\text{eq}} K_{\text{eq}}. \tag{2.57}$$

Because $K_{\text{eq}} = K_1 + K_2 - z_{12} K_1 K_2$ is the equivalent power of the lens pair and because $z_{12} = F_1 + F_2$ is the distance between the lenses, we find that

$$\begin{aligned} u_3 - u_1 &= -h_{\text{eq}}[K_1 + K_2 - (F_1 + F_2)K_1K_2] \\ &= -h_{\text{eq}}\left[K_1 + K_2 - \left\{\frac{1}{K_1} + \frac{1}{K_2}\right\}K_1K_2\right] \\ &= -h_{\text{eq}}\left[K_1 + K_2 - \left\{\frac{K_1 + K_2}{K_1K_2}\right\}K_1K_2\right] = 0, \end{aligned} \tag{2.58}$$

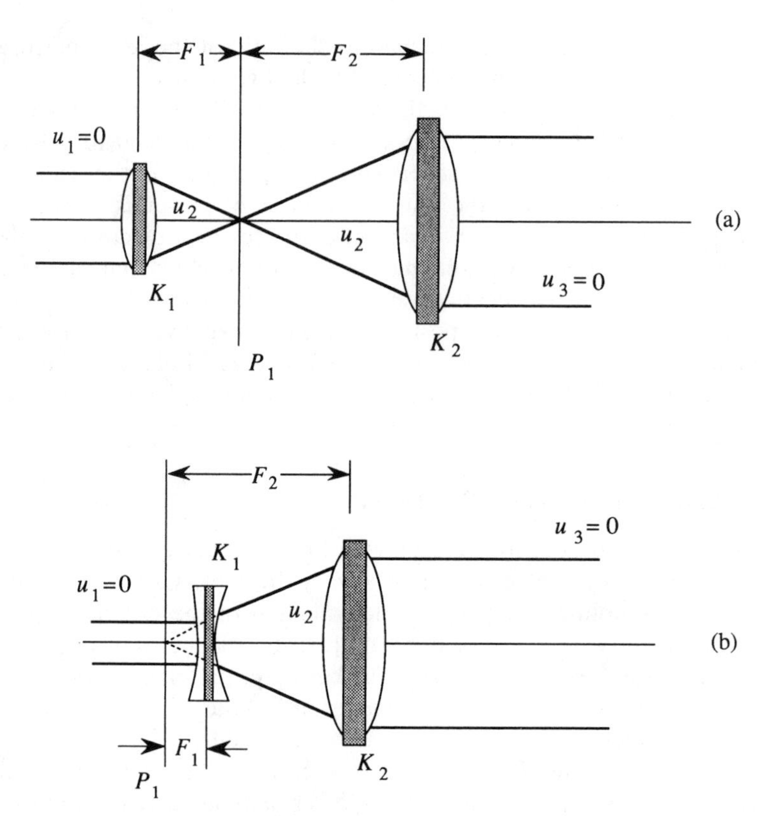

Figure 2.19. Cascaded lenses in afocal arrangements: (a) positive lenses and (b) negative/positive lenses.

so that $u_3 = u_1$. The name *telescopic* derives from the property of a telescope, which accepts parallel light at a given angle and produces parallel light, generally with some beam magnification, at the same angle.

The telescopic condition arises whenever $z_{12} = F_1 + F_2$. The net power of the system is therefore zero, and rays are not bent in traversing the system. There are two generic telescopic configurations that result according to the signs associated with the focal lengths of the two lenses. In the first configuration, shown in Figure 2.19(a), it is clear that the two lenses are separated by the sum of their focal lengths. In the second configuration, shown in Figure 2.19(b), the first lens has a negative power, but the two lenses are still separated by the sum of their focal lengths and plane P_1 is a common focal plane for both lenses. There are two other telescopic configurations that provide magnifications less than one; these configurations are simply those of Figure 2.19(a) and Figure 2.19(b) with the lenses interchanged. The total length of the system and the sign of the magnification depend on the signs of the focal lengths of the two lenses.

2.6. THE GENERAL IMAGING CONDITION

So far we have considered some special conditions in which either the object or the image is at infinity; these are called *infinite conjugate* imaging conditions. A more general condition arises when an object at plane P_1 is imaged at plane P_3, as shown in Figure 2.20. Both planes are at finite distances from the lens, resulting in a *finite conjugate* imaging condition. With the lens as the current origin, z_1 and z_3 indicate the distances of these planes from the lens. A sample in the object plane, a height h_1 above the axis, generates a family of rays that produces a wavefront W_1 normal to all the rays. In a well-corrected system, Fermat's principle states that the optical paths traveled by all rays are equal. Therefore wavefront W_2 is also normal to the ray family on the image side; the image occurs at plane P_3, a height h_3 below the optical axis. For all conjugate-ray pairs the refraction equation states that

$$u_3 = u_1 - hK, \tag{2.59}$$

where h is the height at which the ray intercepts the lens. When we divide all terms in Equation (2.59) by h, we find that

$$\frac{u_3}{h} = \frac{u_1}{h} - K. \tag{2.60}$$

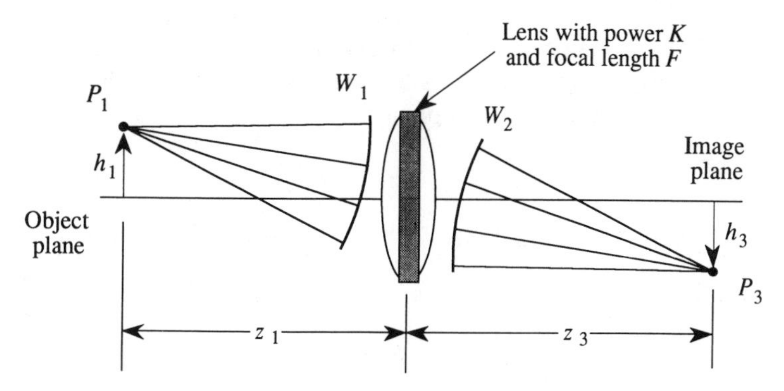

Figure 2.20. General imaging condition.

For the sample located on the optical axis at plane P_1, we find that $u_3 = -h/z_3$ and $u_1 = -h/z_1$, so that

$$-\frac{1}{z_1} + \frac{1}{z_3} = \frac{1}{F}, \tag{2.61}$$

which is commonly referred to as the *lens equation*. Thus, if we know the value of F and the value of either z_1 or z_3, we can determine the remaining unknown distance. Remember that Equation (2.61) is valid only for a thin lens in air. The more general result, as given by Equation (2.39), is always valid.

2.6.1. Ray Tracing

To analyze the image quality produced by an optical system, we trace many rays from an object sample through the system and require that they all fall within a specified distance from the true image sample position. However, only a few rays are required to determine the positions of the object and image planes relative to the lens. The trick is to select the most useful ones.

Throughout this section, we treat axial distances as having positive magnitudes and account for their negative values by compensating for the signs in the equations. We do this because we need to shift the current origin several times to derive general results. The sign of an axial distance may therefore be both positive and negative, depending on the position of the current origin, and ambiguities could arise when making the final calculations. This procedure is not necessary, of course, when we numerically calculate ray positions because the calculations are completed at each surface before we proceed to the next surface.

We begin by selecting a ray parallel to the optical axis, as shown in Figure 2.21(a), for which $u_1 = 0$. This ray starts from an object sample at plane P_1 located a height h_1 above the axis. We use the transfer equation to find the ray height at the lens plane P_3:

$$\begin{aligned} h_3 &= h_1 + (z_{12} + z_{23})u_1 \\ &= h_1. \end{aligned} \tag{2.62}$$

The refraction equation for a lens in air states that

$$\begin{aligned} u_3 &= u_1 - h_3 K \\ &= -h_3 K = -h_1 K, \end{aligned} \tag{2.63}$$

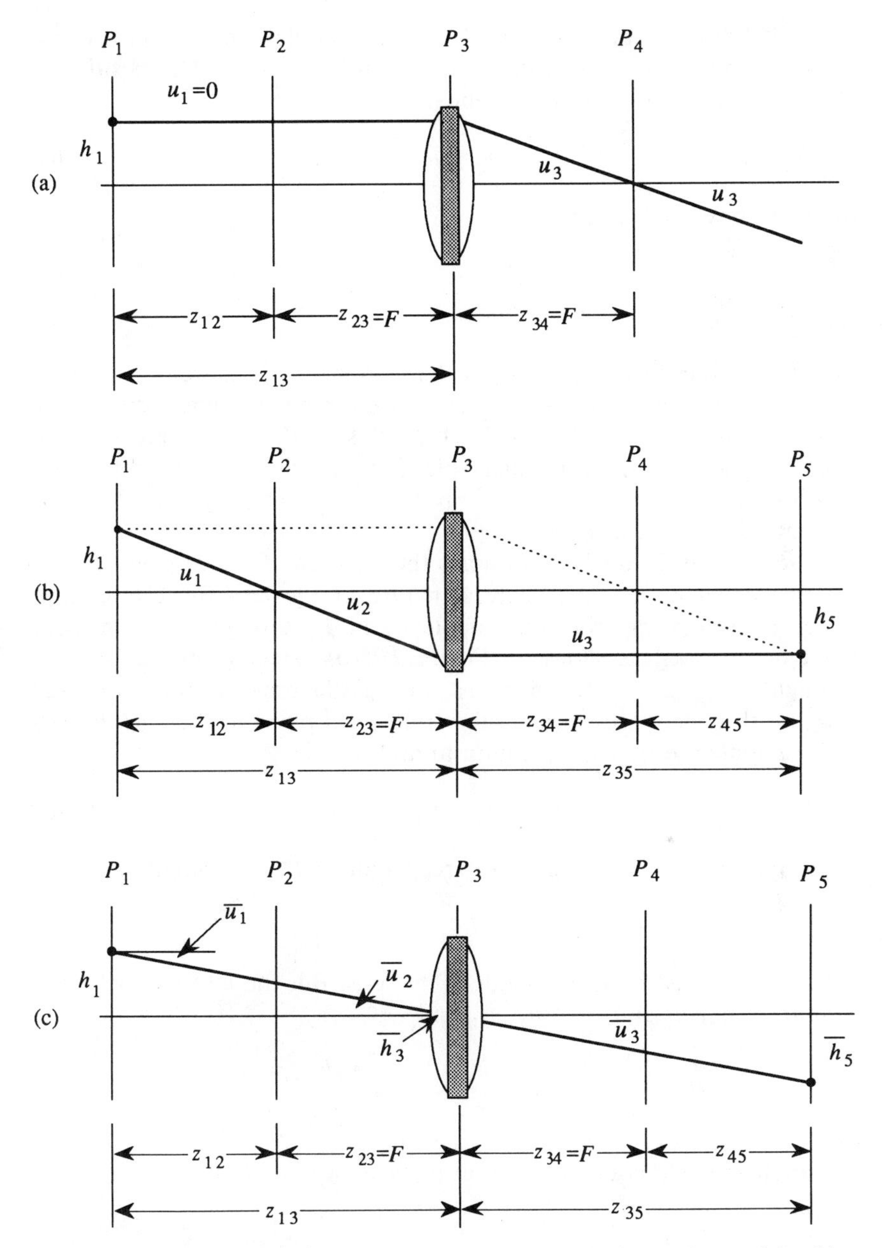

Figure 2.21. Three convenient rays for finding the position of the image plane and the object magnification: (a) input parallel ray, (b) output parallel ray, and (c) principal pupil ray.

where K is the power of the thin lens. This parallel input ray crosses the axis at plane P_4, located a distance z_{34} from the lens. The position of plane P_4 is found by solving the equation

$$h_4 = h_3 + u_3 z_{34} = 0, \tag{2.64}$$

from which we deduce that

$$z_{34} = -\frac{h_1}{u_3} = \frac{1}{K} = F, \tag{2.65}$$

where F is the focal length of the lens. Plane P_4 is therefore the back focal plane of the lens; all parallel rays entering the lens focus at this plane, as we showed in Section 2.5.4. At this stage of the analysis, we do not know the position of the image plane P_5, nor the image height h_5. We therefore temporarily extend the ray from P_3 through plane P_4 for an arbitrary distance to the right of plane P_4.

A second ray uniquely determines the position of image plane P_5 and the image height h_5. The bilateral nature of the lens suggests that we select a ray, passing from the sample h_1 in P_1 through the front focal point of the lens, as shown in Figure 2.21(b). Tracing this ray forward through the system is the same as tracing the previous ray backwards. Because this ray passes through the front focal plane of the lens, $h_2 = 0$, and we apply the transfer equation to find u_1:

$$h_2 = h_1 + u_1 z_{12} = 0, \tag{2.66}$$

so that $u_1 = -h_1/z_{12}$. This ray intercepts plane P_3 at a height

$$h_3 = h_2 + u_2 z_{23}, \tag{2.67}$$

where $z_{23} = F$ is the focal length of the lens. But the intersection height $h_2 = 0$ so that $u_2 = u_1 = -h_1/z_{12}$ and

$$h_3 = \frac{-h_1 z_{23}}{z_{12}} = \frac{-h_1 F}{z_{12}}. \tag{2.68}$$

We apply the refraction equation at plane P_3 to find that

$$\begin{aligned} u_3 &= u_2 - h_3 K \\ &= -\frac{h_1}{z_{12}} - \left[\frac{-h_1 F}{z_{12}}\right] K = 0, \end{aligned} \tag{2.69}$$

so that the second ray is rendered parallel to the optical axis by the collimating action of the lens.

The intersection of the second ray with the extension of the first ray defines the position of the image plane P_5. For the second ray, we find that

$$h_5 = h_4 = h_3 = \frac{-h_1 F}{z_{12}}, \tag{2.70}$$

which gives the height of the image sample.

The distance z_{45} can now be found from the transfer equation

$$\begin{aligned} h_5 &= h_4 + u_3 z_{45} = u_3 z_{45} \\ &= -h_1 K z_{45}. \end{aligned} \tag{2.71}$$

We use Equation (2.70) in Equation (2.71) to find that

$$-\frac{h_1 F}{z_{12}} = -h_1 K z_{45}, \tag{2.72}$$

which is rearranged as

$$z_{12} z_{45} = F^2, \tag{2.73}$$

to obtain *Newton's formula* for relating the distances of the object and image plane from the focal planes of the lens.

2.6.2. Lateral Magnification

The *lateral magnification* M is given by the ratio of the image height to the object height, as shown in Figure 2.21(b). From similar triangles, we find that

$$M = \frac{h_5}{h_1} = \frac{-F}{z_{12}}, \tag{2.74}$$

which is the ratio of the focal length of the lens to the distance from the front focal plane to the object plane. We can also express the magnification in terms of the object to lens distance z_{13}:

$$z_{13} = z_{12} + z_{23} = z_{12} + F \tag{2.75}$$

so that

$$M = \frac{-F}{z_{13} - F}. \tag{2.76}$$

Three special cases are of interest. In the first case, $z_{13} \to F$, which means that the object plane approaches the front focal plane of the lens. In this case, $M \to -\infty$ so that the lateral magnification is infinite and the lens acts as the collimator shown in Figure 2.15(b). In the second case, $z_{13} = 2F$ so that $M = -1$ and the object and the image planes are symmetrically located two focal lengths from the lens. The modulus of the magnification is unity, with the negative sign implying a reversal of the image coordinate system in passing from plane P_1 to plane P_5. The image is therefore rotated 180° with respect to the object. In the third case, $z_{13} \to \infty$ so that $M \to -1/\infty = 0$, and the lens acts as the condenser shown in Figure 2.15(a). The rays on the object side of the lens are therefore parallel and the image is formed at the back focal plane of the lens.

We also express the magnification in terms of the parameters on the image side of the lens. From similar triangles, we find that

$$M = \frac{h_5}{h_1} = \frac{-z_{45}}{F} = \frac{F - z_{35}}{F}. \tag{2.77}$$

Arguments similar to those given in the previous paragraph are applied to Equation (2.77) to determine the position of the image plane for various values of the magnification.

2.6.3. The Principal Pupil Ray

Another easy ray to trace is the *principal pupil ray*, which is midway between the two marginal rays and passes through the center of the lens as shown in Figure 2.21(c). The angles and intersection heights of principal pupil rays are usually indicated by an overbar; they obey, of course, the same laws of propagation as do other rays. For the principal pupil ray we have

$$\bar{u}_1 = \frac{-h_1}{z_{12} + z_{23}} = \bar{u}_2. \tag{2.78}$$

At plane P_3 we have

$$\bar{u}_3 = \bar{u}_2 - \bar{h}_3 K. \tag{2.79}$$

But $\bar{h}_3 = 0$ by construction so that $\bar{u}_3 = \bar{u}_2 = \bar{u}_1$, which shows that the ray is not bent as it passes through the center of the lens. It therefore arrives at plane P_5 at a height

$$\begin{aligned} \bar{h}_5 = h_5 &= \bar{h}_3 + (z_{34} + z_{45})\bar{u}_3 \\ &= 0 + \left[-h_1 \frac{z_{34} + z_{45}}{z_{12} + z_{23}} \right] \\ &= \frac{-h_1 z_{35}}{z_{13}}. \end{aligned} \tag{2.80}$$

The magnification is $M = h_5/h_1 = -z_{35}/z_{13}$, which is an alternative, and perhaps the easiest, way to obtain the magnification. The magnification is simply the ratio of image to object distance. If $z_{35} > z_{13}$, the magnitude of the magnification is greater than 1; if $z_{35} < z_{13}$ the corresponding magnitude is less than 1.

These three rays are useful for determining the image position and magnification when the object position and the lens focal length are known. Any two of the three rays are sufficient; we generally choose whichever pair is most convenient for a particular system. These rays do not, however, reveal anything about the detailed structure of the image in terms of the required sampling distance for the object or about the spatial resolution of the lens. We now show how we can trace some rays that are associated with object and image resolution and combine them with those traced so far to develop an important result: the optical invariant.

2.7. THE OPTICAL INVARIANT

Suppose that the object has a sampling interval d_0, the distance between the delta functions of the sampling function, for a signal bandlimited to α_{co}, as discussed in Chapter 1. Further, each sample in the object between $-h_1$ to h_1, shown in Figure 2.22, diffracts light rays over a range of angles bounded by $\pm u_1$. For the moment we concern ourselves with only those rays produced by the sample located at O_1 at plane P_1 and use the refraction equation to establish the position of the image plane P_5. We begin by finding a relationship between u_1 and u_5. From plane P_2, the

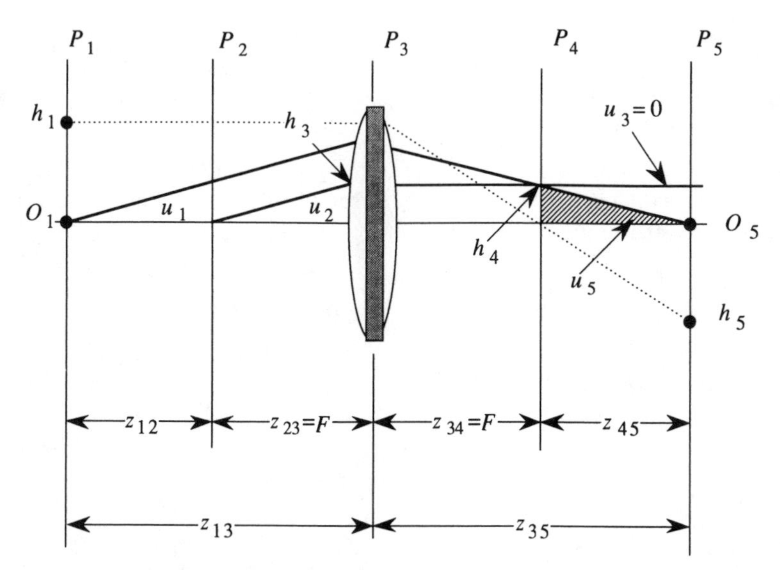

Figure 2.22. Rays necessary to determine the optical invariant.

front focal plane of the lens, we construct a parallel ray with the same angle $u_2 = u_1$ as the marginal ray from sample O_1. As $h_2 = 0$ at plane P_2 for this ray, we find that

$$\begin{aligned} h_3 &= h_2 + u_2 z_{23} \\ &= u_2 z_{23}. \end{aligned} \tag{2.81}$$

The refraction equation requires that

$$\begin{aligned} u_3 &= u_2 - h_3 K \\ &= \frac{h_3}{z_{23}} - h_3 K = 0. \end{aligned} \tag{2.82}$$

This ray is therefore parallel to the optical axis and it intercepts plane P_4 at a height $h_4 = h_3$; as before, we extend this ray to infinity on the image side of the lens. Because parallel rays on the object side of the lens focus at plane P_4, the ray originating from the sample O_1 also intersects plane P_4 at height h_4. From the shaded triangle we see that

$$h_4 = u_5 z_{45}. \tag{2.83}$$

But because $h_4 = h_3 = z_{23}u_1$, we further find that

$$u_1 z_{23} = -u_5 z_{45}, \tag{2.84}$$

which relates the image ray bundle angle u_5 to the object ray bundle angle u_1.

We now find a relationship between h_1 and h_5. By tracing the dotted ray through the lens, we see by inspection that

$$M = \frac{h_5}{h_1} = \frac{-z_{45}}{z_{34}}. \tag{2.85}$$

We solve Equation (2.85) for z_{45} and use this value in Equation (2.84) to find that

$$u_1 z_{23} = u_5 \left[\frac{h_5 z_{34}}{h_1} \right]. \tag{2.86}$$

As $z_{23} = z_{34} = F$ is the focal length of the lens, we further simplify Equation (2.86) to show that $h_1 u_1 = h_5 u_5$. We developed this result on the assumption that $n_1 = n_5$. The more general and complete result, for arbitrary indices of refraction on either side of the lens, is that

$$\boxed{n_1 h_1 u_1 = n_5 h_5 u_5.} \tag{2.87}$$

This relationship is variously known as Helmholtz's equation, Lagrange's equation, or Smith's equation; it is becoming more universally known as the *optical invariant*.

For a given system, the value of the *optical invariant is the same for all image planes*, provided that no information is lost due to aperture stops and that the system is free from aberrations. For example, a periscope has many intermediate image planes between the object plane and the final image plane; the optical invariant has the same value at each of these planes, even though the image sizes may be different.

The optical invariant is valuable for quickly sketching the geometry of an optical system, for double-checking mathematical calculations, and for showing why the brightness at any set of conjugate planes is fixed. The optical invariant is an important and quick way to check whether an optical system is feasible and how difficult it is to design. Some examples of how the optical invariant is used are given in the following sections.

2.7.1. Magnification Revisited

The optical invariant provides a useful alternative method for calculating the magnification of a system. Because the lateral magnification is defined as $M = h_5/h_1$, we use Equation (2.87) to show that

$$M = \frac{n_1 u_1}{n_5 u_5} \tag{2.88}$$

is an alternative expression for the lateral magnification. As h_5 is negative and h_1 is positive, the magnification is negative for the system shown in Figure 2.22, as confirmed by the fact that u_1 is positive and u_5 is negative. We see that $|n_5 u_5| > |n_1 u_1|$ when $|M| < 1$. For the typical case of $n_5 = n_1 = 1$, Equation (2.88) states that the angle u_5 is inversely proportional to the magnification. Because an image that is smaller than the object has smaller sample spacings, the maximum value of the included angle defining the image bundle must increase.

2.7.2. Spatial Resolution

Consider the telescope, shown in Figure 2.23, that focuses an incoming parallel ray bundle at its back focal plane. The plane-wavefront W_1 is focused by the lens to a point on the optical axis at plane P_2. The ray bundle on the image side of the lens does not, of course, form an infinitesimally small spot at plane P_2 because of diffraction phenomena. In Chapter 3 we show that the intensity of the diffraction pattern for a uniformly illuminated one-dimensional aperture is $\text{sinc}^2(\xi L/\lambda F)$, where ξ is the coordinate in the image plane as shown in Figure 2.24(a). The sinc^2

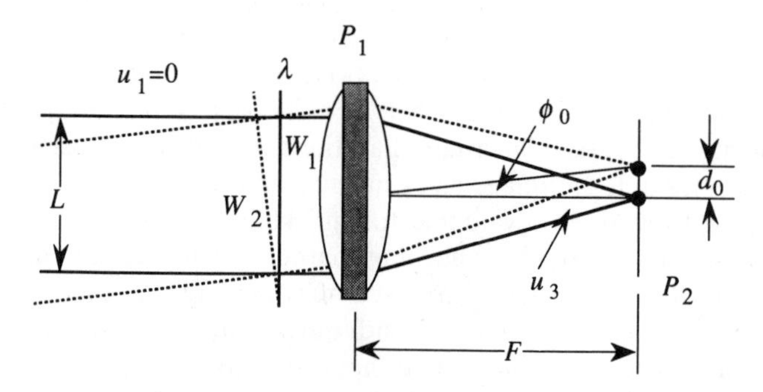

Figure 2.23. Resolution limit of a telescope.

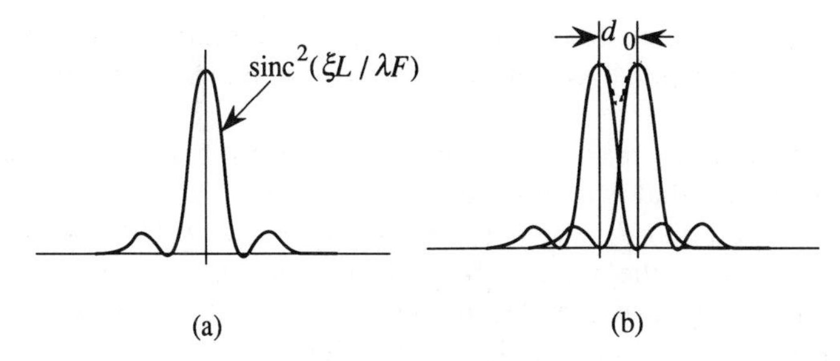

Figure 2.24. Rayleigh resolution criterion.

function is also the impulse response of the system, the impulse in this case being the infinitesimally small star at infinity.

Two samples from an object at infinity are resolved if the wavefront W_2 from the second sample is tilted with respect to W_1 by one wavelength of light over the aperture L of the lens. The *angular resolution* of the telescope is therefore

$$\phi_0 = \frac{\lambda}{L}, \tag{2.89}$$

as shown in Figure 2.23. Thus, the minimum resolvable distance at plane P_2, equivalent to the sampling interval, is

$$d_0 = \left(\frac{\lambda}{L}\right) F = \frac{\lambda}{2\theta_{co}}, \tag{2.90}$$

as shown in Figure 2.24(b), where we use θ_{co} to indicate the maximum value of u_3 as produced by the marginal ray with respect to the central ray. The relationship given by Equation (2.90) establishes the connection between θ_{co} and the interval d_0 between the samples at the image plane, a result that we used in Chapter 1 to describe the optimum sampling interval for a bandlimited analog optical signal.

Consider the spatial resolution of the telescope shown in Figure 2.23. For a lens aperture of $L = 100$ mm and focal length of $F = 1000$ mm, the relative aperture of the telescope is $L/F = 0.1$. A more commonly used measure is the $f/\#$, which is the ratio of the focal length to the aperture; the $f/\#$ for this set of parameters is $f/10$. For $\lambda = 0.5\ \mu$, we use Equation (2.90) to find that the *spatial resolution* of the telescope is $d_0 = 0.5(1000)/100 = 5\ \mu$. From Figure 2.23, we see that the largest angle θ_{co} is equal to $L/2F$ so that $\theta_{co} = 0.5\ \mu/2(5\ \mu) = 0.05 = 2.87°$. This is a surprisingly small angle for a fairly high spatial resolution and supports the notion that most optical systems are accurately analyzed by using paraxial rays.

If we set u_3 to its maximum value of θ_{co}, we have a special case of the optical invariant that we indicate by the symbol J_x. From the optical invariant we find that $J_x = n_3 h_3 \theta_{co}$ or that $J_x = n_3 h_3 \lambda / 2d_0$. We recognize that the height of the image, $2h_3$, divided by the minimum resolvable distance d_0 is a measure of the number of resolvable samples N_x in the image in the x direction. For $n_3 = 1$, we find that

$$\boxed{J_x = \frac{N_x \lambda}{4},} \tag{2.91}$$

so that the optical invariant is equal to the product of the number of samples and a quarter wavelength. Similar comments apply to the optical invariant J_y and the number of samples N_y in the y direction.

2.7.3. Space Bandwidth Product

In Chapter 1 we showed that we need $N = 2TW$ samples to accurately represent a signal that has duration T and highest frequency W. A similar concept applies to optical signals. The *length bandwidth product* of a spatial signal is the product of the object length L and the highest spatial frequency α_{co}: LBP $= \alpha_{co} L$. Thus, $N_x = 2\alpha_{co} L$ is the number of samples necessary to accurately represent the object in the horizontal direction.

To describe a two-dimensional image, we define the *height bandwidth product* as HBP $= \beta_{co} H$. Generally $\beta_{co} = \alpha_{co}$, but sometimes the cutoff frequency is different in the two directions. In a corresponding way, $N_y = 2\beta_{co} H$ is the number of samples necessary to accurately represent the object in the vertical direction. The *space bandwidth product* is simply the product of the length and height bandwidth products with the result that SBP = (LBP)(HBP), and we need $N = N_x N_y$ samples to accurately represent a two-dimensional image.

Finally, as J_x is measured in units of distance, we note that the dimensionless quantities N, TW, LBP, HBP, and SBP are proportional to the optical invariant normalized by the wavelength of light. A typical value of J_x for optical processing systems is of the order of 0.25 mm. As a general rule, lens design starts to get difficult when J_x exceeds 1 mm or so, corresponding to a length bandwidth product of 4000 for green light.

2.7.4. Matching the Information Capacity of System Components

As we showed in Section 2.7.2, the angular resolution of the telescope is dependent only on the aperture of the lens and the wavelength of light. The angular resolution of the telescope, whose parameters are given in Section 2.7.2, is $\phi_0 = \lambda/L = 0.5\ \mu/100(10^3)\ \mu = 5\ \mu$rad. The angular resolution of the eye is only about 500 μrad; we therefore need an eyepiece which, in combination with the objective lens, provides a 100 × magnification to fully appreciate the resolving capability of the telescope. This magnification is more or less consistent with the idea that the magnification of a telescope is equal to the ratio of the entrance-beam diameter to that of the exit beam, as discussed in Section 2.5.8. The pupil of the eye establishes the exit-beam diameter at 3–5 mm, depending on the illumination level; the overall useful magnification of a small-diameter telescope (80–100 mm) is therefore approximately 20–30 × , although magnifications of about 2–3 × these values seem to produce "sharper" images.

Using a significantly more powerful eyepiece does not provide more resolution. Why not? A principle of system design is that we match the bandwidths of all components of a system. All real-time electronic systems have a fixed metric for the time base. In optical systems, however, the spatial bandwidths may not be equal because the image may have a different size than the object. For optical systems, the *optical invariant plays the same role as does bandwidth in a real-time system*, and it ensures that the number of samples required to represent a signal does not change as the signal progresses through the system. An equivalent statement is that the optical invariant ensures that no signal information is lost.

Increasing the bandwidth of the last component of the system, which is equivalent to using a higher-power eyepiece, does not increase the information content of a signal. The angular resolution has already been set by the primary aperture of the telescope, which is why we generally rate the angular resolution of a telescope by citing its aperture size. Other factors, such as atmospheric turbulence, affect resolution, of course, but the aperture sets the theoretical resolution. The use of a higher-power eyepiece than necessary results in an image that has *empty magnification*.

We further illustrate these points by considering the resolving power of an $f/4$ camera lens, shown in Figure 2.25, whose focal length is $F = 100$ mm and whose diameter is $L = 25$ mm. Suppose that the image

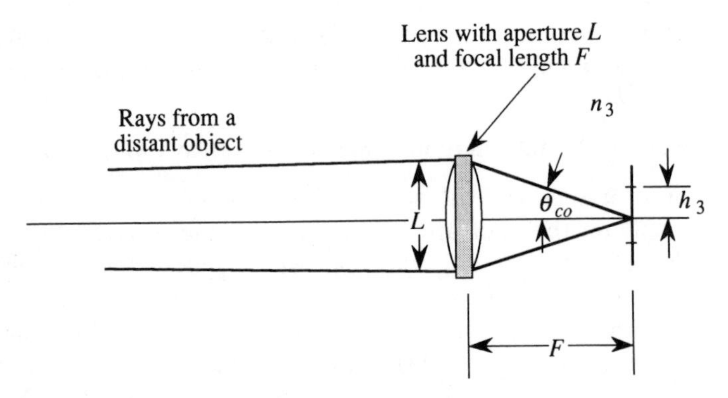

Figure 2.25. Camera with $f/4$ lens.

height is $2h_3 = 25$ mm. At the image plane, we have

$$\theta_{\text{co}} = \frac{L}{2F} = \frac{25 \text{ mm}}{200 \text{ mm}} = 0.125 \text{ rad} = 7.18^\circ,$$

$$J_y = n_3 \theta_{\text{co}} h_3 = 1(0.125)(12.5 \text{ mm}) = 1.56 \text{ mm},$$

$$d_0 = \frac{\lambda}{2\theta_{\text{co}}} = \frac{0.5\ \mu}{0.25} = 2\ \mu,$$

$$N_y = \frac{2h_3}{d_0} = \frac{25{,}000\ \mu}{2\ \mu} = 12{,}500 \text{ samples}. \qquad (2.92)$$

We see that a fairly simple camera lens produces an image of high quality, as shown by the large value of the optical invariant J_y and the large number of samples as shown by N_y.

If the image length is $L = 35$ mm, the number of samples required is $N_x = 17{,}500$ in that direction and the total number of samples produced is $N = N_x N_y = 2.28(10^8)$. At $f/2$, the image has four times the number of resolvable elements because the sample interval is halved in each direction; at $f/1.4$ the image has two times even more samples. Hence, a high-quality image contains a considerable amount of information.

Suppose that we record the image on photographic film. As the sampling theorem requires two samples per cycle of the highest spatial frequency to accurately represent a signal, the highest spatial frequency that the lens produces is

$$\alpha_{co} = \frac{1}{2d_0}, \tag{2.93}$$

which is called the *cutoff frequency* of the lens system. The units of the spatial frequency α_{co} are cycles/mm when d_0 is expressed in millimeters.

The optics community does not have an equivalent designator, such as Hertz (one cycle per second), to indicate spatial frequencies. We suggest the use of the designator *Abbe*, in honor of Earnst Abbe who did pioneering work in describing the relationship of spatial frequency content to image quality, to mean one cycle per millimeter and to abbreviate it as *Ab*. The reader is cautioned to note that spatial frequencies are often expressed as *lines per millimeter* in the literature and that the television industry tends to describe their systems in terms of *scan lines per inch*. This terminology is potentially confusing because a line sometimes refers to a cycle, but in other instances it may refer to a sample. The use of Ab to represent the spatial frequency of an optical signal may help to eliminate this potentially confusing situation.

We connect the cutoff spatial frequency α_{co} to the scattering angle θ_{co} associated with sample size d_0 through the use of Equations (2.90) and (2.93):

$$\boxed{\alpha_{co} = \frac{1}{2d_0} = \frac{\theta_{co}}{\lambda}.} \tag{2.94}$$

We find that $\alpha_{co} = 250$ Ab for the given parameters of the $f/4$ camera lens. The length bandwidth product of the image is therefore LBP $= 2h_3\alpha_{co} = 6250$. Consistent with normal design rules, photographic film must have a frequency response equal to α_{co} to avoid loss of information in the recording process. As the eye can resolve about 4 Ab at normal viewing distances and pupil sizes, we can magnify the resulting image approximately $M = \alpha_{co}/\alpha_{eye} = 62.5\times$. Additional magnification causes more pronounced film grain noise; less magnification results in loss of detail. Matching the optical invariant J at all image planes and for all components in an optical system therefore provides the maximum information content at the output.

It is important to realize that the cutoff angle θ_{co}, in any given situation, may be determined either by the optical system or by the object itself. For example, in the camera system just analyzed, we assumed that the cutoff angle was limited by the relative aperture of the lens. But in some applications the intrinsic resolution available from the object may be the limit; in this case, the object cannot produce a ray bundle that fills the lens aperture. In other applications the recording film or a CCD photodetector array may set the minimum sampling interval. Ideally, then, we first determine which part of a system imposes the limiting cutoff parameter and then match all the components of the system in terms of the optical invariant. In low-light-level television cameras a lens with an excessive $f/\#$ may be used simply to collect more photons—similar to the use of a larger antenna in a microwave system.

2.8. CLASSIFICATION OF LENSES AND SYSTEMS

Recall that the refraction equation for a lens in air is

$$u_3 = u_1 - hK, \tag{2.95}$$

where $K = (n_2 - 1)(c_1 - c_2)$ is the power of the lens. Because the refractive index of the lens is greater than one, the sign of the power of the lens is determined by the signs and magnitudes of c_1 and c_2. If c_1 is positive and numerically larger than c_2, or if c_1 is positive and c_2 is negative, the power of the lens is positive.

2.8.1. The Coddington Shape Factor

The shape of the lens is also determined by the magnitudes of c_1 and c_2. The *Coddington shape factor* σ is defined as

$$\sigma \equiv \frac{c_1 + c_2}{c_1 - c_2}. \tag{2.96}$$

If we add an incremental curvature Δc to both c_1 and c_2, their difference is constant so that the power of the lens is unchanged. Lens shapes for various values of σ are shown in Figure 2.26. We start with $\Delta c = 0$ and with $c_2 = c_1$ so that the Coddington shape factor is $\sigma = 0$ and the lens is called a *biconvex* lens. If we add an incremental curvature $\Delta c = -|c_1|$ to each surface, we find that $\sigma = -1$; and the lens is called a *plano convex* lens. If $\Delta c > -|c_1|$, we find that $\sigma < -1$; and the lens is called a *positive meniscus* lens. Adding amounts $\Delta c = |c_2|$ and $\Delta c > |c_2|$ results in plano convex and meniscus lenses, as shown in Figure 2.26. The following points

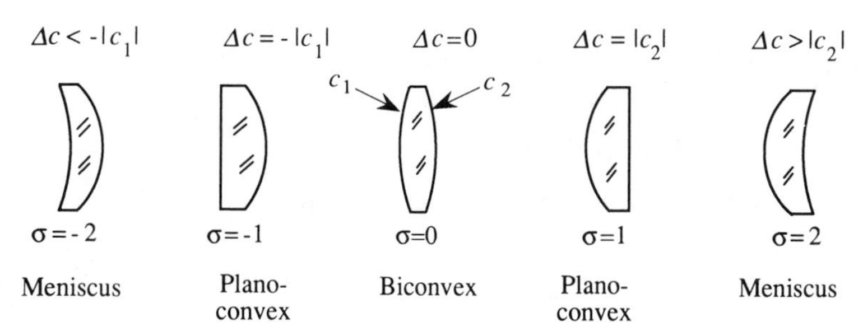

Figure 2.26. Types of positive lenses.

are noted:

- All the lenses have the *same power*.
- All the lenses are *positive* lenses.
- Although not shown in the figure, the principal planes of the lenses are in different places relative to the vertices of the lens surfaces.
- The process of adding a given value to both curvatures of the lens is called *bending the lens*. Bending is easily visualized by keeping the center of the lens fixed and pushing on the rim of the lens in either direction.

A similar set of lens shapes is obtained for lenses with negative power, a condition that arises when c_1 is negative and c_2 is positive, or if c_1 is negative and numerically larger than c_2. These lens shapes, shown in Figure 2.27, are called *biconcave*, *plano concave*, and *negative meniscus*

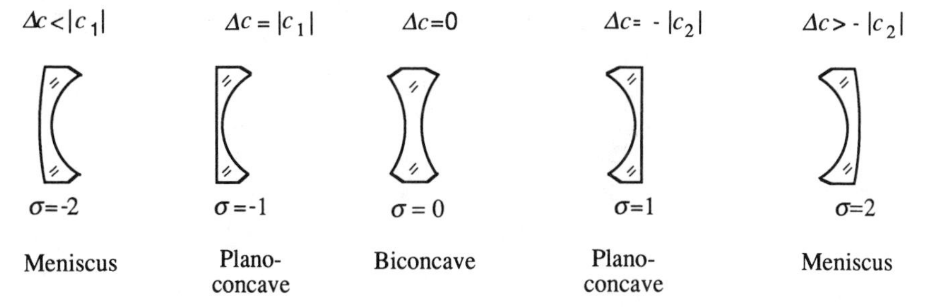

Figure 2.27. Types of negative lenses.

lenses. Comments similar to those made previously apply to these lens shapes.

2.8.2. The Coddington Position Factor

The *Coddington position factor* π is defined as

$$\pi \equiv \frac{u_3 + u_1}{u_3 - u_1}. \tag{2.97}$$

The various imaging geometries are shown in Figure 2.28. The condensing lens configuration shown in Figure 2.28(a) has $\pi = +1$, the unity magnification arrangement shown in Figure 2.28(b) has $\pi = 0$, and the collimating lens configuration shown in Figure 2.28(c) has $\pi = -1$. As $M = u_1/u_3$,

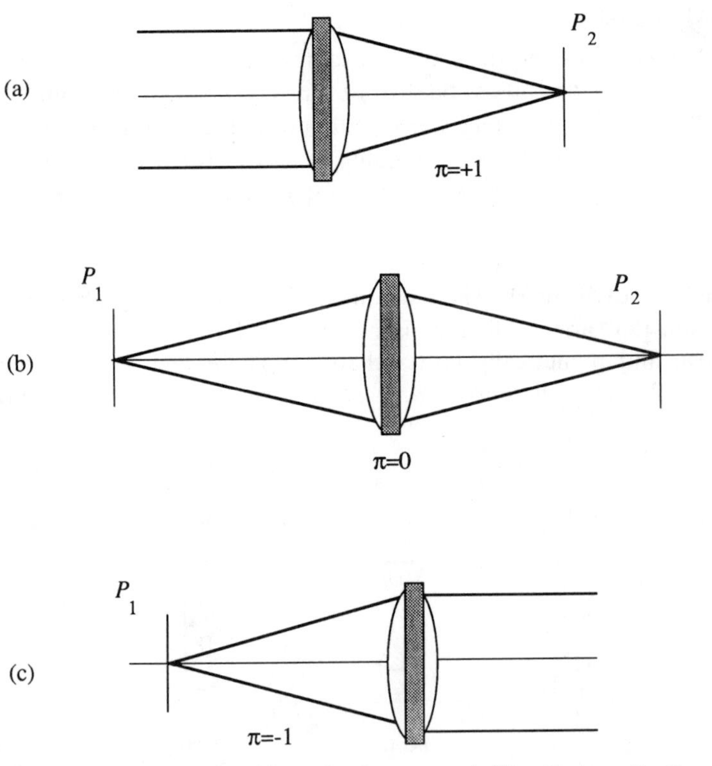

Figure 2.28. Imaging geometries: (a) condensing, $\pi = +1$; (b) unity magnification, $\pi = 0$; and (c) collimating, $\pi = -1$.

we relate the Coddington position factor to the magnification by

$$\pi = \frac{1 + M}{1 - M}, \tag{2.98}$$

so that we easily find the position factor required for any magnification.

2.9. ABERRATIONS

Aberrations result from the imperfect way in which rays are bent as they pass through an optical system. One way to characterize aberrations is to trace rays from selected object samples and to determine where they intersect the image plane. The resulting ray pattern is called the *point spread function*, equivalent to the *impulse response* in linear system theory. Ray tracing is at the heart of geometrical optics design. With present-day computers and lens-design programs, it is relatively easy to arrive at such a well-designed lens that all the rays pass through a region smaller than the diffraction limited resolution of the lens. The ray-tracing program then no longer accurately describes the light distribution near focus because diffraction effects, as we discuss in Chapter 3, dominate.

As an alternative to ray tracing, we calculate the wavefront that is normal to the ray bundle and express the aberrations in terms of a polynomial function. From the wavefront surface, the nature of the aberrations is easier to visualize, and, in conjunction with diffraction theory, the exact form of the light distribution near the focal point is obtained. The aberration polynomial function is equivalent to the *system response* $H(f)$ in linear system theory, except that the system response changes as a function of the position of a sample in the object. The optical system is therefore *space variant* so that a unique system response does not generally exist. Aberration theory recognizes the space-variant property of the system response and accurately predicts the system response to all samples in the object. The intent of this brief discussion of aberrations is to familiarize the reader with the nature of the aberrations, how to recognize them, and how to select a lens shape or the proper system configuration to minimize them in laboratory setups. More extensive treatments of aberrations and methods for correcting them are found in various texts and monographs (10–13). Photographs of the effects of lens aberrations are found in the first three of these references and in a more recent text (14).

Consider the generalized imaging system shown in Figure 2.29, where the curved surface S images an object sample from O_1 at the focal point O_2. Fermat's principle requires that all rays from O_1 to O_2 have the same

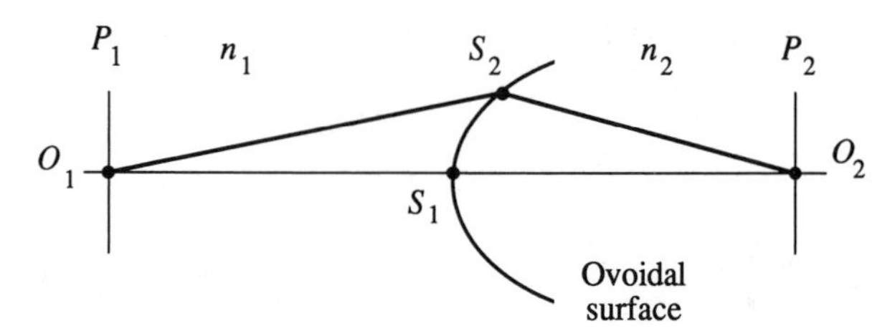

Figure 2.29. Surface to provide aberration-free performance for a single axial point.

optical path length so that the ideal wavefront is the ovoidal surface represented by

$$O_1S_2 + S_2O_2 = O_1S_1 + S_1O_2 = \text{constant}. \tag{2.99}$$

In general we cannot fabricate such surfaces and, even if we could, the imagery is perfect only at one object sample. The lens designer's task, then, is to control aberrations over an extended region, while using surface shapes that are easily manufactured.

We use the coordinate system, shown in Figure 2.30, that defines a sample in the image plane by the distance ρ; the coordinates at the lens plane are r and ϕ. The *monochromatic wavefront aberration polynomial*

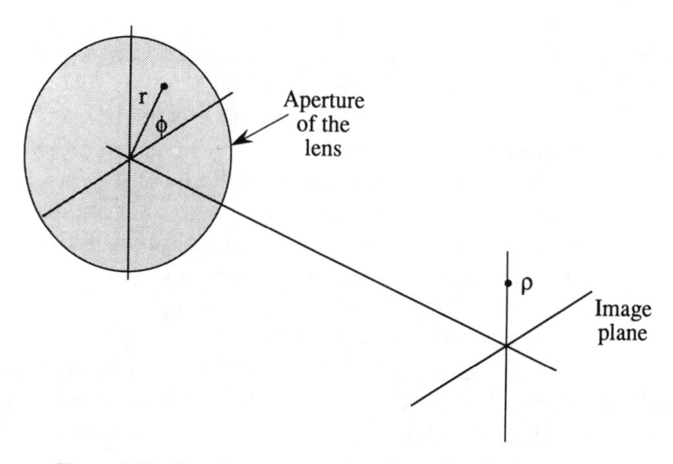

Figure 2.30. Coordinate system for characterizing aberrations.

for third-order aberrations is given by

$$W(\rho, r, \phi) = a_1 r^4 + a_2 \rho r^3 \cos\phi + a_3 \rho^2 r^2 \cos^2\phi + a_4 \rho^2 r^2 + a_5 \rho^3 r \cos\phi, \tag{2.100}$$

where the coefficients determine the magnitude of the aberration. The wavefront aberration $W(\rho, r, \phi)$ is the optical path difference between the actual wavefront and a reference wavefront that is a perfect sphere, with the paraxial focal point as its center. Each of the five primary aberration terms are in the fourth power of combinations of the variables ρ and r.

Because the lateral displacements of the rays at the focal plane are proportional to the derivatives of $W(\rho, r, \phi)$, the aberrations given by Equation (2.100) are called *third-order aberrations*. We have ignored first-order aberrations, such as a defect of focus or a lateral focal shift, which do not affect image quality. We compensate for these "aberrations" by adjusting the axial position of the focal plane or the lateral position of the optical axis. Fifth- and higher-order aberrations are generally not important unless the relative apertures of the lenses are high. This situation does not normally arise in optical signal-processing systems because the optical invariant is held to a reasonably low value by currently available input/output devices. Chromatic aberrations are ignored because the illumination is monochromatic.

2.9.1. Spherical Aberration

In Figure 2.31(a), we show the reference wavefront W_r whose curvature is proportional to r^2 that would produce a perfect image of an object sample located at infinity. The first term of the aberration polynomial is *spherical aberration*, for which the actual wavefront is given by $a_1 r^4$. When spherical aberration is present, the wavefront is curved too much if a_1 is positive, or too little if a_1 is negative. For the case shown, a_1 is positive and we see that the marginal rays are too highly bent so that they cross the optical axis on the object side of the point where the paraxial rays cross. The clue for correcting spherical aberration is to recall that the deviation angle for a prism is a function of the incident ray angle. We consider a lens as a set of prisms, each with a different angle, as shown in Figure 2.31(b); and we attempt to equalize the angles of incidence and refraction by bending the lens.

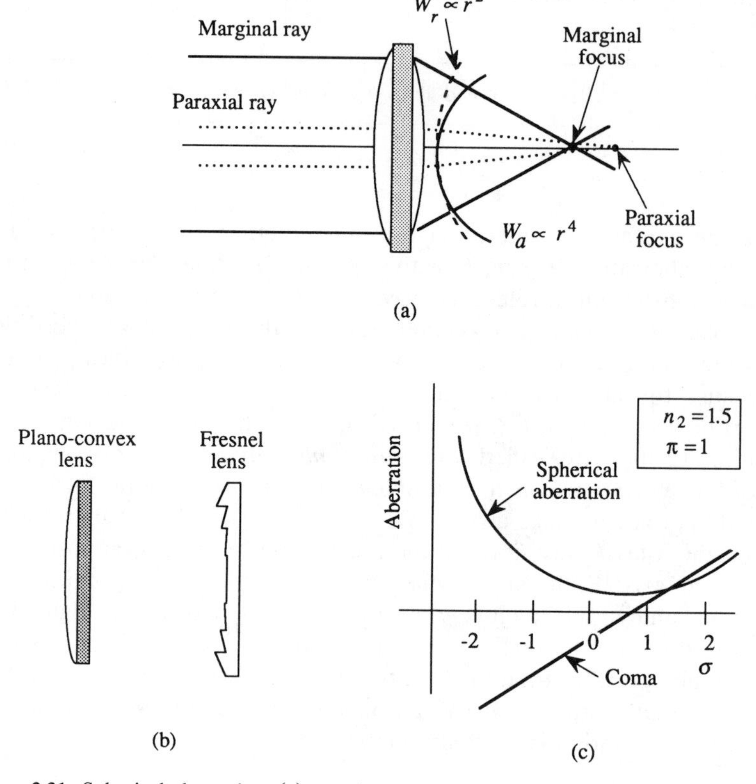

Figure 2.31. Spherical aberration: (a) wavefront representation, (b) equivalent Fresnel lens, and (c) plot of spherical aberration and coma.

A measure of the spherical aberration as a function of the lens shape factor σ is shown in Figure 2.31(c). The proper shape factor to correct spherical aberration is dependent on the object/lens geometry. For a glass lens in air, the optimum shape factor σ for minimizing spherical aberration as a function of the position factor π is (13)

$$\sigma = \frac{2(n_2^2 - 1)}{n_2 + 2}\pi. \tag{2.101}$$

We find that $\sigma = 0.7\pi$ for a typical lens of refractive index $n_2 = 1.5$. When $\pi = 1$, the minimum spherical aberration occurs at $\sigma = 0.7$ as shown in Figure 2.31(c). This σ value is close to that of a plano convex lens, oriented so that its curved surface is toward the plane wavefront produced by the object. With this lens orientation we see that the angle of incidence with respect to the first surface and the angle of refraction with respect to the second surface are nearly equal. When the magnification is unity, the shape factor is zero, independent of n_2; in this case, a biconvex lens minimizes the spherical aberration because the angles of incidence and refraction are then equal. Although spherical aberration is partially corrected by selecting the proper lens shape, it is better controlled by using a cemented doublet. The primary reason for using cemented doublets is to correct chromatic aberrations, but spherical aberration is also significantly reduced.

The *Hartman test* is a simple laboratory test to quickly estimate the amount of spherical aberration and to determine if a lens is overcorrected or undercorrected. As illustrated in Figure 2.32, we construct a mask containing two small apertures, one near the edge of the lens and the second at the center of the lens. We place this mask at the lens plane, and begin by allowing light to pass only through the central aperture to establish the paraxial focal position. This aperture is small enough to produce a well-formed diffraction pattern, sometimes called the *Airy disc*, but large enough to accurately determine the paraxial focus position.

We then cover the central aperture, allow light to pass through the edge aperture, and determine if the second bundle of rays crosses the axis before or after the paraxial focal plane. For example, if the spherical aberration is positive and the edge aperture is above the central aperture, the bundle crosses the axis before the paraxial focal plane and the second Airy disc is below the first Airy disc at the focal plane. Spherical aberration is negative if the relative positions are reversed.

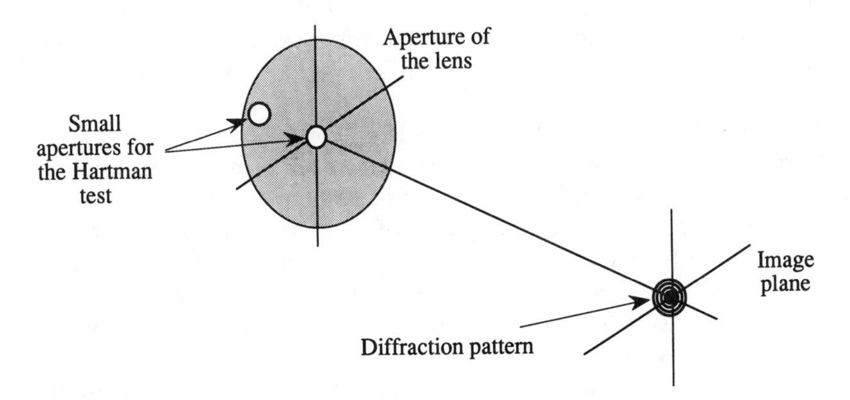

Figure 2.32. Hartman test for spherical aberration.

2.9.2. Coma

From Equation (2.100), we note that spherical aberration is not a function of the position of a sample in the object plane; that is, it is not a function of ρ. It is, therefore, the only primary aberration that produces a space-invariant response. The first aberration that is a function of ρ is *coma*, given by $a_2 \rho r^3 \cos \phi$. Coma arises when the principal planes are curved surfaces, a condition that occurs only in high-performance systems. As a result, the magnification of the system is greater, or less, for marginal rays than for paraxial rays, even in the absence of spherical aberration. When coma is present, the image of a sample has a shape like a comet. Lens designers minimize coma by making the magnification equal for all rays. From the optical invariant, with the sines of the angles retained, we find that $n_3 h_3 \sin u_3 = n_1 h_1 \sin u_1$. For $n_1 = n_3 = 1$ we have

$$M = \frac{\sin u_1}{\sin u_3}, \tag{2.102}$$

known as the *sine condition*, which must hold for all finite angles u_1 and u_3 to ensure that coma is zero.

Coma is also a function of the lens shape and is minimized for a glass lens in air when (13)

$$\sigma = \frac{(2n_2 + 1)(n_2 - 1)}{n_2 + 1} \pi. \tag{2.103}$$

In Figure 2.31(c) we plot the relative value of coma as a function of σ. Coma is minimized when $\sigma = 0.8\pi$ for $n_2 = 1.5$. For $\pi = +1$, both coma and spherical aberration are therefore minimized at nearly the same shape factor. Because the minimum value for spherical aberration is a fairly broad function of σ, we control both aberrations fairly well when we minimize coma.

A simple laboratory test for coma is to construct a set of masks, as illustrated in Figure 2.33, containing two small apertures with various distances between them. We place a mask in the plane of the lens so that the two apertures are at opposite edges of the lens. As the mask is rotated clockwise, the light distribution at the focal plane rotates counterclockwise, tracing out the locus of the extreme tail of the coma which passes through the paraxial focal point. Although the magnification is roughly the same for all rays from either small aperture, it is quite different for the two apertures taken together. As other masks with smaller distances between the apertures are inserted into the system, the light distribution

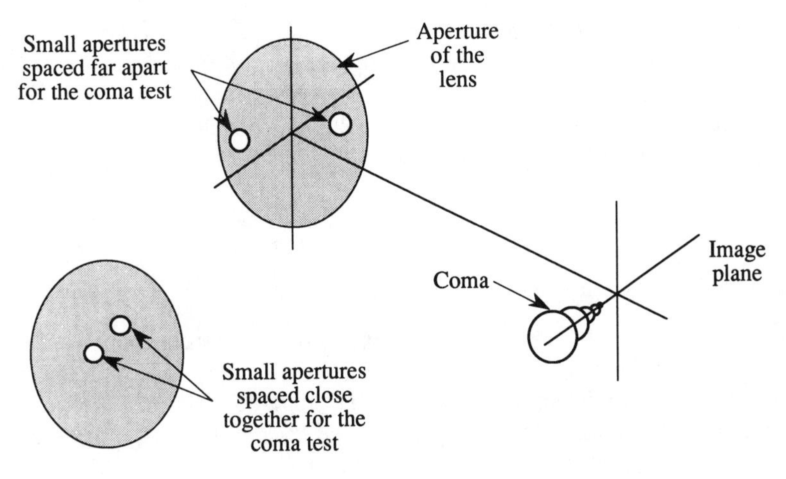

Figure 2.33. Test for coma.

at the focal plane becomes more compact; when only one aperture is present, the head of the comet is produced.

2.9.3. Astigmatism

Astigmatism, the third term in the aberration polynomial (2.100), is given by $a_3\rho^2 r^2 \cos^2 \phi$. Astigmatism results when rays in the *tangential plane* do not focus at the same point as those from the orthogonal plane, called the *sagittal plane*. Astigmatism is illustrated in Figure 2.34, where we see that the curvature of the wavefront is not the same in the two orthogonal

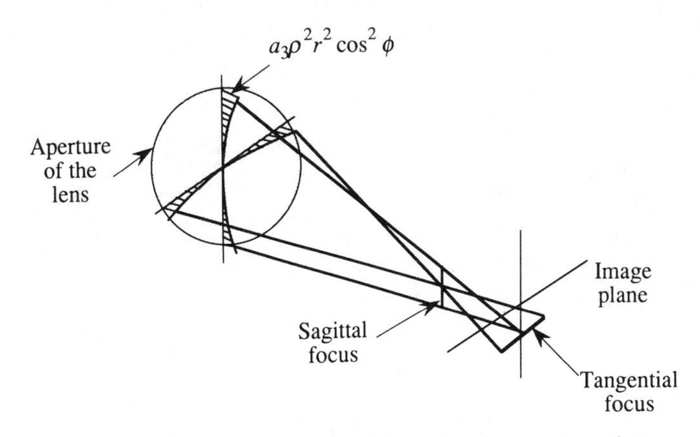

Figure 2.34. Sagittal and tangential focus lines due to astigmatism.

planes, leading to the two focal planes. As astigmatism is not a function of the lens shape, we control it by keeping ρ and r within bounds relative to the lens focal length. A simple laboratory test for astigmatism is to image a test target that looks like a spoked wheel. The rim is in focus at the tangential focal plane, and the spokes are in focus at the sagittal focal plane.

2.9.4. Curvature of Field

A simple lens works best when it focuses an object from a curved surface onto a curved image surface. When a lens is forced to work with flat object and image planes, an aberration called *curvature of field* appears, as illustrated in Figure 2.35. Curvature of field, given by $a_4\rho^2r^2$, is independent of all other aberrations. The coefficient a_4 depends on the ratios of the powers of the lens elements to their refractive indices, summed over all the elements in the system. This value is called the *Petzval sum*:

$$J^2\sum_i \frac{K_i}{n_i}, \tag{2.104}$$

where J is the optical invariant, and K_i and n_i are the powers and the refractive indices of the lens elements. Because a system containing only positive lenses has a large positive Petzval sum, we control curvature of field by introducing negative lenses where they have little effect on image quality. One possibility is to use negative lenses near the object or image planes; in the latter case the lens is called a *field flattener*.

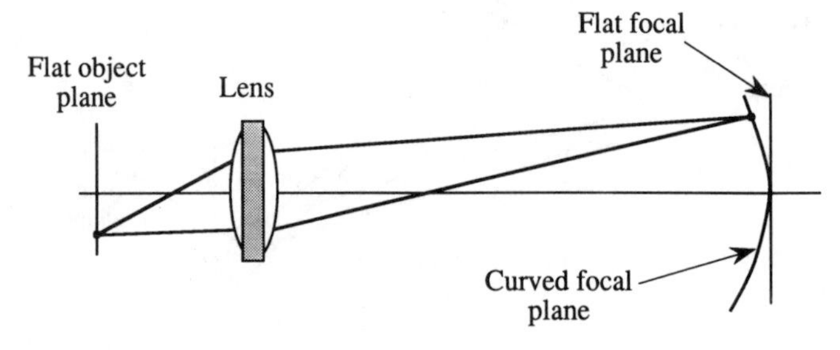

Figure 2.35. Curvature of field.

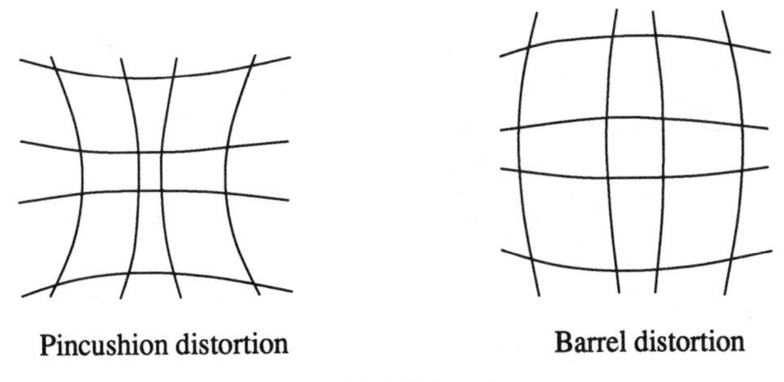

Figure 2.36. Distortion.

2.9.5. Distortion

The fifth primary aberration is *distortion*, described by $a_5\rho^3 r \cos\phi$. Distortion affects the shape of an image, not its sharpness. Distortion is due to a variation in the magnification for off-axis object samples. Distortion causes a regular grid of squares to take on a pincushion or barrel shape, as shown in Figure 2.36, according to the sign of a_5. Distortion is zero for longitudinally symmetric systems, whose aperture is in the middle of the system. As optical signal-processing systems are often configured symmetrically, distortion is generally not a problem.

2.9.6. Splitting the Lens

Using more than one lens is a simple method for reducing aberrations in a laboratory setup. Consider the unity magnification imaging system, shown in Figure 2.37(a), in which a lens of power K is placed a distance $2F$ from the object plane. As $M = -1$, we find that $\pi = 0$ and we might be tempted to use a biconvex lens. Suppose, however, that we split the lens into two lenses, each with power $K/2$ as shown in Figure 2.37(b). The object and image planes are still a distance $2F$ from the two lenses, but the position factors are now $\pi = -1$ and $\pi = +1$ for the two lenses.

The major benefit of lens splitting is that the relative apertures of the lenses have been reduced by a factor of 2 so that aberrations such as spherical aberration, coma, astigmatism, and curvature of field are reduced, some rather dramatically. Distortion is identically equal to zero due to symmetry for the special case of unity magnification. Spherical aberration and coma are minimized by using suitable lens shapes, such as a plano convex lens, configured so that the curved surfaces face each other.

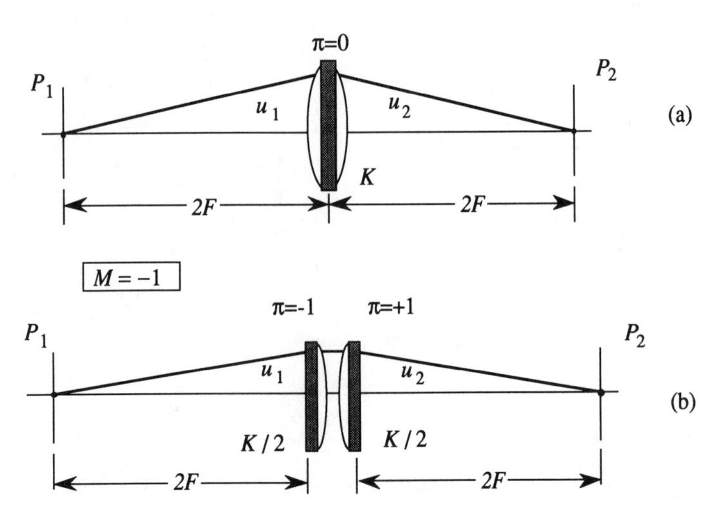

Figure 2.37. Splitting an imaging lens to control aberrations: (a) original system and (b) equivalent system.

Almost any imaging setup benefits from splitting the lens, but the advantages are greatest when the magnification is near unity. To illustrate the more general case, suppose that $M > -1$. The immediate question is how to split the lens of power K into two others. As we would like each of the resulting lenses to operate with $|\pi| = 1$, we find that

$$K_1 = \frac{MK}{M - 1} \tag{2.105}$$

and

$$K_2 = \frac{-K}{M - 1}. \tag{2.106}$$

As a sanity check, we note that $K_{eq} = K_1 + K_2 = K$. We can therefore split the lens for any system configuration.

For large M, we find that the relative aperture of the first lens is not reduced significantly because the object is already near its front focal plane. The aberrations of the second lens are, however, significantly reduced relative to those of either the original lens or to those of the first lens. Finally, the focal lengths of available lenses in a laboratory may not be what we need to achieve a given magnification. We can then depart slightly from the $\pi = \pm 1$ condition to fine-tune the final magnification without seriously changing the aberrations.

Some final thoughts about aberrations. Be aware of these aberrations—know how to identify them and what causes them. Based on this knowledge, make laboratory setups for preliminary experiments by proper choice of the lens types and the geometric arrangements (the values of σ and π). Consult with a lens designer when you are about to build a system. Learn to specify, but not to overspecify, the operational requirements, and let the lens designer do the rest.

PROBLEMS

2.1. You have a prism for which $n_2 = 1.6$ and $\gamma = 30°$.

(a) For an incidence angle of $I_1 = 20°$ (on the base side of the normal of the prism), calculate the deviation angle. Provide a sketch.

(b) What value of I_1 produces total internal reflection? Provide a sketch of the situation.

2.2. A ray is incident at an angle of 10° relative to the normal of a prism whose index of refraction is 1.6. Calculate the apex angle for which the ray is internally reflected by the prism (the critical angle). Plot the angle as a function of the index of refraction of the prism for n_2 ranging from 1.4 to 2.0.

2.3. A parallel beam of light 20 mm high enters a prism normal to its first surface. The prism has an index of refraction of 1.8 and an apex angle of 27 degrees. Calculate:

(a) the magnification of the prism (state clearly the direction in which the light is traveling),

(b) the deviation angle (show clearly by a sketch), and

(c) the smallest apex angle that produces total internal reflection.

2.4. You have a prism whose apex angle is 30° and whose index of refraction is 1.7. Design a second prism (i.e., find its apex angle and index of refraction) if the magnification of the pair must be equal to 0.3. If we were to use two identical prisms whose index of refraction is 1.5, calculate the apex angle needed to produce a magnification of 0.3.

2.5. Derive a general solution for the displacement h of a ray passing through a wobble plate of index n_2 and thickness z_{12} as a function of the angle of incidence I_1. Show that your result reduces to

$$h = \frac{z_{12} I_1 (n_2 - 1)}{n_2},$$

for small I_1. What is the maximum possible displacement? Find the approximate angle at which the exact and approximate solutions differ by 1%.

2.6. Is it possible to image a point object (for simplicity, assume that the object is a star at infinity) on the back surface of a glass sphere whose radius is R? If so, what is the required index of refraction? Hint: Use the refraction equation and remember that it represents the paraxial approximation, i.e., the entering rays should be ones near the optical axis.

2.7. We want to expand and collimate a beam of light from an ideal laser whose beam divergence is zero, using a pair of telescopic lenses whose powers are $K_1 = 0.05\ \text{mm}^{-1}$ and $K_2 = 0.005\ \text{mm}^{-1}$. The light fills the first lens whose diameter is 5 mm. Upon measurement, we find that we made an error in the spacing between the lenses so that the beam has a convergence angle of 10 mrad. How far, and in what direction, should we move the second lens? Remember that the magnification is negative, i.e., that an entrance ray above the optical axis crosses over and becomes an exit ray below the optical axis. Having found the correct position of the lenses, sketch the system and calculate the magnification.

2.8. Consider a spherical surface that separates media of refractive indices n_1 and n_2. In terms of the power K of the surface, calculate the positions of the front and back focal planes.

2.9. We have an optical system that contains a single lens whose focal length is 100 mm. We want to change its effective focal length to 200 mm by using a second lens. Because of mechanical constraints, we must place the second lens 40 mm beyond the first lens. Find the focal length of the second lens and provide a sketch for the final system.

2.10. You need to image an object with a magnification of $M = -4$. You have been given a blob of glass, whose refractive index is $n = 1.62$, that is just enough to make a single lens element. You want the image to be 100 mm from the lens. Provide a sketch of the general geometry. You melt this glass and give it to a lens maker as a blank circular piece from which he must grind the lens. Calculate (1) the Coddington position factor π, (2) the optimum Coddington shape factor σ that will minimize the spherical aberration, (3) the curvatures c_1 and c_2 necessary to achieve this shape, and (4) the optimum shape factor and the curvatures c_1 and c_2 needed to minimize coma. Hint: Use the refraction equation and the formula for M involving the angles, along with the relationship for π and σ. Please watch the signs and be numerically accurate.

2.11. You have just bought a Camcorder for making home videos. The manufacturer states that the sensor in the camera has 400 resolution elements (equivalent to 400 samples) in the vertical direction, which is 10 mm high. Assume that the wavelength of light is 0.5 μ. Calculate (a) the optical invariant (remember that the optical invariant applies to image planes—use the version involving N), (b) the maximum spatial frequency that can be resolved at the sensor, and (c) the $f/\#$ necessary to match the resolution of the lens to that of the sensor (assume an infinite conjugate imaging condition).

2.12. You want to make a logo for your newest video production, using the Camcorder from Problem 2.11. You identify a useful picture in a magazine that has been printed with a 100 sample/mm halftone screen. If you set the system magnification to record a 250-mm-high logo onto the sensor, what will be the maximum resolvable frequency in cycles/mm (referenced to the plane of the logo) that can be preserved? If you wish to preserve all of the detail (or resolution) available in the logo, how much of its height can you capture?

2.13. You have a TV set with 330 scan lines distributed uniformly over 15 inches in the vertical direction. If the angular resolution of the eye is 500 μrad, at what distance from the set must you be to just resolve the scan lines according to the Rayleigh-resolution criterion? (Consider each scan line to be equivalent to a sample or a smallest resolvable detail.)

2.14. Suppose that we have a CCD detector array with 400 photodetector elements along a 10-mm line. We have an LED array with 250 elements in an 8-mm length that we use as the object. Find the focal lengths of a two-lens system, with each lens operating at infinite conjugates (the π factors are ± 1) such that the resolution is matched, given that the diameter of the lenses must be 5 mm.

2.15. You want to use an injection laser diode having dimensions of $10\ \mu \times 50\ \mu$ as a light source. You need a collimated beam of 30 mm in each direction and decide to use a spherical lens, followed by a beam-expanding prism. Calculate the aperture and focal length of the lens, as well as the apex angle for a prism whose index of refraction is 1.55. Do a sketch of the top and side views. Hint: Think of the laser dimensions as the size of a single sample of a generalized signal.

3

Physical Optics

3.1. INTRODUCTION

The basic theory of geometrical optics as given in Chapter 2 describes how optical elements such as lenses, mirrors, and prisms modify the direction of light rays. Physical optics extends the theoretical treatment of optical systems by incorporating the wave nature of light. We take a direct approach and refer the reader to various texts that provide the details of the development from Maxwell's equation (10, 15–18).

We begin with the basic assumption that light waves propagate in an isotropic media with simple harmonic motion and satisfy the scalar wave equation

$$\nabla^2 \phi(z,t) = \frac{1}{c^2} \frac{\partial^2 \phi(z,t)}{\partial t^2}, \tag{3.1}$$

where t is time and z is distance in the direction that the wave travels. The representation for free-space electromagnetic radiation is a real-valued function of the form $\cos(\omega t - kz)$, where $k = 2\pi/\lambda$ and ω is the radian frequency of light.

The transfer functions for lenses, prisms, and other optical elements are usually represented by complex-valued functions. As an example, Figure 3.1 shows an arbitrary optical element, illuminated by monochromatic light at wavelength λ that propagates parallel to the z axis. The light wave at plane z_0, represented by $\sqrt{2}\cos(\omega t - kz_0)$, is spatially modulated by the optical element whose magnitude transmittance is $|a(x)|$ and whose phase is $\phi(x)$. The phase difference between planes z_0 and z_1 is the product of the optical path difference and k:

$$\begin{aligned}\phi(x) &= \frac{2\pi}{\lambda}[n_1 d(x) + \{\Delta z - d(x)\}] \\ &= \frac{2\pi}{\lambda}[(n_1 - 1)d(x) + \Delta z],\end{aligned} \tag{3.2}$$

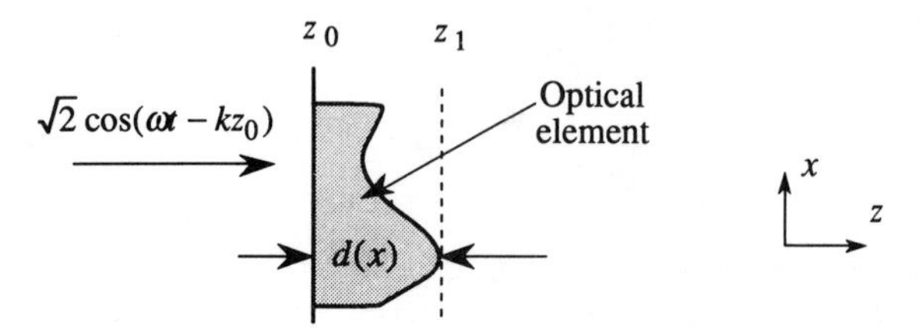

Figure 3.1. A general optical element.

where $\Delta z = z_1 - z_0$. Light at plane z_1 is therefore represented by

$$\sqrt{2}\,|a(x)|\cos[\omega t - kz_0 - \phi(x)], \tag{3.3}$$

which shows that light is modulated in both magnitude and phase.

It is often more convenient to use the phasor notation $\exp[j(\omega t - kz)]$ as a solution of the wave equation. In this sign convention, the wave propagates in the positive spatial z direction and temporal frequencies are positive if, when represented by a phasor diagram, they rotate in a counterclockwise fashion. This time/space sign convention is consistent with those of geometrical optics, physical optics, and electrical engineering. Using this notation, we represent light at plane z_1 of Figure 3.1 as

$$|a(x)|e^{j[\omega t - kz_0 - \phi(x)]}. \tag{3.4}$$

We generally suppress the temporal part of the electromagnetic wave, because no detector has sufficient bandwidth to respond directly to the amplitude fluctuations at light frequencies, and we also generally ignore the relative phase kz_0 at plane z_0. With these conventions, we represent the complex transmittance of the optical element by

$$a(x) = |a(x)|e^{-j\phi(x)}. \tag{3.5}$$

The complex transmittances of several optical elements in series multiply, as we expect. For example, three cascaded elements have an effective transmittance $a_4(x)$:

$$\begin{aligned} a_4(x) &= a_1(x)a_2(x)a_3(x) \\ &= |a_1(x)|\,|a_2(x)|\,|a_3(x)|e^{-j[\phi_1(x)+\phi_2(x)+\phi_3(x)]}, \end{aligned} \tag{3.6}$$

and the associated real-valued representation of the light is

$$\sqrt{2}\,|a_1(x)|\,|a_2(x)|\,|a_3(x)|\cos[\omega t - kz_0 - \phi_1(x) - \phi_2(x) - \phi_3(x)], \tag{3.7}$$

which illustrates that the magnitude transmittances are multiplicative and the phases are additive. Throughout this book we describe optical elements by complex-valued functions of the form given by Equation (3.6), with the understanding that the real-valued representations of propagating light are given by functions represented by Equation (3.7).

Any physical detector senses the *intensity* of light, defined as

$$I(x, y, z) \equiv a(x, y, z, t)a^*(x, y, z, t), \tag{3.8}$$

where $*$ indicates complex conjugate. To illustrate that dropping the time dependence of the wave is valid when the postdetection bandwidth does not extend to the frequency of light, we obtain the intensity of the light at plane z_1 by two different methods. First, we use the real-valued version of the plane wave from Equation (3.3) and find that the intensity at plane z_1 becomes

$$\begin{aligned} I(x,t) &= 2|a(x)|^2\cos^2[\omega t - kz_0 - \phi(x)] \\ &= 2|a(x)|^2\{\tfrac{1}{2} + \tfrac{1}{2}\cos[2\omega t - 2kz_0 - 2\phi(x)]\}. \end{aligned} \tag{3.9}$$

A photodetector responds to the time average of $I(x,t)$ to produce a current $g(x)$ that is a function of only the space variable:

$$g(x) = \langle I(x,t)\rangle = |a(x)|^2. \tag{3.10}$$

The second method is to calculate the current in a direct fashion, as though the temporal component is not present, from Equation (3.5):

$$\begin{aligned} g(x) &= \langle a(x)a^*(x)\rangle \\ &= |a(x)|^2, \end{aligned} \tag{3.11}$$

which confirms that the time average notation is not needed explicitly and that we can ignore the temporal frequency of light in our basic phasor notation.

As an example of phasor notation, consider the complex-valued representation of a plane wave, propagating upward and to the right at an angle

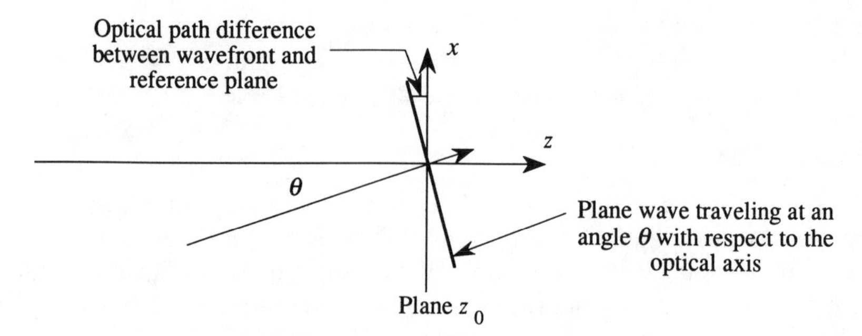

Figure 3.2. Plane-wave representation.

θ, as shown in Figure 3.2. The wave has magnitude $|a(x)|$ and phase $\exp[-j\phi(x)]$. We measure the phase at any value of the variable x relative to the phase at $x = 0$. As the index of refraction in air is unity, the phase is given by the linear function

$$\phi(x) = \frac{2\pi}{\lambda}\theta x. \tag{3.12}$$

We use the relationship established in Chapter 2 between ray angles and spatial frequencies, namely, that $\alpha = \theta/\lambda$, to find that the plane-wave representation becomes

$$a(x) = e^{-j2\pi\alpha x}. \tag{3.13}$$

We see then that we associate a spatial frequency α with a plane wave that propagates at a ray angle θ with respect to the optical axis. The signs of both α and θ are reversed if the wave travels downward and to the right.

The linear phase representation given by Equation (3.13) is also used to represent the complex transmittance of a prism. Recall from Chapter 2 that a prism operates on a plane wave, described there in terms of a set of rays normal to the wavefront, and deflects the wave so that it travels in a new direction. Thus, if θ_1 is the incidence angle of a ray with respect to the optical axis and if θ_2 is the corresponding angle of the refracted ray, the deviation angle is $\theta_3 = \theta_2 - \theta_1$. The prism is therefore represented by $\exp[-j2\pi\alpha_3 x]$, where $\alpha_3 = \pm\theta_3/\lambda$, and the $\pm$ sign indicates whether the wave has an upward or downward component.

From Equations (3.10) and (3.11) we see that the intensity reveals no information regarding either the temporal or the spatial frequency of a plane wave. To strengthen the idea that a plane wave traveling with a ray angle with respect to the optical axis has an associated spatial frequency, we render that frequency visible. Suppose that we add a plane wave $r(x)$, traveling parallel to the optical axis, to the inclined wavefront shown in Figure 3.2. The total amplitude at plane z_0 is then the sum of the two plane waves; the intensity, for $|r(x)| = |a(x)| = 1$, is

$$\begin{aligned} I(x) &= \left| r(x) + a(x)e^{-j2\pi\alpha x} \right|^2 \\ &= \left| 1 + e^{-j2\pi\alpha x} \right|^2 \\ &= 2[1 + \cos(2\pi\alpha x)]. \end{aligned} \tag{3.14}$$

From Equation (3.14) it is clear that α is the spatial frequency associated with the light distribution at plane z_0. As the angle θ between the two waves increases, the spatial frequency increases correspondingly. A firm link between spatial frequencies and the angle between two wavefronts is therefore established.

3.2. THE FRESNEL TRANSFORM

Fresnel transforms relate the complex-valued light distributions located at two planes separated by free space. In holography, for example, it is the Fresnel transform of an object that is recorded for subsequent reconstruction. Fresnel transforms are used in synthetic aperture radar processing to determine the appropriate range and azimuth processing operations. In this chapter, we use the Fresnel transform to illustrate interference and diffraction phenomena and to develop the more familiar Fourier transform.

Fresnel extended Young's principle of interference to cases where the light is polarized. His work did much to confirm the transverse nature of light waves. In a key development, Fresnel modified Huygens' principle for relating the complex-valued light distribution at two separated planes in an optical system. Suppose, for example, that we know the light distribution $f(x, y)$ at plane P_1, as shown in Figure 3.3. We want to calculate the light distribution $g(\xi, \eta)$ at plane P_2, a distance D from plane P_1.

Fresnel's idea, stated here in a somewhat revised form, is that the elemental contribution δg at a point (ξ, η) in the observation plane, due to an elemental region $dx\,dy$ near a point (x, y) in the input plane, is proportional to several factors. First, we find that $f(x, y)$ represents the complex-valued amplitude in the neighborhood of a point (x, y) at plane

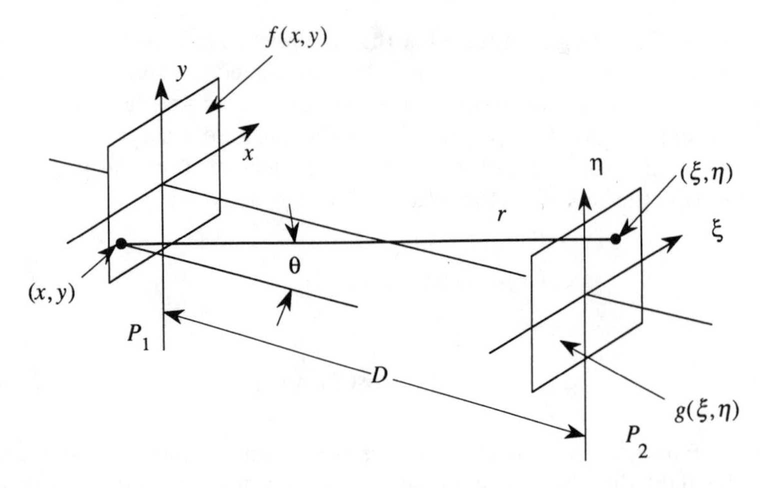

Figure 3.3. Fresnel diffraction.

P_1. Second, the exponential factor $\exp[j(\omega t - kr)]$ represents the phase change in the light as it propagates between planes P_1 and P_2, where r is the length of the ray connecting the points (x, y) and (ξ, η). Third, two factors determine how the magnitude changes as light propagates from plane P_1 to plane P_2. The first factor is $1/r$, which accounts for the fact that intensity is reciprocally related to the square of the propagation distance; the magnitude is therefore inversely proportional to r. The second factor is the *obliquity factor* $(1 + \cos\theta)/2$, where θ is the ray angle with respect to the optical axis. Finally, a fixed phase factor j is needed to obtain the correct results in Fresnel diffraction calculations, as we show in Section 3.2.4, and λ is a scaling factor to account for the wavelength of light. We combine all these factors to find the total contribution at (ξ, η):

$$g(\xi, \eta) = \frac{je^{j\omega t}}{\lambda r} \iint_{-\infty}^{\infty} \left[\frac{1 + \cos\theta}{2}\right] f(x, y) e^{-jkr}\, dx\, dy. \tag{3.15}$$

We begin the solution of Equation (3.15) by considering the obliquity factor $(1 + \cos\theta)/2$. In Chapter 2 we showed that the maximum ray angle $\theta_{co} = \lambda/(2d_0)$, where d_0 is the sample interval required to support a signal bandlimited to the cutoff frequency α_{co}. The distance d_0 is no less than a few wavelengths, even for rather wideband signals. For example, if the sample interval is $d_0 > 3\lambda$, all ray angles are such that $\theta < \frac{1}{6}$ and the minimum value of $(1 + \cos\theta)/2$ is therefore always greater than 0.993. This result, based on a wide-bandwidth signal for which $\alpha_{co} = \theta_{co}/\lambda = 333.3$ Ab, allows us to ignore the obliquity factor and to calculate the Fresnel diffraction integral for arbitrary values of the propagation distance D, including $D \to 0$.

The next step in the solution of Equation (3.15) is to consider the value of r:

$$r^2 = D^2 + (x - \xi)^2 + (y - \eta)^2. \tag{3.16}$$

By use of the binomial expansion, we find that

$$r = D\left[1 + \frac{(x - \xi)^2}{2D^2} + \frac{(y - \eta)^2}{2D^2} + \text{higher-order terms}\right]. \tag{3.17}$$

We neglect the higher-order terms because their effects are dominated by the first-order terms for all practical signal-processing applications. Furthermore, we replace r by D in the denominator of Equation (3.15), with an error in the magnitude that is generally less than 1%; in fact, replacing r by D partially compensates for ignoring the obliquity factor and increases the accuracy of the results. In any event, magnitude weighting has relatively little effect on diffraction; the phase is much more important.

With these factors accounted for, the diffraction integral of Equation (3.15) becomes

$$g(\xi, \eta) = \frac{je^{-jkD}}{\lambda D} \iint_{-\infty}^{\infty} f(x, y) e^{-j(\pi/\lambda D)[(x-\xi)^2 + (y-\eta)^2]}\, dx\, dy. \tag{3.18}$$

We ignore the time exponential factor $e^{-j\omega t}$ because we are mostly concerned with the spatial relationship between $g(\xi, \eta)$ and $f(x, y)$. To further explore the Fresnel transform, we also ignore the exponential phase factor e^{-jkD}, which simply represents the phase accumulation as the light travels from plane P_1 to P_2.

We now switch to a one-dimensional notation and find that free space behaves as an operator that produces a *Fresnel transform* defined by

$$\boxed{g(\xi) \equiv \sqrt{\frac{j}{\lambda D}} \int_{-\infty}^{\infty} f(x) e^{-j(\pi/\lambda D)(x-\xi)^2}\, dx,} \tag{3.19}$$

where the exponential function represents the free-space response to an impulse. The scaling factor in the one-dimensional case is the square root of that for the two-dimensional case because the wave is diverging cylindrically instead of spherically. Because the magnitude does not change in the y direction the magnitude decreases only as $\sqrt{1/D}$.

The formulation of the Fresnel transform as given by Equation (3.19) has the following features.

- Shows the convolutional process between the free-space operator and the input signal.
- Explicitly displays the impulse response of free space as the function $\sqrt{j/\lambda D}\,\exp[-j(\pi/\lambda D)x^2]$.
- Produces a continuous transition from the Fresnel, or near-field, diffraction pattern, to the Fraunhofer, or far-field, diffraction pattern as a function of the distance D.
- Retains the necessary phase factors to facilitate the analysis of optical systems, using additional lenses and free-space intervals to achieve other processing operations, as discussed in Section 3.6.

3.2.1. Convolution and Impulse Response

To further illustrate the nature of the Fresnel transform, we define a function

$$\psi(x;d) \equiv e^{j\pi x^2/\lambda D}, \tag{3.20}$$

where $d = 1/D$ is a reciprocal distance. From Equations (3.19) and (3.20), we see that $g(\xi)$ is the convolution of $f(x)$ with the *free-space impulse response* $\sqrt{d}\,\psi^*(x;d)$, where $*$ indicates complex conjugate. We therefore describe the output $g(\xi)$ in terms of the input signal $f(x)$ that has passed through a black box whose impulse response is $\sqrt{d}\,\psi^*(x;d)$, as shown in Figure 3.4. The output of the black box is

$$g(\xi) = \sqrt{d}\int_{-\infty}^{\infty} f(x)\psi^*(\xi - x;d)\,dx, \tag{3.21}$$

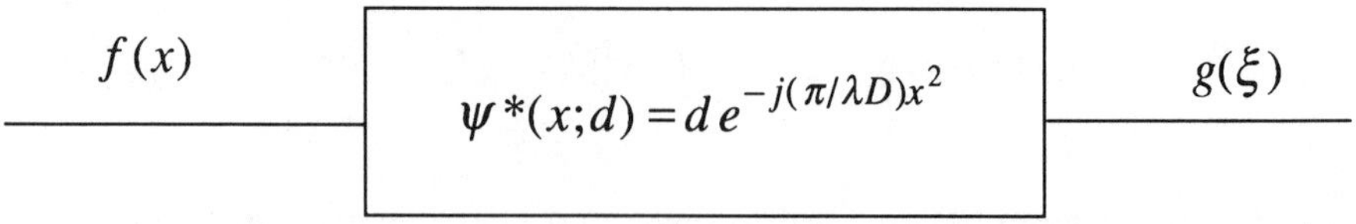

$f(x)$ $\longrightarrow$ $\boxed{\psi^*(x;d) = d\,e^{-j(\pi/\lambda D)x^2}}$ $\longrightarrow$ $g(\xi)$

Figure 3.4. Equivalent block diagram for propagation of light through free space.

which is similar to Equation (3.19) in its form. In general, we drop scaling constants when developing major concepts; we restore them when needed to calculate actual light intensities.

To illustrate these concepts, we represent one sample of the input signal by an impulse function so that $f(x) = \delta(x)$. The output of the system, by the sifting property of the delta function, is

$$g(\xi) = \sqrt{d}\,\psi^*(\xi; d) = \frac{1}{\sqrt{D}} e^{-j\pi\xi^2/\lambda D}, \tag{3.22}$$

so that $g(\xi)$ is a cylindrically diverging wave of radius D, as shown in Figure 3.5. Light from the point source is therefore *dispersed* spatially as it propagates through free space. The magnitude of the wave at plane P_2 is uniform, whereas the phase is quadratic in ξ. The phase is proportional to the optical path difference between the wavefront and plane P_2, as a function of the coordinate ξ. As $D \to \infty$, the magnitude of $g(\xi)$ tends to zero while its phase approximates that of a plane wave.

As another example, suppose that the sample is centered at $x = x_0$ so that $f(x) = \delta(x - x_0)$. We again use the sifting property of the delta function to find that

$$g(\xi) = \frac{1}{\sqrt{D}} e^{-j(\pi/\lambda D)(\xi - x_0)^2}. \tag{3.23}$$

This result represents a diverging cylindrical wave of radius D, translated

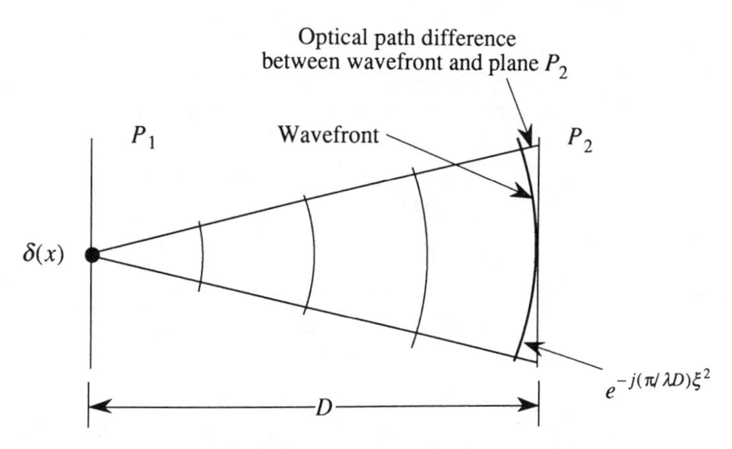

Figure 3.5. Spherical waves propagating from a point source.

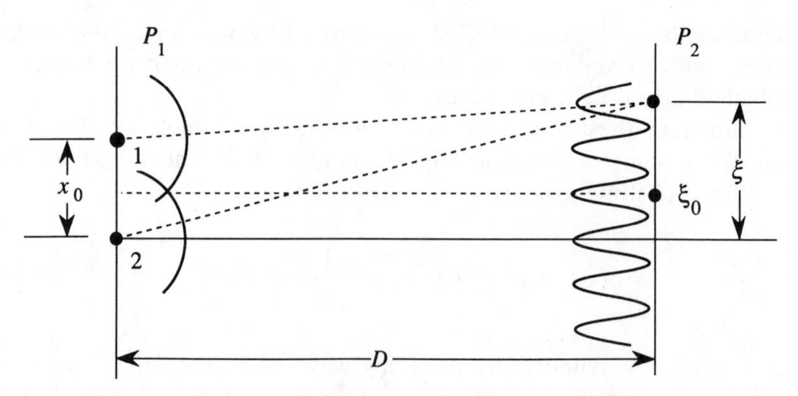

Figure 3.6. One-dimensional representation of the two-source geometry.

a distance x_0 in the x direction throughout the space from P_1 to P_2. Therefore, if an arbitrary signal $f(x)$ is displaced a distance x_0 in plane P_1 of Figure 3.5, so too is its Fresnel transform $g(\xi)$ displaced a distance x_0 in plane P_2.

3.2.2. Diffraction by Two Sources

Figure 3.6 shows two light sources at plane P_1, separated by a distance x_0. The light amplitude at plane P_2, due to the two sources when they are in phase, is given by Equations (3.22) and (3.23):

$$g(\xi) = C_0 e^{-j(\pi/\lambda D)\xi^2} + C_1 e^{-j(\pi/\lambda D)(\xi - x_0)^2}, \tag{3.24}$$

where C_0 and C_1 are the magnitudes of the waves at plane P_2 due to the two sources. The observable quantity at plane P_2 is the intensity of light:

$$\begin{aligned} I(\xi) &= g(\xi)g^*(\xi) = |g(\xi)|^2 \\ &= C_0^2 + C_1^2 + 2\,\mathrm{Re}\left[C_0 C_1 e^{-j(\pi/\lambda D)\xi^2} e^{+j(\pi/\lambda D)(\xi - x_0)^2}\right] \\ &= C_0^2 + C_1^2 + 2C_0 C_1 \cos\left[\frac{2\pi}{\lambda D} x_0 \xi - \frac{\pi}{\lambda D} x_0^2\right]. \end{aligned} \tag{3.25}$$

The first two terms of $I(\xi)$ are intensities produced by the individual sources; their sum is a spatially uniform intensity called the *bias*. The third term, a spatially varying cosine distribution called the *fringe pattern*, has a *spatial frequency* α given by the partial derivative of the phase with respect to the spatial variable ξ:

$$\alpha = \frac{1}{2\pi} \frac{\partial}{\partial \xi} \left(\frac{2\pi}{\lambda D} x_0 \xi \right) = \frac{x_0}{\lambda D}. \tag{3.26}$$

The spatial frequency of the fringe pattern therefore increases as the separation x_0 between the sources increases and as the observation distance D decreases. Note that the ratio x_0/D is simply the angular size of the entire source as seen from the receiving screen.

The phase of the fringe pattern is also given by Equation (3.25):

$$\varphi = -\frac{\pi x_0^2}{\lambda D}. \tag{3.27}$$

The physical interpretation of the phase is that it specifies the position of the maximum intensity in the observation plane. The *principal maximum* occurs where the argument of the cosine is equal to zero, i.e., where $2\pi \xi x_0/\lambda D - \pi x_0^2/\lambda D = 0$. The principal maximum therefore occurs at $\xi_0 = x_0/2$, which is directly opposite the midpoint between the two sources, as shown in Figure 3.6. Other intensity maxima occur where the phase is an integer multiple of 2π. If the light from the two sources have arbitrary phases ϕ_0 and ϕ_1, the entire fringe pattern at plane P_2 is shifted according to the phase difference, $\phi_1 - \phi_0$, and the principal maximum moves to

$$\xi_1 = \frac{x_0}{2} - \frac{\phi_1 - \phi_0}{2\pi x_0/\lambda D}. \tag{3.28}$$

The variation of the fringe intensity is measured by the *fringe visibility*:

$$\boxed{V = \frac{I_{\max} - I_{\min}}{I_{\max} + I_{\min}} = \frac{2C_0C_1}{C_0^2 + C_1^2}.} \tag{3.29}$$

We see that visibility is not a function of the phase difference between the two sources and that maximum fringe visibility is achieved when the two sources have equal magnitudes.

3.2.3. Fresnel Zones, Chirp Functions, and Holography

Suppose that we add a plane-wave *reference beam* to a unit magnitude wave produced by a sample function as shown in Figure 3.7(a). The plane wave, which may be thought of as the limiting form of $\exp(-j\pi\xi^2/\lambda D)$ as $D \to \infty$, is also assigned unit magnitude. The intensity at plane P_2 due to the sum of the reference and signal waves is

$$I(\xi) = |g(\xi)|^2 = |1 + e^{-j\pi\xi^2/\lambda D}|^2$$
$$= 2\left[1 + \cos(\pi\xi^2/\lambda D)\right]. \qquad (3.30)$$

This fringe pattern is generally called a *Fresnel zone pattern*, as shown in Figure 3.7(b). The first intensity maximum occurs at $\xi = 0$, a point at plane P_2 directly opposite the source at plane P_1. The first minimum occurs where $\pi\xi^2/\lambda D = \pi$ or at $\xi_0 = \sqrt{\lambda D}$. Successive nulls occur when

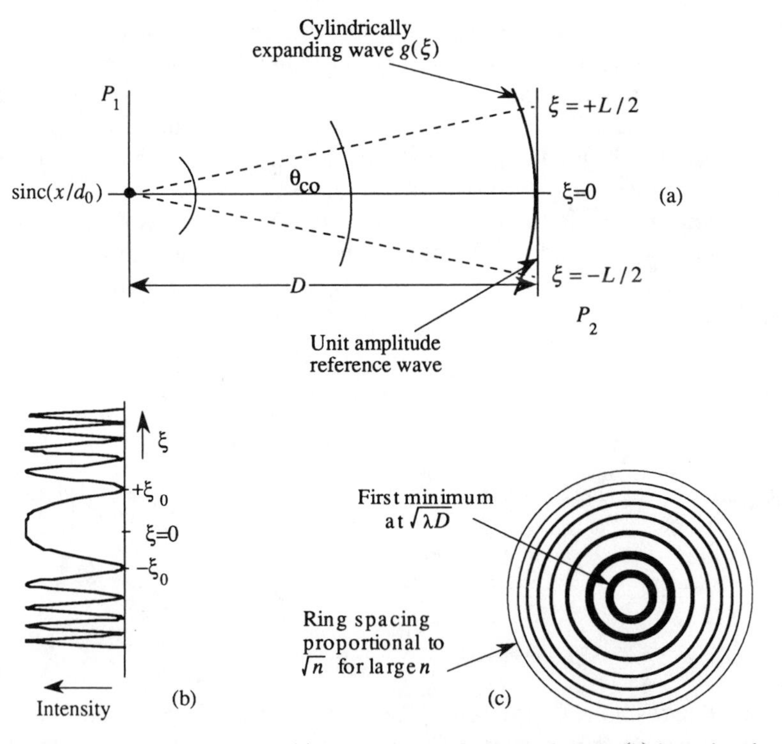

Figure 3.7. Fresnel zone pattern: (a) plane wave and spherical wave, (b) intensity along a radial line, and (c) two-dimensional fringes.

$\pi\xi^2/\lambda D = (2n+1)\pi$ or at $\xi = \sqrt{(2n+1)\lambda D}$, where $n = 0, 1, 2, \ldots$. For large values of n, the nulls are spaced according to $\sqrt{n}$ so that the areas of the rings approach a constant as shown in Figure 3.7(c) for the two-dimensional case.

The spatial frequency is low at the center of the pattern, as shown in Figure 3.7(b), and increases at increasing distances from the center. The spatial frequency of the Fresnel zone pattern along a horizontal line is

$$\alpha = \frac{1}{2\pi}\frac{\partial}{\partial\xi}\left[\frac{\pi\xi^2}{\lambda D}\right] = \frac{1}{2\pi}\left[\frac{2\pi\xi}{\lambda D}\right] = \frac{\xi}{\lambda D}, \tag{3.31}$$

from which we see that the spatial frequency is a linear function of the position variable ξ at plane P_2. Because the spatial frequency increases linearly as the distance increases from the origin, the Fresnel zone pattern is called a *chirp function* in signal processing. From Equation (3.31), we see that the slope of the chirp function is $1/\lambda D$, so that a small value of D implies a more rapid change in spatial frequency as a function of spatial position. The slope is called the *chirp rate* and is expressed in Ab/mm.

From Equation (3.31), we see that the highest spatial frequency occurs at the point $\xi_{\max} = L/2$. The maximum spatial frequency is therefore $\alpha_{co} = L/2\lambda D$, where L is the total length of the chirp at plane P_2. In Chapter 2 we defined the length bandwidth product as LBP $= L\alpha_{co}$, so that the length bandwidth product of the chirp is

$$\text{LBP} = L\alpha_{co} = L\left[\frac{\xi_{\max}}{\lambda D}\right] = \frac{L^2}{2\lambda D}. \tag{3.32}$$

From the Nyquist sampling theorem, we know that the number of samples needed to accurately represent the chirp function is equal to twice the length bandwidth product:

$$\boxed{N = 2\text{LBP} = \frac{L^2}{\lambda D}.} \tag{3.33}$$

This important relationship relates the values of LBP and N to the purely geometrical properties of the chirp; we use this result frequently in subsequent developments.

We also relate the chirp function to θ_{co}, the maximum ray angle produced by diffraction from the sample function whose form is $\text{sinc}(x/d_0)$. Again, we use the important fact that a spatial frequency is associated with every ray angle; in particular, the highest spatial frequency α_{co} associated with the cutoff ray angle θ_{co} does not change as the wave propagates. From Equation (3.31), we find that $\alpha_{co} = L/2\lambda D$ and, from Figure 3.7(a), that $\theta_{co} = L/2D$, so that

$$\boxed{\alpha_{co} = \frac{\theta_{co}}{\lambda} = \frac{1}{2d_0},} \tag{3.34}$$

as we established in Chapter 2 by appealing to purely geometric arguments. This curious result shows that the maximum spatial frequency of a chirp at *all* planes from P_1 to P_2 is α_{co}. As a direct consequence, the sample spacing d_0 is also constant for all planes.

The number of samples needed to describe the chirp is $N = L/d_0$, where L is the length of the chirp waveform at an arbitrary plane. The value of N is therefore a function of the propagation distance D, a result that is easily seen by noting that $L = 2\theta_{co}D$. Only one sample is needed to characterize the chirp at plane P_1, whereas many samples are required when D is large. This result seems to conflict, at first, with the notion that the value of N and the optical invariant are connected by $N = 4J/\lambda$. But the optical invariant applies only to *conjugate image planes*, a condition not satisfied here.

The Fresnel zone pattern is fundamental to holography in which the magnitudes of diverging wavefronts caused by secondary scattering samples from a signal are added together and combined with a reference beam. Gabor originally conceived holography as a way to correct the aberrations of an electron beam microscope (19). He recognized that a two-dimensional interference pattern represented by the Fresnel zone, as shown in Figure 3.7(c), contains information about the position of the object sample in three-dimensional space. Because an arbitrary signal consists of many such samples, it should be possible to record the information on film so that the image could be reconstructed at visible wavelengths after the aberrations were corrected. The basic problem with the Gabor hologram, however, is that using a reference beam colinear with the signal beam does not allow the signal beam to be cleanly reconstructed because the desired information spatially overlaps other terms, such as the bias.

In the early 1960's, Leith and Upatnieks demonstrated the benefits of using an off-axis reference wavefront (20). Figure 3.8 shows an arrange-

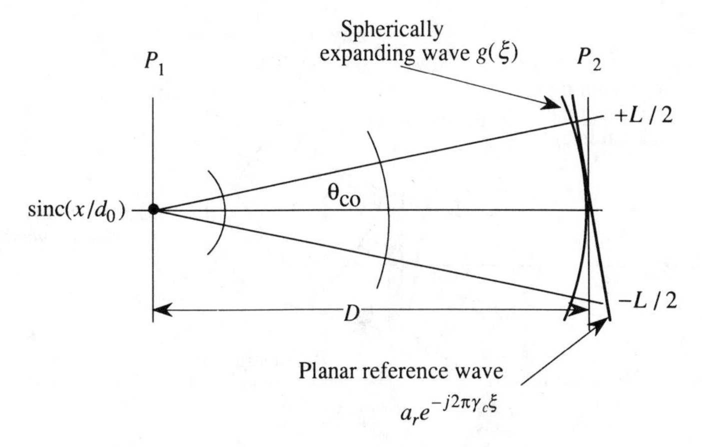

Figure 3.8. Setup for constructing an off-axis hologram.

ment for recording the simplest possible off-axis hologram, a one-dimensional sample of the form $\mathrm{sinc}(x/d_0)$. This arrangement is similar to that shown in Figure 3.7, except that the reference wave, represented by $r(\xi) = a_r \exp(-j2\pi\gamma_c\xi)$, now propagates upward and to the right. The signal waveform is represented by $g(\xi) = a_s \exp[-j\pi\xi^2/\lambda D]$ so that the intensity at plane P_2 is

$$\begin{aligned} I(\xi) &= \left| a_r e^{-j2\pi\gamma_c\xi} + a_s e^{-j(\pi/\lambda D)\xi^2} \right|^2 \\ &= a_r^2 + a_s^2 + 2a_r a_s \,\mathrm{Re}\left[e^{-j2\pi\gamma_c\xi} e^{+j(\pi/\lambda D)\xi^2} \right] \\ &= a_r^2 + a_s^2 + 2a_r a_s \cos\left[2\pi\gamma_c\xi - \frac{\pi}{\lambda D}\xi^2 \right]. \end{aligned} \tag{3.35}$$

The hologram is recorded by illuminating a photosensitive medium, such as photographic film, for a time t_0; the *exposure* is defined as $E(\xi) \equiv t_0 I(\xi)$.

The hologram is reconstructed by illuminating it with a replica of the reference beam; this beam is now called the *reconstruction beam*, as shown in Figure 3.9. If the hologram is recorded so that its amplitude transmittance is linearly proportional to $E(\xi)$, the total amplitude to the right of the hologram is $h(\xi) = t_0 r(\xi) I(\xi)$. The nature of the three wavefronts released from the hologram becomes evident when we expand the cosine

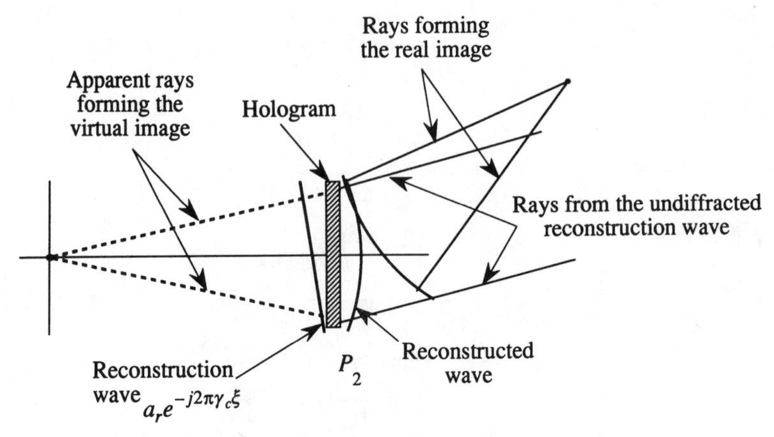

Figure 3.9. Setup for reconstructing an off-axis hologram.

in Equation (3.35) using the Euler formula:

$$\begin{aligned} h(\xi) &= t_0 a_r e^{-j2\pi\gamma_c\xi}\Big[a_r^2 + a_s^2 + a_r a_s e^{j2\pi\gamma_c\xi} e^{-j(\pi/\lambda D)\xi^2} \\ &\qquad + a_r a_s e^{-j2\pi\gamma_c\xi} e^{+j(\pi/\lambda D)\xi^2}\Big] \\ &= t_0 a_r\left(a_r^2 + a_s^2\right) e^{-j2\pi\gamma_c\xi} + t_0 a_r^2 a_s e^{-j(\pi/\lambda D)\xi^2} \\ &\quad + t_0 a_r^2 a_s e^{-j4\pi\gamma_c\xi} e^{+j(\pi/\lambda D)\xi^2}. \end{aligned} \tag{3.36}$$

The first term of Equation (3.36) is simply the reconstruction beam which continues to propagate upward and to the right, with some attenuation as introduced by the hologram. The second term of Equation (3.36) has the same form as the original signal beam and it propagates to the right of the hologram as though it had never been intercepted. This wavefront produces a *virtual image* whose apparent position, as seen from the right of the hologram, is that of the original signal; the rays associated with this image are shown dashed in Figure 3.9. The last term of Equation (3.36) represents a wave, propagating toward the right at twice the reconstruction beam angle, whose spherical phase factor has a positive sign produced by the conjugation operation. This *conjugate* wavefront represents a convergent wave that forms a *real image* of the original signal as shown by the solid rays. At some distance to the right of the hologram, the three beams no longer overlap, thus producing a clean reconstruction of the signal.

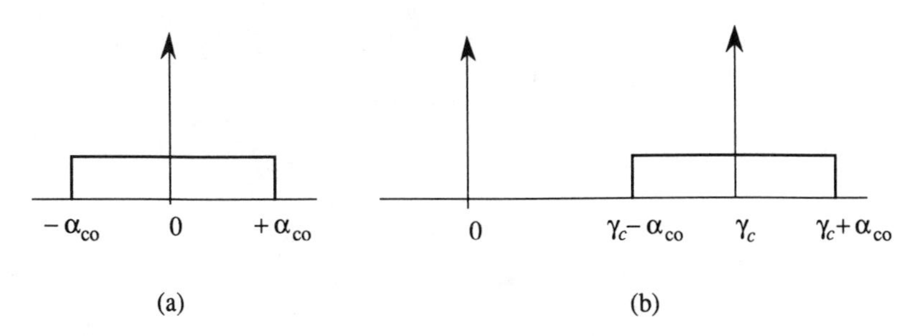

Figure 3.10. Spatial frequencies of (a) a Gabor hologram and (b) a Leith-Upatnieks hologram.

An off-axis hologram requires more spatial frequency response from the recorded hologram. In Figure 3.10(a), we see that the frequency response of a Gabor hologram ranges from zero to α_{co}. In a Leith-Upatnieks hologram, the off-axis reference beam shifts the frequency range from baseband to a bandpass region, as shown in Figure 3.10(b), with γ_c as the center frequency. The frequency range is therefore from $\gamma_c - \alpha_{co}$ to $\gamma_c + \alpha_{co}$. If the signal subtends a large angle, the center frequency required of the hologram is correspondingly large because γ_c must be proportional to the angle subtended by the object.

From Equation (3.35) we see that all the information about the point signal is encoded in the hologram by using a combination of magnitude, frequency, and phase modulation. If the signal contains several samples, each sample produces its own Fresnel zone pattern at the hologram plane. The amplitude responses of each sample are added at the plane of the hologram by the principle of superposition. The hologram reconstruction distinguishes each sample *position* because its associated spatial frequency shifts upward or downward, within the passband, according to whether the sample produces a larger or smaller angle with respect to the reference beam. Three-dimensional signals are recorded by encoding the *distance* of the samples from the hologram through phase modulation of the type shown in Equation (3.35). Finally, we do not require a planar reference wave; holograms can be recorded under a wide range of geometrical conditions. The interested reader can consult one of several books on holography for more details (21–23).

The values of the spatial frequencies associated with the exact Fresnel transform, when the higher-order terms in the expansion of the radius vector r are retained, differ somewhat with the approximate result obtained so far. The difference arises because of the approximations in the expansion for the ray length r as given by Equation (3.17); the approximate form is

$$r \cong D\left[1 + \frac{(x-\xi)^2}{2D^2}\right], \tag{3.37}$$

but the exact form is

$$r = \sqrt{D^2 + (x-\xi)^2}. \tag{3.38}$$

As the value of the spatial frequency is dependent only on the ray angle and is independent of x, we set x equal to zero for convenience. The exact frequency is therefore

$$\alpha = \frac{1}{2\pi}\frac{\partial}{\partial \xi}\left[\frac{2\pi}{\lambda}\sqrt{D^2 + \xi^2}\right] = \frac{\xi}{\lambda D}\left[1 + \frac{\xi^2}{D^2}\right]^{-1/2}, \tag{3.39}$$

so that the approximate spatial frequency value $\alpha = \xi/\lambda D$ must be divided by a correction factor $(1 + \xi^2/D^2)^{1/2}$ to obtain the exact spatial frequency value. Recall that $\alpha_{co} = 1/(2d_0) = 1/(6\lambda)$ for the case where $d_0 = 3\lambda$. In this event, the maximum error in frequency is 1.4% at α_{co}. In most signal-processing systems we find that $d_0 \approx 10\lambda$, for which the maximum frequency error is of the order of 0.5%. The difference between the approximate and the exact solution therefore is often of little consequence in most signal processing applications.

In holography, however, the ray angles between the signal and reference beams can sometimes approach 90°. From the exact solution, we find that the maximum spatial frequency, as $\xi \to \infty$, is bounded by $\alpha = 1/\lambda$. In contrast, the approximate solution for r leads to a spatial frequency that goes to infinity as $\xi \to \infty$. This difference between the exact and approximate spatial frequency is important when calculating Fresnel transforms on a digital computer, because the Fresnel kernel as given by Equation (3.20) will produce alias frequencies if the approximate solution is used, no matter how small the sample interval at the input. When using the exact solution, the Fresnel kernel will not alias, provided that the input signal is sampled with $d_0 \leq \lambda/2$.

3.2.4. The Fresnel Transform of a Slit

Consider the Fresnel transform of a slit of length L, shown in Figure 3.11, and represented by $f(x) = \text{rect}(x/L)$. The Fresnel transform at plane

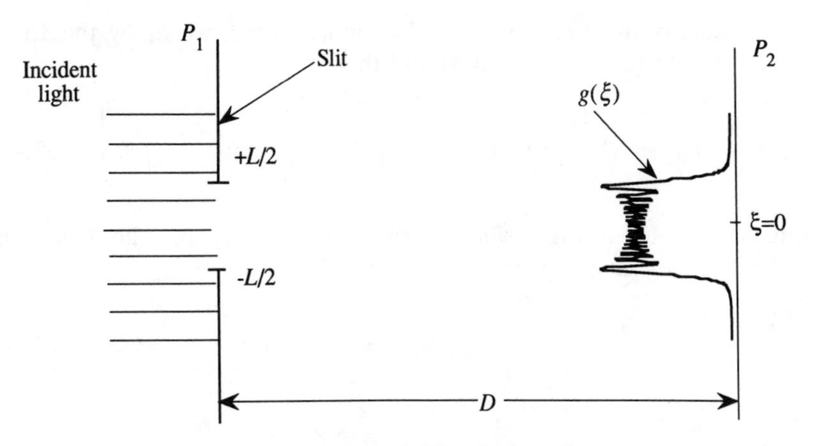

Figure 3.11. Fresnel diffraction by a slit.

P_2 is

$$\begin{aligned} g(\xi) &= \sqrt{\frac{jI_0}{\lambda D}} \int_{-\infty}^{\infty} f(x) e^{-j(\pi/\lambda D)(x-\xi)^2}\, dx \\ &= \sqrt{\frac{jI_0}{\lambda D}} \int_{-\infty}^{\infty} \operatorname{rect}\left(\frac{x}{L}\right) e^{-j(\pi/\lambda D)(x-\xi)^2}\, dx \\ &= \sqrt{\frac{jI_0}{\lambda D}} \int_{-L/2}^{L/2} e^{-j(\pi/\lambda D)(x-\xi)^2}\, dx, \end{aligned} \tag{3.40}$$

where I_0 is the intensity at $x = 0$ at plane P_1. By a change of variables in which $x - \xi = \sqrt{\lambda D/2}\, z$, the differential becomes $dx = \sqrt{\lambda D/2}\, dz$ and the upper and lower limits become

$$\begin{aligned} z_2 &= \sqrt{2/\lambda D}\,(L/2 - \xi), \\ z_1 &= \sqrt{2/\lambda D}\,(-L/2 - \xi). \end{aligned} \tag{3.41}$$

With this change of variables, Equation (3.40) becomes

$$g(\xi) = \sqrt{\frac{jI_0}{2}} \int_{z_1}^{z_2} e^{-j\pi z^2/2}\, dz, \tag{3.42}$$

which is, aside from the scale factor, in the standard form of a Fresnel *integral*.

To examine the Fresnel integral in more detail, we apply the Euler relationship to the exponential to find that

$$g(\xi) = \sqrt{\frac{jI_0}{2}}\left[\int_{z_1}^{z_2} \cos\left(\frac{\pi z^2}{2}\right) dz - j\int_{z_1}^{z_2} \sin\left(\frac{\pi z^2}{2}\right) dz\right], \quad (3.43)$$

which are related to the cosine and sine Fresnel integrals. The definitions of the *Fresnel cosine and sine integrals* are that

$$C(z) \equiv \int_0^z \cos(\pi u^2/2)\, du, \quad (3.44)$$

and

$$S(z) \equiv \int_0^z \sin(\pi u^2/2)\, du. \quad (3.45)$$

Neither $C(z)$ nor $S(z)$ can be solved in closed form. For numerical calculations, the Fresnel sine and cosine integrals are approximated by the relationships that

$$C(z) = \tfrac{1}{2} + f(z)\sin\left(\frac{\pi}{2}z^2\right) - g(z)\cos\left(\frac{\pi}{2}z^2\right) \quad (3.46)$$

and

$$S(z) = \tfrac{1}{2} - f(z)\cos\left(\frac{\pi}{2}z^2\right) - g(z)\sin\left(\frac{\pi}{2}z^2\right), \quad (3.47)$$

where we use the rational approximations that (24)

$$f(z) = \frac{1 + 0.926z}{2 + 1.792z + 3.104z^2} + \varepsilon(z); \qquad |\varepsilon(z)| \le 2(10^{-3}) \quad (3.48)$$

and

$$g(z) = \frac{1}{2 + 4.142z + 3.492z^2 + 6.670z^3} + \varepsilon(z); \qquad |\varepsilon(z)| \le 2(10^{-3}). \quad (3.49)$$

The error $\varepsilon(z)$ in these approximations is small and the rational approximations are much easier and faster to compute than the integrals given by Equation (3.44) or Equation (3.45). The rational approximations are valid

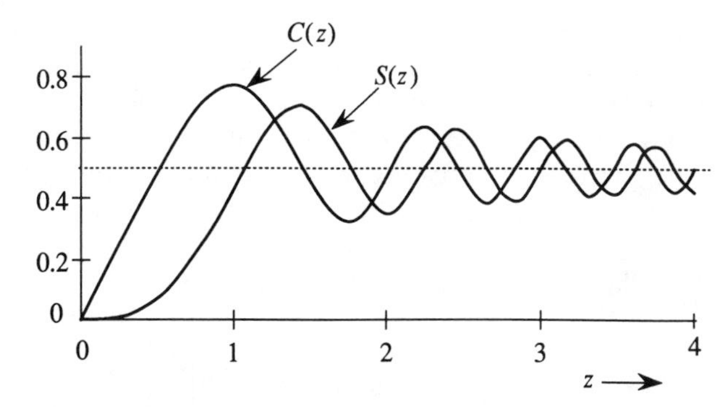

Figure 3.12. Fresnel sine and cosine integrals: $C(z)$ and $S(z)$.

only for positive values of z; and the symmetry relationships for $C(z)$ and $S(z)$ provide the values for all z. The Fresnel sine and cosine integrals are plotted in Figure 3.12. The oscillatory behavior of both functions is obvious and, for large values of z, both functions are asymptotically equal to $\frac{1}{2}$.

One way to better understand the Fresnel integral is through the use of the Cornu spiral, as shown in Figure 3.13. The *Cornu spiral* is a plot of the

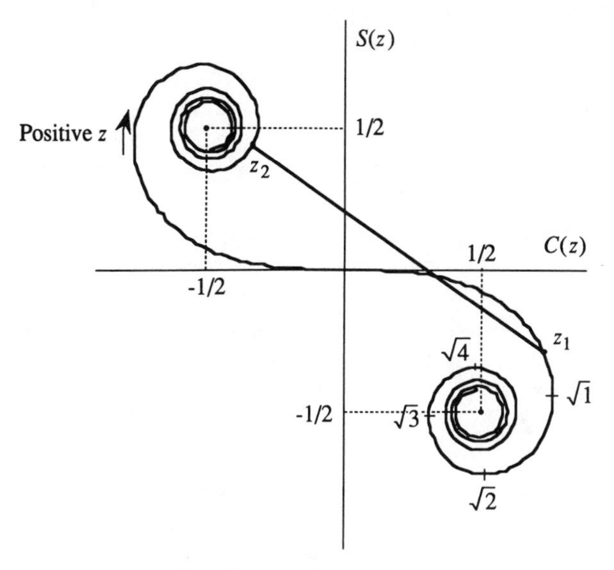

Figure 3.13. Cornu spiral.

real part of the Fresnel integral from Equation (3.43) on the horizontal axis and the imaginary part on the vertical axis; this is equivalent to the parametric plotting of $S(z)$ as a function of $C(z)$. Because the exponential factor in Equation (3.42) is negative as a result of our fundamental sign convention, the Cornu spiral must lie in the second and fourth quadrants. The parameters z_1 and z_2 represent points on the spiral and $z_{12} = z_2 - z_1$ is a measure along the arc of the spiral.

A tangent to the Cornu spiral has zero slope when

$$\frac{\partial}{\partial z} S(z) = \frac{\partial}{\partial z}\left[\int_0^z \sin\left(\frac{\pi u^2}{2}\right) du\right] = \sin\left(\frac{\pi z^2}{2}\right) = 0, \qquad (3.50)$$

from which we conclude that $\pi z^2/2 = n\pi$ so that $z = 0, \sqrt{2}, \sqrt{4}, \ldots$ are the horizontal tangent points. In a similar fashion, the tangent has infinite slope when

$$\frac{\partial}{\partial z} C(z) = \frac{\partial}{\partial z}\left[\int_0^z \cos\left(\frac{\pi u^2}{2}\right) du\right] = \cos\left(\frac{\pi z^2}{2}\right) = 0, \qquad (3.51)$$

so that the vertical tangent points occur when $z = \sqrt{1}, \sqrt{3}, \sqrt{5}, \ldots$. Thus, we see that the quarter turning points on the Cornu spiral occur when the values of the parameter z are the square roots of successive integers.

The parameters z_2 and z_1 represent normalized distances at plane P_1 from a perpendicular line that intersects plane P_2 at the point of observation to the edges of the slit, as shown in Figure 3.14. When the observation point is at a large value of ξ, both z_1 and z_2 are large in magnitude. Depending on the sign of ξ, the line length z_{12} shown in Figure 3.13 is tightly wrapped into one end or the other of the spiral; and the intensity of the Fresnel transform is low.

Suppose that the observation point is initially at $\xi = +\infty$, where the intensity is low. As the observation point moves toward $\xi = 0$, a point is reached when $z_1 = 0$ and $z_2 = -\sqrt{2/\lambda D}\,L$. The intensity in this region increases rapidly because it is the transition region from the geometric shadow to the clear aperture region. As the observation point moves to where $z_2 = 0$, we transition back into the shadow region at the side of the slit where $\xi = -L/2$.

Suppose that the slit is infinitely long so that $L \to \infty$. In this case, $z_1 = \infty$ and $z_2 = -\infty$ and the ends of the vector are located at the centers of each end of the spiral. From the Cornu spiral, we find that the magnitudes of Equations (3.44) and (3.45) are equal to $\sqrt{2}/2$, and the phase of the vector is given by $\exp[-j\pi/4]$. The complex amplitude at

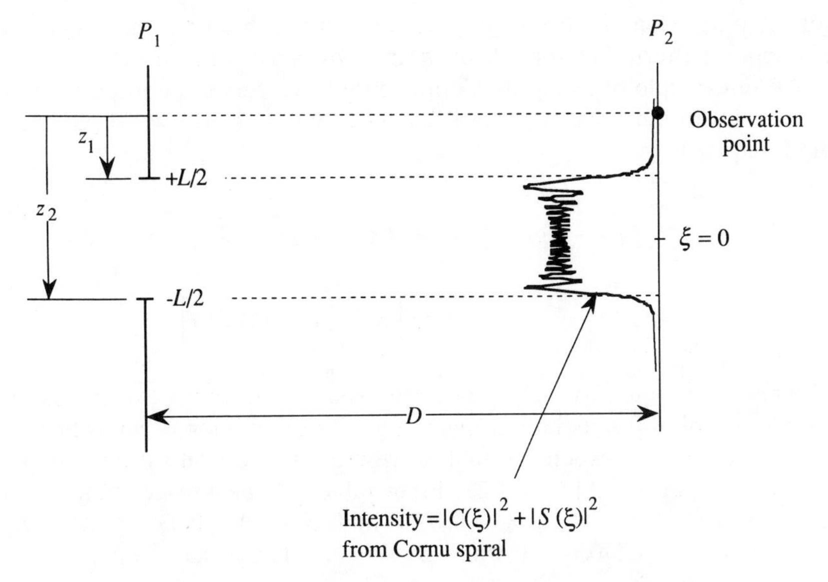

Figure 3.14. Intensity pattern produced by diffraction from a slit.

$\xi = 0$ is therefore

$$g(\xi) = \sqrt{\frac{jI_0}{2}}\left[\sqrt{2}\,e^{-j\pi/4}\right]$$

$$= \sqrt{\frac{I_0}{2}}\,e^{+j\pi/4}\left[\sqrt{2}\,e^{-j\pi/4}\right] = \sqrt{I_0}\,, \qquad (3.52)$$

so that the magnitude is exactly the same at plane P_2 as it is at plane P_1 and intensity is equal to I_0 as expected. From Equation (3.52) we see that the $\exp[-j\pi/4]$ phase factor from the calculation of the Fresnel integral exactly cancels the $\sqrt{j}$ factor from Equation (3.19) that Fresnel included at the onset; the Fresnel transform therefore predicts accurately both the magnitude and the phase of the propagating diffraction pattern.

The key information regarding the behavior of the Fresnel integral is in the limits of integration, because they characterize the transitions into and out of the shadow regions. When $|z_{12}|$ is large, the contributions to the Fresnel integral from the two edges do not interfere and the distance between the half-magnitude points in the Fresnel transform is nearly the same as the slit length. But if $|z_{12}|$ is small so that the slit is narrow, contributions from the two edges interfere with each other. The result is a diffraction pattern that spreads over larger distances at plane P_2.

As an example of using the Cornu spiral to calculate the intensity at a given point in the Fresnel pattern, suppose that z_1 is at $+\infty$. At plane P_2 the intensity is

$$I(\xi) = \frac{I_0}{2}|g(\xi)|^2 = \frac{I_0}{2}|C(\xi) + jS(\xi)|^2 = \frac{I_0}{2}\left[|C(\xi)|^2 + |S(\xi)|^2\right]. \tag{3.53}$$

The maximum intensity occurs when the vector connecting two points on the Cornu spiral has its largest magnitude. This condition occurs when z_2 is about halfway between the first horizontal and vertical turning points, say at $z_2 = [\sqrt{1} + \sqrt{2}]/2 \approx 1.22$. From tables of the Fresnel transforms (24) we find that $C(1.22) = -0.7021$ and $S(1.22) = 0.6383$. Also, for $z_1 = \infty$, we have $C(\infty) = -0.5$ and $S(\infty) = 0.5$. The maximum intensity is

$$I_{\max} = \frac{I_0}{2}\left[(-0.5 - 0.7021)^2 + (0.5 + 0.6383)^2\right] = \frac{2.74 I_0}{2} = 1.37 I_0. \tag{3.54}$$

The intensity at this point is therefore 37% larger than if the slit were not present. To find the physical coordinate at which the intensity is a maximum, recall that z_2 is measured in units of $\sqrt{2/\lambda D}[L/2 - \xi]$. For the example at hand, the distance from the edge of the half-plane

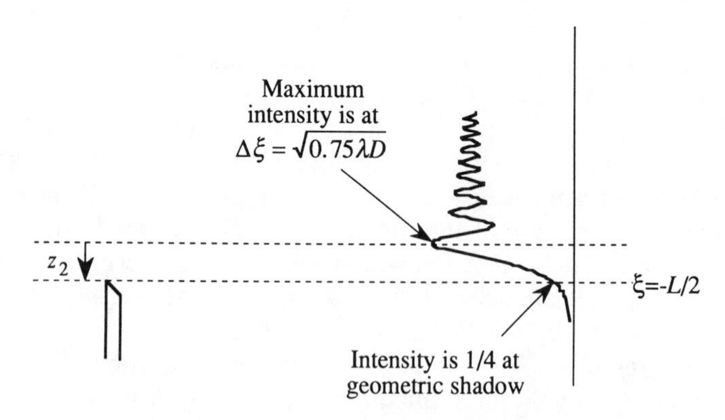

Figure 3.15. Diffraction produced by a half-plane aperture.

slit to the intensity maximum is $\Delta\xi = [-L/2 - \xi] = z_1\sqrt{\lambda D/2}$. For $z_2 = 1.22$, the maximum intensity is located about $\sqrt{0.75\lambda D}$ units away from the geometric shadow of the edge. The resulting intensity pattern is sketched in Figure 3.15.

3.3. THE FOURIER TRANSFORM

The Fourier transform is a widely used tool in the physical sciences for signal analysis. Its principal value is that it generates a function that displays the frequency content of a signal. As a result, certain signal features are more easily analyzed or detected in the frequency domain than in the spatial domain. For example, the presence of a weak sinusoidal signal may be masked by noise in the spatial domain. In the frequency domain, however, this signal component may be easily detected because all the signal energy is concentrated at one frequency, but the noise energy remains spread over the bandwidth of the total signal. In other applications, signals are characterized by a combination of spectral components —they have a "signature" or a "fingerprint" that is more easily detected in the frequency domain than in the time or space domains.

3.3.1. The Fourier Transform of a Periodic Function

The Fourier transform has its roots in a method used by Fourier to represent a periodic signal by a set of weighted sinusoidal components. The basic idea is that if $f(x)$ is a periodic signal so that $f(x + L) = f(x)$, where L is the period of the signal, the frequency content is revealed if we expand $f(x)$ into a series of the form

$$f(x) = a_0 + \sum_{n=1}^{\infty} [a_n \cos(2\pi nx/L) + b_n \sin(2\pi nx/L)], \quad (3.55)$$

which is known as a *Fourier series*. The coefficients a_n and b_n are obtained by multiplying both sides of Equation (3.55) by $\cos(2\pi nx/L)$ or $\sin(2\pi nx/L)$ and integrating the products over one period of the signal $f(x)$. This method leads directly to expressions for the coefficients because the cosine and sine functions are orthogonal. We find that

$$a_n = \frac{2}{L}\int_c^{c+L} f(x)\cos\left(\frac{2\pi nx}{L}\right) dx \quad (3.56)$$

and

$$b_n = \frac{2}{L}\int_c^{c+L} f(x)\sin\left(\frac{2\pi nx}{L}\right) dx, \tag{3.57}$$

where c is an arbitrary starting point for the integration. The value of a_0 is given by Equation (3.56) with $n = 0$, but the coefficient is halved. From Equation (3.56), it is clear that a_0 gives the average value of $f(x)$. We identify $\alpha_0 = 1/L$ as the fundamental spatial frequency whose dimensions are reciprocal to those of L. As n increases, we obtain the coefficients a_n and b_n associated with the harmonics of α_0.

3.3.2. The Fourier Transform for Nonperiodic Signals

If the period of the signal $f(x)$ becomes large or if $f(x)$ is not periodic, we must develop alternatives to the Fourier-series representation of a signal. One alternative is to induce periodicity by replicating $f(x)$ at regular intervals and to proceed with the frequency decomposition as described above. The other alternative is to extend the basic interval L to infinity so that $\alpha_0 \to 0$ and the discrete coefficients become a continuous function of the spatial variable α. In this case, the summation becomes an integral and the discrete frequency components become the *Fourier transform* of the nonperiodic signal:

$$F(\alpha) = \int_{-\infty}^{\infty} f(x)e^{j2\pi\alpha x}\, dx, \tag{3.58}$$

and the corresponding *inverse Fourier transform* is

$$f(x) = \int_{-\infty}^{\infty} F(\alpha)e^{-j2\pi\alpha x}\, d\alpha. \tag{3.59}$$

The integral relationships of Equations (3.58) and (3.59) are called *Fourier-transform pairs*. We treat the value of the continuous transform $F(\alpha)$ as the limiting form of the weights a_n and b_n at discrete frequencies $2\pi n/L$. That is, if we integrate $|F(\alpha)|^2$ over a small interval $d\alpha$ centered at one of the discrete frequencies $2\pi n/L$, its value is equal to $(a_n^2 + b_n^2)$.

We do not venture into the mathematical subtleties of the Fourier transform nor the conditions for which it exists. As we always consider only those signals that have finite total energy, we can safely say that the Fourier transform $F(\alpha)$ of any spatial signal exists.

3.3.3. The Fourier Transform in Optics

We develop the Fourier transform in optics using the basic system shown in Figure 3.16. To simplify the mathematics, we use a one-dimensional notation to find the light distribution $F(\xi)$ at the back focal plane of the lens. To further simplify the mathematics, we place $f(x)$ at the front focal plane of the lens; the more general conditions under which the Fourier transform exists are treated in Section 3.6. We begin by recognizing that the light distribution $g(u)$ just before the lens is the Fresnel transform of $f(x)$:

$$g(u) = \sqrt{\frac{j}{\lambda F}} \int_{-\infty}^{\infty} f(x) e^{-j(\pi/\lambda F)(u-x)^2}\, dx. \tag{3.60}$$

Next, we find the amplitude $h(u)$ at plane P_3 on the other side of the lens. In Chapter 2 we showed that a lens collimates light from a point source located at its front focal plane. Because collimated light is represented as a plane wave of unit magnitude, the lens must render a cylindrically diverging wavefront, of the form $\exp(-j\pi u^2/\lambda F)$, into a wave of the form $\exp(j0) = 1$:

$$(\text{lens function}) \times e^{-j\pi u^2/\lambda F} = e^{j0} = 1, \tag{3.61}$$

from which we conclude that the lens function must be $\psi(u; K)$, as given by Equation (3.20), where $K = 1/F$ is the power of the lens. The light distribution at plane P_3, just beyond the lens, is therefore

$$h(u) = g(u) e^{j\pi u^2/\lambda F}, \tag{3.62}$$

which is the product of $g(u)$ and the lens function. The light distribution

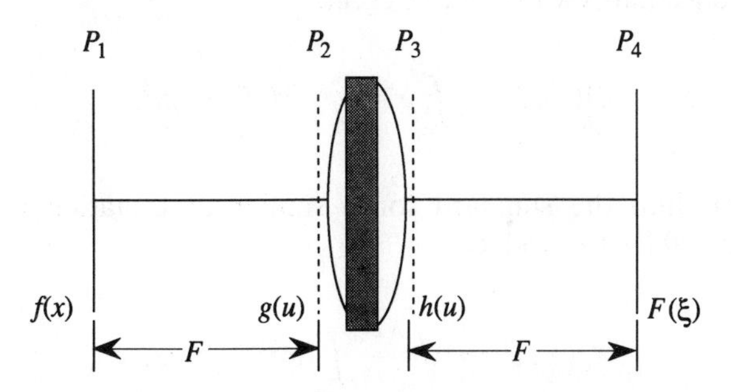

Figure 3.16. Basic Fourier-transform system.

$F(\xi)$ at plane P_4 is obtained by a second application of the Fresnel transform:

$$F(\xi) = \sqrt{\frac{j}{\lambda F}} \int_{P_3} h(u) e^{-j(\pi/\lambda F)(\xi - u)^2} \, du. \tag{3.63}$$

We use Equations (3.62) and (3.60) in Equation (3.63) to find that

$$F(\xi) = \sqrt{\frac{j}{\lambda F}} \int_{P_3} \left[\sqrt{\frac{j}{\lambda F}} \int_{-\infty}^{\infty} f(x) e^{-j(\pi/\lambda F)(u-x)^2} \, dx \right] \times e^{j\pi u^2/\lambda F} e^{-j(\pi/\lambda F)(\xi - u)^2} \, du. \tag{3.64}$$

We collect all the exponential terms and evaluate the resulting kernel function:

$$-j(\pi/\lambda F) \Big[\underbrace{(u-x)^2}_{\text{first free space}} - \underbrace{u^2}_{\text{lens}} + \underbrace{(\xi - u)^2}_{\text{second free space}} \Big] = -j(\pi/\lambda F)\left[u^2 - 2u(x+\xi) + x^2 + \xi^2\right]. \tag{3.65}$$

We complete the square in the variable u, and arrange the integral in the form

$$F(\xi) = \frac{j}{\lambda F} \int_{-\infty}^{\infty} f(x) \left\{ \int_{P_3} e^{-j(\pi/\lambda F)(u - x - \xi)^2} \, du \right\} e^{j2\pi x\xi/\lambda F} \, dx. \tag{3.66}$$

We almost have the desired Fourier-transform relationship. The final step is to show that the integral in braces is not a function of x or ξ. The integral in question is a Fresnel integral:

$$Q(x, \xi) = \int_{P_3} e^{-j(\pi/\lambda F)(u - x - \xi)^2} \, du, \tag{3.67}$$

which is put into the standard form, similar to Equation (3.42), by a change of variables to produce

$$Q(x, \xi) = \sqrt{\frac{\lambda F}{2}} \int_{z_1}^{z_2} e^{-j\pi z^2/2} \, dz, \tag{3.68}$$

where $z_1 = \sqrt{2/\lambda F}[-A/2 - x - \xi]$, $z_2 = \sqrt{2/\lambda F}[A/2 - x - \xi]$ and A is the aperture of the lens. From the results of Section 3.2, we know that the most rapid change in $Q(x, \xi)$ occurs when either z_1 or z_2 approaches zero. Recall from the discussion associated with Equation (3.52) that $Q(x, \xi)$ is approximately equal to $\sqrt{\lambda F/j}$ when the magnitude of either z_1 or z_2 is greater than zero; the value of the integral settles exactly to $\sqrt{\lambda F/j}$ as the magnitudes of z_1 and z_2 become large.

To satisfy the constraint that the integral is constant, we require that $|A/2| > (x + \xi)_{\text{max}}$. The signal $f(x)$ is limited in space to the region $|x|_{\text{max}} < L/2$. To find the maximum value of ξ, we note that light from $f(x)$ is spread over a region equal to $\theta_{\text{co}} F$ at plane P_3 so that $\xi_{\text{max}} = \theta_{\text{co}} F = \lambda F \alpha_{\text{co}}$, because the light is then collimated. Because the sample interval must be d_0 throughout $f(x)$, the condition under which $Q(x, \xi)$ is constant is

$$\left|\frac{A}{2}\right| \geq \left|\frac{L}{2} + \lambda F \alpha_{\text{co}}\right|. \tag{3.69}$$

The required lens aperture is therefore a function of both the object size and its frequency content. In Sections 3.6 and 3.7 we more fully explore how finite lens apertures affect system performance. For now, we assume that Equation (3.69) is satisfied so that we can set $Q(x, \xi) = \sqrt{\lambda F/j}$ in Equation (3.66) to express $F(\xi)$ as

$$\boxed{F(\xi) = \sqrt{\frac{j}{\lambda F}} \int_{-\infty}^{\infty} f(x) e^{j2\pi x \xi/\lambda F}\, dx,} \tag{3.70}$$

which is a Fourier transform of the signal $f(x)$ in terms of the physical coordinate ξ of the Fourier plane. To satisfy the scalar wave equation, the sign of the kernel in the *spatial* Fourier transform must be opposite to the kernel of the *temporal* Fourier transform. Note that the kernel function has a positive sign, a direct result of the sign convention we have used throughout.

If we want to emphasize that the Fourier transform is a function of the spatial frequency variable α, we express the result as

$$F(\alpha) = \int_{-\infty}^{\infty} f(x) e^{j2\pi\alpha x}\, dx, \tag{3.71}$$

where $\xi = \lambda F \alpha$, and we generally drop the scaling constant when we use this form of the transform. The two-dimensional version of the Fourier transform, in terms of the orthogonal spatial frequencies α and β, is

$$F(\alpha, \beta) = \iint_{-\infty}^{\infty} f(x, y) e^{j2\pi(\alpha x + \beta y)} \, dx \, dy, \tag{3.72}$$

where β is the spatial frequency in the vertical direction and the vertical coordinate at the frequency plane is $\eta = \lambda F \beta$. The two-dimensional Fourier transform is obtained by a similar line of analysis and is useful in the image-processing applications treated in the following chapters.

3.4. EXAMPLES OF FOURIER TRANSFORMS

In this section we calculate the Fourier transforms of a few functions that illustrate some of the basic principles. Gaskill's book (18) is a thorough and readable discussion of the Fourier transforms of many other functions and their application in optics.

3.4.1. Fourier Transforms of Aperture Functions

Fourier transforms of aperture functions are important in applications such as spectrum analysis, which we discuss in Chapter 4. We use the *aperture function* $a(x)$ to describe the amplitude weighting due to the laser illumination and the truncation effects due to lenses or other elements in an optical system. The effective signal is then the product $a(x)f(x)$ which, by the convolution theorem, produces a Fourier transform that is the convolution of $A(\alpha)$ and $F(\alpha)$.

In the mathematical developments associated with optical signal processing, we can usually calculate integral equations in closed form if we use a uniform aperture function. We therefore examine the form of $A(\alpha)$ when $a(x) = \text{rect}(x/L)$, which represents a clear aperture of length L, centered on the optical axis. From the Fourier transform we find that

$$\begin{aligned} A(\alpha) &= \int_{-\infty}^{\infty} \text{rect}\left(\frac{x}{L}\right) e^{j2\pi\alpha x} \, dx \\ &= \int_{-L/2}^{L/2} e^{j2\pi\alpha x} \, dx \\ &= L \frac{\sin(\pi\alpha L)}{\pi\alpha L} = L \,\text{sinc}(\alpha L). \end{aligned} \tag{3.73}$$

We see that $A(\alpha)$ attains its maximum value when $\alpha = 0$, and that zeros of $A(\alpha)$ occur when the argument of the sinc function is a nonzero integer. The first zeros are at the spatial frequencies $\alpha_0 = \pm 1/L$, or at the physical distances

$$\xi_0 = \pm \frac{\lambda F}{L}. \tag{3.74}$$

From Equation (3.74) we see that the scale of the transform is inversely related to the scale of the aperture; that is, when the aperture length L is large, $A(\alpha)$ is compact because the first zero occurs at a small value of ξ. Conversely, when L is small, $A(\alpha)$ is spread over a large region in the Fourier domain. This relationship satisfies our intuitive notion that a signal with coarse or fine sample intervals has a Fourier transform that contains low or high frequencies as evidenced by the spectral spread in $A(\alpha)$.

In Chapter 1 we showed that the inverse Fourier transform of a $\mathrm{rect}(\alpha/\alpha_{co})$ function in the spatial frequency plane produced a sample function $\mathrm{sinc}(x/d_0)$ in the space plane. Here we note the dual relationship: a $\mathrm{rect}(x/L)$ function in the space domain produces a sample function of the form $\mathrm{sinc}(\alpha L)$ in the frequency plane. We use this sample function in Chapter 4 to characterize the Fourier transforms of arbitrary signals.

A Gaussian-weighted illumination beam is intrinsically produced by gas lasers and injection laser diodes. For our purposes, we define the Gaussian beam as $a(x) = \exp[-2A(x/L)^2]$ so that the intensity response at $x = \pm L/2$ is $1/e^{-A}$. We find that the Fourier transform of a Gaussian beam remains Gaussian:

$$\int_{-\infty}^{\infty} e^{-2A(x/L)^2} e^{j2\pi\alpha x}\, dx = L\sqrt{\frac{\pi}{2A}}\, e^{-(1/2A)(\pi\alpha L)^2}. \tag{3.75}$$

The Gaussian function is therefore one of several that retain their functional identity under Fourier transformation. The Fourier transforms of truncated Gaussian beams cannot be calculated in closed form; several computer solutions are given in Chapter 4.

3.4.2. A Partitioned Aperture Function

Suppose that we have a partitioned aperture function of the form

$$f(x) = \begin{cases} 1; & -2.5L \le x \le -1.5L \\ 1; & +1.5L \le x \le +2.5L \\ 0; & \text{elsewhere,} \end{cases} \tag{3.76}$$

as shown in Figure 3.17(a). This function represents an optical system that has a clear aperture with a central occlusion. We can calculate the Fourier transform directly, if we so choose, but a useful trick is to note that $f(x)$ is equal to $\text{rect}(x/L)$ convolved with $[\delta(x + 2L) + \delta(x - 2L)]$, as suggested in Figure 3.17(b). By the convolution theorem, we find that $A(\alpha)$ is equal to the product of the individual Fourier transforms of the rect function and of the delta functions. We find that

$$\int_{-\infty}^{\infty} \text{rect}(x/L) e^{j2\pi\alpha x}\, dx = L\, \text{sinc}(\alpha L), \tag{3.77}$$

and that

$$\int_{-\infty}^{\infty} [\delta(x + 2L) + \delta(x - 2L)] e^{j2\pi\alpha x}\, dx = 2\cos(4\pi\alpha L). \tag{3.78}$$

The Fourier transform of $f(x)$ is then $F(\alpha) = 2L[\text{sinc}(\alpha L)]\cos(4\pi\alpha L)$, the product of a cosine function established by the delta functions and a sinc function established by the rect function. There are exactly four cycles of the cosine function under the main lobe of the sinc function, as shown in Figure 3.17(c). As the distance between the two apertures increases, the

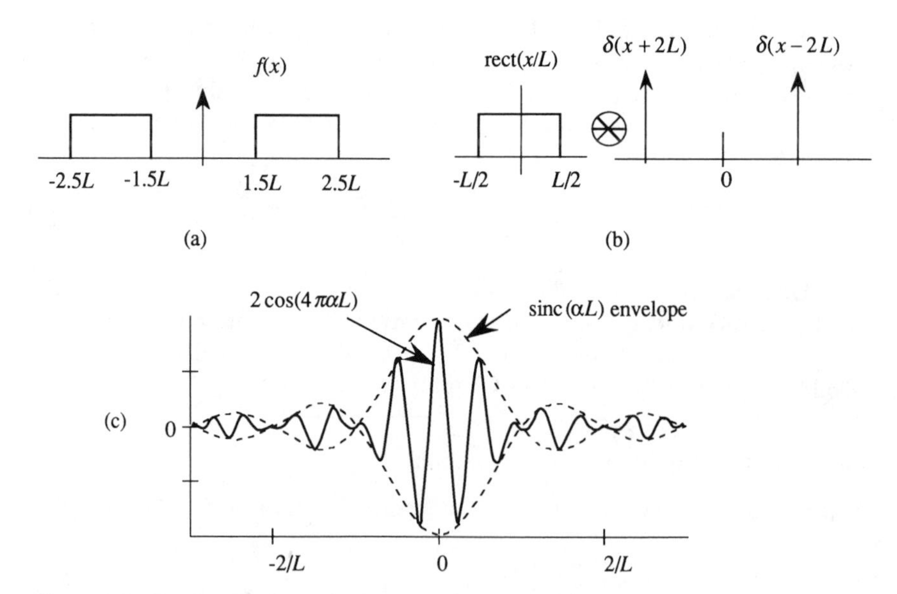

Figure 3.17. Fourier transform of a partitioned aperture: (a) the partitioned aperture, (b) an equivalent representation of the aperture, and (c) the Fourier transform.

frequency of the cosine increases so that more cycles of the cosine appear under the central lobe of the sinc function.

3.4.3. A Periodic Signal

The convolution theorem also helps to calculate Fourier transforms of a train of short pulses represented by

$$f(x) = \mathrm{rect}(x/L) * \sum_{n=-\infty}^{\infty} \delta(x - nx_0), \tag{3.79}$$

where x_0 is the separation between the pulses. The impulse train in Equation (3.79) is called a *comb function*. The Fourier transform of the comb function is

$$\begin{aligned} \int_{-\infty}^{\infty} \sum_{n=-\infty}^{\infty} \delta(x - nx_0) e^{j2\pi\alpha x}\, dx &= \sum_{n=-\infty}^{\infty} e^{j2\pi\alpha n x_0} \\ &= 2\alpha_o \sum_{n=-\infty}^{\infty} \delta(\alpha - 2n\alpha_o), \end{aligned} \tag{3.80}$$

where $\alpha_o = 1/2x_0$. We note that the Fourier transform of the comb function produces many plane waves in the Fourier domain whose phases are arranged so that the spectral distribution is also a comb function. The final result, then, is that

$$F(\alpha) = 2\alpha_o L\, \mathrm{sinc}(\alpha L) \sum_{n=-\infty}^{\infty} \delta(\alpha - 2n\alpha_o), \tag{3.81}$$

which shows that the comb function is weighted by a sinc function which is the Fourier transform of a single short pulse. In this fashion, we see that the Fourier transforms of seemingly complicated signals are found by a combination of the Fourier transforms of their component parts.

3.5. THE INVERSE FOURIER TRANSFORM

Suppose that we cascade two Fourier-transform systems of the type shown in Figure 3.16, with the second following the first, as shown in Figure 3.18. The second system generates the Fourier transform from the output of the

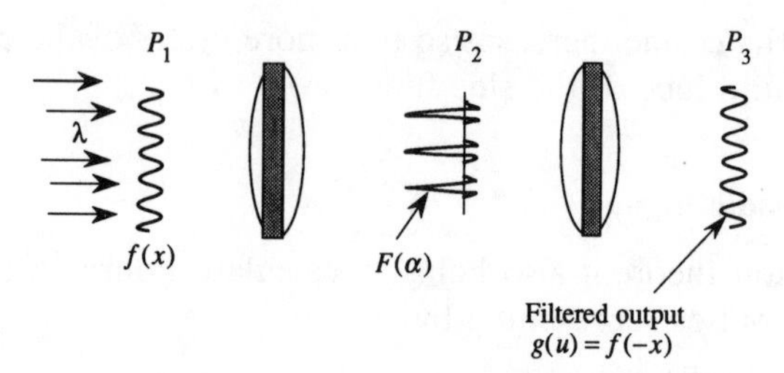

Figure 3.18. Inverse Fourier-transform system.

first system:

$$g(u) = \int_{P_2} F(\alpha) e^{j2\pi\alpha u}\, d\alpha. \tag{3.82}$$

Note that the second spatial Fourier transform also has an exponential whose sign is positive, similar to that of the forward-going transform. We use Equation (3.71) and substitute for $F(\alpha)$:

$$g(u) = \int_{P_2} \left[\int_{-\infty}^{\infty} f(x) e^{j2\pi\alpha x}\, dx \right] e^{j2\pi\alpha u}\, d\alpha. \tag{3.83}$$

If the aperture at plane P_2 does not stop any light rays, we can extend the aperture limits to infinity and perform the integration on α to find that

$$\int_{-\infty}^{\infty} e^{j2\pi\alpha(x+u)}\, d\alpha = \delta(x+u). \tag{3.84}$$

We now use the sifting property of the delta function:

$$\begin{aligned} g(u) &= \int_{-\infty}^{\infty} f(x)\delta(x+u)\, dx \\ &= f(-u), \end{aligned} \tag{3.85}$$

and we see that $g(u)$ is identical to $f(x)$ but with a reversed coordinate, as indicated by the negative sign associated with the space variable.

Space, as contrasted to time, does not have a preferred sense of direction; and geometrical optics requires *two forward-going transforms* to produce the required spatial inversion of the signal. A true "inverse transform" can, in fact, be created by using a negative focal length lens in the second of the two systems shown in Figure 3.18, but the image does not then exist anywhere in the space to the right of the lens. The true inverse transform leads to "virtual" images instead of to "real" images.

3.5.1. Bandlimited Signals

Suppose that $f(x)$ is bandlimited to spatial frequencies less than or equal to α_{co}. Furthermore, suppose that we have a frequency plane weighting function $A(\alpha) = \text{rect}[\alpha/2\alpha_m]$, where $\alpha_m \geq \alpha_{co}$. The light leaving the Fourier plane is then $F(\alpha)A(\alpha)$; and the Fourier transform of this product, by use of the convolution theorem, is

$$\begin{aligned} g(x) &= \int_{-\infty}^{\infty} F(\alpha)A(\alpha)e^{j2\pi\alpha x}\,d\alpha \\ &= 2\alpha_m \int_{-\infty}^{\infty} f(u)\text{sinc}[2\alpha_m(x+u)]\,du \\ &= f(-x). \end{aligned} \tag{3.86}$$

The transition from the second to the third step is made without appeal to complicated mathematical arguments because the function $A(\alpha)$ does not affect, or in any way further limit, the already bandlimited signal $f(x)$ if $\alpha_m \geq \alpha_{co}$. The general rule is this: *if we encounter a convolution between any signal $f(x)$ whose maximum frequency is equal to α_{co} and a function of the form* $\text{sinc}(2\alpha_m x)$, *we replace the* sinc *function by a delta function if* $\alpha_m \geq \alpha_{co}$. The sifting theorem is then used with confidence that the spatial frequency content of $f(x)$ has not been altered.

A function of the form $2\alpha_m\text{sinc}(2\alpha_m x)$ is, of course, a suitable form of a delta function, in the limit as $\alpha_m \to \infty$. This development shows that the conditions defining a suitable delta function are relaxed significantly when dealing with signals or systems that are bandlimited. The property of bandlimited functions as given by Equation (3.86) is used frequently in analytical developments throughout this book.

3.5.2. Rayleigh-Resolution Criterion

In our development of the Fresnel transform we ignored the obliquity factor, based on the argument that the angular spreading of light produced by a signal sample is small. We now support this conclusion and

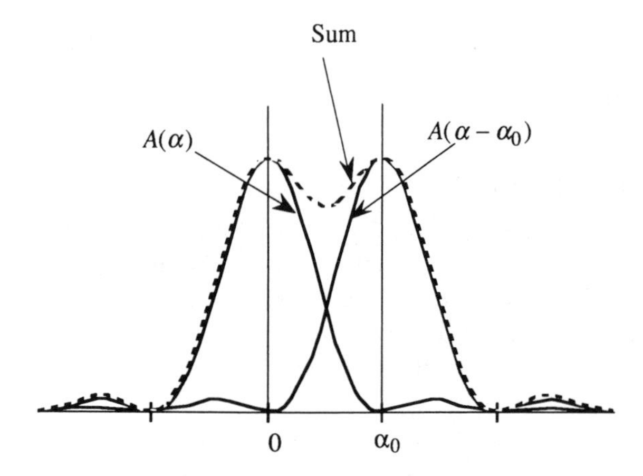

Figure 3.19. Rayleigh-resolution criterion for a sinc^2impulse response.

return to the Rayleigh-resolution criterion for a telescope as stated initially in Chapter 2. Suppose, for example, that $f(x) = 1$ so that the product $a(x)f(x)$ is simply $a(x)$. This situation might arise when light from a distant scene, such as an isolated star, enters a lens whose aperture function is $a(x) = \text{rect}(x/L)$. The image of the star appears at the back focal plane of the lens; thus the image is identical to the Fourier transform $A(\alpha)$ of the aperture function $a(x)$ as just derived. The equivalence of $A(\alpha)$ to the image of a star implies that the light distribution at plane P_1 of Figure 3.16 is a plane wave of infinite extent that is rendered to a finite extent by the action of $a(x)$.

The Rayleigh-resolution requirement, based on visually detecting a dip between the peaks of the two stars, is that the peak of the response due to the second star, $A(\alpha - \alpha_0)$, falls at the first zero of $A(\alpha)$, as shown in Figure 3.19. The physical distance between the peaks is $\xi_0 = d_0 = \lambda F/L$, and the angular resolution is therefore $\phi_0 = \lambda/L$, which is just equal to the difference in the angle that the wavefront has as it enters the lens.

3.5.3. Abbe's Resolution Criterion

Abbe showed that, under certain illumination conditions, false details in the image are generated if the frequency components of the signals are altered (25). He showed, by a somewhat different line of analysis from Rayleigh's, that coherently illuminated systems have a resolution limit d_0 which is related to the cutoff spatial frequency: $d_0 = 1/(2\alpha_{co})$. For example, consider the coherently illuminated optical system shown in

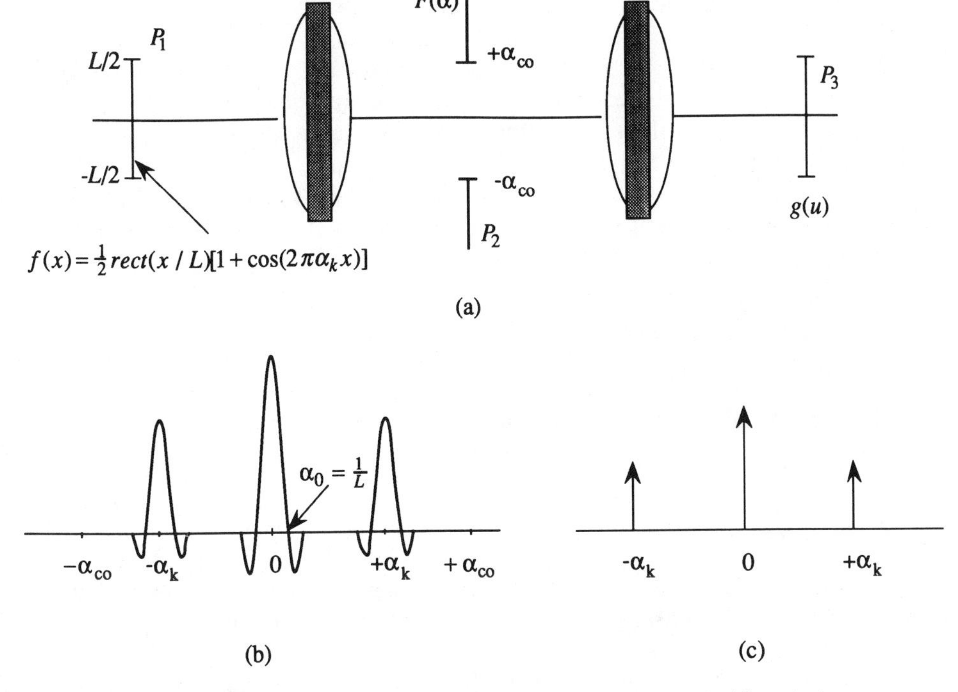

Figure 3.20. Coherently illuminated system to illustrate sharp cutoff: (a) optical system, (b) Fourier transform of a biased cosine, and (c) alternative representation of the Fourier transform.

Figure 3.20(a), in which a sinusoidal function

$$f(x) = 0.5\,\mathrm{rect}(x/L)[1 + \cos(2\pi\alpha_k x)]$$

is the input signal. The Fourier transform of $f(x)$ is

$$F(\alpha) = \int_{-\infty}^{\infty} f(x)e^{j2\pi\alpha x}\,dx = \int_{-L/2}^{L/2} \tfrac{1}{2}[1 + \cos(2\pi\alpha_k x)]e^{j2\pi\alpha x}\,dx$$

$$= \int_{-L/2}^{L/2} \tfrac{1}{2}\left[1 + \tfrac{1}{2}e^{j2\pi\alpha_k x} + \tfrac{1}{2}e^{-j2\pi\alpha_k x}\right]e^{j2\pi\alpha x}\,dx$$

$$= \int_{-L/2}^{L/2} \tfrac{1}{2}\left[e^{j2\pi\alpha x} + \tfrac{1}{2}e^{j2\pi(\alpha+\alpha_k)x} + \tfrac{1}{2}e^{j2\pi(\alpha-\alpha_k)x}\right]dx$$

$$= \frac{L}{2}\left\{\mathrm{sinc}(\alpha L) + \tfrac{1}{4}\,\mathrm{sinc}[(\alpha + \alpha_k)L] + \tfrac{1}{4}\,\mathrm{sinc}[(\alpha - \alpha_k)L]\right\}, \quad (3.87)$$

where L is the length of the signal at plane P_1. As shown in Figure 3.20(b), $F(\alpha)$ consists of three sinc functions whose first zeros are at $\alpha_0 = \pm 1/L$; one sinc function is centered at $\alpha = 0$, one is centered at $\alpha = \alpha_k$, and one is centered at $\alpha = -\alpha_k$. The sinc functions are typically so much narrower than those shown in the figure that we sometimes represent them by δ functions as shown in Figure 3.20(c).

We place an aperture at plane P_2 whose extent is $\pm\alpha_{co}$. If the spatial frequency is low so that $|\alpha_k| \ll \alpha_{co}$, all the energy in $F(\alpha)$ passes through the aperture and the amplitude at image plane P_3 is

$$\begin{aligned} g(u) &= \int_{-\infty}^{\infty} F(\alpha) e^{j2\pi\alpha u}\, d\alpha \\ &= \int_{-\infty}^{\infty} \operatorname{rect}\left[\frac{\alpha}{2\alpha_{co}}\right]\left\{\frac{L}{2}\operatorname{sinc}(\alpha L) + \frac{L}{4}\operatorname{sinc}[(\alpha + \alpha_k)L]\right. \\ &\qquad \left. + \frac{L}{4}\operatorname{sinc}[(\alpha - \alpha_k)L]\right\} e^{j2\pi\alpha u}\, d\alpha, \end{aligned} \tag{3.88}$$

where $\operatorname{rect}[\alpha/2\alpha_{co}]$ represents the frequency response of the system. The output therefore consists of three terms, a typical one being

$$g_1(u) = \int_{-\infty}^{\infty} \operatorname{rect}\left[\frac{\alpha}{2\alpha_{co}}\right]\frac{L}{2}\operatorname{sinc}(\alpha L) e^{j2\pi\alpha u}\, d\alpha. \tag{3.89}$$

The easiest way to evaluate Equation (3.89) is to recognize that if we express $\operatorname{sinc}(\alpha L)$ in terms of its Fourier transform, we have

$$g_1(u) = \int_{-\infty}^{\infty} \frac{1}{2}\left\{\operatorname{rect}\left[\frac{\alpha}{2\alpha_{co}}\right]\int_{-\infty}^{\infty}\operatorname{rect}\left[\frac{x}{L}\right] e^{j2\pi\alpha x}\, dx\right\} e^{j2\pi\alpha u}\, d\alpha. \tag{3.90}$$

We interchange the order of integration to find that

$$\begin{aligned} g_1(u) &= \int_{-\infty}^{\infty} \frac{1}{2}\operatorname{rect}\left[\frac{x}{L}\right]\left\{\int_{-\infty}^{\infty}\operatorname{rect}\left[\frac{\alpha}{2\alpha_{co}}\right] e^{j2\pi\alpha(x+u)}\, d\alpha\right\} dx \\ &= \int_{-\infty}^{\infty} \frac{1}{2}\operatorname{rect}\left[\frac{x}{L}\right] 2\alpha_{co}\operatorname{sinc}[2\alpha_{co}(x+u)]\, dx. \end{aligned} \tag{3.91}$$

Because $|\alpha_k| \ll \alpha_{co}$, the sinc function in the integral on the second line of Equation (3.91) is sufficiently narrow that it behaves as a delta function, as we show in Section 3.5.1. The convolution is therefore easily performed by using the sifting property of the delta function to provide

$$g_1(u) = \frac{1}{2}\,\mathrm{rect}\left[\frac{-u}{L}\right], \tag{3.92}$$

which is simply the image of the average value of the original signal $f(x)$. Each of the other terms of Equation (3.88) is evaluated in a similar fashion and we find that the entire output is

$$g(u) = 0.5\,\mathrm{rect}(-u/L)[1 + \cos(-2\pi\alpha_k u)] = f(-x),$$

so that the image has exactly the same form as the signal, but with a coordinate reversal.

When we increase α_k so that $|\alpha_k| \approx \alpha_{co}$, the sinc functions centered at $\alpha = \pm\alpha_k$ are just at the edges of the aperture in the Fourier plane, but the image is still accurately related to the object. A further increase in the frequency, so that $|\alpha_k| > \alpha_{co}$, results in a sudden change in the structure of the image. Instead of being a sinusoidal function $g(u) = 0.5\,\mathrm{rect}(-u/L)[1 + \cos(-2\pi\alpha_k u)]$, the image becomes a constant $g(u) = 0.5\,\mathrm{rect}(-u/L)$, because all information about the magnitude and frequency of the sinusoid is lost.

As the spatial frequency response is constant for all frequencies below the cutoff frequency and zero thereafter, we represent the normalized *coherent modulation transfer function* of the system as

$$\boxed{H(\alpha) = \mathrm{rect}\left(\frac{\alpha}{2\alpha_{co}}\right),} \tag{3.93}$$

as shown in Figure 3.21(a). Because the image intensity is the magnitude squared of the image amplitude, the *incoherent modulation transfer func-*

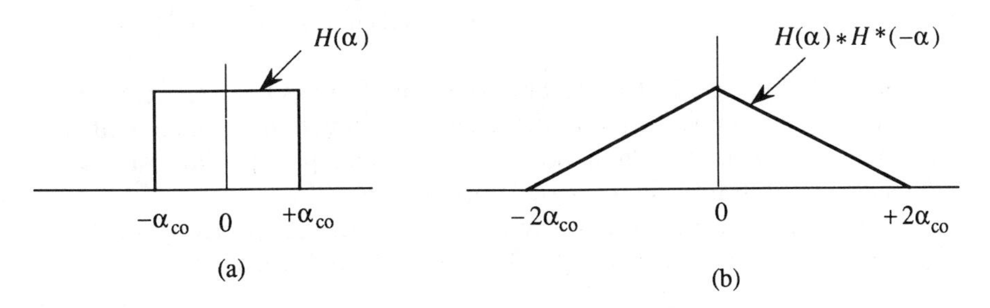

Figure 3.21. Normalized modulation transfer functions: (a) coherently illuminated system and (b) incoherently illuminated system.

tion of an optical system is the normalized autocorrelation of the coherent modulation transfer function:

$$\boxed{\mathcal{H}(\alpha) = \int_{-\infty}^{\infty} H(\gamma) H^*(\gamma + \alpha)\, d\gamma = \mathrm{tri}\left(\frac{\alpha}{2\alpha_{\mathrm{co}}}\right),} \tag{3.94}$$

where $\mathrm{tri}(\alpha/2\alpha_{\mathrm{co}}) = 1 - |\alpha|/2\alpha_{\mathrm{co}}$ for $|\alpha| < 2\alpha_{\mathrm{co}}$ and is equal to zero elsewhere, as shown in Figure 3.21(b). The incoherent and coherent modulation transfer functions are therefore related by the autocorrelation function. As a result, the image contrast in an incoherently illuminated system changes gradually as $|\alpha|$ increases, falling to zero when $|\alpha| = 2\alpha_{\mathrm{co}}$. Although an incoherently illuminated system can resolve spatial frequencies twice as high as its coherently illuminated counterpart, the contrast ratio for the sinusoid is uniformly better, over its bandpass region, for the coherently illuminated system.

Abbe also noticed that the image of a coherently illuminated signal changed its appearance when the angle of the illumination was off axis. We now consider the more general case of *oblique illumination* as shown in Figure 3.22(a). Light leaving plane P_1 is now expressed as

$$f(x) = \tfrac{1}{2}\,\mathrm{rect}\left(\frac{x}{L}\right) e^{-j2\pi\alpha_i x}\left[1 + \cos(2\pi\alpha_k x)\right], \tag{3.95}$$

where $\theta_i = \lambda\alpha_i$ is the oblique illumination angle. The Fourier transform of this signal is found by a procedure similar to that used to derive the result given by Equation (3.87):

$$F(\alpha) = \frac{L}{2}\{\mathrm{sinc}[(\alpha + \alpha_i)L] + \tfrac{1}{2}\,\mathrm{sinc}[(\alpha + \alpha_i + \alpha_k)L]$$
$$+ \tfrac{1}{2}\,\mathrm{sinc}[(\alpha + \alpha_i - \alpha_k)L]\}. \tag{3.96}$$

As shown in Figure 3.22(b), $F(\alpha)$ has the same form as before, except that the entire spectrum is shifted in the negative α direction by an amount α_i. We note that oblique illumination provides a nice physical illustration of the shift theorem from Fourier-transform theory.

As α_i increases, we see that eventually some of the spectrum is cut off by the finite aperture at plane P_2. This event occurs when

$$\mathrm{sinc}[(\alpha + \alpha_i + \alpha_k)L]\,\mathrm{rect}(\alpha/2\alpha_{\mathrm{co}}) = 0. \tag{3.97}$$

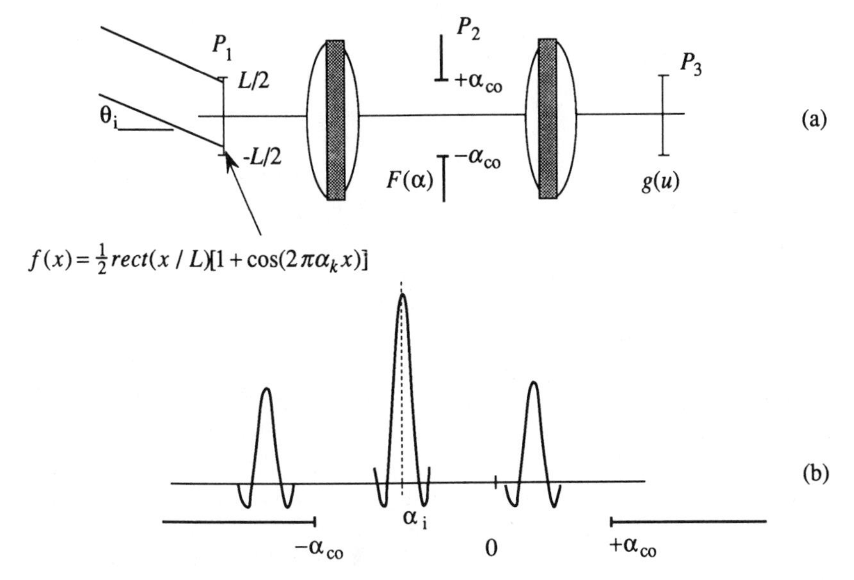

Figure 3.22. Oblique illumination: (a) optical system and (b) shifted spectrum.

From Equation (3.97), we see that the center of the sinc function is within the passband aperture whenever $0 < \alpha_i < (\alpha_{co} - \alpha_k)$. For larger values of the angle of illumination the sinc function is cut off, and the output of the system becomes

$$g(u) = \int_{-\infty}^{\infty} \text{rect}\left[\frac{\alpha}{2\alpha_{co}}\right] \frac{L}{2} \{\text{sinc}[(\alpha + \alpha_i)L) + \tfrac{1}{2} \text{sinc}[(\alpha + \alpha_i - \alpha_k)L]\} e^{j2\pi\alpha u} \, d\alpha, \quad (3.98)$$

so that the complex amplitude at plane P_3 is

$$g(u) = \tfrac{1}{2} e^{-j2\pi\alpha_i u}\left[1 + \tfrac{1}{2} e^{j2\pi\alpha_k u}\right] \text{rect}(u/L). \quad (3.99)$$

This filtering operation generates false information as is seen by examining the intensity of the output signal for these two cases. When $\alpha_i = 0$ and $|\alpha_k| \ll \alpha_{co}$, the intensity is $I_0(u) = |g(u)|^2$:

$$I_0(u) = \tfrac{1}{4}\left[\tfrac{3}{2} + 2\cos(2\pi\alpha_k u) + \tfrac{1}{2}\cos(4\pi\alpha_k u)\right] \text{rect}(u/L), \quad (3.100)$$

whereas when one of the diffracted orders is cut off, the intensity is the squared magnitude of Equation (3.99):

$$I_1(u) = \tfrac{1}{4}\left[\tfrac{5}{4} + \cos(2\pi\alpha_k u)\right]\mathrm{rect}(u/L). \tag{3.101}$$

By comparing Equation (3.101) with Equation (3.100), we conclude that the intensity of the fundamental frequency component has been reduced by a factor of 2 because of the loss of half the energy in the fundamental, and that the harmonic of the fundamental is missing. The output therefore contains false information about the input signal because of filtering the spatial frequencies.

Oblique illumination has the advantage, however, that it increases the bandwidth of a spectrum analyzer. If the input signal is real valued so that the spectrum has polar symmetry, negative spatial frequencies have the same magnitude as positive ones and measuring the spectrum from zero frequency to α_{co} is sufficient to extract all the information. Oblique illumination displays the spatial frequencies from the signal that fall into the range from zero to α_{co} in the interval $(-\alpha_{co}, 0)$ at the Fourier plane, thus making the interval $(0, \alpha_{co})$ at the Fourier plane available for displaying spatial frequencies ranging from α_{co} to $2\alpha_{co}$. The frequency analysis range is therefore doubled. Oblique illumination is also useful in certain special cases of correlation. It is not, however, generally used when accurate imaging is required.

We see that Equation (3.100) reveals a spatial frequency $2\alpha_k$ that was not displayed at the Fourier plane P_2 of the coherent optical system. This apparent discrepancy is explained by noting that the harmonic $2\alpha_k$ arises only when we *observe*, *measure*, or *record* the intensity of an amplitude signal that is a pure sinusoidal waveform. In the example given, the harmonic was, in fact, already present in the intensity stored on the spatial light modulator at the input of the system. The coherently illuminated system produced the Fourier transform of the *amplitude* of the light at plane P_2, and, because the intensity of the input signal was, in fact, $I(x) = |f(x)|^2$, the system properly calculates the Fourier transform of $f(x) = \sqrt{I(x)}$.

3.5.4. The Sample Function, Sampling Theorem, and Decomposition

Both Rayleigh and Abbe studied resolution for optical systems, but from different viewpoints. We now summarize their results and relate them to the sampling theorem and to methods for representing signals. If we follow Rayleigh's resolution criterion, we represent a signal $f(x)$, bandlimited to the frequency range $|\alpha| \leq \alpha_{co}$, by a set of weighted sample

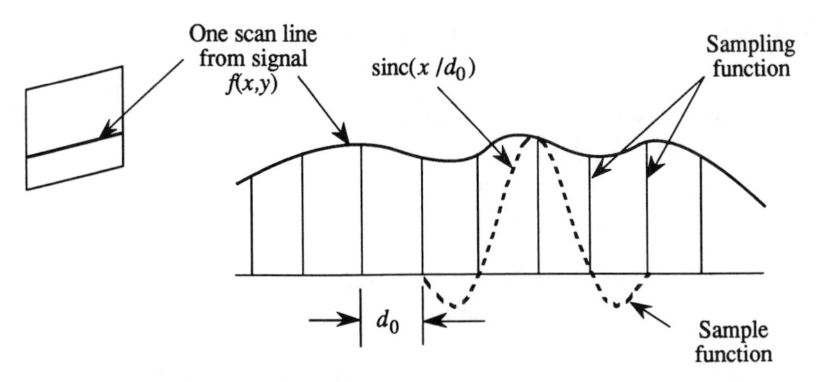

Figure 3.23. The sample (interpolation) function for a sampled bandlimited signal.

functions:

$$f(x) = \underbrace{\operatorname{rect}\left(\frac{x}{L}\right)}_{\text{sets limit on signal length}} \sum_{n=-\infty}^{\infty} \underbrace{a_n e^{j\phi_n}}_{\text{weights in magnitude and phase}} \underbrace{\operatorname{sinc}[2\alpha_{co}(x - nd_0)]}_{\text{sample function}}, \tag{3.102}$$

where the a_n are the sample magnitudes and the ϕ_n are the sample phases at the midpoints of the sinc functions. In general, ϕ_n takes on the value $\pm\pi$ for a real-valued signal, but we also allow for the fact that the signal may become complex-valued after some filtering operations. The sample function $\operatorname{sinc}(2\alpha_{co}x)$ is wrapped around each element of the sampling function as shown in Figure 3.23. The rect function in Equation (3.102) shows that the number of samples in $f(x)$ is limited to $N = L/d_0$.

The signal representation given by Equation (3.102) is equivalent to the Nyquist criterion for representing a bandlimited signal by a sequence of weighted sample functions. This criterion states that the highest spatial frequency must be sampled at least twice per cycle of the highest frequency to accurately represent the signal; the relationship $\alpha_{co} = 1/(2d_0)$ fulfills this requirement. By applying the shift theorem to the Fourier transform, we find that all light produced by all samples of $f(x)$ pass through an aperture in the Fourier plane P_2 bounded by $|\alpha| \leq \alpha_{co}$; the entire signal, as well as each sample, is therefore strictly bandlimited.

If N degrees of freedom completely specify $f(x)$, so too will N degrees of freedom completely specify $F(\alpha)$. As a result of Equation (3.102), we find that the Fourier transform of $f(x)$ is

$$F(\alpha) = \underbrace{\text{rect}\left(\frac{\alpha}{2\alpha_{\text{co}}}\right)}_{\substack{\text{due to the}\\ \text{sample}\\ \text{function}}} \underbrace{\{\text{sinc}(\alpha \text{L}) *}_{\substack{\text{due to}\\ \text{aperture}\\ \text{function}}} \sum_{n=-\infty}^{\infty} \underbrace{a_n e^{j\phi_n}}_{\substack{\text{weights in}\\ \text{magnitude}\\ \text{and phase}}} \underbrace{e^{j2\pi n d_0 \alpha}\}}_{\substack{\text{due to}\\ \text{sample}\\ \text{positions}}}, \tag{3.103}$$

where $*$ indicates convolution. The spectrum therefore consists of a set of weighted plane waves, each wave due to a sample contained in $f(x)$, that interfere to produce the spectrum $F(\alpha)$. The rect function shows that the signal is bandlimited to $|\alpha| \leq \alpha_{\text{co}}$. The convolution of the spectrum with $\text{sinc}(\alpha L)$ is included for completeness; it does not alter the shape of the spectrum by an argument similar to that developed in Section 3.5.1.

Alternatively, we might follow Abbe's approach and represent $f(x)$ as a sequence of weighed sinusoids whose frequencies are separated by $\alpha_0 = 1/L$. A more convenient representation is to replace the cosines by exponentials, using the Euler expansion, to find that

$$f(x) = \underbrace{\text{rect}\left(\frac{x}{L}\right)}_{\substack{\text{aperture}\\ \text{function}}} \underbrace{\{\text{sinc}(2\alpha_{\text{co}} \text{x}) *}_{\substack{\text{due to the band}\\ \text{limit in the}\\ \text{Fourier plane}}} \sum_{n=-\infty}^{\infty} \underbrace{b_n e^{j\phi_n}}_{\substack{\text{weights in}\\ \text{magnitude}\\ \text{and phase}}} \underbrace{e^{j2\pi n \alpha_0 x}\}}_{\substack{\text{exponential}\\ \text{function}}}, \tag{3.104}$$

where the b_n are the magnitudes and the ϕ_n are the phases of the exponential functions at the signal plane. The sinc function reveals that $2\alpha_{\text{co}}/\alpha_0 = 2L\alpha_{\text{co}} = N$ complex exponentials completely describe $f(x)$. In turn, we express $F(\alpha)$ as

$$F(\alpha) = \underbrace{\text{rect}\left(\frac{\alpha}{2\alpha_{\text{co}}}\right)}_{\substack{\text{frequency}\\ \text{plane cutoff}}} \sum_{n=-\infty}^{\infty} \underbrace{b_n e^{j\phi_n}}_{\substack{\text{weights in}\\ \text{magnitude}\\ \text{and phase}}} \underbrace{\text{sinc}[(\alpha - n\alpha_0)L]}_{\substack{\text{samples in the}\\ \text{Fourier plane}}}, \tag{3.105}$$

where $\alpha_0 = 1/L$ is the minimum resolvable spatial frequency, and the b_n and the ϕ_n provide the weights of the sinc functions in the Fourier plane.

This exercise shows that, since $f(x)$ and $F(\alpha)$ are both functions of space coordinates, we can represent them in either of the two forms listed above. An observer of a sinusoidal spatial signal is hard pressed to know whether they are looking at a space signal, a spatial frequency function, or neither (they may be looking at a Fresnel transform). We can therefore interchange our notions of space and frequency planes as we please, useful for helping to visualize the operation of some signal-processing systems.

Again, we remind the reader that signals cannot, in theory, be both space limited and bandlimited, as we have assumed here. In practice, however, the impact of such an assumption is usually small because the limits of observation are set by noise levels at both the space and spatial frequency planes. In any event, Equations (3.102)–(3.105) are useful to help visualize the general nature of the signals and are accurate, except possibly near the edges of the signals.

3.6. EXTENDED FOURIER-TRANSFORM ANALYSIS

The Fourier-transform result was developed in Section 3.3, under some mild restrictions that simplified the analysis considerably. We assumed that the signal was illuminated by a plane wave of light and was at the front focal plane of the lens. The Fourier-transform relationship exists, however, for a wide range of geometries. To explore these possibilities further, we introduce an operational method of analysis that reveals the richness of the conditions under which the transform exists. In the process, we show that the fundamental results from geometrical optics are obtained from analyses involving only the principles of physical optics. As examples, we generate two key results: the fundamental lens equation and the Fourier-transform relationship.

To facilitate the analysis and to draw analogies to linear system theory, we represent each optical system by a block diagram. By using an operational notation, a basic set of optical elements is synthesized into a system for either imaging or Fourier transforming a signal. Once the basic systems are synthesized, we cascade them to produce more complex ones.

3.6.1. The Basic Elements of an Optical System

Figure 3.24 illustrates that light from the source is represented by $a(x, y) = |a(x, y)|\exp[j\phi(x, y)]$, where $|a(x, y)|$ is the magnitude and $\phi(x, y)$ is the phase of the light. The amplitude transmittance of the spatial light modulator is $f(x, y) = |f(x, y)|\exp[j\theta(x, y)]$ so that the modulated light wave is given by

$$g(x, y) = |a(x, y)||f(x, y)|e^{j[\phi(x, y)+\theta(x, y)]}. \tag{3.106}$$

The block diagram element for a spatial light modulator is therefore a *multiplier*, as shown in Figure 3.24(b).

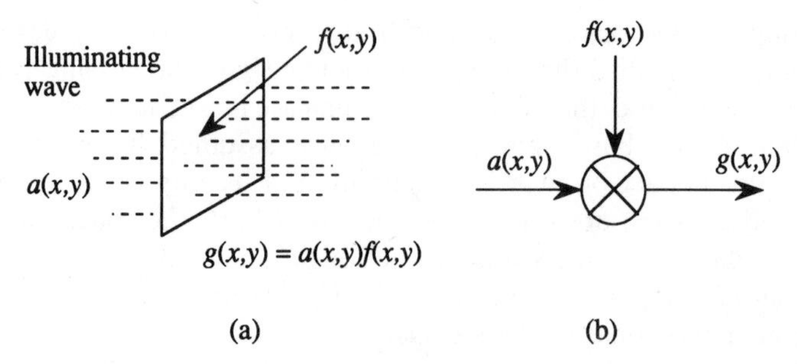

Figure 3.24. Effect of a spatial light modulator: (a) optical system and (b) block diagram.

Lenses are important elements in optical signal-processing systems. In Section 3.3.3 we showed that a spherical lens is represented by the phase function $\exp[j(\pi/\lambda F)(x^2 + y^2)]$, where F is the focal length of the lens. Cylindrical lenses are represented as functions of only one variable: $\exp[j\pi x^2/\lambda F]$ or $\exp[j\pi y^2/\lambda F]$, depending on the direction of the power of the lens. The focal length of cylindrical lenses in the orthogonal direction is infinite. The operation of a lens and its block diagram is shown in Figure 3.25.

The next important step is to represent how light propagates through free space. We showed in Section 3.2 that, if $f(x, y)$ is a light distribution in a given plane, the propagation of light through a distance D produces the Fresnel transform

$$g(\xi, \eta) = \frac{C}{D} \iint_{-\infty}^{\infty} f(x, y) e^{-j(\pi/\lambda D)[(x-\xi)^2+(y-\eta)^2]} \, dx \, dy, \quad (3.107)$$

where ξ, η are coordinates in the new plane and C is a complex-valued constant that is independent of all the variables.

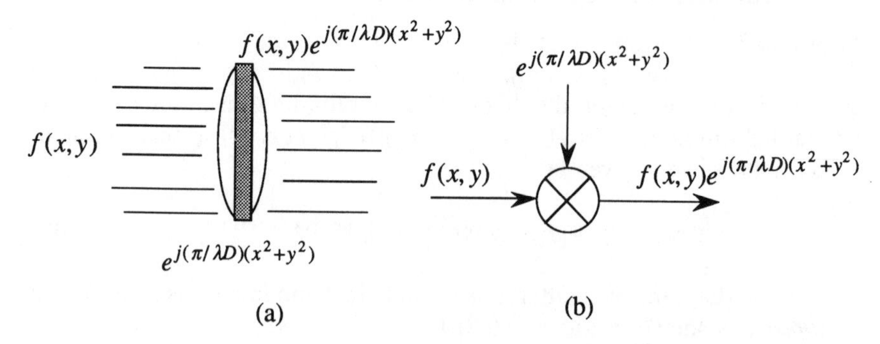

Figure 3.25. Effect of a lens: (a) optical system and (b) block diagram.

3.6.2. Operational Notation

As the lens function and the free-space impulse response have similar forms, we represent both by a ψ function (27):

$$\boxed{\psi(x, y; d) \equiv e^{j(\pi/\lambda D)(x^2+y^2)}.} \tag{3.108}$$

Because the distance between two planes occurs in the denominator of the argument of the exponential function, we use a lower-case letter to represent the reciprocal of the propagation distance as indicated by an upper-case letter. With this notation, we describe the propagation of light through a distance D, shown in Figure 3.26, as though it passed through a black box whose impulse response is $d\psi^*(x, y; d)$, where $*$ indicates complex conjugate and $d = 1/D$. A lens of focal length F is represented by $\psi(x, y; K)$, where K is the power of the lens. A cylindrical lens is represented by $\psi(x; K)$ or $\psi(y; K)$, according to which axis of the cylinder has the focal power.

Some of the more useful properties of the ψ function are

P1 $\qquad \psi(x, y; d) = \psi^*(x, y; -d),$

P2 $\qquad \psi(-x, -y; d) = \psi(x, y; d),$

P3 $\qquad \psi(x, y; d_1)\psi(x, y; d_2) = \psi(x, y; d_1 + d_2),$

P4 $\qquad \psi(x, y; d) = \psi(x; d)\psi(y; d),$

P5 $\qquad \psi(cx, cy; d) = \psi(x, y; c^2 d),$

P6 $\qquad \psi(x - u, y - v; d) = \psi(x, y; d)\psi(u, v; d)e^{-jkd(ux+vy)},$

P7 $\qquad \lim_{K \to 0} \psi(x, y; K) = 1.$

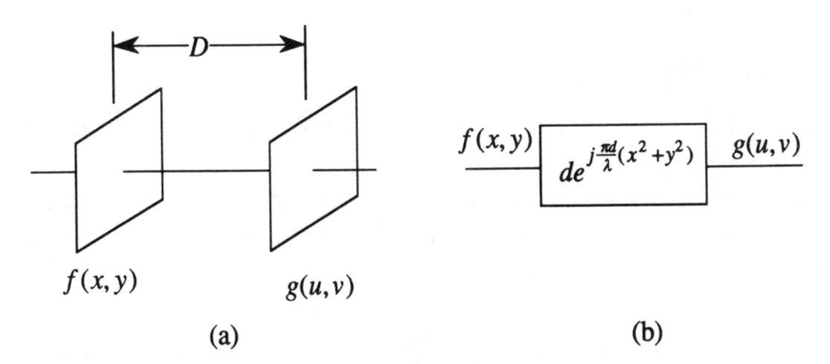

Figure 3.26. Propagation through free space: (a) optical system and (b) block diagram.

Property P1 shows asymmetry along the optical axis; that is, a spherical wave appears to diverge or converge according to the direction of observation. Property P2 shows symmetry normal to the optical axis; an example is that the power of a spherical lens has polar symmetry. Property P3 gives the rule for multiplication and property P4 shows the separability of the ψ function. Property P5 gives the effect of a scale change, which is useful when we solve certain integral equations; and property P6 is useful in expanding convolution integrals involving the ψ function. Property P7 states that a lens of infinite focal length has no effect on a light distribution.

Another useful property of the ψ function is that the Fourier transform of the function $\psi(x, y; d_1)$ with respect to a parameter d_2 is also a ψ function:

$$\text{P8} \qquad \iint_{-\infty}^{\infty} \psi(x, y; d_1) e^{j(2\pi/\lambda)d_2(ux+vy)}\, dx\, dy = \frac{c}{d_1}\psi*(u, v; d_2^2/d_1), \qquad (3.109)$$

where c is a complex-valued constant that is generally neglected. Property P7 shows that the argument of the integral becomes a constant when $d_1 \to 0$ and P8, in turn, shows that

$$\text{P9} \qquad \lim_{d_1 \to 0} \frac{1}{d_1}\psi^*(x, y; d_2^2/d_1) = \delta(x, y) \qquad (3.110)$$

for any value of d_2. Property P9 is independent of d_2, in the limit, because the Fourier transform of a constant over an infinite interval is a δ function. In Section 3.5.1, we showed that the finite aperture of a lens leads to a sample function in the image plane that is, in an operational sense, equivalent to a δ function, provided that the signal is bandlimited and that the aperture limit does not remove information.

3.6.3. A Basic Optical System

A fundamental combination of the elements described in Section 3.6.1 consists of a spatial light modulator, free space, a spherical lens, and more free space, as shown in Figure 3.27. The spatial light modulator is illuminated by a unit magnitude monochromatic light wave with an arbitrary spherical phase factor. The illumination is represented by $\psi^*(x, y; d_1)$ if the spherical wave is divergent, or by $\psi(x, y; d_1)$ if the wave is convergent; the radius of curvature is $D_1 = 1/d_1$. According to P1 we could also

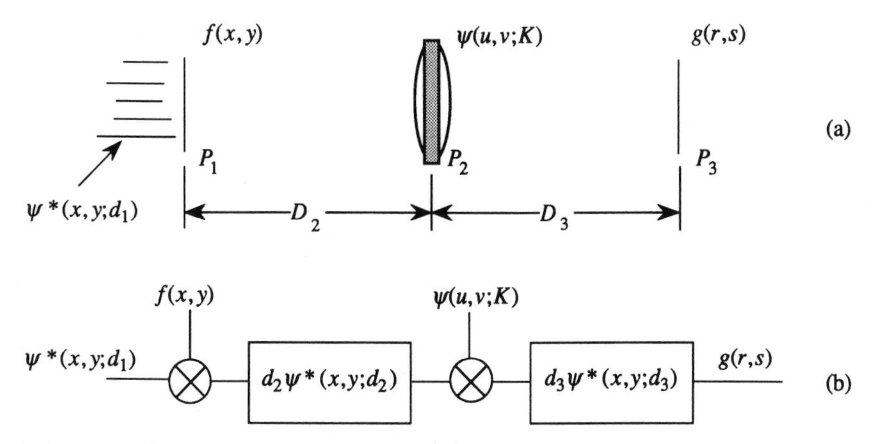

Figure 3.27. General Fourier-transform module: (a) optical system and (b) block diagram.

use $\psi(x, y; -d_1)$ to represent a diverging spherical wave; this notation is consistent with the sign convention adopted in geometrical optics, where we consider D_1 negative when plane P_1 is the origin of the current coordinate system.

From the block diagram we express the output of the system, after using P6, as

$$g(r,s) = d_2 d_3 \iint_{P_1} \iint_{P_2} \psi^*(x,y;d_1) f(x,y) \psi^*(x,y;d_2) \times \psi^*(u,v;d_2) e^{j(2\pi/\lambda)d_2(ux+vy)} \psi(u,v;K) \psi^*(u,v;d_3) \times \psi^*(r,s;d_3) e^{j(2\pi/\lambda)d_3(ur+vs)}\, dx\, dy\, du\, dv, \tag{3.111}$$

where d_2 and d_3 are the reciprocals of the distances D_2 and D_3. Using P3 to collect terms, we have

$$g(r,s) = d_2 d_3 \psi^*(r,s;d_3) \iint_{P_1} \iint_{P_2} \psi^*(x,y;d_1+d_2) f(x,y) \times \psi^*(u,v;d_2-K+d_3) \times e^{j(2\pi/\lambda)[u(d_2x+d_3r)+v(d_2y+d_3s)]}\, dx\, dy\, du\, dv. \tag{3.112}$$

We first carry out the (u, v) integration, with the aid of P8, to obtain

$$g(r,s) = \frac{d_2 d_3}{d_2 - K + d_3} \psi^*(r, s; d_3) \iint_{P_1} \psi^*(x, y; d_1 + d_2) f(x, y) \times \psi\left(x + \frac{d_3 r}{d_2}, y + \frac{d_3 s}{d_2}; \frac{d_2^2}{d_2 - K + d_3}\right) dx\, dy. \tag{3.113}$$

Equation (3.113) is the *central result* in this operational notation. It relates the complex-valued light distribution in the output plane of a basic optical system to the input distribution in terms of the parameters d_1, d_2, d_3, and K.

3.6.3.1. The Imaging Condition. Suppose we wish to synthesize the basic elements shown in Figure 3.27 into an imaging system. First, we note that Equation (3.113) is a convolution operation so that if we find the relationship among d_2, d_3, and K necessary to convert the ψ function into a delta function, the output will be an image of the input. By property P9, we find that

$$\frac{d_2^2}{d_2 - K + d_3} \psi\left(x + \frac{d_3 r}{d_2}, y + \frac{d_3 s}{d_2}; \frac{d_2^2}{d_2 - K + d_3}\right) = \delta\left(x + \frac{d_3 r}{d_2}, y + \frac{d_3 s}{d_2}\right) \tag{3.114}$$

when $d_2 - K + d_3 = 0$. We neglect the multiplicative constants and use the sifting property of the delta function so that Equation (3.113) becomes

$$g(r,s) = \psi^*(r, s; d_4) f\left(-\frac{d_3 r}{d_2}, -\frac{d_3 s}{d_2}\right), \tag{3.115}$$

where $d_4 = d_3 + (d_1 + d_2) d_3^2 / d_2^2$. Equation (3.115) shows that, aside from a spherical phase factor, $g(r, s)$ is an inverted and scaled image of $f(x, y)$; the scaling factor $-d_3/d_2 = -D_2/D_3$ accurately represents the lateral magnification M of the system. As the imaging condition is valid only when $d_2 + d_3 = K$, we find that $1/D_2 + 1/D_3 = 1/F$. This condition is recognized as the fundamental lens equation or, by suitable modification, as the refraction equation, whose general form is given by $n_2 u_2 = n_1 u_1 - hK$.

The phase factor that modifies $f(r, s)$ in Equation (3.115) is not important if $g(r, s)$ is recorded or otherwise detected because physical detectors are insensitive to phase. Therefore the intensity $|g(r, s)|^2$ is independent of d_1; i.e., the phase curvature of the illuminating wave does not affect recording of the image. The phase factor provides the information needed to determine where the next image plane occurs if further operations are performed on $g(r, s)$. Thus, all the important features of an imaging system are readily obtained from Equation (3.113).

3.6.3.2. The Fourier-Transform Condition. Suppose that we wish to synthesize the elements of the basic optical system to provide a Fourier-transform relationship between $g(r, s)$ and $f(x, y)$. By using P6 and collecting terms, we rewrite Equation (3.113) as

$$g(r, s) = \psi^*(r, s; d_5) \iint_{P_1} \psi^*\left(x, y; d_1 + d_2 - \frac{d_2^2}{d_2 - K + d_3}\right) f(x, y) \times e^{j(2\pi/\lambda)[d_2 d_3/(d_2 - f + d_3)](xr + ys)}\, dx\, dy, \tag{3.116}$$

where $d_5 = d_3 - d_3^2/(d_2 - K + d_3)$. This relationship is almost in the form of a Fourier transform. We apply P7 so that the ψ function in the integral is equal to unity. This condition implies that

$$\boxed{d_1 + d_2 - \frac{d_2^2}{d_2 - K + d_3} = 0} \tag{3.117}$$

must be satisfied. First, suppose that $d_1 = 0$ so that the input signal is illuminated by a plane wave. If Equation (3.117) is satisfied for *any* value of d_2, it is satisfied for *every* value of d_2 so that the distance from the input plane to the lens is of no consequence, aside from ensuring that all the rays pass through the lens. Finally, we find that $d_3 = K$, which implies that the Fourier transform of $f(x, y)$ occurs in the back focal plane of the lens. Under these conditions Equation (3.116) becomes

$$g(r, s) = \psi^*\left(r, s; K - \frac{K^2}{d_2}\right) \iint_{P_1} f(x, y) e^{j(2\pi/\lambda F)(xr + ys)}\, dx\, dy. \tag{3.118}$$

The Fourier-transform relation is made "exact" by setting the ψ function in Equation (3.118) equal to one. To do so, we set $K - K^2/d_2 = 0$ which requires that $d_2 = K$; thus, we place $f(x, y)$ in the front focal plane of the lens. This is the usual result, derived in many texts and papers in the literature. The operational notation used here makes it easy to see what other values of the parameters lead to a useful Fourier-transform relationship—something not easy to do by conventional techniques.

The physical meaning of Equation (3.117) is that the Fourier transform always occurs at the *image plane of the source*. This result is useful in helping us visualize where, for example, the Fourier plane occurs relative to the image plane. The steps necessary to prove this assertion are to observe that Equation (3.117) is rewritten in a sequence of equations as

$$(d_1 + d_2)(d_2 - K + d_3) = d_2^2,$$

$$d_1(d_2 - K + d_3) + d_2(d_3 - K) = 0,$$

$$d_1 d_2 + d_1(d_3 - K) + d_2(d_3 - K) = 0,$$

$$\frac{d_1 d_2}{d_1 + d_2} + d_3 = K,$$

$$\frac{1}{D_1 + D_2} + \frac{1}{D_3} = \frac{1}{F}, \tag{3.119}$$

which is the condition necessary for imaging the source at plane P_3.

The physical meaning of d_5 in Equation (3.116) is that it represents the phase curvature associated with the Fourier transform. This claim is supported by noting that, if $f(x, y) = \delta(x, y)$ in Equation (3.116),

$$g(r, s) = \psi^*(r, s; d_5), \tag{3.120}$$

which is either a convergent or a divergent spherical wave depending on the sign of d_5. Note that $d_5 = 0$ whenever $d_2 = K$; that is, the Fourier transform has no residual phase curvature whenever the signal is in the front focal plane of the lens, *independently of the curvature of the illuminating wave*. We therefore create an "exact" Fourier transform under a wide range of conditions, contrary to the popular belief that such a transform occurs only when $d_2 = K$ *and* the illumination is a plane wave.

It is not necessary for the Fourier transform to be "exact" in optical signal-processing systems, a point often misunderstood. A residual phase factor in no way affects the spatial filtering operations, to be fully discussed in Chapter 5, other than to determine the position of the output plane.

In a similar fashion, the ψ function associated with the image as given by Equation (3.115) represents a converging or diverging spherical phase function according to the sign of d_4; this result is most easily seen if we set $f(x, y) = 1$. Setting $f(x, y)$ to unity or to $\delta(x, y)$ often provides a quick assessment of how an optical system works. These two functions are a dual pair in the sense that the Fourier transform of a delta function is a constant, and vice versa, as we showed in Section 3.4. These two functions are especially useful for the analysis of anamorphic systems containing cylindrical lenses; often the system is constructed to produce a delta function in one direction while it produces a constant value (in the absence of a signal) in the orthogonal direction.

3.6.3.3. A Variable-Scale Fourier Transform. Equation (3.116) shows that the scale of the Fourier transform is a function of d_2 when $d_1 \neq 0$ and that the transform does not appear in the back focal plane of the lens. From this result we learn that the scale of the Fourier transform is a function of the axial position of the signal if the input signal is placed in nonparallel light. We use this fact to synthesize a variable-scale Fourier transform system, shown in Figure 3.28. From the block diagram we

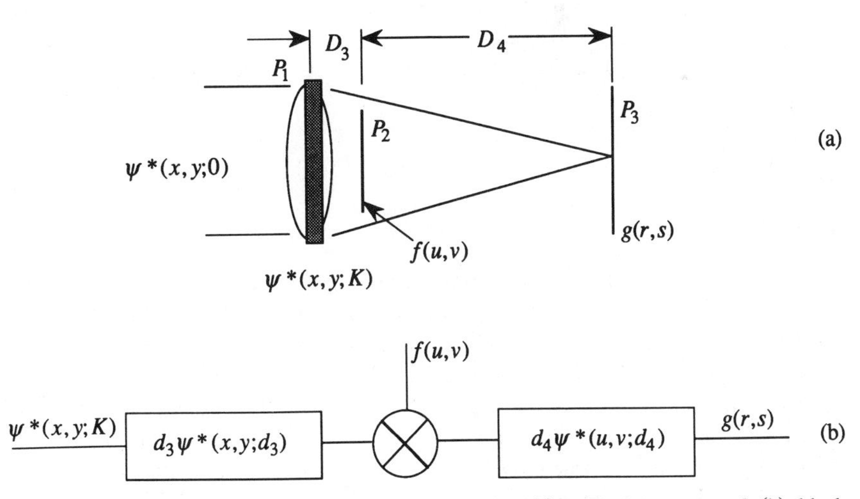

Figure 3.28. Variable-scale Fourier-transform system: (a) optical system and (b) block diagram.

express

$$g(r,s) = d_3 d_4 \psi^*(r,s;d_4) \iint_{P_1} \iint_{P_2} \psi^*(x,y;K)\psi^*(x,y;d_3)\psi^*(u,v;d_3) \times f(u,v) e^{j(2\pi/\lambda)d_3(ux+vy)} \psi^*(u,v;d_4) e^{j(2\pi/\lambda)d_4(ur+vs)}\, dx\, dy\, du\, dv. \tag{3.121}$$

Applying the same methods as before, we find that

$$g(r,s) = \psi^*(r,s;d_4) \iint_{P_2} \psi(r,s;d_5) f(u,v) e^{j(2\pi/\lambda)d_4(ur+vs)}\, du\, dv, \tag{3.122}$$

where $d_5 = -d_3 - d_4 + d_3^2/(d_3 - K)$. Again, $g(r,s)$ and $f(u,v)$ are Fourier-transform pairs if the ψ function in the integrand is equal to one. From P7, we find that d_5 must be set equal to zero, which results in the requirement that $d_3^2/(d_3 - K) = d_3 + d_4$ or that

$$\frac{d_3 d_4}{d_3 + d_4} = K. \tag{3.123}$$

This equation is satisfied for *all* values of d_3 and d_4 because $D_3 + D_4 = F$ by the initial constraints imposed on the optical system. Hence, Equation (3.122) becomes

$$\boxed{g(r,s) = \psi^*(r,s;d_4) \iint_{P_2} f(u,v) e^{j(2\pi/\lambda)d_4(ur+vs)}\, du\, dv.} \tag{3.124}$$

The Fourier transform is now carried out using the variable d_4 rather than the parameter K; by varying d_4, we *vary the scale* of the transform of $f(u,v)$. The presence of the ψ function serves to indicate where the inverse transform occurs, as discussed in the next section.

We close this section by noting that the basic optical system shown in Figure 3.27 can either image or Fourier transform a signal, depending on the configuration. We ask whether this system performs both simultaneously. The reader can show, as an exercise, that a Fourier transform must be created somewhere in the system, not necessarily to the right of the lens, if the system produces an image of the input signal $f(x,y)$. The existence of a Fourier transform in an optical system does not, however, guarantee the existence of an image of the signal.

3.6.4. Cascaded Optical Systems

An optical filtering system is most conveniently analyzed by repeated application of the Fourier-transform relationship or, in some cases, by repeated use of the imaging relationship. Hence, the detailed discussion of a basic optical system simplifies the analysis of cascaded optical systems. We often want to perform an operation described by the general linear integral operator

$$g(r,s) = \iint_{-\infty}^{\infty} f(u,v)h(r-u,s-v)\,du\,dv, \tag{3.125}$$

which can also be expressed in the frequency domain as

$$G(\alpha,\beta) = F(\alpha,\beta)H(\alpha,\beta). \tag{3.126}$$

The advantages of being able to perform this operation by realizing the *spatial filter* $H(\alpha, \beta)$ in the frequency plane, where $H(\alpha, \beta)$ is the Fourier transform of $h(x, y)$, will become apparent in Chapter 5.

A typical filtering system consists of a Fourier-transform operation, a mask that introduces $H(\alpha, \beta)$, and a second Fourier-transform operation. Such a system, with a scale searching capability, is shown in Figure 3.29.

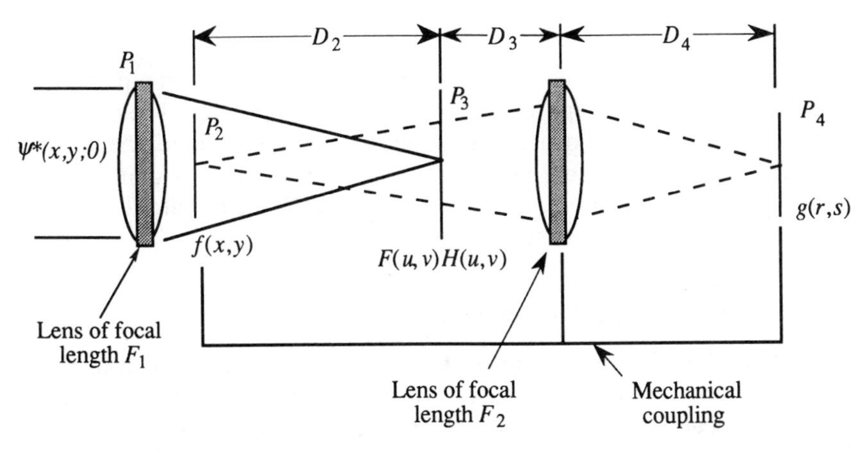

Figure 3.29. Variable-scale correlator.

The input function is illuminated by a converging wave, as before. From the block diagram, we replace the system up to plane P_3 by the equivalent diagram shown in Figure 3.28. Therefore, the distribution at plane P_3 in Figure 3.29 is

$$F(u,v) = \psi^*(u,v;d_2) \iint_{P_1} f(x,y)e^{j(2\pi/\lambda)d_2(ux+vy)}\,dx\,dy. \quad (3.127)$$

The light distribution emerging from P_3 is $F(u,v)H(u,v)$. A second lens, with focal length F_2, takes the Fourier transform of this product and images the filtered signal at the output image plane P_4. This Fourier transform is obtained by applying the general result given in Equation (3.116):

$$g(r,s) = \psi^*(r,s;d_5) \iint_{P_2} \psi^*\left(u,v;d_2+d_3-\frac{d_3^2}{d_3-d_2+d_4}\right)$$
$$\times F(u,v)H(u,v)e^{j(2\pi/\lambda)[d_3d_4/(d_3-K_2+d_4)](ur+vs)}\,du\,dv, \quad (3.128)$$

where $d_5 = d_4 - d_4^2/(d_3 - K_2 + d_4)$. The condition for making $g(r,s)$ the Fourier transform of $F(u,v)H(u,v)$ is that

$$d_2 + d_3 - d_3^2/(d_3 - K_2 + d_4) = 0$$

or that

$$\frac{1}{D_2 + D_3} + \frac{1}{D_4} = \frac{1}{F_2}. \quad (3.129)$$

Equation (3.129) is the condition for imaging plane P_2 into plane P_4, which satisfies our concept about how the system operates. The variable-scale correlator is implemented by connecting the input plane, the second lens, and the output plane together so that they move as a unit. Hence, $D_2 + D_3$ is constant and Equation (3.129) shows that the output at plane P_3 always represents a focused image of the filtered data.

3.6.5. The Scale of the Fourier Transform

In Section 3.6.3.3, we developed the variable-scale Fourier transform under the condition that the signal $f(x, y)$ is illuminated with convergent light. The scale of the Fourier transform is governed simply by the distance between the signal and the Fourier plane. The signal might,

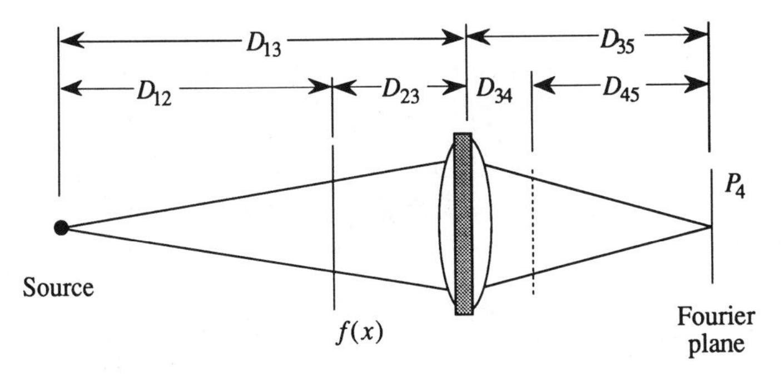

Figure 3.30. Scale of the Fourier transform.

however, by placed in the divergent beam as shown in Figure 3.30, in which case the scaling factor for the Fourier transform seems to be more difficult to find. In this section, we give a simple method for finding this scaling factor. Instead of using the operational notation, we appeal to simple geometrical optics tools and thereby strengthen the relationships between geometrical and physical optics.

The simplest way to find the scale of the Fourier transform of $f(x, y)$ is to temporarily reverse the direction in which the light travels. With light traveling from right to left, we find that the signal is now in *convergent* light. We can therefore calculate the Fourier transform *referenced to the source plane*, using the distance D_{12} as the scaling parameter. The final step is to use the magnification, $M = -D_{35}/D_{13}$, between the source and Fourier planes to get the final scale of the transform.

An alternative technique is to *project* the signal to the lens plane, using the axial point in either the source plane or the Fourier plane as the central projection point. By simple geometrical arguments, we see that the scale of the signal, when projected to the plane of the lens, is $f(Qx, Qy)$, where

$$Q = \frac{D_{12} + D_{23}}{D_{12}}. \tag{3.130}$$

This scaling procedure is equivalent to determining the scale of a signal that would produce the Fourier transform if the signal were illuminated by collimated light and the lens were assigned the focal length D_{35}. Therefore, after the scaling of $f(x, y)$ is done, we can replace the optical system of Figure 3.30 with one in which the signal is illuminated by collimated light at the front focal plane of a lens whose focal length is D_{35}. The usual

Fourier-transform relationships are then applied to find the scale of the transform. A similar procedure can be used when the signal is placed in the convergent beam, replacing D_{12} by D_{45} and D_{23} by D_{34}.

3.7. MAXIMUM INFORMATION CAPACITY AND OPTIMUM PACKING DENSITY

The wide range of configurations, given in Section 3.6, under which a lens produces an image or a Fourier transform, suggests that we seek additional criterion for determining the best configuration of an optical system. A common criterion in communication systems is to maximize the information capacity in terms of bits per unit time. An equivalent criterion in optics is to maximize the packing density in terms of samples per unit length or per unit area. In turn, the size of the optical system determines the total information capacity. We now consider these issues for a coherently illuminated optical system.

Recall that $N = 4(\text{LBP})(\text{HBP})$ is a measure of the number of independent samples required to form a two-dimensional image. The amount of information, on a per-sample basis, is equal to the number of resolvable states, such as magnitudes, polarizations, or wavelengths. The minimum detectable increment in any of these quantities is a function of noise; the noise characteristics therefore set the ultimate rate at which information is transmitted. At this point, we are interested only in the relationship between the geometry of the optical system and the number of samples that is transmitted in a noise-free system. Our results are based on an assumed zero/one state for each sample; we multiply the results by the number of independent states to get the total information capacity.

3.7.1. Maximum Information Capacity

Consider a one-dimensional signal $f(x)$ whose highest frequency is α_{co}, located at the front focal plane of the lens shown in Figure 3.31. The signal has length L and the number of samples necessary to describe the signal is $N = 2LBP = 2L\alpha_{co}$. A sample function $\text{sinc}(x/d_0)$ in $f(x)$ diffracts light within the angle $\theta_{co} = \lambda\alpha_{co}$. If the illumination is collimated, the marginal ray from a sample located at $x = 0$ intercepts the lens plane at $h_2 = \lambda\alpha_{co}F$. The lens must also capture all the rays from samples located at the extreme edges of the signals.

The *relative aperture* R of the lens is the ratio of the clear aperture A to the focal length F of the lens: $R = A/F$. For the conditions shown in Figure 3.31, the aperture is the sum of the aperture $2h_2$ needed to capture

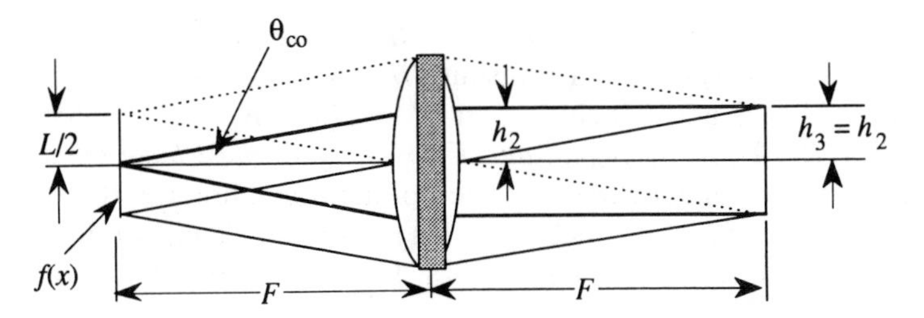

Figure 3.31. Ray diagram for finding the capacity of a lens system.

all the light from an on-axis sample and the aperture L needed to capture all the light diffracted by extreme off-axis samples. The relative aperture therefore is

$$R = \frac{|L| + |2h_2|}{F} = \frac{|L| + |2\lambda a_{co} F|}{F}. \qquad (3.131)$$

Because L and h_2 have the same sign at the lens plane, whether we deal with the upper or the lower half of the aperture, we remove the absolute-value signs to obtain

$$R = \frac{L + 2\lambda\alpha_{co} F}{F} = R_0 + R_t, \qquad (3.132)$$

where R_0 is the relative aperture of the signal and R_t is the relative aperture of the Fourier-transform plane. This relationship suggests that, for a lens of given relative aperture, we could use a small signal with high resolution or a large signal with low resolution. Which is best? The answer is provided by multiplying R by L/λ and solving for $N = 2\text{LBP} = 2L\alpha_{co}$:

$$N = \frac{RL}{\lambda} - \frac{L^2}{\lambda F}. \qquad (3.133)$$

We now find the signal length that maximizes the system capacity by differentiating N with respect to the signal length L. We find that

$$\frac{\partial N}{\partial L} = \frac{R}{\lambda} - \frac{2L}{\lambda F} = 0, \qquad (3.134)$$

from which we conclude that $L = RF/2$ maximizes the number of samples that the system can transmit. We substitute this value of L into Equation (3.132) and conclude that $R = 2R_0 = R_0 + R_t$, which implies that $R_t = R_0$. *Thus, the signal and Fourier planes must have the same size to maximize the system capacity*. The optimum condition is therefore one that strikes a compromise between signal size and signal cutoff frequency so that the number of samples, which is the product of the two quantities, is maximized.

The maximum capacity is found by substituting the value of L into Equation (3.133) to find that

$$\boxed{N_{\text{max}} = \frac{R^2F}{4\lambda} = \frac{RA}{4\lambda}.} \tag{3.135}$$

This result shows a lens with a high relative aperture R and a large clear aperture A yields a system with a high capacity. As $N = 2L\alpha_{\text{co}}$, we also see that

$$\alpha_{\text{co}} = \frac{R}{2\lambda} \tag{3.136}$$

is the maximum frequency that passes through a lens of relative aperture R. These results show that α_{co} is not dependent on the focal length of the lens. Since $\lambda \approx 0.5\ \mu$, a rule of thumb is that $\alpha_{\text{co}} \approx 1000R$ so that an f/2 lens, which has a relative aperture of 0.5, has a coherent cutoff frequency of 500 Ab.

The maximum information capacity was derived under the condition that the lens is operating at infinite conjugates. An image of the signal is easily created by using a second lens, also working at infinite conjugates. If we use finite conjugates, the distance from the signal to the lens is greater than F and the lens aperture would have to increase somewhat to accommodate the diffracted light, thereby lowering the capacity of the system. The results given in this section have application to the design of holographic memories (28) and show that the system capacity is generally set by the optical invariant, not by the capacity of the recording material.

3.7.2. Optimum Packing Density

In applications such as optical storage and retrieval of information, we want to maximize the *packing density*, defined as (29)

$$\rho = \frac{N}{Q} \text{ samples/mm}, \tag{3.137}$$

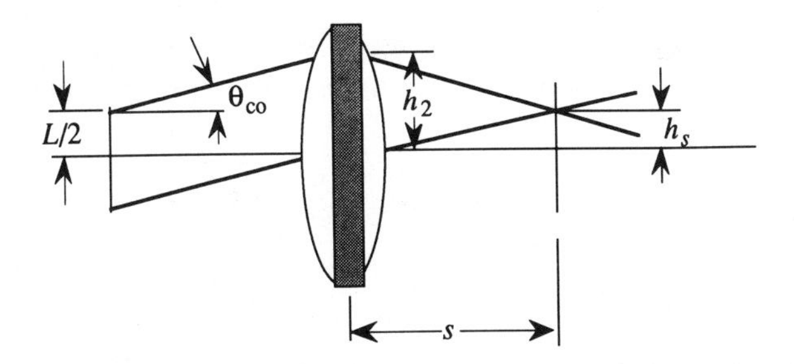

Figure 3.32. Ray diagram for finding the maximum packing density.

where Q is the length of the aperture. We examine all planes in the system shown in Figure 3.32 to find the smallest one through which all information passes. For collimated illumination, the smallest aperture to the left of the lens must be $Q = L$ because no additional aperture is needed to accept diffraction caused by the signal. On the signal side of the lens, we therefore find that the packing density is

$$\rho = \frac{N}{Q} = \frac{2L\alpha_{co}}{L} = 2\alpha_{co} = \frac{1}{d_0}. \tag{3.138}$$

This result shows that the maximum packing density ρ is *independent of the size of the signal when we are on the signal side of the lens.*

Consider the rays to the right of the lens, in which the distance from the lens to a candidate plane is s. The smallest aperture that contains all rays occurs where the marginal ray from a sample at the upper edge of the signal and the parallel ray from the lower edge of $f(x)$ intersect. Because these parallel rays intersect at the back focal plane of the lens, $s = F$. Note that the same result is obtained by summing the heights of these two rays at an arbitrary plane to form a relative aperture; we then calculate the partial derivative of ρ with respect to s, following the same procedure as in handling Equation (3.131).

The minimum value of Q on the image side of the lens occurs at the Fourier plane, or when $s = F$ (29). *The highest packing density on the image side of the lens is therefore at the Fourier plane.* As $\theta_{co} = \lambda\alpha_{co}$, we find that $Q = 2\lambda F\alpha_{co}$ so that the packing density at the Fourier plane is

$$\rho = \frac{2L\alpha_{co}}{2\lambda\alpha_{co}F} = \frac{L}{\lambda F} = \frac{R_0}{\lambda}, \tag{3.139}$$

which is independent of the frequency content of the signal.

The *gain* in packing density is given by the ratio of the maximum density on the image side of the lens to that on the signal side:

$$G = \frac{R_0/\lambda}{1/d_0} = \frac{R_0 d_0}{\lambda} = \frac{R_0}{2\lambda\alpha_{co}}. \tag{3.140}$$

As an example, suppose that the lens has a relative aperture $R_0 \approx 0.5$ and that $\lambda = 0.5\ \mu$ so that $G = 500/\alpha_{co}$. If $\alpha_{co} = 5$ Ab, which is typical for textual material at normal reading distances, $G = 100$. Thus, we store information in an area 10^4 smaller than that occupied by the signal itself; that is, we store an 8×11-in page in an area of about 0.1×0.1-in. When α_{co} is greater than 500 Ab, the gain is less than unity; there is no advantage to storing information at the Fourier plane when the spatial frequencies of the signal are higher than 500 Ab.

3.7.3. Convergent Illumination

We briefly show how to improve the results from Section 3.7.2 by using convergent illumination and a two-lens solution as shown in Figure 3.33. The first lens serves as a field lens and focuses parallel bundles of rays through an aperture of length L that contains the signal $f(x)$. The signal is placed just to the right of the first lens whose aperture is just large enough to fully illuminate the signal with a convergent wave. The aperture of this lens is clearly *independent of the frequency content* of the signal and is dependent only on the signal length.

The aperture required of the second lens is *independent of the length* of the signal and is dependent only on the frequency content of the signal. Each of these lens apertures are therefore smaller than those of lenses operating at infinite conjugates. If G is fairly large, the aperture of the second lens is much smaller than that of the first lens; for example, if

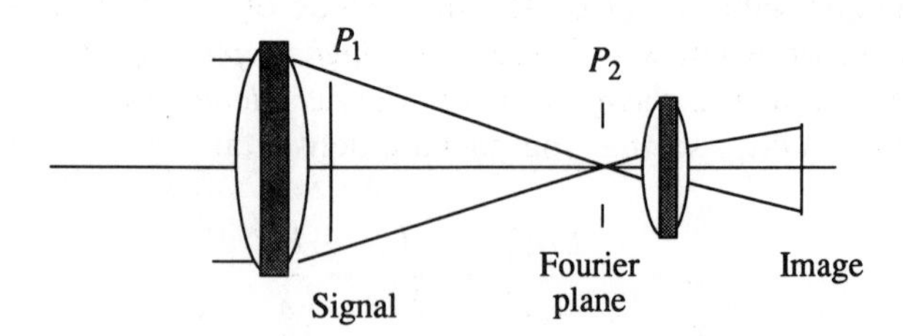

Figure 3.33. Two-lens solution for maximizing the system capacity.

$G = 100$, the aperture of the second lens is 100 times smaller than that of the first lens. Note that the aberrations of the first lens do not affect the imagery; because the second lens has a low relative aperture, it produces high-quality imagery because its aberrations tend to be small.

3.7.4. The Chirp-Z Transform

The configuration shown in Figure 3.33 is suggestive of the configuration needed to create the chirp-Z transform. This transform is sometimes used in electronic systems to convert a time coordinate to a frequency coordinate. Recall that the one-dimensional Fourier transform is

$$F(\xi) = \int_{-\infty}^{\infty} f(x) e^{j(2\pi/\lambda F)\xi x}\, dx. \tag{3.141}$$

The basic idea of the chirp-Z transform is that

$$2\pi\xi x/\lambda F = (2\pi/\lambda F)\left[(\xi + x)^2 - (\xi - x)^2\right]$$

so that we can also express the Fourier transform as

$$F(\xi) = e^{j(\pi\xi^2/\lambda F)} \int_{-\infty}^{\infty} \left[f(x) e^{j(\pi x^2/\lambda F)}\right] e^{-j(\pi/\lambda F)(\xi - x)^2}\, dx. \tag{3.142}$$

This result shows that we obtain the Fourier transform if we premultiply the signal $f(x)$ by a chirp function, perform a Fresnel transform on the resultant product, and postmultiply the integral by a chirp. The optical equivalent of these operations is shown in Figure 3.34. The first lens

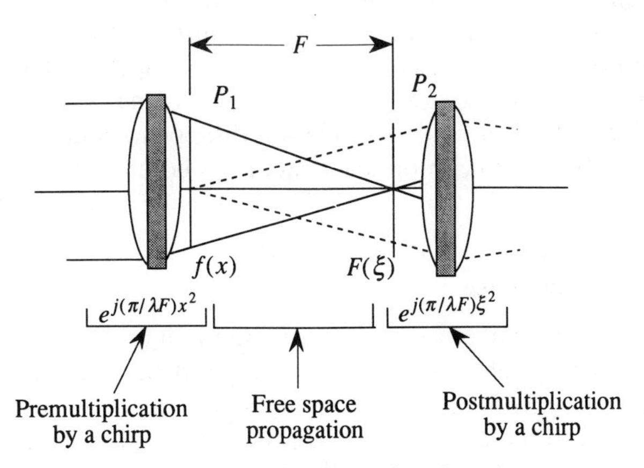

Figure 3.34. Chirp-Z transform in optics.

provides the premultiplication by illuminating $f(x)$ with a converging spherical wave, equivalent to a chirp function. Propagation through free space provides the required Fresnel transform between the signals at planes P_1 and P_2. The second lens provides the postmultiplication of the Fourier transform $F(\xi)$ by a chirp. The similarity of the optical layout in Figure 3.34 to that of Figure 3.33 is obvious and we see that the chirp-Z configuration is useful for minimizing the apertures of the lenses and their aberrations.

3.8. SYSTEM COHERENCE

Coherence theory, which deals with the statistical fluctuations of light, is important in physical optics. For our purposes, we want to establish a few working rules that help us understand some of the phenomena we observe and to settle some issues relating to system linearity.

Any system with transfer function T is linear if

$$T[af_1(x) + bf_2(x)] = ag_1(x) + bg_1(x), \tag{3.143}$$

where $g_1(x) = T[f_1(x)]$ and $g_2(x) = T[f_2(x)]$. In optical systems we are faced with some choices that tend to blur the usual concepts of linearity. Optical systems may be linear in terms of amplitude (equivalent to voltage), in terms of intensity (equivalent to power), or in terms of neither. The subject of linearity cannot be resolved until we assess the role of coherence in optics.

Coherence is a measure of how well light from a source is correlated over space and time. If light is nearly monochromatic, which represents the situation of greatest interest to us, we visualize that wave trains of light arrive at an observation screen with the same average frequency but with slowly varying magnitudes and phases. When the source is strictly monochromatic, the amplitude modulation disappears and the fringes are stable in time. In general, an optical system has both spatial and temporal coherence properties.

3.8.1. Spatial Coherence

We quantify spatial coherence by considering the intensity pattern created at plane P_3 from two elemental apertures Q_1 and Q_2 located at plane P_2, as shown in Figure 3.35. At the secondary source plane P_2 the intensities are proportional to the elemental source area $d\sigma$ so that the magnitude is proportional to the square root of the primary source area at plane P_1.

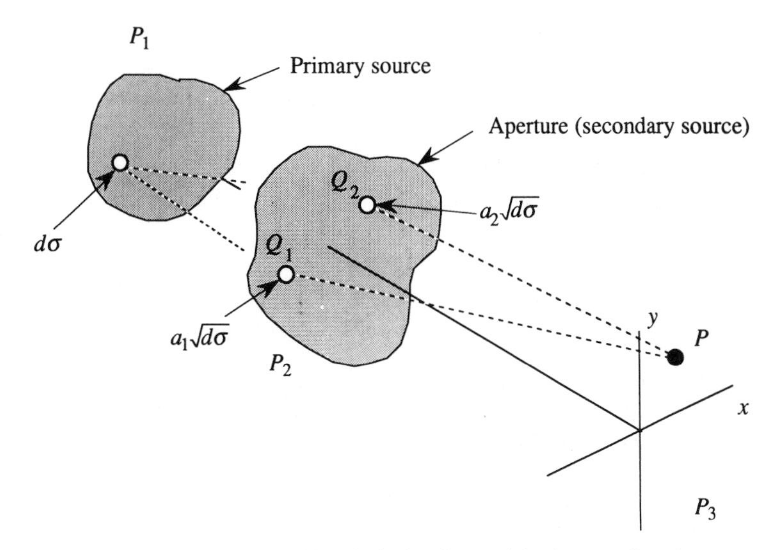

Figure 3.35. Geometry for calculating the spatial coherence function.

Thus, the complex amplitudes $a_1\sqrt{d\sigma}$ and $a_2\sqrt{d\sigma}$ define the secondary sources. If unit value sources at Q_1 and Q_2 produce complex amplitudes u_1 and u_2 at an observation point P at plane P_3, the elemental intensity at P is

$$\delta I_p = |a_1u_1\sqrt{d\sigma} + a_2u_2\sqrt{d\sigma}|^2. \tag{3.144}$$

The total intensity at P due to the entire primary source is given by the integral of δI_p over the primary source. We define the integral of $|a_1u_1|^2\, d\sigma$ as I_1 and that of $|a_2u_2|^2\, d\sigma$ as I_2 to obtain

$$\begin{aligned} I_p &= \int_{-\infty}^{\infty} |a_1u_1|^2\, d\sigma + \int_{-\infty}^{\infty} |a_2u_2|^2\, d\sigma + 2\,\mathrm{Re}\left\{\int_{-\infty}^{\infty} a_1u_1a_2^*u_2^*\, d\sigma\right\} \\ &= I_1 + I_2 + I_{12}, \end{aligned} \tag{3.145}$$

where

$$I_{12} = 2\,\mathrm{Re}\left\{\int_{-\infty}^{\infty} a_1u_1a_2^*u_2^*\, d\sigma\right\} \tag{3.146}$$

is the *mutual intensity* due to the two sources at plane P_2. As the first integral of Equation (3.145) is proportional to I_1 and the second integral is proportional to I_2, we can express the mutual intensity as

$$I_{12} = 2\sqrt{I_1 I_2}\,\gamma_{12}, \tag{3.147}$$

where the function γ_{12} is called the *complex degree of coherence*. Note that γ_{12} is a function of the primary source geometry, the secondary source geometry, and the paths traversed by the light from P_2 to P_3. We express γ_{12} as

$$\gamma_{12} = |\gamma_{12}| e^{j(\phi_1 - \phi_2 + \phi_{12})}, \tag{3.148}$$

where ϕ_1 and ϕ_2 are the phases associated with the path lengths from the sources at Q_1 and Q_2 to the point P and ϕ_{12} is the phase difference between points Q_1 and Q_2.

Finally, we find that the intensity at P is

$$\boxed{I_p = I_1 + I_2 + 2\sqrt{I_1 I_2}\,|\gamma_{12}| \cos(\phi_1 - \phi_2 + \phi_{12}).} \tag{3.149}$$

Generally the argument of the cosine function is unimportant because it simply establishes the principal maximum of the interference fringes. Of greater importance is the factor $|\gamma_{12}|$, which is called the *degree of spatial coherence*. The degree of coherence is bounded to the interval between zero and one because the mutual intensity has been normalized according to Equation (3.147). We see that when $|\gamma_{12}| = 0$, the intensity $I_p = I_1 + I_2$ is simply the sum of the individual intensities of the two sources. As a result, the optical system is *incoherent* and is linear in *intensity*.

We connect the fringe visibility from Equation (3.29) with the degree of coherence by using Equation (3.149) in Equation (3.29):

$$\begin{aligned} V &= \frac{I_{\max} - I_{\min}}{I_{\max} + I_{\min}} \\ &= \frac{I_1 + I_2 + 2\sqrt{I_1 I_2}\,|\gamma_{12}| - \left[I_1 + I_2 - 2\sqrt{I_1 I_2}\,|\gamma_{12}|\right]}{I_1 + I_2 + 2\sqrt{I_1 I_2}\,|\gamma_{12}| + \left[I_1 + I_2 - 2\sqrt{I_1 I_2}\,|\gamma_{12}|\right]} \\ &= \frac{2\sqrt{I_1 I_2}\,|\gamma_{12}|}{I_1 + I_2}. \end{aligned} \tag{3.150}$$

When $I_1 = I_2$, we find that $V = |\gamma_{12}|$ so that the visibility of the fringes is a direct measure of the degree of coherence. When $I_1 \neq I_2$, of course, we must measure I_1 and I_2 independently and use Equation (3.150) to calculate $|\gamma_{12}|$.

When $|\gamma_{12}| = 1$, the intensity at the observation point ranges from $(I_p)_{\max} = I_1 + I_2 + 2\sqrt{I_1 I_2}$ to $(I_p)_{\min} = I_1 + I_2 - 2\sqrt{I_1 I_2}$. When $I_1 = I_2$, the fringe visibility in a region about the point I_p is unity. When $I_1 \neq I_2$, there is an intensity bias and the fringe visibility is not unity. In either case, if $|\gamma_{12}| = 1$, the system is *coherent* and is linear in *amplitude*.

When $0 < |\gamma_{12}| < 1$, the source is *partially coherent*. There is some evidence of fringes over small areas but the fringe visibility is never unity, even if $I_1 = I_2$. Such systems are linear in neither amplitude nor intensity. Although they can be analyzed using linear systems theory, the analysis is complex and rarely offers much physical insight into the performance of the system. We avoid them like the plague.

3.8.2. Temporal Coherence

In an interferometer, light from a single source is divided into two beams and recombined after traversing different paths. If the light is monochromatic, a wave train has an infinitely long duration so that the *coherence distance* ΔD and the *coherence time* $\Delta t = \Delta D/c$ are infinite. As the source becomes increasingly polychromatic, the average length of the wave train shortens, as do coherence time and distance.

Temporal coherence is the property of light that allows interference between a wave train from a source and a delayed wave train from the same source. Consider the Michelson interferometer shown in Figure 3.36, in which we represent the signal from the light source by $s(t)$ at some

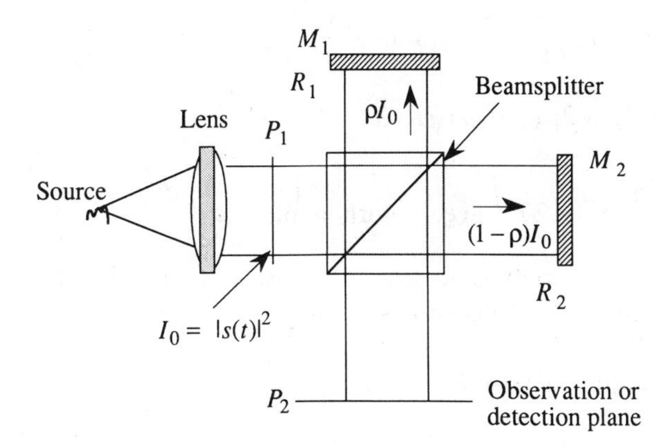

Figure 3.36. Geometry for calculating the temporal coherence function.

arbitrary plane P_1 before the beamsplitter. The intensity at this plane is $I_0 = |s(t)|^2$. Suppose that a fraction ρ of the light intensity is reflected by the beamsplitter and directed toward mirror M_1. If the reflectivity of the mirror is R_1, the light intensity reflected back toward the beamsplitter is ρR_1. If the beamsplitter is nonabsorbing, a fraction $(1-\rho)$ of the intensity reflected from mirror M_1 reaches the observation plane. The total light intensity reaching plane P_2 due to the first branch of the interferometer is therefore $I_1 = \rho(1-\rho)R_1 I_0$.

A similar argument is applied to the second branch of the interferometer, in which light is transmitted by the beamsplitter, reflected by mirror M_2, and then reflected by the beamsplitter to reach plane P_2. The total light intensity reaching plane P_2 due to the second branch of the interferometer is therefore $I_2 = \rho(1-\rho)R_2 I_0$. At the observation plane P_2, the signal is

$$f(t) = \sqrt{\rho(1-\rho)R_1}\, s(t - z_1/c) + \sqrt{\rho(1-\rho)R_2}\, s(t - z_2/c), \quad (3.151)$$

where $z_1/2$ and $z_2/2$ are the distances measured along the optical axis from plane P_1 to plane P_2. The bandwidths of all physical detectors are too low to detect the frequency of light directly. Such detectors have low-pass characteristics, equivalent to integrating the intensity over a time period T. The *effective* intensity, aside from a scaling factor, is therefore given by the time average of the square of the resultant amplitude signal:

$$\begin{aligned}
I_d &= \langle f(t) f^*(t) \rangle \\
&= \left\langle \left| \sqrt{\rho(1-\rho)R_1}\, s\left(t - \frac{z_1}{c}\right) \right|^2 \right\rangle + \left\langle \left| \sqrt{\rho(1-\rho)R_2}\, s\left(t - \frac{z_2}{c}\right) \right|^2 \right\rangle \\
&\quad + 2\,\mathrm{Re}\left\{ \left\langle \rho(1-\rho)\, s\left(t - \frac{z_1}{c}\right) s^*\left(t - \frac{z_2}{c}\right) \right\rangle \right\} \\
&= I_1 + I_2 + 2\rho(1-\rho)\sqrt{R_1 R_2} \\
&\quad \times \mathrm{Re}\left\{ \lim_{T\to\infty} \frac{1}{2T} \int_{-T}^{T} s\left(t - \frac{z_1}{c}\right) s^*\left(t - \frac{z_2}{c}\right) dt \right\}. \quad (3.152)
\end{aligned}$$

We now change variables so that $t - z_1/c = q$ and find that the third term of Equation (3.152) becomes

$$\begin{aligned}
r_{ss}(\tau) &= 2\rho(1-\rho)\sqrt{R_1 R_2} \\
&\quad \times \mathrm{Re}\left\{ \lim_{T\to\infty} \frac{1}{2T} \int_{-T-z_1/c}^{T-z_1/c} s(q)\, s^*\left(q + \frac{z_1}{c} - \frac{z_2}{c}\right) dq \right\}. \quad (3.153)
\end{aligned}$$

If $s(t)$ is a stationary process, the integral is a function only of the time difference $(z_1 - z_2)/c = \tau$ so that the time origin is immaterial. We therefore write the integral as

$$r_{ss}(\tau) = 2\rho(1-\rho)\sqrt{R_1 R_2}\,\mathrm{Re}\left\{\lim_{T\to\infty}\frac{1}{2T}\int_{-T}^{T} s(q)s^*(q+\tau)\,dq\right\}, \tag{3.154}$$

which we recognize as the autocorrelation function of $s(t)$.

In the optics literature, $r_{ss}(\tau)$ is normally indicated by $\Gamma_{12}(\tau)$ and is called the *mutual intensity*. If we normalize $\Gamma_{12}(\tau)$ by $\Gamma_{11}(0)$, we find that the intensity at plane P_2 is

$$\boxed{I_d = I_1 + I_2 + 2\sqrt{I_1 I_2}\,\gamma_{12}(\tau),} \tag{3.155}$$

where $\gamma_{12}(\tau)$ is called the *temporal coherence function*. As $s(t)$ must be real valued, $\gamma_{12}(\tau)$ is also real valued and we normally speak of the magnitude $|\gamma_{12}(\tau)|$ as the *degree of temporal coherence*.

The source is completely incoherent when we use white light and $\gamma_{12}(\tau) = \delta(\tau)$. In this case we see that, for $\tau \neq 0$, the intensity at plane P_2 is $I_d = I_1 + I_2$, which is the ordinary splitting case which would result from two independent sources. When the two paths of the interferometer are exactly balanced so that $\tau = 0$, the intensity at the output increases to $I_d = I_1 + I_2 + 2\sqrt{I_1 I_2}$.

To further illustrate the concept of temporal coherence and to relate it to communication theory, suppose that $s(t) = a(t)\cos(2\pi f_l t)$, where $a(t)$ is a baseband modulation signal and f_l is the frequency of an assumed monochromatic light source. Equation (3.154) then becomes

$$r_{ss}(\tau) = 2\rho(1-\rho)R_1 R_2 \times \mathrm{Re}\left\{\lim_{T\to\infty}\frac{1}{2T}\int_{-T}^{T} a(q)\cos(2\pi f_l q)a^*(q+\tau)\cos[2\pi f_l(q+\tau)]\,dq\right\}. \tag{3.156}$$

We expand the cosine product to form sum and difference frequencies. The sum frequency, when integrated over a long period of time, does not contribute to the integral so that we have

$$\begin{aligned} r_{ss}(\tau) &= 2\rho(1-\rho)R_1 R_2 \cos(2\pi f_l \tau)\mathrm{Re}\left\{\lim_{T\to\infty}\frac{1}{2T}\int_{-T}^{T} a(q)a^*(q+\tau)\,dq\right\} \\ &= 2\sqrt{I_1 I_2}\,r_{aa}(\tau)\cos(2\pi f_l \tau), \end{aligned} \tag{3.157}$$

where we have noted that the integral is the autocorrelation function of the modulation function $a(t)$. The total intensity at plane P_2 for this example is

$$I_d = I_1 + I_2 + 2\sqrt{I_1 I_2}\, r_{aa}(\tau)\cos(2\pi f_l \tau), \tag{3.158}$$

and the temporal degree of coherence is

$$\gamma_{12}(\tau) = r_{aa}(\tau)\cos(2\pi f_l \tau), \tag{3.159}$$

so that, although the light is monochromatic, the degree of temporal coherence is determined by the baseband signal $a(t)$. When $r_{aa}(\tau) = 1$, Equation (3.158) shows that $I_d = I_1 + I_2 + 2\sqrt{I_1 I_2}$ when the cosine function is at its maximum value. When the cosine term has its minimum value, we have that $I_d = I_1 + I_2 - 2\sqrt{I_1 I_2}$. In general, Equation (3.158) shows that the output of the interferometer, as one of the mirrors is moved to change the value of τ, follows the shape of the autocorrelation function of the baseband signal $a(t)$, but with rapid fluctuations due to the cosine function of τ. The autocorrelation function $r_{aa}(\tau)$ is the envelope of the cosine because it is a slowly varying function.

3.8.3. Spatial and Temporal Coherence

In some systems the temporal coherence may further modify spatial coherence. For example, in Figure 3.35 any point on the observation plane that is equally distant from the two sources will show fringes, even in white light, because $\tau = 0$ and $\gamma_{12}(\tau) = \delta(\tau) = 1$. As we move to a position away from the bisector, the degree of temporal coherence decreases because the path lengths become unequal. The fringe visibility in the observation plane is, to the first order, the product of the degree of spatial coherence and the temporal coherence function.

The coherence length $\Delta L = z_2 - z_1$ is the physical path difference corresponding to the coherence time duration of the source. The coherence length indicates how accurately the two path lengths in an interferometer must be balanced. The coherence length is related to the frequency spread in the source in the following way:

$$f = \frac{c}{\lambda},$$

$$|\Delta f| = \frac{c\Delta\lambda}{\lambda^2} = \frac{f\Delta\lambda}{\lambda},$$

$$\frac{\Delta f}{f} = \frac{\Delta\lambda}{\lambda}, \tag{3.160}$$

so that the wavelength spread is proportional to the frequency spread. Furthermore, since $\Delta L = c\Delta t$, we have

$$\Delta L = c\Delta t = \frac{c}{\Delta f} = \frac{\lambda^2}{\Delta\lambda}. \tag{3.161}$$

PROBLEMS

3.1. Calculate the observed spatial frequency of the fringe pattern generated in a plane P_2 by applying the Fresnel transform to two sample functions of the form $\text{sinc}(x/d_0)$, spaced 4 mm apart in plane P_1. Let the distance D between the planes be 1000 mm. Assume completely coherent light and a wavelength of 0.5 μ. If the sources each have dimensions of $d_0 = 0.05$ mm, what is the approximate extent of the fringe pattern? Hint: First calculate the diffraction angle produced by each source; then calculate the width (or extent) of the beam in plane P_2; then calculate the extent of the overlap. Recall that interference fringes can occur only when two or more beams of light overlap.

3.2. Suppose that the signal now consists of three sample functions of the form $\text{sinc}(x/d_0)$ in the (x, y) plane: one at $(0, 0)$, one at $(4, 0)$, and one at $(0, 3)$. Sketch the input signal as it would appear if you looked at the x-y plane from the positive z direction. If the other parameters are as in Problem 3.1, provide a two-dimensional sketch of the spatial frequency fringe pattern in the Fresnel plane, calculate the values of the spatial frequencies, and show the regions of overlap. Hint: Treat the sample functions in a pairwise fashion and use the principle of superposition.

3.3. You observe a Fresnel zone present on a viewing screen. You measure the diameter of the first dark ring as 2 mm. If the wavelength is 0.5 μ how far away from the screen is the point source? Assume that the reference beam is collimated and that the intensity of the zone is a maximum in the center. What is the *diameter* of the seventh dark ring?

3.4. Calculate the observed spatial frequency of the fringe pattern at the Fourier-transform plane when the input is the set of two sources given in Problem 3.1; the focal length of the lens is 1000 mm. Is the value of the spatial frequency different from that of Problem 3.1? How about the extent of the fringes; i.e., is the amount of overlap in the Fourier plane less than, the same as, or greater than that of the Fresnel plane? Support your answer with calculations and sketches.

3.5. A source of the form sinc(x/d_0), where $d_0 = 5\ \mu$, is located a distance of 700 mm from a screen. Another source is located 900 mm from the screen and displaced laterally from the first source by 20 mm. Calculate the size that the second source must have so that it completely overlaps light from the first source at the screen and calculate the highest spatial frequency in the interference pattern.

3.6. A plane-wave reference beam is added to the wave produced by a point source of the form sinc(x/d_0). The maximum frequency measured in the interference pattern is 25 Ab. The source is 275 mm from the observation screen. Calculate (a) the chirp rate, (b) the length of the chirp, (c) the number of samples needed to accurately represent the chirp function, and (d) the source size d_0.

3.7. A one-dimensional Fresnel zone is recorded on a strip of film 220 mm long. The spatial frequency at one end is zero and the spatial frequency at the other end is 430 Ab. The strip of film is illuminated by collimated light and is moved through an aperture that is 30 mm long. Calculate where the light is focused when the lowest-frequency portion of the strip is in the aperture. Repeat the calculation for when the highest-frequency portion is in the aperture. Describe what happens to the size and location of the focused spot when the film is moved through the aperture at a constant velocity v_0. Illustrate your results with a sketch. Assume the wavelength to be 500 nm.

3.8. You have a prism whose index of refraction is 1.55 and whose apex angle is 24°. The prism is 14 mm high and is illuminated, normal to one of its faces, by a plane wave of collimated light whose extent is infinite in the plane of the prism. Calculate the distance from this plane to the plane where the refracted light completely overlaps the undisturbed light. Calculate the form of the resultant intensity pattern if the prism absorbs a fraction 0.4 of the light (do not forget to calculate the effect of the prism magnification on the amplitude of the transmitted light). Calculate the spatial frequency. Assume the wavelength to be 500 nm. Do a sketch.

3.9. Calculate the two-dimensional Fourier transform of a rectangular aperture that is 4 mm long in the x direction and 6 mm high in the y direction; the focal length of the lens is 200 mm and the wavelength of light is 0.5 μ. Provide labeled sketches of both the object and transform plane light distributions. If the object plane were stretched by a factor of 2 in the x direction, how would the transform change in shape and amplitude?

3.10. The amplitude of a simple one-dimensional signal is given by $f(x) = [1 + \cos(2\pi\alpha_1 x)]\text{rect}(x/L)$. Suppose that the magnitude response of a spatial filter in the region of $\pm\alpha_1$ in the Fourier plane is 0.4, and that its phase response is $e^{j\beta}$; let the transmittance be 1.0 in the region where the undiffracted light is located ($\alpha = 0$). Sketch the magnitude and phase of the filter. Calculate the intensity $|g(u)|^2$ at the output of the filtering system and compare it with the intensity of the object $|f(x)|^2$. Hint: It may be useful to let $L \to \infty$ to solve for the inverse transform. Comment on the effect that the phase β has on the intensity of the output. Make a similar comparison if the magnitudes are as given but the phase response is $e^{j\beta}$ at $+\alpha_1$ and $e^{-j\beta}$ at $-\alpha_1$?

3.11. A function $f(x) = 1 + \cos(2\pi\alpha_1 x) + \cos(2\pi\alpha_2 x)$ is placed in a Fourier-transforming system for which $\lambda = 0.5\ \mu$, the focal length is $F = 200$ mm, and the input aperture is $L = 50$ mm. Assume normal incidence illumination and let $\alpha_1 = 10$ Ab and $\alpha_2 = 15$ Ab. Suppose that the function is moving with velocity $v = 2$ mm/sec in the positive x direction. Part (a): Write the general form of $f(x, t)$ that incorporates $f(x)$ and the velocity v. Part (b): Derive the Fourier transform $F(\alpha, t)$ of $f(x, t)$ in terms of the general variables and parameters. Part (c): Using the specified values of α_1, α_2, λ, F, and v, calculate the locations and the temporal frequencies in the Fourier plane associated with each spectral component of the input signal.

3.12. Consider the convolution in the space domain:

$$g(u) = \int_{-\infty}^{\infty} \text{sinc}(ax)\text{sinc}[b(u - x)]\, dx.$$

Find $g(u)$ for (1) $a > b$, (2) $a = b$, and (3) $a < b$. Hint: There is no need to solve the convolution directly; the solution is apparent when you consider what happens in the Fourier domain.

3.13. You place a lens whose focal length is 100 mm a distance of 400 mm from a source of the form $\text{sinc}(x/d_0)$, where $d_0 = 5\ \mu$. You place an aperture whose diameter is 36 mm a distance of 300 mm from the source; the signal is $f(x) = 1 + \cos(2\pi\alpha_1 x)$, where $\alpha_1 = 50$ Ab. Calculate (a) the position of the Fourier-transform plane, (b) the function $F(\xi)$, (c) the function $F(\xi)$ if the signal and aperture are placed in the plane of the lens, and (d) the distance between the undiffracted light and the diffracted light in the Fourier domain if the signal is placed 10 mm away from the lens on the Fourier-transform side of the lens.

3.14. We enter a laboratory and discover that someone has set up an interferometric system that causes a time delay between two beams of light derived from a common source. We block one of the beams and find that a power meter reads 3 mW/mm^2 at the output plane. We block the other beam and find that the meter reads 1 mW/mm^2. With both beams falling on the power meter, we read 6 mW/mm^2. What is the magnitude of the degree of coherence for this particular set up? Hint: This problem deals with temporal coherence.

3.15. In an adjacent laboratory, we notice a different interferometric arrangement in which two small apertures in a plane appear to be illuminated by a common light source. We notice that a set of spatial fringes exists at an output plane. When we block one of the apertures, we measure the intensity as 2 mW/mm^2. When we block the other aperture, we measure the intensity as 9 mW/mm^2. When both apertures are open, we note that the maximum intensity in the fringes is 12.7 mW/mm^2. Calculate (1) the magnitude of the complex degree of coherence, (2) the minimum intensity in the fringe pattern, and (3) the fringe visibility. Hint: This problem deals with spatial coherence.

3.16. We observe fringes from two apertures located at $x = \pm 4$ mm in plane P_1 at a screen located 100 mm away from plane P_1. The intensities of the two sources are $I_1 = 10$ mW/mm^2 and $I_2 = 5$ mW/mm^2. The visibility of the fringes is measured as $V = 0.72$ at the observation screen opposite the midpoint between the two sources. For this setup, the visibility drops linearly, being zero at ± 20 mm from the center: (a) Calculate the degree of spatial coherence $|\gamma_{12}|$ and (b) the degree of temporal coherence $\gamma_{12}(\tau)$ at $\tau = 2$ picoseconds (i.e., for a path difference of 0.6 mm).

3.17. Part (a): In a spatial interferometer the intensity due to one beam is 16 times that due to the other (when each is measured separately). In terms of one of the unknown intensities, calculate I_{max}, I_{min}, and the fringe visibility when (1) $\gamma_{12} = 0.4$ and (2) when $\gamma_{12} = 0$. Part (b): Under what general conditions could you adjust the path-length difference of a temporal interferometer and find that the maximum and minimum intensities are periodic? Give at least one specific example to support your claim.

4

Spectrum Analysis

4.1. INTRODUCTION

Spectrum analysis is the most widely used signal-processing technique in the physical sciences for gaining information about unknown signals. It is used in applications such as pattern recognition, cloud-cover analysis, inspection of manufactured items, particle-size analysis, measurements of turbulence, sea state analysis, characterization of the electromagnetic spectrum, determining direction of arrival of emitters, and structural analysis. The Fourier transform, as developed in Chapter 3, plays a central role in optical spectrum analysis. Signal features, such as periodic structures, are more easily detected in the Fourier domain than in the space domain because the energy from each frequency in the signal is concentrated at a particular point in the Fourier plane.

In the optical system shown in Figure 4.1, a signal stored on a spatial light modulator at plane P_1 is illuminated by coherent light. In Chapter 3 we discussed the range of geometrical conditions for which the Fourier transform occurs; for convenience, we use a system in which the signal and Fourier planes are at the front and back focal planes of the lens. The complex-valued light at plane P_2, the image plane of the primary source, is the Fourier transform of the light at plane P_1:

$$S(\alpha, \beta) = \iint_{-\infty}^{\infty} a(x, y) s(x, y) e^{j2\pi(\alpha x + \beta y)} \, dx \, dy, \tag{4.1}$$

where $s(x, y)$ is the amplitude of the signal, $a(x, y)$ is an aperture function, x and y are the spatial coordinates of plane P_1, and α and β are spatial frequencies. The spatial coordinates ξ and η at plane P_2 are related to the spatial frequencies α and β by

$$\begin{aligned} \xi &= \lambda F \alpha, \\ \eta &= \lambda F \beta, \end{aligned} \tag{4.2}$$

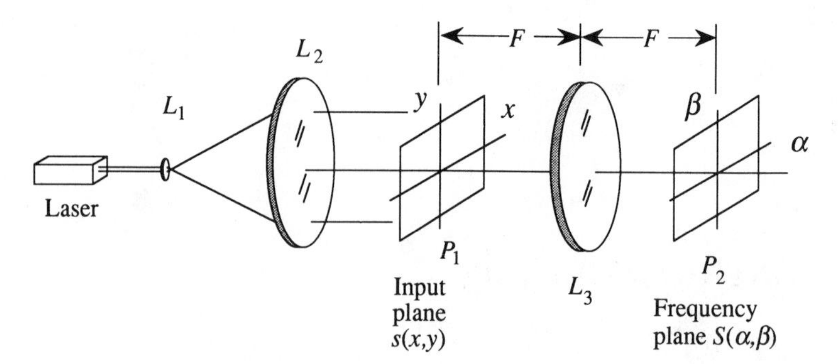

Figure 4.1. The Fourier transform is basic to spectrum analysis.

where F is the focal length of the lens and λ is the wavelength of light. A photodetector array, placed in the Fourier plane, measures the spectral content of the signal.

In this chapter we discuss the key active elements of a typical spectrum analyzer: the light source, the spatial light modulator, and the photodetector array. We then discuss the key performance parameters and develop some design guidelines for maximizing the dynamic range of the analyzer. We conclude this chapter with a discussion of a two-dimensional spectrum analyzer that provides excellent frequency resolution over an extremely wide signal bandwidth.

4.2. LIGHT SOURCES

The usual source of coherent light in optical signal processing is a laser. Water-cooled argon-ion lasers have strong spectral lines at 488.0 and 514.5 nm and power outputs in the 0.2–20-W range. Helium-neon lasers, emitting light at 632.8 nm, are more often used because they are air cooled, compact, and reliable. They are generally less powerful, however, with outputs in the 1–50 mW range.

Semiconductor lasers are the most useful lasers for signal-processing applications because they provide high optical power levels, help to significantly reduce the size of the processing system, and efficiently convert electrical power into optical power. Laser diodes provide output powers comparable to those of much bulkier gas lasers; single-element lasers have been developed, with both spatial and temporal single-mode operation, at powers of greater than 50 mW at 830 nm (using GaAlAs layers of various fractional compositions) and at 1300 nm (using InGaAsP layers). Large arrays of laser diodes have been developed that produce watts of cw optical power (30).

4.3. SPATIAL LIGHT MODULATORS

To create the spectrum of a signal, we structure the input data in an optical format. *Spatial light modulators* are devices that format electronic or incoherent optical information so that it can be processed using coherent light. Historically, the most common spatial light modulator was photographic film as used in pattern recognition and radar processing applications. Photographic film has the attractive feature that its space bandwidth product most nearly matches that of both optical sensors and optical processing systems. In an interesting study, Kardar has shown that film also has a high information channel capacity, even when we account for the relatively long exposure times (31).

For modern signal-processing applications, we need spatial light modulators that have several distinctive features.

- A large space band-width product to provide a high level of performance.
- Adequate bandwidth and response time for use in computationally intensive applications.
- Wide dynamic range for applications such as spectrum analysis and radar processing.
- Good linearity so that intermodulation products are controlled.
- Good efficiency so that optical sources with high power outputs are not required.
- Good phase control so that aberrations do not limit system performance.
- Good geometric fidelity so that the processed signals are not distorted spatially.

We briefly review the current state of two-dimensional spatial light modulators and assess which are most useful for real-time signal processing (32–34). One-dimensional acousto-optic spatial light modulators needed for real-time signal processing, as discussed in the later chapters of this book, are treated separately in Chapter 7.

4.3.1. Light Valve Spatial Light Modulators

An early real-time modulator was a modified *light valve* developed for theater projection television. An electron beam gun is used to deposit charge onto a viscous fluid supported on a rotating disc substrate as shown in Figure 4.2(a). Electrostatic forces deform the fluid to produce a path-length variation that phase modulates light passing through the fluid film. The thickness variation is represented by $f(x, y, t) = s(x, y, t)\cos(2\pi\alpha_c x)$, where $s(x, y, t)$ is the TV baseband signal and α_c is a spatial carrier frequency introduced in the x direction so that we can separate diffracted light, containing the information, from undiffracted light.

The complex-valued amplitude transmittance is a pure phase function $\exp[jmf(x, y, t)]$, where m is a scaling constant, that can be expanded into a power series. If the argument of the exponential is small, we retain just the first two terms and express the transmittance as $1 + jmf(x, y, t)$. To use the light valve in a coherently illuminated system, we want the amplitude transmittance of the device to be proportional to the amplitude of the applied signal. We use the *Schlieren imaging technique* shown in Figure 4.2(b) in which we coherently illuminate the light valve to produce

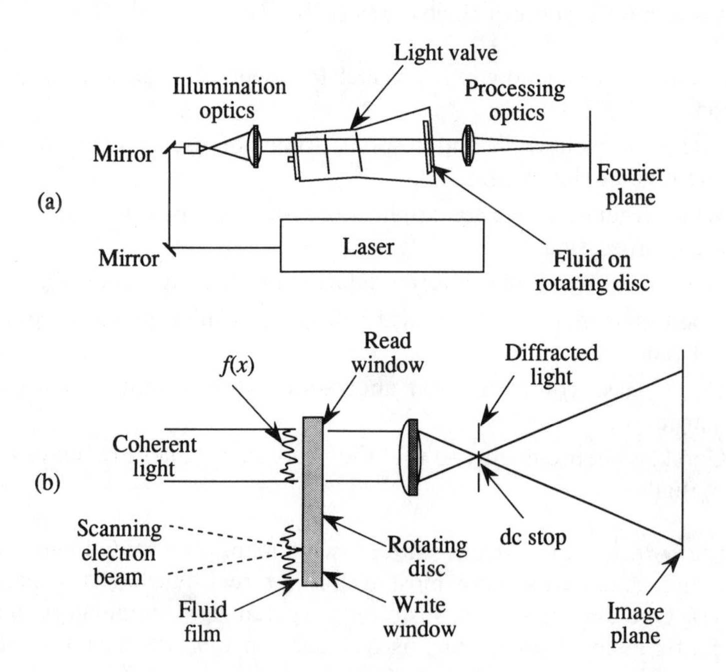

Figure 4.2. Light valve as a spatial light modulator: (a) optical system and (b) details of the recording and reading functions.

the Fourier transform of the input signal and use a stop in the Fourier plane to remove the undiffracted light and one diffracted order. The effective amplitude of the transmitted light is therefore proportional to $jmf(x, y, t)$ so that the light amplitude is now proportional to the applied signal voltage.

The light valve operates at TV frame rates, has a bandwidth in the 30-MHz range, and has a space bandwidth product of approximately 10^6. The maximum diffraction efficiency is 33.8% as dictated by the diffraction theory associated with thin phase modulating media. The light valve is a good real-time display for raster-scanning spectrum analysis, which we discuss in Section 4.7. Its major disadvantages are its large size and the use of high-voltage electron beam tubes.

4.3.2. Optically Addressed Electro-Optic Spatial Light Modulators

One of the first electro-optic spatial light modulators was the Pockels effect Readout Optical Modulator (PROM). The basic concept is that a voltage, applied across a sandwich device as shown in Figure 4.3, produces an electrostatic charge pattern that changes the transmittance of the electro-optic crystal. Incoherent light, striking the device from the left, activates a bismuth silicon oxide photosensitive layer to produce a spatially varying charge pattern that controls the transmittance. Readout light, at a different wavelength from the readin light, reflects from the dichroic mirror and makes two passes through the electro-optic material. By changing the voltage across the device, we can perform useful operations such as contrast reversal or subtraction of two successive images. The

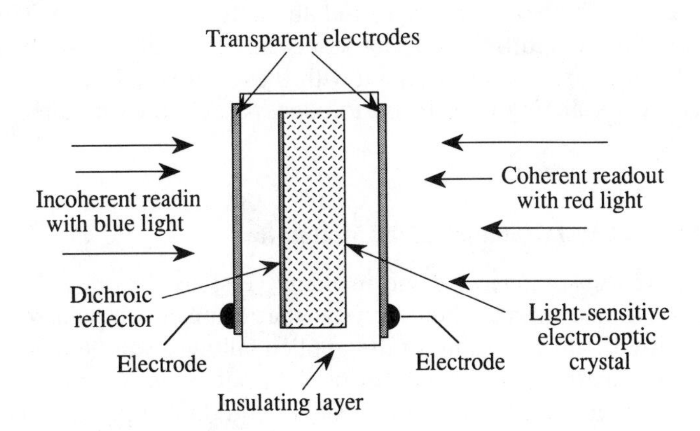

Figure 4.3. Bismuth silicon oxide spatial light modulator.

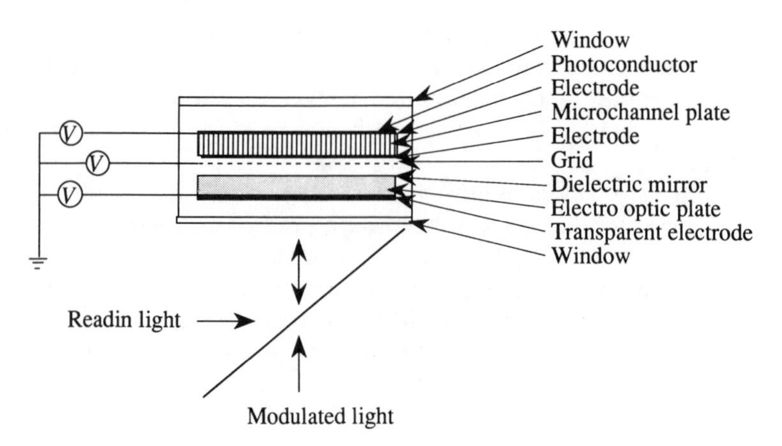

Figure 4.4. A microchannel plate signal light modulator.

space bandwidth products of these devices are of the order of 10^7, with $\alpha_{co} \approx 150$–500 Ab and contrast ratios of 5000 : 1 or so. The major problem is the uniformity of response across the aperture of the device.

A *microchannel plate* is similar to the PROM. The key difference is that the electro-optic material is replaced by one consisting of several small hollow cylinders, as shown in Figure 4.4. The surfaces of the cylinders are coated with a material that has an electron secondary emission coefficient greater than one. Light strikes a photocathode, which releases electrons into the microchannel tubes. The electrons are amplified and strike an electro-optic plate, as with the electron beam addressed devices, or they strike a fluorescent screen to provide a brighter optical image. These devices have 40–200-Hz frame rates and about 10-Ab resolution. They are sometimes used in nighttime spotting scopes and binoculars; extremely low light level scenes are readily observed with these devices. They can also be addressed by a scanning laser beam to process real-time one-dimensional signals.

4.3.3. Liquid-Crystal Spatial Light Modulators

Liquid-crystal display devices come in many versions. In one version the active material is a *nematic* liquid crystal that changes the polarization of the transmitted light according to the applied voltage. The field is applied by means of a matrix of electrodes or by light falling onto a photoconductive surface. Readout is by means of a single or a double pass, as shown in Figure 4.5. Another version operates in the *variable grating modulation*

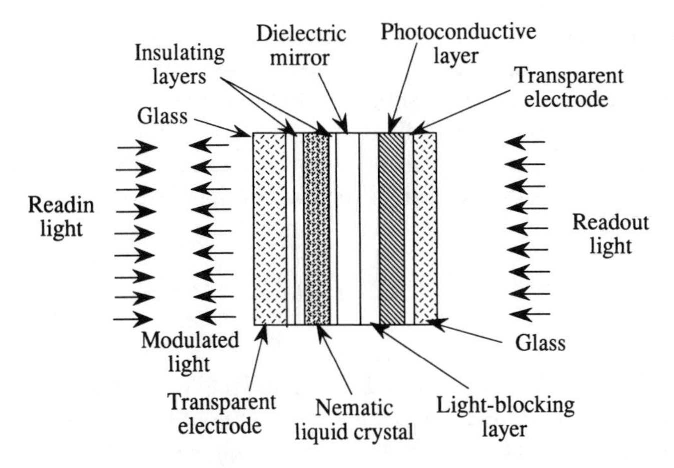

Figure 4.5. Nematic liquid crystal display spatial light modulator.

mode, wherein the application of a dc voltage causes the liquid-crystal display to assume a grating structure. Light is thereby diffracted according to the amplitude of the applied voltage. Typical liquid-crystal displays have frame rates of 10–20 frames per second, contrast ratios of 10 : 1, resolutions in the 8–12-Ab range, and fairly poor throughput efficiencies. Their performance parameters are improving rapidly and they are finding increasing use as flat panel television and computer displays.

4.3.4. Magneto-Optic Spatial Light Modulators

Magneto-optic devices are based on an epitaxial garnet film grown on a nonmagnetic substrate and are addressed by a matrix conductor array. The Faraday effect causes a rotation of the plane of polarization. The time required to switch one element is about 10 μsec so that frame rates range from 3 msec for 48×48-element arrays to 20 ms for 128×128-element arrays. Efficiency is poor, of the order of 0.1%. Other magneto-optic phenomena include optically writing on a magnetic medium and then heating it to the Curie point at which the magnetic flux rotates appreciably. After cooling, the magnetic domains are rotated such that the amplitude from an analyzer changes according to the exposure; again, the amount of rotation is small, yielding low diffraction efficiencies.

Two-dimensional spatial light modulators have been under development for many years but progress has been relatively slow. The issue is finding a magic material that records analog signals and can be repeatedly erased and recorded without memory of past activity. The material must also have high resolution and a high dynamic range. As materials with these properties are also useful for optical memories to implement a read/write capability, that field may develop alterable materials also suitable for optical processing. A review of two-dimensional light modulators, including deformable mirror technology, has recently been given (34).

4.4. THE DETECTION PROCESS IN THE FOURIER DOMAIN

The outputs of optical processing systems were initially detected, recorded, and displayed using photographic film; we might say that the first two-dimensional "photodetector array" was photographic film. In common with other photodetector devices, film requires a certain number of photons per unit time, is sensitive to light in certain spectral ranges, and has a dynamic transfer function, a modulation transfer function, and a noise floor (often expressed in terms of granularity, grain size, or Selwyn's number). Although the terminology is different, the concepts are similar to those associated with modern photodetector arrays.

A key advantage of photographic film is that its high spatial resolution matches that of the optical system. The major disadvantage is the time delay in developing the film; as a result, it cannot support real-time operations. Electron beam tubes, such as vidicons or image orthicons, have limitations such as inadequate dynamic range and geometric fidelity. One- and two-dimensional photodetector arrays, based on photosensitive charge-coupled device (CCD) structures, provide a new flexibility of operation and have many desirable features.

4.4.1. A Special Photodetector Array

If a signal $s(x, y)$ contains regular features, such as the street pattern of a city, the spacing and width of the sidelobe structure in the spectrum $S(\alpha, \beta)$ indicate the period of the street spacings. In contrast, the spectrum associated with natural terrain is generally more uniformly distributed over all spatial frequencies with no predominant peaks. As an illustration of two-dimensional spectrum analysis, we show, in Figure 4.6, a photograph that contains a variety of ground textures. From the spectra shown in the inserts, we note that strong diffraction occurs in directions normal to the ground texture; this is most notable in the spectra of the bridge and of the streets. Images of the surface of the ocean produce spectra that give information about wave direction and ocean depth. Regions containing clouds and natural terrain have more uniform angular spectra but still show variations in the spatial frequency distribution. If we

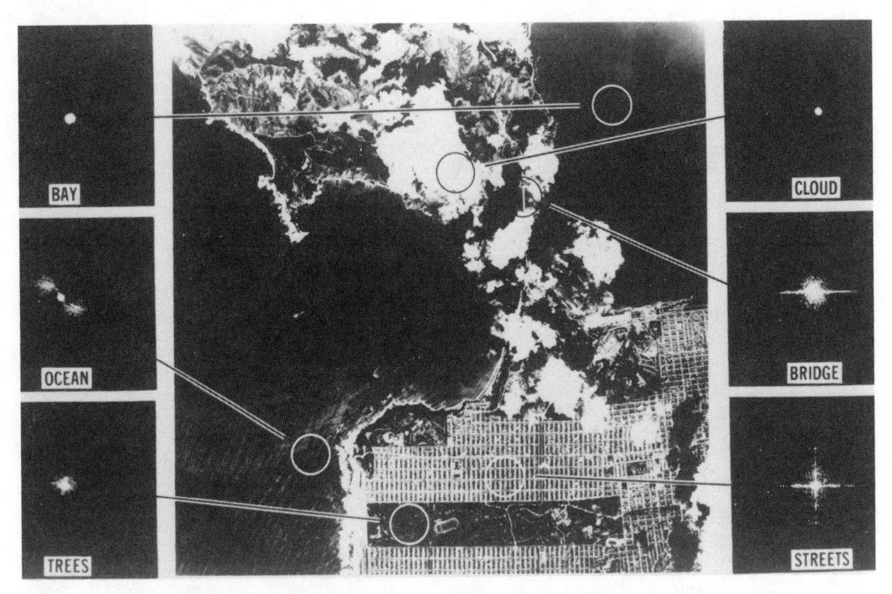

Figure 4.6. Examples of two-dimensional spectrum analyses (courtesy of Robert Leighty, U.S. Army Engineering Topographic Laboratory).

sequentially scan small portions of the input, the spectra of these subregions can be detected in the Fourier plane to give an indication of texture and its variation from one region to another.

To properly characterize the spectral information in a signal, we generally need to illuminate a subregion of the signal with a light beam that is the right size. If the light beam is too large, the spectral characteristics of the region are not well defined; if the light beam is too small, there is insufficient information on which to form a spectral estimate. A typical illumination subsystem is shown in Figure 4.7, in which the laser beam of diameter A_1 is expanded to a diameter A_2 by means of the beam-

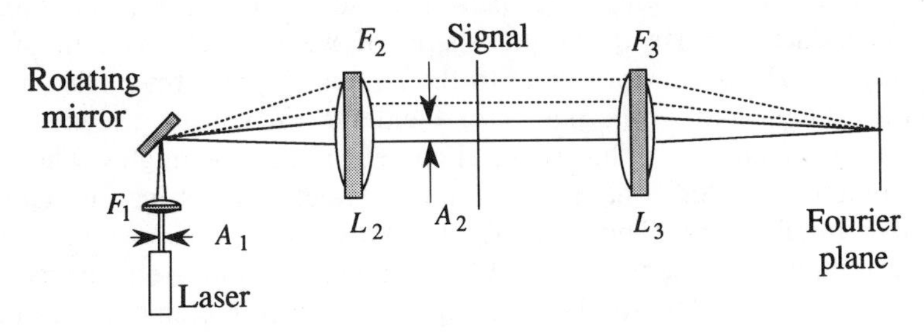

Figure 4.7. Telecentric scanning system.

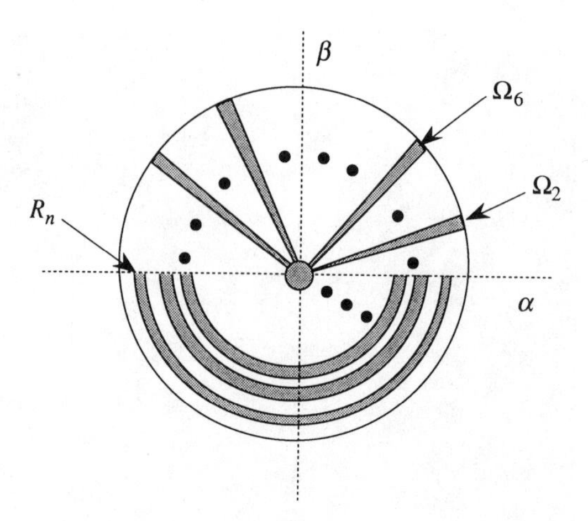

Figure 4.8. Special ring/wedge photodetector.

expanding telescope. The beam size ratio A_1/A_2 is the same as the focal length ratio F_1/F_2. At the common focal plane, we place a rotating mirror to provide the beam-scanning action at the signal plane.

It is important that the light strike the signal normal to its surface so that the Fourier transform, as produced by lens L_3, is always centered on the optical axis in the Fourier plane. At each position of the telecentric illuminating beam, the photodetector array is read out to produce the spectral content of the region being illuminated. Scanning in the orthogonal direction can be provided by a second scanning stage, similar to the one described. Sometimes the first stage is a fast scanning subsystem, using an acousto-optic scanner (see Chapter 7), combined with a slow scanning stage provided by a galvanometer mirror. Multifaceted rotating mirrors may also be used when the scanning velocity is high.

A useful array for detecting these spectral features consists of a set of wedge and annular photosensitive areas (35, 36), as shown in Figure 4.8. In this device, called a *ring/wedge detector*, one half of the area contains N photoconductive surfaces in the shape of wedges. The output signals $\Omega_1, \Omega_2, \ldots, \Omega_N$ are proportional to the amount of light that falls on each of the wedge-shaped photodetector elements to indicate the degree to which the signal $s(x, y)$ has spectral content at specific angles. The other half of the detector consists of N photodetector areas in the shape of narrow annular rings. The ring output signals $R_1, R_2, \ldots, R_N$ indicate the relative energy present in $s(x, y)$ at various spatial frequencies. The output from R_1, due to the circular photodetector element located in the center of the array, is proportional to the integrated amplitude of the signal and is used to normalize the spectrum.

As the spectrum $S(\alpha, \beta)$ is symmetric about the origin when the signal $s(x, y)$ is real valued, no information is lost in the detection process other than that which falls between the active areas. The signal is classified by postprocessing the Ω_i and R_i values according to algorithms developed for specific applications. Details of the postprocessing are not important here; requirements on the photodetector are. We now consider some of these requirements.

4.4.2. Spectral Responsivity and Typical Power Levels

The operating wavelength is dictated by factors such as the diffraction efficiency of the spatial light modulator, scattering in the optical system, and the availability of compact, efficient sources. We need photodetector elements whose *spectral responsivity* is in the 450–850-nm range to operate effectively with commonly available light sources. The photodetector must have a high responsivity at the wavelength generated by the laser but need not have a broad spectral response.

Photometry is a science whose terminology is strange. In the photometry literature we find quantities such as lumens, lux, phots, stilbs, apostilbs, and candelas, which are divided by feet, square feet, steradians, centimeters, and the like, to get even stranger quantities such as a nit (a candela per square meter). Because we deal largely with systems that use monochromatic light, photometry is considerably simplified. We use watts as the measure of optical power and millimeters as the unit of distance. The responsivity S of a discrete photodetector element is the ratio of the photocurrent to the incident light power, expressed in units of amps/watt.

The amount of optical power at the output of a spectrum analyzer is dependent on the efficiency of various components in the system and the collection efficiency of the photodetector. In Section 4.6, we discuss the optical power budget in detail; here we summarize just the main points. In a spectrum analyzer, we generally operate the spatial light modulator at a diffraction efficiency of no more than 1% per frequency to contain intermodulation products at an acceptable level. The rest of the system is typically about 10% efficient, primarily due to beam-shaping losses. The photodetector element collects about 30% of the light power associated with any given frequency because we generally use three detectors per frequency to achieve the required system performance. The maximum power that a photodetector intercepts for a laser with output power in the 10–30-mW range is of the order of 3–9 μW. The weakest signal is often 60–70 dB below this level.

4.4.3. The Number of Photodetector Elements

The ring/wedge photodetector, while useful for feature analysis, does not preserve all the information available in the Fourier plane because it does not have enough elements. The space bandwidth product of the input signal establishes the number of elements required of the photodetector array. The sampling theorem, as stated in Chapter 1, requires that we have $N \geq 4\text{SBP}$ elements in a two-dimensional photodetector array to avoid loss of information.

The number of photodetector elements required is large when processing high-quality signals. For example, if the highest frequency in a two-dimensional signal is 100 Ab and if the signal is 100 mm on a side, we require a $20{,}000 \times 20{,}000$-element array; this is at least an order of magnitude larger, in each dimension, than existing photodetector arrays. Therefore either bandwidth or frequency resolution must be sacrificed until larger photodetector arrays become available.

The situation is more favorable when processing two-dimensional signals with low information content, such as those produced by two-dimensional collection systems that also use CCD technology. Low-resolution video cameras may produce images in a 300×400-sample array, requiring a fairly simple photodetector array. The photodetector-array requirements are also more easily achieved in some one-dimensional signal-processing applications. For example, we often use acousto-optic cells to convert a wideband time signal to one that is a function of both space and time. (See Chapter 7 for more details.) Typical values for the time bandwidth product of acousto-optic cells are of the order of $\text{TW} = 1000$, which is almost independent of the interaction material used. Linear photodetector arrays, with up to 4096 elements on a single chip and coupled to a CCD readout structure, are available to satisfy these sampling requirements.

4.4.4. Array Geometry

The appropriate photodetector array for a one-dimensional spectrum analyzer is a linear array as shown in Figure 4.9. The array consists of M active elements shown shaded, each of width d' and height h. Each

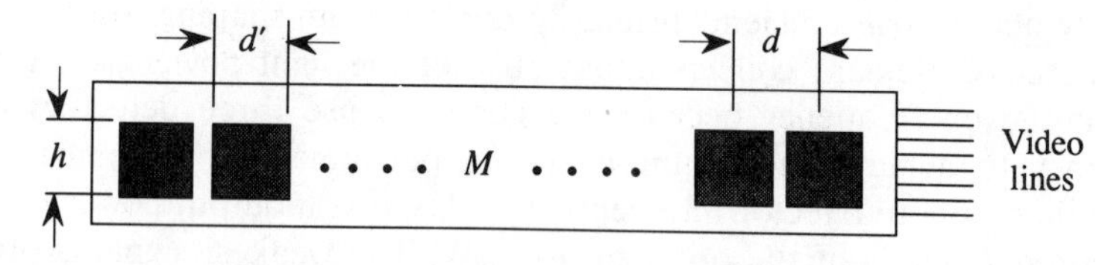

Figure 4.9. One-dimensional linear photodetector array.

element behaves as a photocapacitor on which electric charge accumulates in proportion to the incident light intensity and to the integration time. The array is read out serially by transferring the charge to a CCD structure and shifting the information onto multiple video lines, whose readout rates are each in the 1–10-MHz range.

The center-to-center spacing between photodetector elements is d and $c = d'/d$ is the spatial duty cycle of the array elements. The angular separation between adjacent frequencies and the focal length of the transform lens must be related to the dimensions of the array. For a uniformly illuminated signal of length L, the angular separation $\delta\theta$ between minimum resolvable frequencies $\delta\alpha$ is simply $\delta\theta = \lambda/L$ so that the frequencies are separated by a distance $\lambda F/L$, at plane P_2.

We need at least two detector elements per frequency to satisfy the sampling theorem in the Fourier domain; we therefore conclude that

$$2d \leq \frac{\lambda F}{L}. \tag{4.3}$$

In Section 4.5, we refine these calculations to account for signals that have weighted illumination to control sidelobe levels in spectral analysis applications.

Aberrations in an optical system are most easily kept under control if the relative aperture, which is approximately equal to $2L/F$, is less than $\frac{1}{10}$. Thus, an ideal center-to-center spacing for the photodetector elements is $d \approx 10\lambda$, which is of the order of 6–8 μ. The center-to-center spacing on currently available linear arrays is typically 12 μ or more; as a result, the focal length of the Fourier-transform lens and, therefore, the overall length of the optical system is somewhat greater than the optimum length. The size of the optical system is not a concern for ground-based systems because the volume of the optical system is generally a small fraction of that of the electronics. In airborne systems, however, the electronics are often packaged using very large scale integration (VLSI) techniques so that the volume and weight of the optical spectrum analyzer becomes relatively more important.

For one-dimensional applications, the height h of the photodetector is set by the signal height H in a fashion similar to Equation 4.3. If $H \neq L$, we use cylindrical lenses to match the vertical height of the spectrum to that of the array. For a two-dimensional application the photodetector center spacing and duty cycle are generally the same in both directions.

The required geometric accuracy for most spectrum analysis and correlation applications is that center spacings are within $\pm 1\%$ of their nominal positions, with a cumulative error of less than $\pm d/10$ at any position in the array. Semiconductor fabrication precision is usually better than the optical distortion in the system.

4.4.5. Readout Rate

In two-dimensional applications, the frame rate is a function of how rapidly the signal information changes. The spatial light modulator frame rate, in turn, determines how rapidly the spectrum changes; the photodetector array is typically read out once per frame. Integration on the array may range from a few milliseconds to a few seconds, provided that the detector does not saturate so that a linear response is maintained. The contents of the array are read out just before saturation, digitized, stored, and accumulated in postdetection memory. The accumulated values represent the desired spectrum and are available for further postprocessing operations.

In other applications, such as those involving real-time processing of one-dimensional signals as discussed more fully in Chapters 8 and 10, the spectrum may change rapidly so that we must read the array once every T seconds, where T is of the order of several microseconds. Unfortunately, the resultant temporal sampling rates are often not sustained by CCD transfer rates or by the digital postprocessing system, even when multiple video lines are used.

In both one-dimensional and two-dimensional spectrum analysis, we would benefit from the development of *smart arrays*, in which some postprocessing functions are included on the photodetector chip. Implementing logic functions at the array element level reduces the transfer data rates and the complexity of the subsequent electronics drastically. For example, suppose that we transfer information only if a photodetector element exceeds some preselected threshold. Or, suppose that we transfer information only if the instantaneous intensity exceeds some preset value. These operations require built-in circuitry for each element or groups of elements in the array. The implications of developing such a capability are enormous because the transfer rates associated with processing a wideband received signal may then be reduced by factors of 10^2–10^3 or more.

4.4.6. Blooming and Electrical Crosstalk

When we use long integration times to detect weak signals, we must ensure that the excess charge produced by strong signals is properly drained away so that spillover into adjacent elements does not mask weak signals. The spillover is sometimes optical in origin, with photons migrating from one photodetector element to another through the substrate of the array; this spillover is called *blooming*. Blooming is a more severe problem at longer wavelengths because of increased penetration distances in the substrate.

If the spillover is electrical in origin it is called *crosstalk*. We typically want the effects of crosstalk and blooming to decrease at a rate of at least 10 dB per element, referenced to the saturated element, to a level of at least -70 dB for all elements farther away than the seventh element.

4.4.7. Linearity and Uniformity of Response

At low intensity levels, most discrete photodetector elements have a linear response. At higher intensities the response of the element becomes nonlinear and saturation eventually sets in. Because the dynamic range at the output of a spectrum analyzer is large, we often introduce a compression scheme to facilitate the readout and display of the information. If the response of the photodetector is monotonic, a high degree of linearity is not required, provided that we can establish an inverse mapping that allows us to measure the spectrum to the required accuracy.

The saturation phenomenon is abrupt for some CCD arrays. The charge accumulates until the well is full; additional charge does not accumulate so that there is a discontinuity in the derivative of the transfer curve. As a unique inverse mapping of the signal intensities is not available for this case, we must avoid saturating these CCD arrays. Other CCD arrays offer a more gradual approach to saturation in an effort to gain more dynamic range. Again, an inverse mapping can be used to determine the true magnitude of the spectrum.

The *uniformity of response* of an array is defined as the variation in the output voltage or current due to manufacturing imperfections. A uniformity of response of $\pm 10\%$ is generally adequate; this degree of uniformity is easily met with current technology or can be corrected by using lookup tables. Differences in odd/even channel output levels may be corrected by postdetection gain compensation.

4.5. SYSTEM PERFORMANCE PARAMETERS

The key parameters that affect performance of spectrum analyzers at the system level are summarized as

- Total spatial frequency bandwidth (Section 4.5.1)
- Sidelobe levels/crosstalk (Section 4.5.2)

- Frequency resolution/dip between photodetector elements (Section 4.5.3)
- Array spacing and number of photodetector elements (Section 4.5.4)
- Dynamic range (Section 4.6)
- Intermodulation products (Section 4.6.1)
- Signal-to-noise/minimum signal level (Section 4.6.2)
- Integration time/bandwidth (Section 4.6.3)

We discuss the relationships among these performance parameters in the following paragraphs. Because the relationships are coupled in some cases, an iterative procedure may be needed to satisfy all the specifications simultaneously.

4.5.1. Total Spatial Frequency Bandwidth

The spatial frequency bandwidth for a baseband signal is simply equal to the highest spatial frequency α_{co} contained in the signal. If the signal is real valued, the spectrum is redundant about zero frequency and the photodetector need cover only half the Fourier plane, from 0 to α_{co}, as explained in Section 4.4.1. For a bandpass signal, centered on a carrier frequency α_c, the bandwidth extends from $\alpha_c - \alpha_{co}$ to $\alpha_c + \alpha_{co}$. The spectrum is redundant about α_c if the signal represents a baseband signal modulating a carrier frequency. In the more general case, the spectrum is nonredundant and the total spectral range is $2\alpha_{co}$.

4.5.2. Sidelobe Control and Crosstalk

In Section 4.4.2, we calculated the photodetector element spacing on the assumption that the signal is uniformly illuminated. We now consider the more typical situation in which an *aperture weighting function* is used to control the sidelobe levels of a cw frequency response in the Fourier plane. Aperture functions, a subset of the general class of apodization functions used in optics, are also used in digital signal processing, where they are generally called *window functions*. Candidate aperture functions are the rectangular, Bartlett, Hamming, Hanning, Chebyshev, Dolph-Chebyshev, Gaussian, Kaiser-Bessel, and Blackman window functions. Figure 4.10 shows four representative aperture functions defined over an aperture of length L. We indicate aperture functions by $a(x)$; the specific

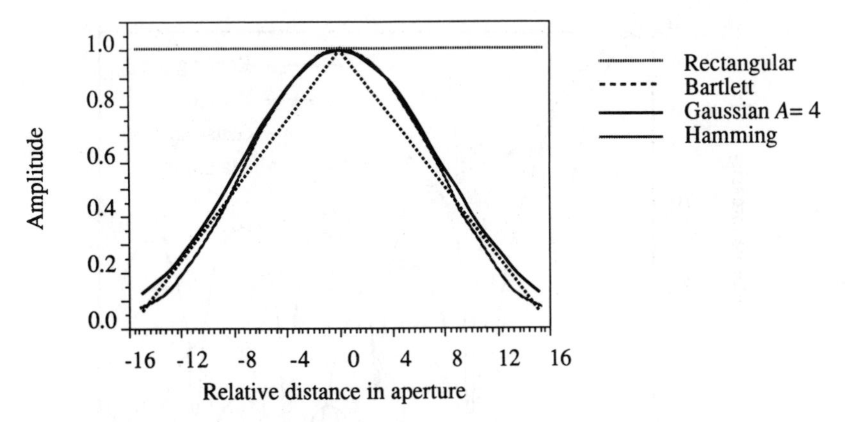

Figure 4.10. Four aperture functions.

functions we consider are

1. *Rectangular*

$$a(x) = \text{rect}(x/L) \equiv \begin{cases} 1, & |x| \leq L/2 \\ 0, & \text{elsewhere.} \end{cases} \tag{4.4}$$

2. *Bartlett*

$$a(x) \equiv \text{rect}\left(\frac{x}{L}\right)\left[1 - \frac{2|x|}{L}\right]. \tag{4.5}$$

3. *Gaussian*

$$a(x) \equiv \text{rect}(x/L)e^{-(A/2)(2x/L)^2}. \tag{4.6}$$

4. *Hamming*

$$a(x) \equiv \text{rect}(x/L)[0.54 + 0.46\cos(2\pi x/L)]. \tag{4.7}$$

The response in the Fourier plane to an aperture function is

$$A(\xi) = \int_{-\infty}^{\infty} a(x)e^{j(2\pi/\lambda F)\xi x}\,dx. \tag{4.8}$$

Figure 4.10 shows that, aside from the rect function, which has a uniform

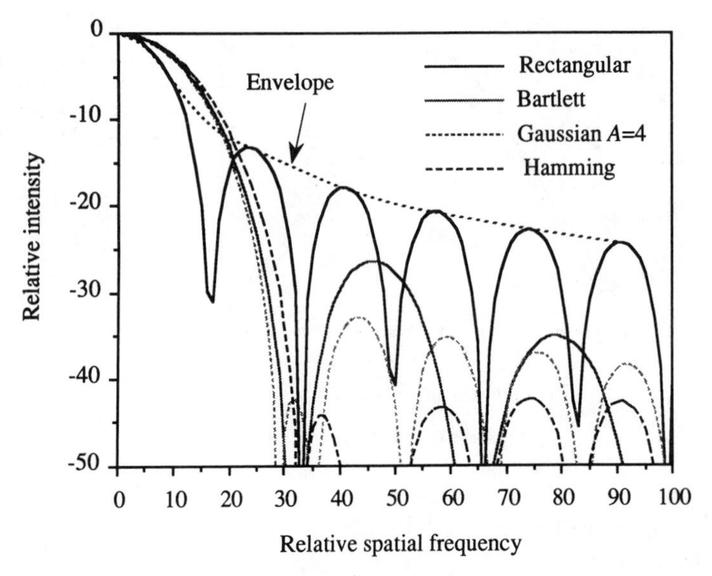

Figure 4.11. Frequency response (in decibels) of aperture functions.

magnitude over the entire aperture, the aperture functions have remarkably similar magnitude responses. Figure 4.11, however, shows that the sidelobe response levels in the Fourier plane are distinctly different for these four aperture functions; all responses have been normalized to unity at zero spatial frequency.

The first aperture function is the rect function for which

$$\begin{aligned} A(\xi) &= \int_{-\infty}^{\infty} \operatorname{rect}\left(\frac{x}{L}\right) e^{j(2\pi/\lambda F)\xi x}\, dx \\ &= L \operatorname{sinc}\left(\frac{\xi L}{\lambda F}\right). \end{aligned} \tag{4.9}$$

The sinc function provides good frequency resolution because its *mainlobe* is narrow, but it has high *sidelobes* that decrease slowly as a function of distance from the mainlobe. The first sidelobe of $|A(\xi)|^2$ is only 13 dB down from the mainlobe, and the sidelobe intensity falls off as $1/\xi^2$. High sidelobe levels may obscure weak signals that we want to detect and contribute energy to adjacent photodetector elements as optical crosstalk. Because the sidelobes originate from large, sharp discontinuities in the signal, we can reduce them by using other aperture functions.

The Bartlett aperture function is a triangular function that we can view as the autocorrelation of two rect functions whose widths are just half that of the full aperture. As the Fourier transform of the convolution of two functions is the product of the Fourier transforms of the individual functions, the magnitude response in the Fourier plane is a sinc^2 function whose nulls are spaced twice as far apart as those in the sinc function produced by $\text{rect}[x/L]$. The first sidelobe due to the Bartlett aperture function is therefore located at about the same spatial frequency as the second sidelobe due to the rect function. Although the magnitude of the sidelobes falls off at twice the rate of those due to the rect function, they are still too high for most applications. Note that the rect function has a discontinuity in its magnitude and in all its derivatives; the Bartlett aperture function does not have a discontinuity in its magnitude, but it does have discontinuities in its derivatives.

The Gaussian aperture function produces lower sidelobe levels than either the rect or the Bartlett aperture functions; its sidelobe level is controlled by selecting the value of the parameter A, which determines the magnitude of the illumination at the aperture edges. The rate at which the sidelobe magnitudes decrease depends on the value of A.

The Hamming aperture function has the interesting property that all its sidelobes are about 45 dB down relative to the mainlobe, even though it has discontinuities in both its magnitude and its derivatives.

A typical specification for a spectrum analyzer is that the sidelobe levels are no higher than -50 dB relative to the peak mainlobe level at a position equivalent to five resolvable frequencies away from the centroid of the mainlobe. The measurement is always made to the *envelope* of the sidelobes, shown in Figure 4.11, to avoid any pathological cases where the required measurement position might fall in a null between two sidelobes. The trick is to determine what is meant by "five resolvable frequencies away." To do so, we discuss frequency resolution in greater detail.

4.5.3. Frequency Resolution / Photodetector Spacing

The price to pay for using an aperture function that produces lower sidelobes is a wider mainlobe width, as measured at the -3-dB intensity response point, which leads to a lower frequency resolution. Figure 4.12 shows, on a linear scale, the mainlobes of $|A(\xi)|^2$ for the four aperture functions under consideration. We see that the rect function provides the most narrow mainlobe and that the Hamming function produces the widest mainlobe. The Gaussian aperture function has the

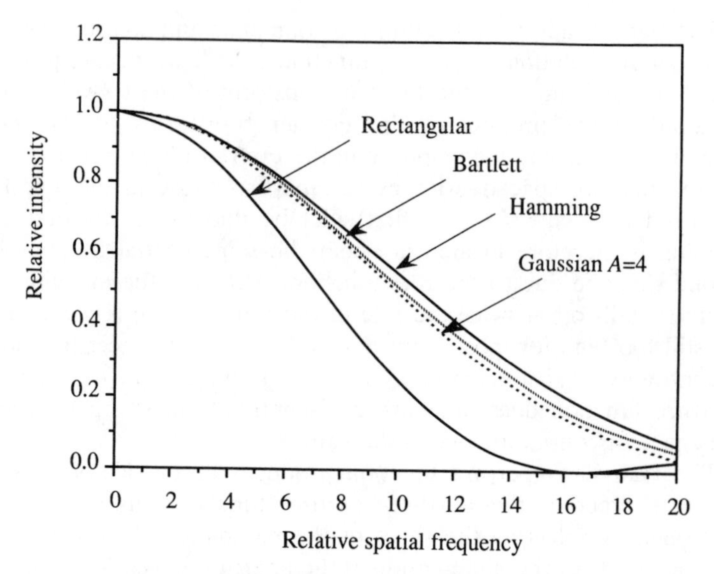

Figure 4.12. Mainlobes plotted on a linear scale.

second-narrowest mainlobe of this set of aperture functions, representing a reasonable compromise between sidelobe control and frequency resolution.

In Figure 4.13 we capture these two key features of the aperture response function by plotting the mainlobe width, relative to that provided by the rect function, on the vertical axis and the highest sidelobe level on the horizontal axis. The performance of the rectangular, Bartlett, and Hamming functions are represented by single points, but the performance of the Gaussian function is given for several values of the parameter A. We also include three other aperture functions for comparison: the Hanning, the Kaiser, and the Chebyshev. These functions are defined as

5. *Hanning*

$$a(x) \equiv \mathrm{rect}(x/L)[0.5 + 0.5\cos(2\pi x/L)] \tag{4.10}$$

and

6. *Kaiser*

$$a(x) \equiv \mathrm{rect}\left(\frac{x}{L}\right)\frac{I_0\left[\beta L\sqrt{1-(2|x|/L)^2}\right]}{I_0(\beta L)}, \tag{4.11}$$

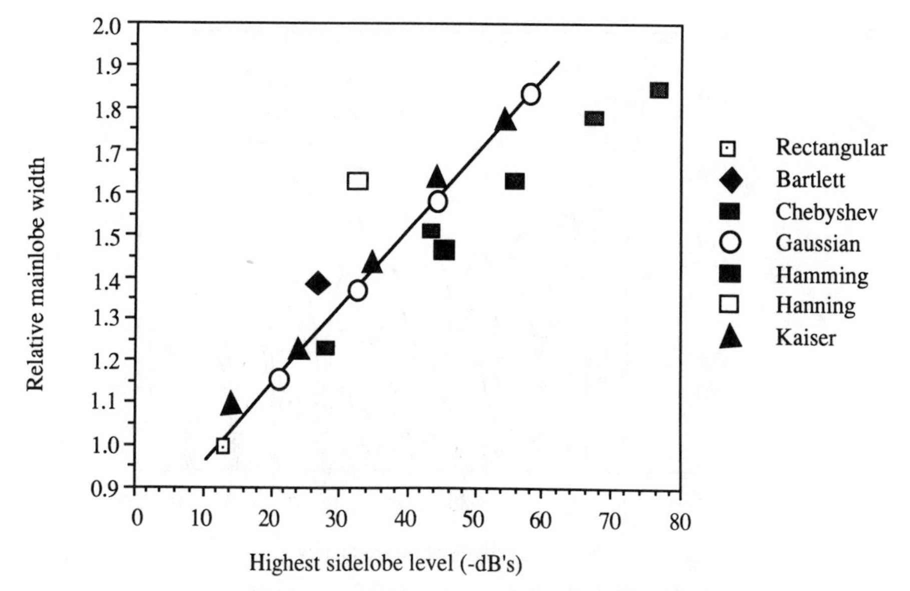

Figure 4.13. Mainlobe width vs. highest sidelobe level.

where $I_0(\cdot)$ is a modified Bessel function and β is a scaling parameter that determines the sidelobe levels.

The last aperture weighting function, the *Chebyshev* function, is described in terms of integrals of the product of polynomials and orthogonal functions. As both the Kaiser and Chebyshev aperture functions are difficult to generate optically, we do not seriously consider them for use in signal-processing systems. Their responses are included here for comparison.

The main conclusion drawn from the data in Figure 4.13 is that the Gaussian aperture function, while not providing the ultimate in performance in terms of sidelobe levels, is the aperture function of choice because it occurs naturally as the intensity profile of illumination from lasers. Attempts to improve on its performance, using additional masks or apertures, generally add noise to the system. We therefore use the Gaussian aperture function as our baseline for optical spectrum analyzers.

Having settled on the Gaussian aperture function as the most practical, we use the parameter A to control the truncation points needed to get the desired sidelobe suppression. Figure 4.14 shows the response in the Fourier plane for $A = 2$, 4, 6, and 8; the Gaussian function degenerates to the rectangular function when $A = 0$. By referring to Equation (4.6), we see that the illumination is truncated at the $1/e^{-A}$ *intensity* points at the

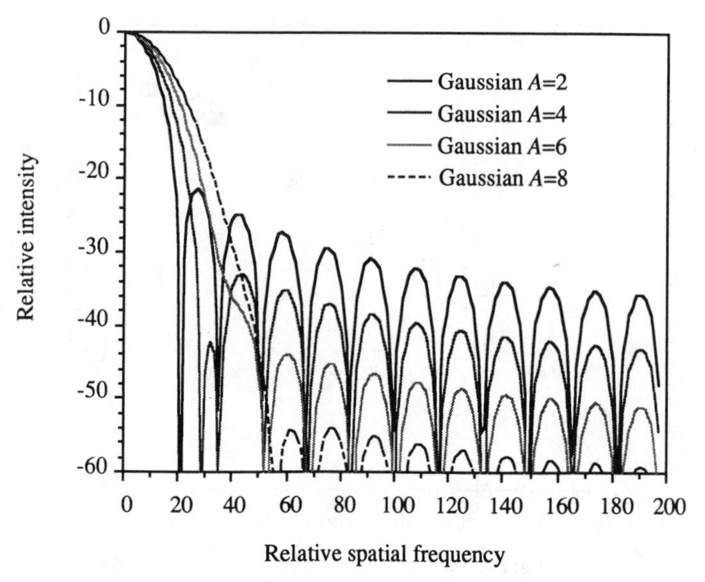

Figure 4.14. Frequency response of four Gaussian aperture functions.

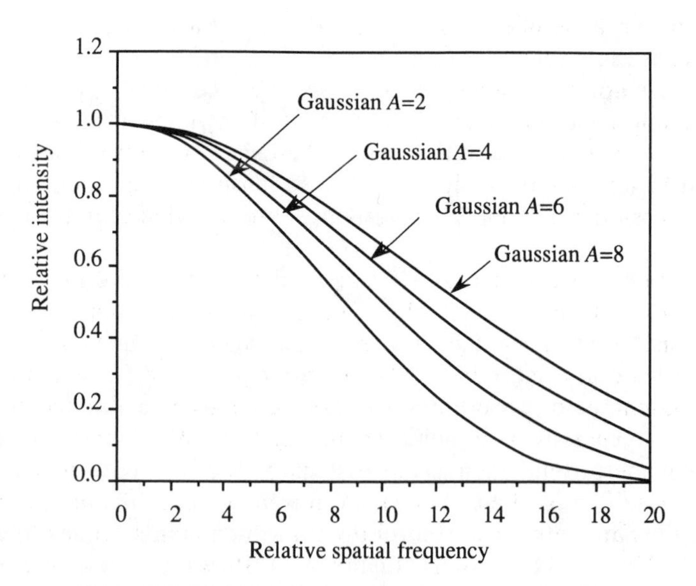

Figure 4.15. Mainlobes for four Gaussian aperture functions.

edges of the signal, where $|x| = L/2$. Figure 4.15 gives the relative mainlobe widths of these aperture functions; we use Figures 4.14 and 4.15 to balance the loss of resolution due to the increased mainlobe width, as a function of a given sidelobe reduction level.

The amount of available laser power used is also a function of the parameter A. The collected power is obtained by integrating the *intensity* of the light over the input plane:

$$\text{fraction of power collected} = \frac{\int_0^{L/2} e^{-A(2x/L)^2}\, dx}{\int_0^{\infty} e^{-A(2x/L)^2}\, dx} = \operatorname{erf}(\sqrt{A}\,), \quad (4.12)$$

where $\operatorname{erf}(\cdot)$ is the error function. For $A > 3$, at least 98.5% of the laser power is used. The Gaussian aperture function, in addition to providing reasonable frequency resolution and sidelobe control, is efficient in terms of the amount of optical power used.

With the sidelobe level under control, we turn our attention to a more complete discussion of frequency resolution. *Frequency resolution* is generally defined in terms of the dip in the intensity response in the Fourier plane produced by two cw signals; the measurement is taken midway between the peak response of the frequencies. A typical specification is that the dip must be 2–3 dB down from the peak response. Figure 4.16(a) shows the response in the Fourier-transform plane to a frequency at ξ_0 and to a frequency spaced $\delta\xi$ away; these frequencies produce the intensity responses $|A(\xi - \xi_0)|^2$ and $|A(\xi - \xi_0 - \delta\xi)|^2$. We sum the intensities from these two functions and find the dip between them relative to the peak response, as shown in Figure 4.16(a).

We cannot, however, simply read the dip value from Figure 4.16(a) because the width of the photodetector element affects the intensity measurements at the Fourier plane. If the photodetector elements were infinitesimally small and the spacing between them were nearly zero, the measured dip would be accurately measured. A photodetector element of finite size d', however, tends to smooth the detected spectrum somewhat, thereby reducing the dip between frequencies.

A few of the elements from the photodetector array are shown in Figure 4.16(a). One way to account for the finite size of the photodetector elements is to integrate the light intensity over the photodetector elements and to then measure the dip relative to the maximum value. Unfortunately, we would need to do this for all possible input frequencies because we do not know *a priori* where the frequencies occur relative to the

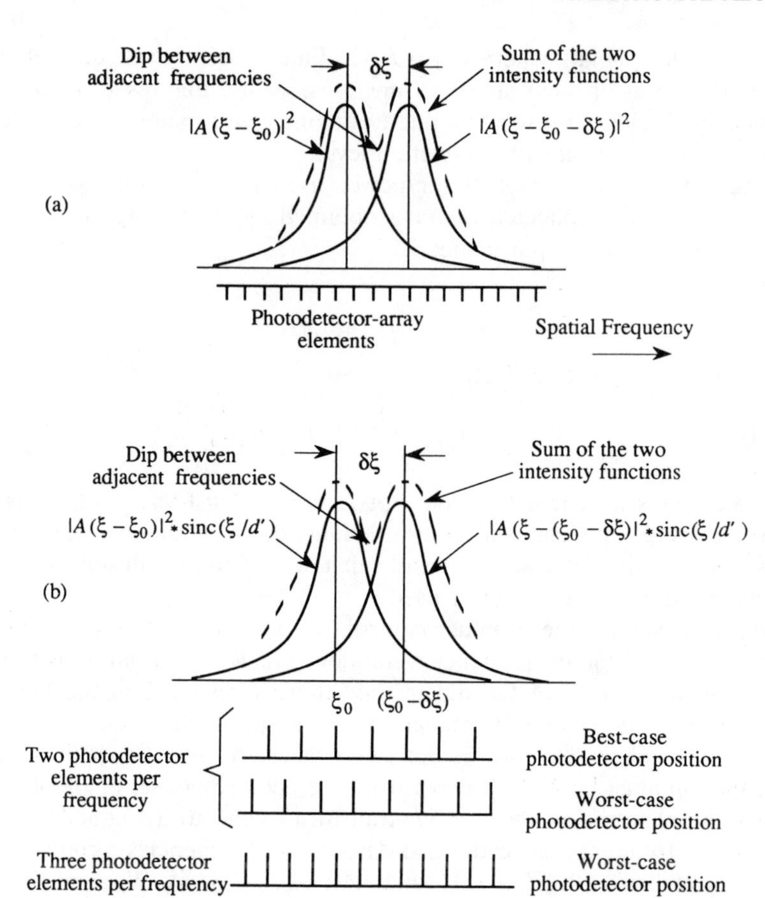

Figure 4.16. Frequency resolution and number of photodetectors: (a) sum of intensity of two frequencies after convolution with a photodetector element and (b) method for determining the dip response.

photodetector elements. An easier way to determine the effect of finite photodetector elements is to start by convolving $|A(\xi)|^2$ with a photodetector element of size d'. The question is "what value do I initially assign to d'?". We now describe an iterative procedure that produces a photodetector-element separation that satisfies all the constraints.

4.5.4. Array Spacing and Number of Photodetector Elements

The first issue in determining the array spacing is to find the required number of photodetector elements per frequency. One might argue, of course, for using as many photodetector elements as possible to provide for the best possible frequency resolution under all conditions. But an unnecessarily large number of photodetector elements leads to excessive output data rates from the array, as discussed in Section 4.4.5. The trick is to use the smallest number of elements in the photodetector array that meets the frequency-resolution specification for an aperture function that meets, in turn, the sidelobe level specification. An iterative calculation may be needed in which we increase the value of $\delta\xi$ until all the specifications are met simultaneously. When this process is finished, we must ensure that the length L of signal history is sufficient to provide the frequency resolution and that the sidelobe level still meets specification for that condition.

In Section 3.5.4, we argued that the number of samples in the spatial and Fourier domains are equal. It might appear, therefore, that one detector per resolvable frequency satisfies the sampling theorem. In a spectrum analyzer application, however, we generally need to resolve closely spaced frequencies. To handle the worst-case frequency-resolution specification, we need about three photodetector elements per frequency, as we show shortly.

A starting point is to use the specification for $\delta\alpha$ and the selected $|A(\xi)|^2$ to provide the required sidelobe control to make an estimate of the photodetector size d'. Here is where some judgment is exercised; experience helps, too. As the physical distance in the Fourier plane between resolvable frequencies is $\delta\xi = \delta\alpha\lambda F$, a reasonable starting point is to require three photodetectors per frequency so that $d' \approx (\delta\xi)/3$. We then convolve the photodetector element, represented by $\text{rect}(\xi/d')$, with $|A(\xi)|^2$ to get the responses for the two frequencies as shown in Figure 4.16(b). As we have now accounted for the finite photodetector size, we can replace the $\text{rect}(\xi/d')$ photodetectors by delta functions as shown in the lower part of Figure 4.16(b). The advantage of this method is that we can now simply slide the delta-function representation of the photodetector array along the spatial frequency axis and read the relative detected values from the dotted function representing the sum of the two intensities. This can be done for any position of the frequencies relative to the photodetector array to find the worst-case response.

The danger of using only two photodetector elements per frequency is illustrated in Figure 4.16(b) in which the delta functions sample the sum of the two intensity responses from the two frequencies. In the best-case condition, one of the photodetector elements is located exactly on the minimum value between frequencies and the dip specification is met. But if the two frequencies shift, relative to the photodetector array, as shown for the worst-case condition, the dip may completely disappear so that the frequencies are not resolved at all. Using three photodetector elements per frequency corrects this situation because at least one photodetector element is always near the maximum dip position.

To illustrate this iterative process, suppose that we use three photodetector elements per frequency and initially estimate that a spacing of 18 units between the frequencies will provide a dip of 3 dB. The spacing between photodetector elements is therefore 6 units. We assign the photodetectors a length of 5 units so that the duty cycle is $c = \frac{5}{6}$. The next step is to convolve the photodetector, represented by rect($\xi/5$), with the response $|A(\xi)|^2$; the result is shown in Figure 4.17(a). We then shift the convolved response 18 units and add it to itself to simulate two frequencies, spaced by 18 units, detected by a photodetector array in which the elements are 5 units wide on 6 unit spacings. The sum of the two responses is shown in Figure 4.17(b).

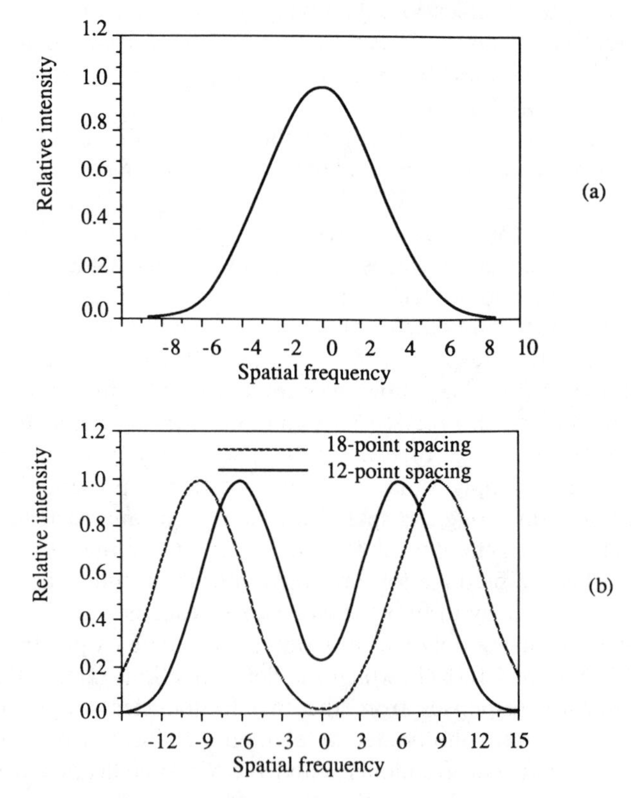

Figure 4.17. Effect of detector width on frequency resolution: (a) convolution of detector with Gaussian response ($A = 4$) and (b) sum of intensities from two frequencies.

We are interested primarily in the region between the two peaks. The worst-case condition for resolving two frequencies occurs when two elements of the photodetector array straddle the null between the frequencies. For this case, the photodetector elements are at the points -9, -3, 3, and 9. We see that the responses at -9 and 9 are equal to one and the responses at -3 and 3 are equal to 0.15. The dip is therefore about 8 dB, which is too high because of our overly conservative decision to space the two frequencies by 18 units.

The next iteration is to space the frequencies by 12 units; the photodetector elements are then chosen as 3 units wide, on 4 unit spacings for a duty cycle of $c = 3/4$. The result of summing the responses from two frequencies is also shown in Figure 4.17(b) for this case. Note that the photodetector elements are located at -6, -2, 2, and 6 for the worst-case situation. The responses at -6 and 6 are equal to one and the responses at -2 and 2 are equal to 0.4, which yields a dip of about 4 dB. This spacing is still slightly too large, but convergence to the desired dip of 3 dB will probably take only one more iteration.

All these calculations are performed without regard to the actual physical size of the photodetector element or the focal length of the Fourier-transform lens. The final step is to calculate the focal length of the lens so that the interval between two resolvable frequencies is equal to three photodetector intervals on the chosen array.

At first glance, it seems that the source size affects the mainlobe width and, therefore, the frequency resolution. Since $a(x)$ is the Fourier transform of the source, the required aperture function $a(x)$ is provided only if the source has the correct size at the onset. Spectral purity of the source may also spread the mainlobe, thereby reducing the frequency resolution. This spreading is rarely a problem with gas lasers where the fractional spectral spread is on the order of $\Delta\lambda/\lambda \approx 10^{-7}$. A high-pressure source has a $\Delta\lambda/\lambda \approx 10^{-3}$, and a typical injection laser diode has $\Delta\lambda/\lambda \approx 10^{-2}$. In all cases, the increased spreading of the mainlobe due to the source wavelength spread is small. Spectral spreading in the source also causes the sidelobes to smear somewhat, filling in the nulls, but this does not significantly change the calculations because we always use the envelope of the sidelobes to determine the required aperture function.

4.6. DYNAMIC RANGE

Dynamic range is the most important performance parameter of a spectrum analyzer. We may require a dynamic range of 60–80 dB in many applications. Dynamic range is defined as the ratio of the largest to the smallest signal supported by the system, *referenced to the input of the system*. The smallest signal is generally determined when the signal-to-noise ratio is unity, although some users prefer to establish a signal-to-noise ratio of 3–6 dB as the minimum signal level that is meaningful. The maximum signal level is set by intermodulation products, caused by nonlinearities in the response of the spatial light modulator.

4.6.1. Intermodulation Products

All spatial light modulators have nonlinearities that distort the amplitude of the input signal. The input/output curve is generally linear at small input signal levels; at higher signal levels all spatial light modulators eventually saturate and distort the signal amplitudes. This distortion results in intermodulation products that produce false signals in the Fourier plane. For example, suppose that the input/output relationship, shown in Figure 4.18(a), is a sinusoidal function, defined between 0 and

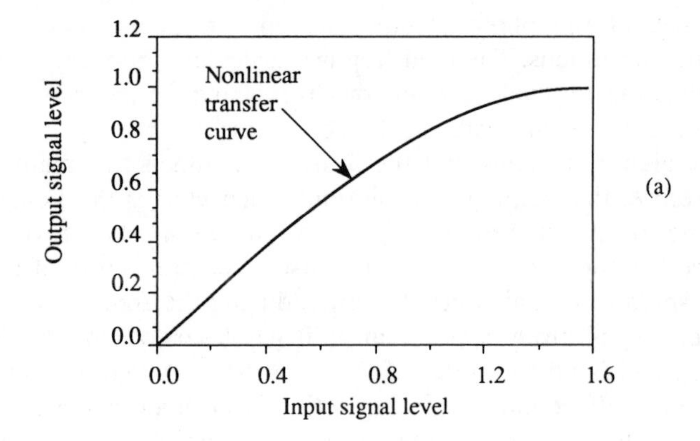

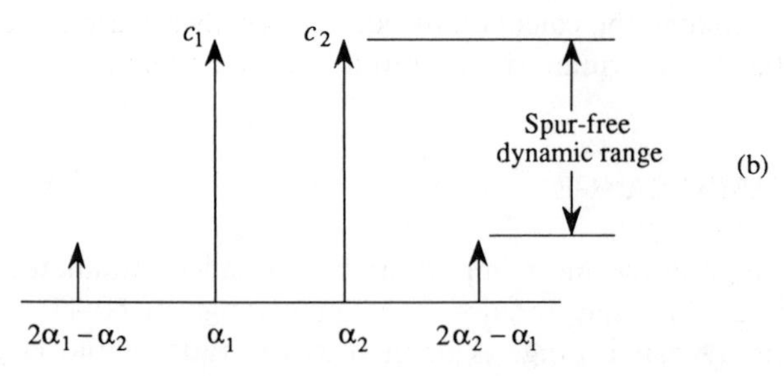

Figure 4.18. Effects of nonlinearities: (a) dynamic transfer curve and (b) spurious frequencies.

$\pi/2$, so that the two leading terms of the power-series expansion are

$$O(x, y) = I(x, y) + \tfrac{1}{6}I^3(x, y), \tag{4.13}$$

where $I(x, y)$ is the input signal and $O(x, y)$ is the output signal from the spatial light modulator. Suppose that the input is the sum of a bias and two cosine waves of the form

$$I(x, y) = \tfrac{1}{3}[1 + \cos(2\pi\alpha_1 x) + \cos(2\pi\alpha_2 x)]. \tag{4.14}$$

After substituting Equation (4.14) into Equation (4.13), we find frequency components at α_1 and α_2, due to the linear part of the input/output curve, and several new frequency components, including those at $2\alpha_1 - \alpha_2$ and at $2\alpha_2 - \alpha_1$, generated by the cubic nonlinearity term in the expansion of the input/output curve.

The new frequencies are *spurious signals*, generally shortened to spurs. A sketch of the frequency components in the Fourier plane is shown in Figure 4.18(b). We cannot, in general, distinguish the spurs from true signal frequencies. An important system specification, therefore, is one that requires a certain *spur-free dynamic range* so that weak signals can be distinguished from the spurs produced by strong signals. The spur-free dynamic range is defined as the ratio of the signal power at α_1 or α_2 to that at $2\alpha_1 - \alpha_2$ or $2\alpha_2 - \alpha_1$. It is met only by keeping the amplitudes of the sine waves below some maximum value by controlling the efficiency ε_m of the spatial light modulator; this efficiency level is stated as the *diffraction efficiency per frequency*.

Another factor in the detection of small signals is the amount of light scattered by various optical elements in the system. Multiple reflections before the spatial light modulator are particularly bad because they produce replicas of the spectrum, displaced from the primary spectrum according to the angles of reflection. Scratches and digs in optical elements, particularly those near the Fourier plane, may also cause scattered light to mask weak signals. Finally, regular structures such as the sample interval in a liquid-crystal display produce spurious noise patterns in the Fourier plane that may obscure signals. The solution to these problems is to ensure that the optical design does not contain flat surfaces which produce multiple reflections and to use antireflecting coatings on all surfaces. Specify minimum scratch and dig tolerances on all optical components and keep the components clean to reduce scattering.

4.6.2. Signal-to-Noise Ratio and the Minimum Signal Level

The minimum detectable signal level is determined by the signal-to-noise ratio available at the output of the system. In a well-designed spectrum

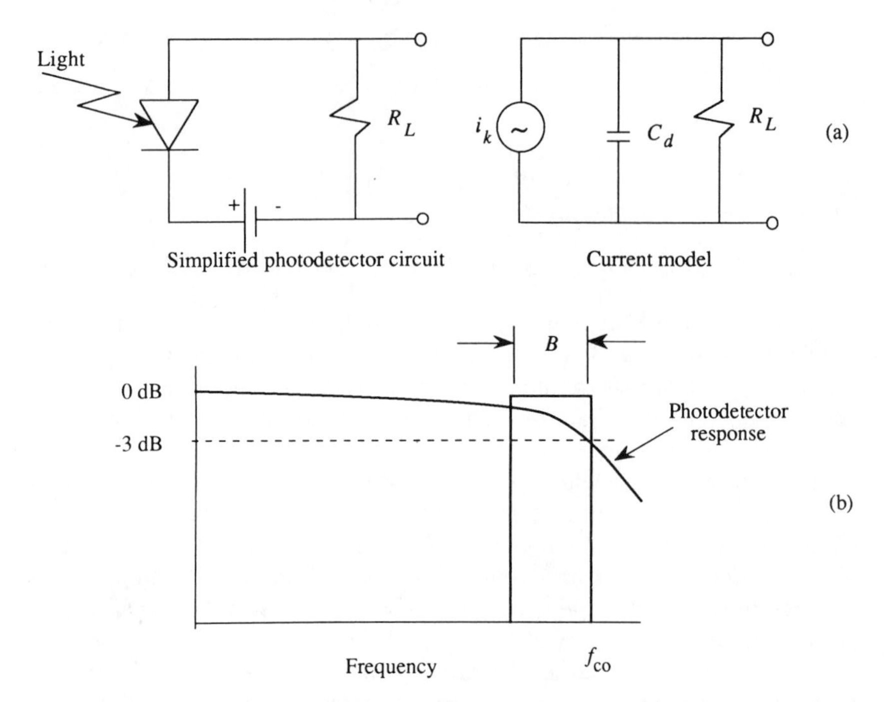

Figure 4.19. Photodetector model for signal-to-noise calculations: (a) photodetector circuit model and (b) frequency response of photodetector.

analyzer, the principal source of noise at the output is from the photodetector and its associated circuitry. A photodetector is modeled by the circuit shown in Figure 4.19(a), in which light falls on a semiconductor junction to generate electrons. The responsivity of the detector is a function of the nature of the material (e.g., silicon or gallium arsenide) and of the wavelength of light. The output current i_k from the kth frequency is the product of the responsivity of the photodetector and the incident optical power: $i_k = SP_k$, where S is the responsivity and P_k is the optical power collected by the photodetector element. In terms of the light intensity I_k, we find that $P_k = I_k\,dA$, where dA is the area of the photodetector. Thus the photocurrent is

$$i_k = SI_k\,dA. \tag{4.15}$$

The *signal-to-noise ratio* is calculated as

$$\text{SNR} = \frac{\text{signal power}}{\text{sum of the noise sources}}. \tag{4.16}$$

For the kth photodetector in the array, the signal-to-noise ratio is

$$\mathrm{SNR} = \frac{\langle i_k^2 \rangle R_L}{2eB(i_d + \bar{i}_k)R_L + 4kTB}, \tag{4.17}$$

where i_k is the signal current, $e = 1.6(10)^{-19}$ Coulomb is the charge on an electron, B is the postdetection bandwidth, i_d is the dark current of the photodetector, $\bar{i}_k$ is the average signal current, $k = 1.38(10)^{-23}$ J/K is Boltzmann's constant, and T is the temperature in degrees Kelvin. The first term in the denominator of Equation (4.17) is the *shot noise*, sometimes called quantum noise, and the second term is the *thermal noise*, sometimes called Johnson noise.

Equation (4.17) shows that we can increase the signal-to-noise ratio arbitrarily by increasing the load resistance until we are shot-noise limited. This procedure, however, may lead to an insufficient photodetector bandwidth. The relationship between the load resistance, capacitance, and cutoff frequency f_{co} is obtained from basic circuit theory as

$$f_{\mathrm{co}} = \frac{1}{2\pi c_d R_L}. \tag{4.18}$$

We solve Equation (4.18) for R_L and use it in Equation (4.17) to generate a more useful form of the signal-to-noise ratio equation:

$$\boxed{\mathrm{SNR} = \frac{\langle i_k^2 \rangle}{2eB(i_d + \bar{i}_k) + 8\pi kTBf_{\mathrm{co}}c_d}.} \tag{4.19}$$

In subsequent discussions we still refer to the terms in the denominator of Equation (4.19) as "shot" and "thermal" noise even though they represent the square of the noise *current* instead of the noise *power*, as they normally do.

The signals we detect may be either baseband signals, whose frequency content ranges from 0 to f_{co}, or a bandpass signal whose frequencies range from $f_{\mathrm{co}} - B$ to f_{co}. In either case, the maximum frequency response required of the photodetector circuit is f_{co}. For bandpass signals, we can improve the postdetection signal-to-noise ratio by using a bandpass filter of width $B < f_{\mathrm{co}}$, as shown in Figure 4.19(b).

To find the signal-to-noise ratio we relate the current to the input signal amplitudes. Consider a cw input signal $s(x) = 0.5[1 + c_k \cos(2\pi\alpha_k x)]$. The Fourier transform of the positive diffracted order is

$$\begin{aligned} S(\xi) &= \sqrt{\frac{P_0\varepsilon\varepsilon_m}{\lambda FL}} \int_{-\infty}^{\infty} \tfrac{1}{2}a(x)\left[\tfrac{1}{2}c_k e^{-j(2\pi/\lambda F)\xi_k x}\right] e^{j(2\pi/\lambda F)\xi x}\, dx \\ &= \tfrac{1}{4}\sqrt{\frac{P_0\varepsilon\varepsilon_m}{\lambda FL}}\, c_k \int_{-\infty}^{\infty} a(x) e^{j(2\pi/\lambda F)(\xi-\xi_k)x}\, dx \\ &= 0.25\sqrt{\frac{P_0\varepsilon\varepsilon_m}{\lambda FL}}\, c_k A(\xi - \xi_k), \end{aligned} \tag{4.20}$$

where $A(\xi)$ is the Fourier transform of the aperture function $a(x)$. For the spectrum analyzer shown in Figure 4.1, P_0 is the laser power, L is the length of the line illumination of the signal, ε_m is the efficiency of the spatial light modulator on a per-frequency basis, and ε is the efficiency of the remainder of the optical system, including the effects of the aperture function as discussed in Section 4.5.3. Recall from Chapter 3 that the Fourier transform has a scaling factor of $\sqrt{j/\lambda F}$ for the one-dimensional transform; this term appears in the scaling factor of Equation (4.20) but we ignore the $\sqrt{j}$ factor.

A photodetector element in the Fourier plane integrates the intensity $I(\xi) = |S(\xi)|^2$ of the light over the area of the photodetector element. The total optical power collected is therefore

$$P_k = \int_{-\infty}^{\infty} 0.0625 \frac{P_0\varepsilon\varepsilon_m}{\lambda FL} c_k^2 |A(\xi - \xi_k)|^2\, d\xi. \tag{4.21}$$

Although we use a Gaussian function to control the sidelobe levels, a rectangular aperture function provides closed-form solutions that illustrate details of the integration process. We therefore consider the case where $a(x) = \text{rect}(x/L)$, for which $|A(\xi - \xi_k)|^2 = L^2 \text{sinc}^2[(\xi - \xi_k)L/\lambda F]$, so that Equation (4.21) becomes

$$P_k = 0.0625 \frac{P_0\varepsilon\varepsilon_m}{\lambda FL} c_k^2 L^2 \int_{-\infty}^{\infty} \text{sinc}^2\left[\frac{(\xi - \xi_k)L}{\lambda F}\right] \text{rect}\left[\frac{\xi - \xi_k}{d'}\right] d\xi, \tag{4.22}$$

where $\text{rect}[(\xi - \xi_k)/d')$ represents the size of the photodetector element centered at ξ_k. Figure 4.20(a) shows the geometry in the Fourier plane for

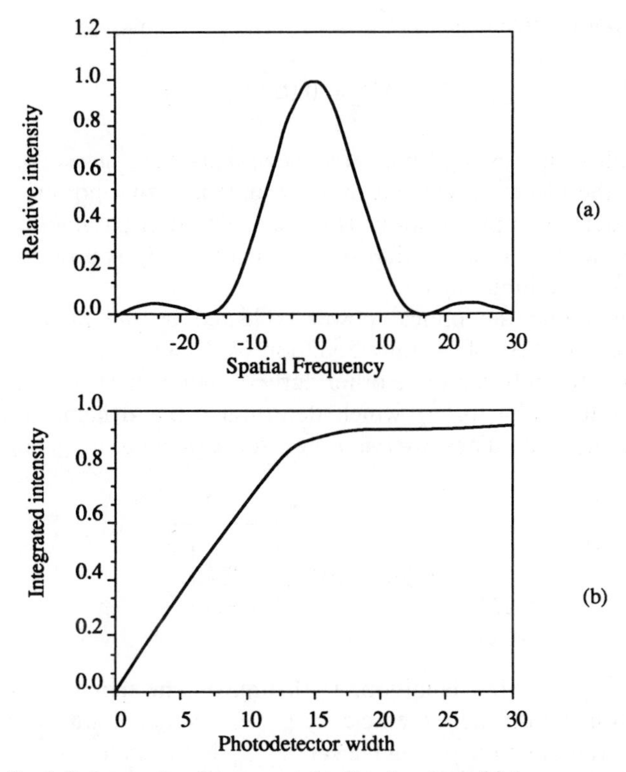

Figure 4.20. Spatially integrating detectors at the Fourier plane: (a) frequency response of a rect function and (b) intensity vs. detector width.

a sinc^2 function produced by the signal. To relate the arbitrary horizontal scale to ξ, we note that the first zero of the sinc^2 function occurs at $\xi = \lambda F/L = 15$ units. When we use three photodetector elements per frequency, the photodetectors are spaced $\Delta\xi = \lambda F/3L = 5$ units apart. The question is how much power the photodetector collects.

Figure 4.20(b) shows the integrated intensity for the sinc^2 function and we note that the integral is nearly a linear function of the photodetector width, for small widths. For a photodetector whose width is 5 units, we find that the value of the integral in Equation (4.22) is equal to $\lambda F/3L$. The total collected power is therefore

$$P_k = 0.0625 \frac{P_0 \varepsilon \varepsilon_m}{\lambda F L} c_k^2 L^2 \left[\frac{\lambda F}{3L} \right] = 0.02 P_0 \varepsilon \varepsilon_m c_k^2, \tag{4.23}$$

and the photocurrent is

$$i_k = SP_k = 0.02SP_0\varepsilon\varepsilon_m c_k^2. \tag{4.24}$$

This result is interesting from three viewpoints: (1) it gives the connection between the electrical current and the optical signal power, (2) it shows that the detected optical power is not a function of parameters such as λ, L, and F, and (3) it shows that the photocurrent i_k is proportional to the *power* c_k^2 of the input signal.

Having found the optical power collected by the photodetector, we substitute Equation (4.24) into Equation (4.19) and set the signal-to-noise ratio equal to unity (or some other agreed upon value). We then solve for the minimum value of c_k which determines the dynamic range of the spectrum analyzer. The *dynamic range*, for a single tone, given in decibels is

$$\boxed{DR = 10\log\left[\frac{c_{k\,\max}^2}{c_{k\,\min}^2}\right],} \tag{4.25}$$

where $c_{k\,\max}^2 = 1$, by definition, is the maximum signal power that, together with the diffraction efficiency per frequency, establishes the maximum intermodulation product level; $c_{k\,\min}^2$ is the minimum signal power obtained from the signal-to-noise ratio calculation.

From Equations (4.19) and (4.24), we find that

$$1 = \frac{\left(0.02SP_0\varepsilon\varepsilon_m c_{k\,\min}^2\right)^2}{2eB\left(i_d + \bar{i}_k\right) + 8\pi kTBf_{\text{co}}c_d}. \tag{4.26}$$

As all the parameters except c_k are specified, we can solve Equation (4.26) for $c_{k\,\min}^2$:

$$c_{k\,\min}^2 = \frac{\sqrt{2eB\left(i_d + \bar{i}_k\right) + 8\pi kTBf_{\text{co}}c_d}}{0.02SP_0\varepsilon\varepsilon_m}. \tag{4.27}$$

It seems, at first, that we need to solve a quadratic equation in $c_{k\,\min}^2$ because $\bar{i}_k$ is also dependent on $c_{k\,\min}^2$. We find, however, that as c_k^2 becomes small, the value of $\bar{i}_k$ becomes much less than i_d so that we replace $(i_d + \bar{i}_k)$ by i_d.

From Equations (4.25) and (4.27) we find that

$$\boxed{DR = 10\log\left[\frac{0.02SP_0\varepsilon\varepsilon_m}{\sqrt{2eBi_d + 8\pi kTBf_{co}c_d}}\right].} \tag{4.28}$$

From Equation (4.28) we see that the dynamic range increases according to the available laser power and decreases according to the square root of the signal bandwidth.

4.6.3. Integration Time / Bandwidth

Equation (4.19) shows that the signal-to-noise ratio and, therefore, the dynamic range is maximized when the bandwidth B is minimized. The bandwidth is the reciprocal of the integration time of light on the photodetector array. The integration time, in turn, is determined primarily by how often the photodetector array must be read to avoid missing important information.

In some applications, the integration time is the same as the frame cycle time of the input spatial light modulator because each frame may contain a completely new signal. In other applications, such as those where photographic film is moving through the aperture, the integration time is on the order of the amount of time that any given sample stays within the aperture. In yet other applications, the integration time is determined by the rate at which the underlying information content is expected to change. As noted above, the cutoff frequency f_{co} is always dictated by the highest frequency that is passed by the system; the bandwidth B is dependent on whether the input is a baseband or a bandpass signal.

4.6.4. Example

Suppose that a spectrum analyzer has these parameters: $P_0 = 10$ mW, $i_d = 10$ nA, $c_d = 4$ pF, $S = 0.4$ A/W, and $T = 300$ K. Suppose that the maximum diffraction efficiency per frequency to meet a -40-dB intermodulation product specification requires that $\varepsilon_m = 0.05$ and that the rest of the system efficiency is $\varepsilon = 0.25$. If the spectrum analyzer processes information at a frame rate of 1000 frames per second, we have the baseband case for which the cutoff frequency and the bandwidth are $f_{co} = B = 1000$ Hz.

The first step is to find the minimum value of c_k^2 when the signal-to-noise ratio is equal to unity. The solution for $c_{k\,\min}^2$ is therefore found from Equation (4.27):

$$c_{K\,\min}^2 = \frac{\sqrt{2(1.6)(10^{-19})1000(10^{-8}) + 8\pi(1.38)(10^{-23})300(1000)1000(3)(10^{-12})}}{0.02(0.4)(10^{-2})0.25(0.05)}. \tag{4.29}$$

We carry out the calculations to find that

$$c_{k\,\min}^2 = \frac{\sqrt{3.2(10^{-24}) + 3.12(10^{-25})}}{10^{-6}} = 2(10^{-6}), \tag{4.30}$$

and because $c_{k\,\max}^2 = 1$, the dynamic range is easily calculated from Equation (4.25) as 57 dB.

4.6.5. Quantum Noise Limit

In the example of Section 4.6.4, the dominant noise source is shot noise so that the system is quantum noise limited. For low-bandwidth applications where shot noise dominates thermal noise, we can simplify the signal-to-noise ratio expression to

$$\text{SNR} = \frac{\langle i_k^2 \rangle}{2eB\left(i_d + \bar{i}_k\right)}. \tag{4.31}$$

The minimum noise occurs when $\bar{i}_k \approx i_d$.

As the required bandwidth increases, we reach a point at which thermal noise dominates shot noise. By comparing the two noise terms in the denominator of Equation (4.19), we find that the system remains shot-noise limited provided that

$$\boxed{f_{\text{co}} \leq \frac{ei_d}{4\pi kTc_d},} \tag{4.32}$$

which is dependent only on the key parameters of the photodetector. Note, in particular, that Equation (4.32) does not contain the bandwidth explicitly so that it is valid for both baseband and passband applications.

Many photodetector manufacturers quote the performance of their products in terms of *noise equivalent power*, which gives an indication of the amount of optical power needed to just overcome the noise in the detector. The quoted noise equivalent power is usually normalized and expressed as $W/\sqrt{Hz}$ so that we simply multiply noise equivalent power by the square root of the predetection bandwidth (f_{co} in this case) to get the required optical power at the photodetector. But this procedure is valid only when the photodetector is shot-noise limited. The noise equivalent power has no provision to include the effects of thermal noise, so be careful when using noise equivalent power in performance calculations.

4.6.5.1. Avalanche Photodiode. When the system is thermal-noise limited, we can improve the system performance by using an avalanche photodiode to provide internal gain. An avalanche photodiode is a normal photodetector element that is operated near its breakdown voltage. When light is collected and current begins to flow, breakdown action is initiated so that additional electrons are generated. This results in amplification, or gain, within the device. The signal is increased by the gain factor G, while the modified shot-noise term becomes (37)

$$SN = 2eB(i_d + \bar{i}_k)G^m, \tag{4.33}$$

where G^m is a gain factor characteristic of avalanche photodiodes. Typical values for m are 2.3–2.5 for silicon devices and 2.7–3.0 for III-V alloy devices. The thermal noise is not affected by operating the photodiode in the avalanche mode. The signal-to-noise ratio then becomes

$$\text{SNR} = \frac{\langle i_k^2 \rangle G^2}{2eB(i_d + \bar{i}_k)G^m + 8\pi kTBf_{co}c_d}, \tag{4.34}$$

and we note that the signal power is increased by the factor G^2.

We want to select the photodetector gain that maximizes dynamic range, using the minimum laser power. As noted before, this maximization occurs when shot noise and thermal noise are approximately equal so that the optimum gain is

$$G^m \approx \frac{4\pi kTf_{co}c_d}{e(i_d + \bar{i}_k)}, \tag{4.35}$$

which reveals that avalanche diodes are most useful when the detection bandwidth is large. As the value of m must be greater than 2, any gain larger than that governed by Equation (4.35) results in a deterioration of the dynamic range, although the loss in performance is not a rapid function of the excess gain. The chief disadvantage of using avalanche photodiodes is that breakdown voltage and gain are sensitive functions of temperature so that they are more difficult and costly to regulate than conventional detectors.

4.6.5.2. Relationship between Signal-to-Noise Ratio and Dynamic Range. The relationship between signal-to-noise ratio and dynamic range is illustrated in Figure 4.21, in which we plot the shot noise, thermal noise, and output signal power as a function of input signal power. The maximum input signal power is $c_k^2 = 1$, which yields a signal level of 0 dB, as shown on the horizontal axis. The signal part of the output, as given by the numerator of Equation (4.19), is then at its maximum value. As the input level decreases, the signal part of the output also decreases with a slope of -2 because the output signal power is proportional to c_k^4.

As thermal noise is not a function of the input signal level, it is shown as a constant in Figure 4.21. Shot noise, on the other hand, is determined mainly by the value of the dark current when $\bar{i}_k \leq i_d$. As c_k increases, we reach the point where $\bar{i}_k > i_d$, after which the shot noise increases linearly with a slope of -1. The dynamic range is determined by the point where the straight line representing the output signal power intercepts the sum of the shot and thermal-noise powers. Note that the dynamic range is

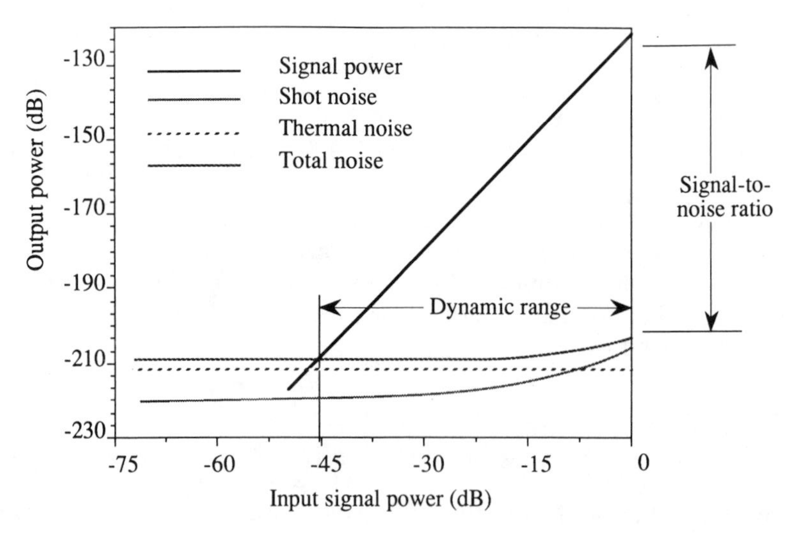

Figure 4.21. Relationship between signal-to-noise ratio and dynamic range.

referenced to the input of the system and is read form the *horizontal* scale in Figure 4.21. The signal-to-noise ratio, on the other hand, is referenced to the output of the system and is read from the *vertical* scale.

4.7. RASTER-FORMAT SPECTRUM ANALYZER

We close this chapter by discussing an important application of two-dimensional optical spectrum analysis, as applied to obtaining high frequency resolution for a wideband one-dimensional time signal. If we want to spectrum analyze a 20-MHz bandwidth signal to a frequency resolution of 20 Hz, we must resolve 10^6 frequencies. The high frequency resolution is obtained only by using a long time history of the signal (at least 50 ms for this example). Given the available resolution capabilities of one-dimensional spatial light modulators, however, we may not be able to process the required signal history.

To overcome this difficulty, we use the full two-dimensional nature of the optical system to process a one-dimensional time signal. The basic idea, according to Thomas (38), is to record a wideband signal of bandwidth W onto photographic film in a raster-scanned format. The raster format is similar to that used for television, in which a wideband time signal of long duration is written onto a cathode-ray tube. Raster-scanning methods are also used in the transmission of images in facsimile systems and in laser printers used as computer peripherals. In all applications, we segment long-duration signals into many shorter ones and record them as a serial set of data lines in the vertical dimension.

4.7.1. The Recording Format

In the raster-scanning format, the wideband time signal modulates a laser beam in magnitude according to the magnitude of the signal $f(t)$, while the laser beam is scanned horizontally across the film by an acousto-optical device as shown in Figure 4.22; these devices are discussed in Chapter 7. For a signal of bandwidth W, the cutoff frequency is $f_{co} = W$ and the sampling rate R_s, expressed in samples/sec, is

$$R_s = 2f_{co}. \tag{4.36}$$

Suppose that the required frequency resolution is f_0 so that we need to resolve a total of $M = f_{co}/f_0$ frequencies in a two-dimensional format. Based on the discussion in Chapter 1, we recognize M_x as the length bandwidth product and M_y as the height bandwidth product for the

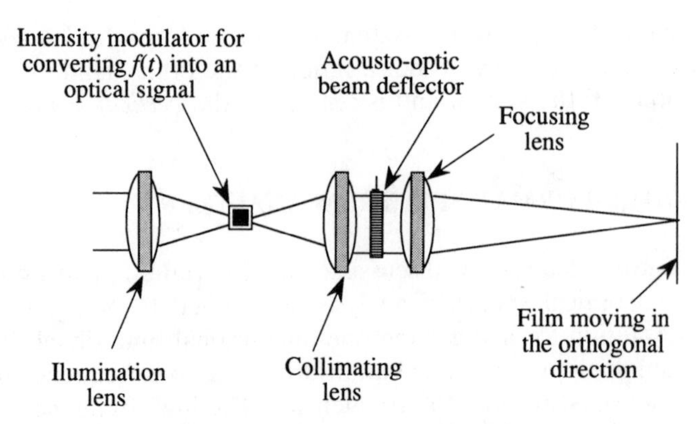

Figure 4.22. Optical wideband recorder.

system, so that $M = M_x M_y$ is the total number of resolvable frequencies. Suitable window functions, needed to control sidelobe levels as discussed in Section 4.5.2, can be introduced later as a refinement to this analysis.

Suppose that the film supports a sample interval of d_0, which means that the cutoff spatial frequency in the horizontal direction is $\alpha_{co} = 1/(2d_0)$ so that the sampling rate at the film is $R_f = 1/d_0$. To record the signal in real time, the laser-beam scanning velocity V_x must be

$$V_x = \frac{R_s}{R_f} = \frac{2f_{co}}{1/d_0} = 2d_0 f_{co}. \tag{4.37}$$

The time duration of a scan line is therefore

$$T_x = \frac{L}{V_x}. \tag{4.38}$$

The minimum resolvable spatial frequency α_0 in the horizontal direction corresponds to a temporal frequency

$$\boxed{f_x = \frac{1}{T_x} = \frac{f_{co}}{M_x},} \tag{4.39}$$

which is generally called the *coarse frequency resolution*.

The film is continuously moving through the scanner, resulting in the scanned line format shown in Figure 4.23. We record the next scan line a

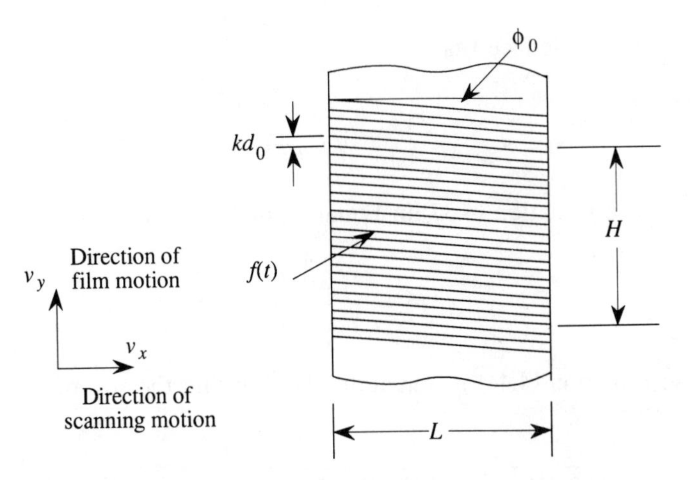

Figure 4.23. Recorded raster format: (a) fine frequencies and (b) full display format.

distance kd_0 from the previous line, where $k \geq 1$ is a parameter whose importance becomes apparent later. The film velocity, based on a zero flyback time for the scanning beam, is

$$V_y = \frac{kd_0}{T_x} = \frac{kd_0 V_x}{L}. \tag{4.40}$$

The height of the processing aperture is given by the product of the film velocity and the recording time T required to provide the frequency resolution f_0. Hence, we find that

$$H = V_y T \tag{4.41}$$

is the height of the processing aperture.

It is tempting to assume that the length and height of the processing aperture should be equal, but this assumption can lead to incompatibilities among other system parameters. We therefore set $L = cH$, where c is a parameter whose value is to be found after other important relationships are developed. From Equations (4.40) and (4.41) we find that

$$L = \frac{kd_0 V_x}{V_y} = cH = cV_y T, \tag{4.42}$$

from which we deduce that

$$V_y = \sqrt{\frac{kd_0V_x}{cT}}. \tag{4.43}$$

We now use Equation (4.37) in Equation (4.43) to yield

$$V_y = \sqrt{\frac{2d_0^2kf_{\text{co}}}{cT}}. \tag{4.44}$$

We use Equation (4.44) in Equation (4.41) to find the size of the processing aperture:

$$H = T\sqrt{\frac{2d_0^2kf_{\text{co}}}{cT}} = \sqrt{\frac{2d_0^2kf_{\text{co}}T}{c}} = \sqrt{\frac{2d_0^2kf_{\text{co}}}{cf_0}} = \sqrt{\frac{2d_0^2kM}{c}},$$
$$L = cH = \sqrt{2d_0^2kcM}. \tag{4.45}$$

We use the values of H and L in Equations (4.38) and (4.39) to find that the coarse frequency f_x is

$$f_x = \frac{1}{T_x} = \frac{V_x}{L} = \frac{f_{\text{co}}}{\sqrt{kcM/2}}. \tag{4.46}$$

We are now in a position to calculate the number of frequencies in the horizontal and vertical directions:

$$\boxed{\begin{aligned} M_x &= \frac{f_{\text{co}}}{f_x} = \frac{\sqrt{2d_0^2kcM}}{2d_0} = \sqrt{kcM/2}\,; \\ M_y &= \frac{f_x}{f_0} = \frac{2d_0M}{\sqrt{2d_0^2kcM}} = \sqrt{2M/kc}\,. \end{aligned}} \tag{4.47}$$

From Equation (4.47) we see that the number of frequencies in the two directions are not equal, in general, but that the product M_xM_y is always equal to M, independent of the values of the parameters k and c.

From the relationships developed so far, we deduce some interesting features about the system that are not intuitively obvious. For example, suppose that we set $k = 1$ to make the most efficient use of the capacity of the film. We might then expect that a square processing aperture, for which $L = H$, leads to an equal number of frequencies in the two directions so that $M_x = M_y$ which, in turn, leads to a square format in the frequency plane. This condition cannot be obtained for $k = 1$, however, as we note by setting $M_x = M_y$ and using the result from Equation (4.47):

$$M_x = \sqrt{kcM/2} = M_y = \sqrt{2M/kc}\,, \tag{4.48}$$

which implies that $kc = 2$. Therefore, to obtain a consistent solution, we find that $c = 2$ when $k = 1$ so that the length of the processing aperture is exactly twice that of the height: $L = 2H$. Furthermore, the minimum resolvable spatial frequency β_0 in the vertical direction, which corresponds to the temporal frequency f_0, and the minimum resolvable spatial frequency α_0 in the horizontal direction, which corresponds to the temporal frequency f_x, are given by

$$\begin{aligned} \alpha_0 &= \frac{\xi_0}{\lambda F} = \frac{1}{L}, \\ \beta_0 &= \frac{\eta_0}{\lambda F} = \frac{1}{H} = \frac{2}{L}. \end{aligned} \tag{4.49}$$

The spot spacings ξ_0 and η_0 in the frequency plane therefore have a two-to-one relationship, as noted in Equation (4.49). The format in the Fourier plane is therefore such that

$$\begin{aligned} \xi_{\max} &= M_x \xi_0 = \frac{M_x \lambda F}{L}, \\ \eta_{\max} &= M_y \eta_0 = \frac{2 M_y \lambda F}{L}, \end{aligned} \tag{4.50}$$

so that the required photodetector array format is rectangular, being twice as high as it is wide. Furthermore, Equation (4.49) reveals that the resolution requirements of the photodetector array are different in the two directions.

Most photodetector arrays are made with equal sampling in the two directions. If we use such a detector, we waste half of its intrinsic bandwidth. Suppose, then, that we set $c = 1$ so that the photodetector array is square. From Equation (4.47) we then find, again with $k = 1$, that $M_y = 2M_x$. In this case we find that $\xi_0 = \eta_0$, which is desirable; but $\xi_{\max}$ is still equal to $2\eta_{\max}$ because $M_y = 2M_x$. The requirements on the overall size of the photodetector array are therefore the same as before.

To avoid wasting photodetector-array capabilities, we remove the constraint that $k = 1$. We want to find a condition, if possible, for which $L = H$ and $M_y = M_x$ so that the photodetector-array format is square. We return to Equation (4.48) and find that $M_y = M_x$ under the general condition that $kc = 2$. As we want to set $c = 1$ so that $L = H$, we find that we must set $k = 2$. This means that successive scan lines are recorded twice as far apart in the vertical direction. Although this procedure wastes half of the film's recording capacity, it is typically a good tradeoff relative to the other options.

A by-product of setting $k = 2$ is that the aperture of the Fourier-transform lens is $\sqrt{2}$ larger than the suboptimum case when $k = 1$ and the film velocity is doubled. Some judgment is therefore needed to find a solution that can be implemented using current technology in the various areas. In the discussions to follow, we assume that the $k = 2$ and $c = 1$ solution is used; appropriate compensations can be made for other conditions as needed.

4.7.2. The Two-Dimensional Spectrum Analyzer

After the recording is finished and the film is developed, it is placed at plane P_1 of the spectrum analyzer as shown in Figure 4.24. One way to understand the spatial frequency display is to follow the locus of the spectrum while we imagine the input temporal frequency to start at zero and finish at f_{co}. First, suppose that the only frequency component present is zero frequency. The recording will therefore consist of $N_y = 2M_y$ scan lines, spaced a distance $2d_0$ apart, with no amplitude modulation. The Fourier transform of the scan lines consists of functions of the form

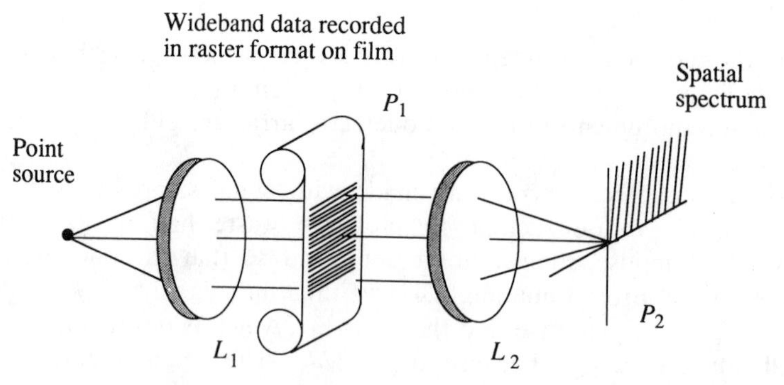

Figure 4.24. Optical spectrum analyzer for raster-scanning applications.

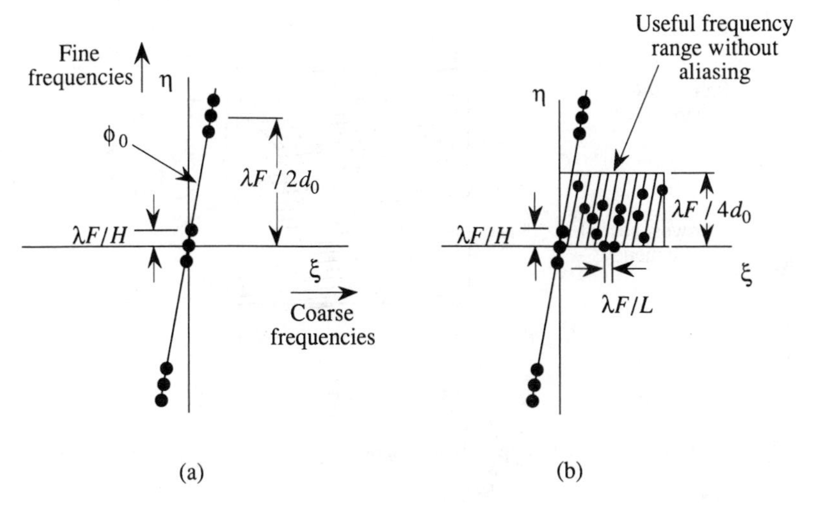

Figure 4.25. Fourier plane.

$\operatorname{sinc}(\xi L/\lambda F)\operatorname{sinc}(\eta H/\lambda F)$, located at $\eta = \pm n\lambda F/2d_0$, as shown in Figure 4.25(a), where F is the focal length of the Fourier-transform lens and n is a positive integer. The spectral components of the scanning pattern do not lie exactly on the $\xi = 0$ axis; instead they lie along a line that is tilted at an angle $\phi_0 = 2d_0/L$. This angle is of the order of $2/N_x$, which is generally small.

The lowest resolvable temporal frequency produces exactly one spatial cycle over the aperture of height H in the vertical direction, as depicted in Figure 4.26. A sinusoidal spatial frequency that has just one full cycle over the aperture produces $\operatorname{sinc}(\xi L/\lambda F)\operatorname{sinc}(\eta H/\lambda F)$ functions, located at $\eta_0 = \pm\lambda F/H$ relative to the sampling spectral points at $\eta = \pm n\lambda F/2d_0$, as shown in Figure 4.25(a). Replicas of the spectrum of the lowest spatial frequency are therefore formed around each of the spectral components, because of the scanning function. We refer to this resolution as the *fine frequency* resolution of the system. The next highest frequency has exactly two cycles over the aperture in the vertical direction, which causes the sinc functions to move a distance $\pm 2\lambda F/H$ with respect to each of the sampling spectral points. This process continues, as the temporal frequency increases, with the spectral components in the Fourier plane moving along the line making an angle ϕ_0 with respect to the vertical axis.

The highest frequency displayed on the first locus is one half the sampling frequency in the vertical direction; the spectrum of this frequency is located at $\eta = \pm\lambda F/4d_0$, as shown in Figure 4.25(b). At this

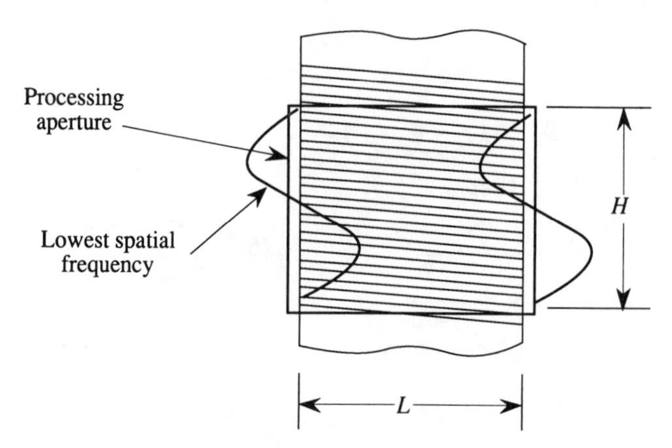

Figure 4.26. Lowest spatial frequency corresponding to lowest temporal frequency.

temporal frequency there is exactly one spatial cycle over the aperture length L in the horizontal direction, in addition to the M_y cycles in the vertical direction. Hence, the spectrum of this frequency has moved a distance $\xi = \lambda F/L$ in the horizontal direction shown in Figure 4.25(b). This coarse frequency resolution, from Equation (4.39), is equal to the cutoff frequency in the vertical direction. As the temporal frequency increases still further, the horizontal spatial frequency remains at one cycle over the aperture, while the vertical spatial frequencies vary from one to M_y cycles; this process forms the second raster scan line in the Fourier plane. At this point, there are two complete spatial cycles in the horizontal direction and this raster movement of the frequency component continues as the temporal frequency increases until we reach f_{co}. This spectral component is located at the upper right-hand corner of the Fourier plane display.

The spectrum at plane P_2 of the optical spectrum analyzer has symmetry about both the ξ and η axes. We generally mask all but one quadrant because the remainder of the information in the Fourier plane is redundant. To completely avoid aliasing, we would like the raster-scanning function on the film to have a sinc-function distribution in the vertical direction so that the spectrum can be cleanly masked. As the film cannot record negative values, we generally shape the raster lines to minimize the overlapping of the spectrum due to aliasing.

To illustrate the performance of a raster-scanning spectrum analyzer, we consider a wideband signal for which $f_{co} = 15$ MHz and want to obtain frequency resolution to 15 Hz. For this example we find that we need

$M = (15\text{ MHz})/15\text{ Hz} = (10^6)$ resolvable frequencies. Suppose that we set $M_x = M_y = \sqrt{M} = 1000$, and use a film for which $d_0 = 5\ \mu$. We begin by using Equation (4.37) to find that the scanning velocity is

$$V_x = 2d_0 f_{co} = 2(5\ \mu)15(10^6)\text{ Hz} = 150\text{ m/sec}. \tag{4.51}$$

The coarse frequency resolution in the horizontal direction is obtained from Equation (4.39):

$$f_x = \frac{15(10^6)}{1000}\text{ Hz} = 15\text{ kHz}, \tag{4.52}$$

and the scan time in the horizontal direction is $T_x = 1/f_x = 66.7$ msec. The film velocity is given by Equation (4.44):

$$V_y = \sqrt{\frac{2d_0^2 k f_{co}}{cT}} = \sqrt{\frac{2(0.005)^2 2(15)(10^6)}{1/15}} = 150\text{ mm/sec}, \tag{4.53}$$

which is reasonable. The aperture size is $L = H = 2d_0\sqrt{M} = 10$ mm.

Suppose that we want to find the location of a cw signal whose frequency is 9,873,705 Hz. We begin by dividing this frequency by 15 kHz to determine the coarse frequency position of the signal; we find that $9{,}873{,}705/15{,}000 = 658.247$. The integer part of this result shows that the frequency response for this signal falls on the 658th coarse frequency locus. We then multiply the fractional part of the result by 15 kHz to find the residual frequency and to find its position on the stated coarse loci: $0.247 \times 15{,}000/15 = 247$. Hence the frequency response occurs at the 247th fine frequency position on the 658th coarse frequency line.

It is interesting to note that the input recording process maps a long-duration, one-dimensional time signal into a raster format. In a similar fashion, the output raster format can be mapped into a continuous one-dimensional spectrum of the input time signal. This process is illustrated in Figure 4.27. A time signal of duration T is subdivided into segments of duration T_x, recorded in the raster format and spectrum analyzed to produce a raster-format spectrum. The spectrum of the signal is reconstructed by taking the spectral segments of bandwidth W_y and assembling them into the one-dimensional spectrum of bandwidth W.

This spectrum analyzer produces a result equivalent to implementing 10^6 narrowband filters, each with a bandwidth of 15 Hz, with their center frequencies ranging from 0 to 15 MHz. Designing and implementing that many filters using discrete electronic elements or by using digital signal

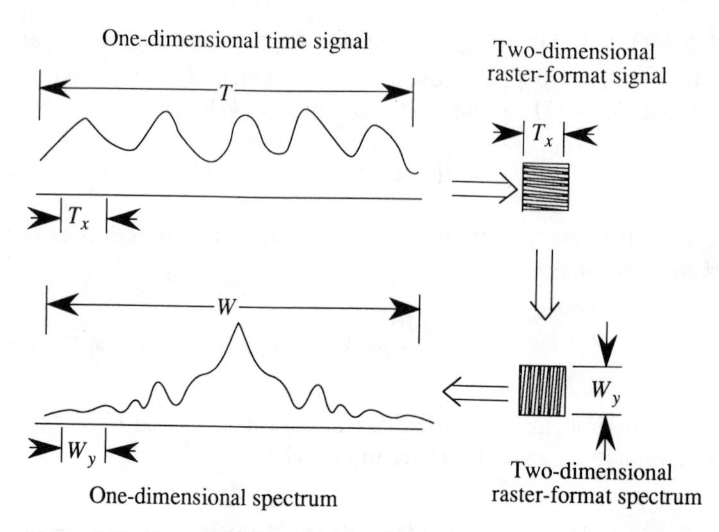

Figure 4.27. Conversion from one-dimensional/two-dimensional functions.

processing is difficult. Furthermore, depending on the beam apodization used, the equivalent optical filters have steep transitions from the passband to the stopband, small inband ripple, and small inband phase variations. None of the performance parameters given here are difficult to achieve optically, even with a relatively modest optical system. It is possible to increase the input signal bandwidth to as high as 1 GHz and to obtain a frequency resolution of about 100 Hz over the entire band.

4.7.3. Illustration of Raster-Format Spectra

In Figure 4.28 we show experimental results for the spectra of several signals displayed in different formats (39). One display format is a two-dimensional format obtained by directly viewing the spectra or by recording the spectra on photographic film. The other display format is a three-dimensional format obtained by plotting the magnitude of the spectrum as a function of the scanned coordinates ξ and η in the Fourier plane. The signal characteristics are given in Table 4.1. The first signal is a sine wave whose time bandwidth product, by definition, is $TW = 1$. The parameter f_c/B gives the ratio of the carrier frequency to the bandwidth of the signal. The number of repeats shows how many times a given signal waveform is repeated within the time interval covered by the Fourier transform; for the sine wave, the value shows that 100 cycles are analyzed.

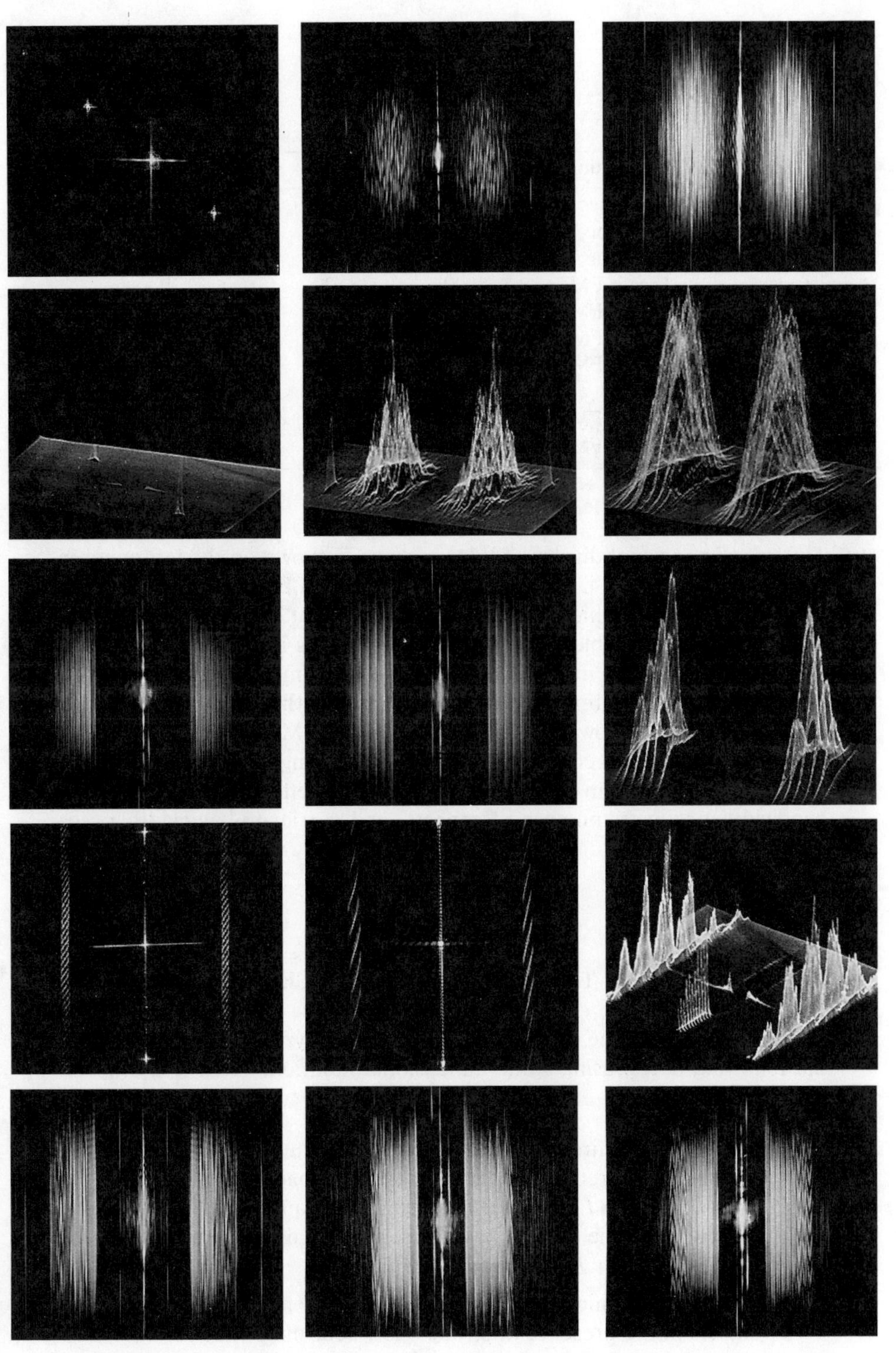

Figure 4.28. Experimental results of raster-scanned data (courtesy G. Lebreton) (39). Top row: two-dimensional spectra of s_1, s_2 and s_3; second row: analog three-dimensional display of spectra s_1, s_2 and s_3; third row: two-dimensional spectra of s_4 and s_5; three-dimensional display for s_5; fourth row: two-dimensional spectra of s_6 and s_7; three-dimensional display for s_7; fifth row: two-dimensional spectra of s_8, s_9 and s_{10}.

Table 4.1

Signal	Modulation	TW	f_c/B	Number of repeats
s_1	Sine wave	1	—	100
s_2	Random	1000	1	9
s_3	Random	1000	1	3
s_4	Linear FM	1000	3/2	9
s_5	Linear FM	1000	3/2	6
s_6	Linear FM	288	17	68
s_7	Linear FM	288	17	34
s_8	V-FM	2×500	1	2×2
s_9	Linear FM	1000	1	6
s_{10}	Linear FM	1000	1	9

The two-dimensional display of this spectrum is shown on the left panel in the top row of Figure 4.28. The other two panels in the top row are for the random signals for which $TW = 1000$; we see that the fine detail in the spectrum is more visible as the number of repeats increases. The second row shows the same information in the three-dimensional format. The remaining signals are linear FM or chirp signals with various combinations of parameters as shown in Table 4.1. The V-FM signal consists of a downchirp FM followed by an upchirp FM, giving the characteristic V shape to the waveform in frequency space. The effects of different time bandwidth products, number of repeats, and carrier-to-bandwidth ratios are apparent.

4.8. SUMMARY OF THE MAIN DESIGN CONCEPTS

The major steps in the design of a spectrum analyzer are summarized in the following notes, using a one-dimensional notation:

1. From the required frequency resolution $\delta\alpha$ and total bandwidth α_{co} to be covered, calculate the required space bandwidth product $M = \alpha_{\text{co}}/\delta\alpha = L\alpha_{\text{co}}$. This number gives a quick assessment of the difficulty of the design task and a preliminary estimate of the required length L of the input signal.
2. From the specification on the sidelobe level needed to detect weak signals in the presence of strong ones, determine the required value

of A for the Gaussian illumination and calculate the Fourier transform $A(\alpha)$ for the chosen truncated Gaussian function.

3. For the chosen Gaussian aperture weighting, calculate the relative loss in resolution and increase the required time bandwidth product by this factor. Since the bandwidth is fixed, we must increase the length of the signal that is processed.
4. From the required dynamic range considerations, determine what type of photodetector subsystem is needed (discrete detectors, a photodetector array, etc).
5. Based on the criterion of using three photodetector elements per resolved frequency, calculate the convolution of the aperture response $A(\alpha)$ and a single detector width.
6. Using the convolved aperture response, determine the value of $\delta\alpha$ needed to satisfy the dip criterion for frequency resolution. Several iterations may be required to achieve the desired result for the worst-case conditions.
7. From the value of $\delta\alpha$ and the photodetector array spacing, calculate the focal length of the Fourier transform lens so that the scale of the displayed spectrum matches that of the detector array.
8. From the required dynamic range, determine the required laser power, select the appropriate laser, and find its Gaussian-beam illumination parameter.
9. Design an illumination subsystem to magnify the laser beam to the plane of the signal so that the required truncation takes place at the edges of the revised signal length.
10. From the spur-free dynamic range specification, determine the maximum value of the input signal level.

PROBLEMS

4.1. The lowest spatial frequency component expected of a signal is 10 Ab and the highest is 200 Ab. Provide an optical layout for a spectrum analyzing system to display the highest and lowest spectral components with a separation of 80 mm in the frequency plane, under the constraint that you have no lenses available whose focal lengths are longer than 100 mm. As an added challenge, see if you can solve this problem using a single 100-mm lens. Sketch and dimension your resulting system.

4.2. Suppose that you have a collection of aerial images. Given a coherently illuminated optical system and a ring/wedge detector, how would you propose to use the outputs of these photodetector elements to classify

(a) wheatfields,

(b) cornfields (or any crop planted in rows),

(c) urban areas, and

(d) an oil tank field (lots of circular objects)?

Provide a rough algorithm for using the outputs to separate these four classes of objects.

4.3. In your laboratory you demonstrate the Fourier transform, using a system similar to the variable-scale correlator shown in Figure 3.29 of the text. You use a 4-μ source that is placed 55 mm in front of a 50-mm focal length lens. You then place a second lens whose focal length is 80 mm a distance of 500 mm behind the first lens. A signal $f(x) = 1 + \cos(2\pi\alpha_1 x)$ is located in contact with the first lens. Part (a): where is the Fourier transform plane located relative to the first and second lenses? Part (b): where is the *image* of the signal located relative to the second lens? Part (c): if $\alpha_1 = 60$ Ab, how far from the optical axis are the centroids of the diffracted light located in the Fourier plane? Solve this problem purely by using ray-tracing methods from geometrical optics, along with the basic connections between physical angles and spatial frequencies.

4.4. Suppose that the aperture of the first lens in Problem 4.3 is 25 mm. We want to create a variable-scale Fourier transform by moving the signal axially, closer to or farther from the first lens. If we need a 20% change in the scale of the transform, over what axial distance must the signal be moved? Can this range of scales be achieved if the object has a diameter of 20 mm? If not, what is the maximum object size that can be illuminated at the extreme end of the scaling range?

4.5. Suppose that the illumination beam in a spectrum analyzer has a Gaussian form $\exp[-(A/2)(2x/L)^2]$. Also, suppose that we want to truncate the beam so that the highest sidelobe is about 30 dB down relative to the peak of the main lobe. Find, from the curves in the text, the value of A required to achieve this performance. What is the penalty in loss of resolution as compared to uniform illumination (use the linear curves for accuracy). Calculate the fraction of laser power available for use.

4.6. From the data in the figures, calculate the relative loss in mainlobe resolution when using a Gaussian illumination beam for which $A = 6$ with that for which $A = 2$. Use the response at the -3-dB level as the criterion for resolution. If the half-width of the mainlobe is r_0 at the -3-dB point, what are the corresponding sidelobe levels at $7r_0$ for each of the same two Gaussian beams?

4.7. Suppose that you have a light source containing a spectral line at 500 nm and a line of equal strength at 550 nm. Sketch the Fourier transform of an aperture function $a(x) = \text{rect}(x/L)$ when using this source. Will you observe any complete nulls? If so, where? If not, why not? Do not forget about the $\cos(2\pi f_1 t)$ and $\cos(2\pi f_2 t)$ terms that denote the frequency of light for these two spectral lines.

4.8. Compare the mainlobe half-power widths and the sidelobe levels when the aperture weighting function $a(x)$ is

$$a(x) = \text{rect}(x/L)$$

with that when

$$a(x) = \text{tri}(x/L)$$

where

$$\begin{aligned} \text{tri}(x/L) &= 1 - |2x/L|, \qquad & |x| \le L/2 \\ &= 0, & \text{otherwise.} \end{aligned}$$

Quantify how rapidly the intensity envelope of the sidelobes falls off in each case. Compute the frequency resolution based on the half-power point for the two cases. Hint: the second weighting function can be obtained from the first through the use of the convolution theorem. But be careful with the scaling! A sketch for both $a(x)$ and $|A(\alpha)|^2$ in each case should keep you out of trouble. Label the half-power points of $|A(\alpha)|^2$ and the positions of the first few nulls of $|A(\alpha)|^2$ in each case.

4.9. We design a power spectrum analyzer using direct detection of the light in the Fourier domain. Assume a signal of the form $f(x) = 0.5[1 + c_k \cos(2\pi\alpha_k x)]$. Assume that the optical system produces a current $i_k = 0.02\varepsilon\varepsilon_m SP_0 c_k^2$. The photodetector parameters are

$$\begin{aligned} i_d &= 10 \text{ nA}, \\ c_d &= 3 \text{ pF}, \\ \varepsilon &= 0.5, \\ \varepsilon_m &= 0.2, \\ S &= 0.4 \text{ A/W} \\ T &= 300 \text{ K, and} \\ f_c &= B = 12 \text{ kHz}. \end{aligned}$$

Calculate the amount of laser power that will make the system shot-noise limited when the signal is such that $c_k = 0.002$. Calculate the minimum signal level. Calculate the dynamic range on the assumption that $c_k = 1$ is the maximum value of the input signal.

4.10. (Double credit) We design a power spectrum analyzer using direct detection of the light in the Fourier domain. To make a first cut, we assume a signal of the form $f(x) = 0.5[1 + c_k \cos(2\pi\alpha_k x)]$. We want the dynamic range to be 40 dB (i.e., we have a SNR = 1 when $c_k^2 = 10^{-4}$). Assume that the optical system produces a current $i_k = 0.02\varepsilon\varepsilon_m SP_0 c_k^2$. The photodetector parameters are

$$\begin{aligned} i_d &= 12 \text{ nA}, \\ c_d &= 5 \text{ pF}, \\ \varepsilon &= 0.5, \\ \varepsilon_m &= 0.2, \\ S &= 0.4 \text{ A/W}, \\ T &= 300 \text{ K, and} \\ B &= 500 \text{ kHz}. \end{aligned}$$

If you follow the questions in the order in which they are posed, you should have no trouble getting some reasonable answers. Calculate:

(a) The cutoff frequency (f_{co}) that makes the shot-noise and thermal-noise terms equal when the signal current $i_k \ll i_d$.

(b) Calculate the required load resistor (assume $f_{co} = 500$ kHz).

(c) Is the system as it stands shot-noise or thermal-noise limited? (Assume that dark current dominates the signal current when the signal is small.) Calculate both $2eBi_d$ and $8\pi kTBf_{co}c_d$ to support your answer and for use later on.

(d) Calculate the laser power P_0 required to achieve the necessary dynamic range.

(e) For this amount of laser power, calculate the minimum signal current and compare it to the dark current as a sanity check. Does this change your mind about whether dark current really does dominate when the signal is at its minimum value?

(f) Calculate the amount of laser power required to make the system shot-noise limited at the minimum signal level. This will be a fairly large value relative to the value calculated in (e).

(g) As an alternative to using lots of laser power, calculate the gain required of an avalanche photodetector (APD) to make the system shot-noise limited. Assume that $m = 2.3$ when you use G_m in the calculation.

(h) Recalculate the laser power required to achieve the necessary dynamic range when using the APD. Compare this value to that obtained in part (d), which uses a PIN detector (an APD whose gain is 1), and to that obtained in part (f), which achieves shot-noise performance by brute force. Of the three, which do you think is the best engineering solution?

4.11. We want to spectrum analyze a 200-MHz bandwidth signal using a falling raster recording format. We have available a spectrum analyzer for which $L = 2H$. The film has a cutoff frequency of $\alpha_{co} = 200$ Ab, and the width of the film is 20 mm. Calculate:

(a) the horizontal velocity V_x,

(b) the number of samples in the horizontal direction N_x,

(c) the minimum resolvable frequency in the horizontal direction (the coarse frequency resolution) f_x,

(d) the film velocity V_y, and

(e) the minimum frequency resolution of the system (the fine frequency resolution) f_0.

4.12. For the spectrum analyzer of Problem 4.11, find the position of a cw signal at 126,290 Hz.

5

Spatial Filtering

5.1. INTRODUCTION

The Fourier transform is key to using the full processing power of optical systems. Although the Fourier-transform relationship was discovered in the early 1800's, strong connections between it and signal processing were not made until the early 1950's. The application of these concepts in optical systems is often direct and elegant because a coherently illuminated optical system produces the Fourier transform of a signal as a physical light distribution, as we showed in Chapter 3. Spectrum analysis is thereby easily achieved by measuring the light intensity as a function of spatial frequency, as discussed in Chapter 4.

In this chapter, we consider signal-processing systems in which a spatial filter directly modifies the Fourier transform of a signal to produce a desired operation, such as pattern recognition. We begin by reviewing some fundamental results from signal processing and communication theory which we use extensively throughout the remainder of this book. Special attention is given to matched filtering as an important tool in pattern-recognition and signal-detection applications. We discuss methods for constructing various types of spatial filters and their applications. The spatial carrier method for constructing filters whose magnitude and phase response are arbitrary is the central theme of the chapter. We often draw analogies between techniques used in optical signal processing and equivalent methods used in communication systems or in electronic signal processing.

5.2. SOME FUNDAMENTALS OF SIGNAL PROCESSING

The fundamental principles of optical signal processing are similar to those developed for digital or electronic signal processing. For example, concepts developed in communication theory and signal processing, such as spectrum analysis and linear filtering, are used almost without modification for analyzing optical signal-processing systems. There are significant differences, however, with respect to the system architecture, the dimensionality of the signals, and the available degrees of freedom. In this section we cover some basic results used in signal processing. The intent is to review the major results with no attempt to derive them; the reader is encouraged to consult appropriate texts to fill in the details (15, 17, 18, 40–42). Our major objective is to state the key results, establish the notation that we use throughout, and delineate the kinds of signals and systems that we study.

Signals are classified as having either finite total power or finite average power. A *deterministic signal* has finite total power as characterized by

$$0 < \int_{-\infty}^{\infty} |f(x)|^2 \, dx < \infty, \tag{5.1}$$

where $f(x)$ is either a real- or a complex-valued signal. An example of a deterministic function with finite total power is $f(x) = \mathrm{rect}(x/L)$. The second class of signals is characterized by

$$0 < \lim_{L \to \infty} \frac{1}{2L} \int_{-L}^{L} |f(x)|^2 \, dx < \infty; \tag{5.2}$$

these signals have *finite average power*. An example of the second class of signals is a random signal that is unbounded in space. We have a meaningful definition of the Fourier transform for signals belonging to the first class, but the concept of spectral density, as defined in Section 5.2.3, is required to characterize the frequency content of signals belonging to the second class.

5.2.1. Linear, Space-Invariant Systems

Suppose that the input to a linear, space-invariant filtering system is a deterministic signal $f(x)$. The output of the system is $g(x) = T[f(x)]$ so that the operator T maps $f(x)$ into $g(x)$. Suppose that $g_1(x) = T[f_1(x)]$ and $g_2(x) = T[f_2(x)]$. The operator is *linear* if, and only if, for any constants a and b,

$$T[af_1(x) + bf_2(x)] = ag_1(x) + bg_2(x). \tag{5.3}$$

The operator is *space invariant* if, and only if, for any value of the displacement variable u,

$$T[f(x-u)] = g(x-u). \tag{5.4}$$

The output of a linear, space-invariant system is characterized by the *convolution integral*

$$g(x) = \int_{-\infty}^{\infty} f(u)h(x-u)\,du. \tag{5.5}$$

If $f(u)$ is a delta function, we use the sifting properties of the δ function in Equation (5.5) to show that $g(x) = h(x)$; $h(x)$ is called the *impulse response* of the system.

If $f(x)$ has finite power, we use the Fourier transform of the output signal $g(x)$ to develop the *convolution theorem*:

$$\begin{aligned} G(\alpha) &= \int_{-\infty}^{\infty} g(x)e^{j2\pi\alpha x}\,dx = \int_{-\infty}^{\infty}\left\{\int_{-\infty}^{\infty} f(u)h(x-u)\,du\right\}e^{j2\pi\alpha x}\,dx \\ &= \int_{-\infty}^{\infty} f(u)e^{j2\pi\alpha u}\,du \int_{-\infty}^{\infty} h(x-u)e^{j2\pi\alpha(x-u)}\,dx \\ &= F(\alpha)H(\alpha). \end{aligned} \tag{5.6}$$

The convolution theorem states that the Fourier transform of the convolution of two signals is equal to the product of the Fourier transforms of the two signals. The function $F(\alpha)$ is the Fourier transform of $f(x)$ and $H(\alpha)$, the Fourier transform of the impulse response $h(x)$, is the *filter function* for the system.

The *double convolution theorem* is a natural extension of the convolution theorem. Suppose that we have two systems in series to produce the output

$$g(x) = \iint_{-\infty}^{\infty} f(u)h_1(v-u)h_2(x-v)\,du\,dv. \tag{5.7}$$

By a calculation similar to that used to obtain Equation (5.6), we can show that

$$G(\alpha) = F(\alpha)H_1(\alpha)H_2(\alpha). \tag{5.8}$$

By an extension of this result, we find that an N-fold convolution of space signals can also be expressed as an N-fold multiplication of spatial frequency functions. Convolution is both associative and commutative.

5.2.2. Parseval's Theorem

A corollary of the convolution theorem is Parseval's theorem, derived for deterministic signals as follows. Let $h(x) = f^*(-x)$ where $*$ indicates complex conjugate. From Equation (5.5) we find that

$$g(0) = \int_{-\infty}^{\infty} |f(u)|^2 \, du. \tag{5.9}$$

We may also express $g(x)$ in terms of $G(\alpha)$ by using the inversion theorem:

$$g(x) = \int_{-\infty}^{\infty} G(\alpha) e^{-j2\pi\alpha x} \, d\alpha. \tag{5.10}$$

From Equation (5.6) we find that, under these conditions, $G(\alpha) = |F(\alpha)|^2$ so that

$$g(0) = \int_{-\infty}^{\infty} |F(\alpha)|^2 \, d\alpha. \tag{5.11}$$

Equating Equations (5.9) and (5.10) gives

$$\boxed{\int_{-\infty}^{\infty} |f(u)|^2 \, du = \int_{-\infty}^{\infty} |F(\alpha)|^2 \, d\alpha,} \tag{5.12}$$

which is known as *Parseval's theorem*. This theorem shows that the total signal power in the space domain is the same as the total signal power in the Fourier domain. Parseval's theorem also applies to Fresnel transforms (see Problem 5.2).

5.2.3. Correlation

The *cross-correlation function* is defined as

$$\boxed{c_{fs}(u) = \int_{-\infty}^{\infty} f(x) s^*(x + u) \, dx,} \tag{5.13}$$

where either $f(x)$ or $s(x)$ is a deterministic signal. Correlation is a measure of the degree of similarity between $f(x)$ and $s(x)$ as a function of the variable u. The Fourier transform of the cross-correlation function is

$$C_{fs}(\alpha) = \int_{-\infty}^{\infty} c_{fs}(u) e^{j2\pi\alpha u}\, du, \tag{5.14}$$

which, from Equations (5.13) and (5.6), is written as

$$C_{fs}(\alpha) = F(\alpha) S^*(\alpha). \tag{5.15}$$

If we let $s(x) = f(x)$ in Equation (5.13), we have the *autocorrelation function*:

$$c_{ff}(u) = \int_{-\infty}^{\infty} f(x) f^*(x+u)\, dx, \tag{5.16}$$

where $f(x)$ is a deterministic signal. For future reference, we note that

$$c_{ff}(0) = \int_{-\infty}^{\infty} |f(x)|^2\, dx, \tag{5.17}$$

so that $c_{ff}(0)$ is proportional to the total power of $f(x)$.

Because the finite apertures of lenses set a limit on the length of spatial signals, these results for deterministic signals generally apply to optical processing systems. Random signals with no time or spatial limits, however, do not satisfy Equation (5.1) so that the Fourier transform of a random process does not exist. To develop the concept of a "spectrum" for such signals, we define the autocorrelation function $r_f(u)$ of a *wide-sense stationary random process* as

$$r_f(u) = \lim_{L \to \infty} \frac{1}{2L} \int_{-L}^{L} f(x) f^*(x+u)\, dx. \tag{5.18}$$

A random process is stationary, in the wide sense, if its autocorrelation function is a function only of the displacement u and if its mean value is not a function of u. As $r_f(u)$ generally satisfies Equation (5.1), we define the *power spectral density* of $f(x)$ as the Fourier transform of $r_f(u)$:

$$\boxed{R_f(\alpha) = \int_{-\infty}^{\infty} r_f(u) e^{j2\pi\alpha u}\, du.} \tag{5.19}$$

The autocorrelation function can also be expressed as the inverse Fourier transform of $R_f(\alpha)$:

$$r_f(u) = \int_{-\infty}^{\infty} R_f(\alpha) e^{-j2\pi\alpha u} \, d\alpha. \tag{5.20}$$

It follows immediately from Equations (5.20) and (5.18) that the integral of the spectral density gives the average intensity of $f(x)$:

$$r_f(0) = \int_{-\infty}^{\infty} R_f(\alpha) \, d\alpha = \lim_{L \to \infty} \frac{1}{2L} \int_{-L}^{L} |f(x)|^2 \, dx. \tag{5.21}$$

5.2.4. Input / Output Spectral Densities

From Equation (5.6) we see that the spectrum of the output of a linear, space-invariant system is the product of the spectrum of the deterministic input signal and the filter of the system. What is the relationship between the spectral densities of the input and output of a system when the input is a random process? A wide-sense stationary random process $f(x)$ that passes through a linear, space-invariant system produces a new random process $g(x)$:

$$g(x) = \int_{-\infty}^{\infty} f(u) h(x-u) \, du, \tag{5.22}$$

where $h(x)$ is the real-valued impulse response of the system. The limits on the integral are safely extended to infinity because the duration of $h(x)$ is finite. We calculate the autocorrelation function for the new random process $g(x)$ using Equation (5.22) in Equation (5.18) to find that

$$\begin{aligned} r_g(q) &= \lim_{L \to \infty} \frac{1}{2L} \int_{-L}^{L} g(x) g^*(x+q) \, dx \\ &= \lim_{L \to \infty} \frac{1}{2L} \int_{-L}^{L} \left\{ \iint_{-\infty}^{\infty} f(u) f^*(y) h(x-u) h^*(x+q-y) \, du \, dy \right\} dx. \end{aligned} \tag{5.23}$$

The spectral density of the output is calculated by using a procedure similar to that used to develop the double convolution theorem given by Equation (5.7):

$$
\begin{aligned}
R_g(\alpha) &= \int_{-\infty}^{\infty} r_g(q) e^{j2\pi\alpha q}\, dq \\
&= \int_{-\infty}^{\infty} \left[\lim_{L\to\infty} \frac{1}{2L} \int_{-L}^{L} \left\{ \iint_{-\infty}^{\infty} f(u) f^*(y) h(x-u) \right.\right. \\
&\quad \left.\left. \times h^*(x+q-y)\, du\, dy \right\} dx \right] e^{j2\pi\alpha q}\, dq \\
&= \lim_{L\to\infty} \frac{1}{2L} \int_{-L}^{L} \left\{ \iint_{-\infty}^{\infty} f(u) f^*(y) h(x-u) \right. \\
&\quad \left. \times \left[\int_{-\infty}^{\infty} h(x+q-y) e^{j2\pi\alpha(x+q-y)}\, dq \right] e^{-j2\pi\alpha(x-y)}\, du\, dy \right\} dx \\
&= \lim_{L\to\infty} \frac{1}{2L} \int_{-L}^{L} \left\{ \int_{-\infty}^{\infty} f(u) f^*(y) H(\alpha) \right. \\
&\quad \left. \times \left[\int_{-\infty}^{\infty} h^*(x-u) e^{-j2\pi\alpha(x-u)}\, dx \right] e^{j2\pi\alpha(y-u)}\, du\, dy \right\} \\
&= \lim_{L\to\infty} \frac{1}{2L} \int_{-L}^{L} \left\{ \int_{-\infty}^{\infty} f(u) f^*(y) |H(\alpha)|^2 e^{j2\pi\alpha(y-u)}\, du \right\} dy.
\end{aligned}
\tag{5.24}
$$

We now make a change of variable by letting $y - u = z$ to obtain

$$
\begin{aligned}
R_g(\alpha) &= \int_{-\infty}^{\infty} \left\{ \lim_{L\to\infty} \frac{1}{2L} \int_{-L}^{L} f(u) f^*(u+z)\, du \right\} |H(\alpha)|^2 e^{j2\pi\alpha z}\, dz \\
&= |H(\alpha)|^2 \int_{-\infty}^{\infty} r_f(z) e^{j2\pi\alpha z}\, dz \\
&= |H(\alpha)|^2 R_f(\alpha).
\end{aligned}
\tag{5.25}
$$

Equation (5.25) provides the justification for our definition of *spectral density* because the intensity at the output of the system is proportional to $R_f(\alpha_0)$ when $f(x)$ is passed through a narrowband filter $H(\alpha)$ whose center frequency is α_0. Note that the phase of the filter $H(\alpha)$ is important in Equation (5.6), but the phase is completely unimportant in

Equation (5.25). The phase of a filter $H(\alpha)$ does not, therefore, change the spectral density of a random process.

5.2.5. Matched Filtering

The matched filter is the optimum linear filter for maximizing the ratio of peak signal to mean-square noise when we process a deterministic signal in the presence of additive, stationary random noise. Because many of the correlation techniques considered in this chapter are developed to give a solution to this important problem, we derive the filter function of a system that optimizes the detection of a deterministic signal under the given conditions. We derive the optimum filter by assuming that no realizability constraints are placed on the filter. Because there are no causality considerations, this assumption is valid for almost all optical signal-processing problems in which the signal is not a function of time. The data we process, $f(x, y) = s(x, y) + n(x, y)$, is the sum of a deterministic signal $s(x, y)$ and stationary random noise $n(x, y)$, the spectral density of which is $R_n(\alpha, \beta)$. We do not impose the usual requirement that the noise spectrum must have a uniform magnitude at all frequencies. We process $f(x, y)$ to maximize the ratio of the intensity of peak signal to mean-square noise at the output of the system. This criterion is equivalent to that used to calculate signal-to-noise ratios in electronic systems.

Suppose that $h(x, y)$ is the impulse response of a linear, space-invariant filter $H(\alpha, \beta)$, the response of which we seek to optimize. As the signal is deterministic and as the signal and noise are additive, the signal part of the output is

$$g_s(x, y) = \iint_{-\infty}^{\infty} S(\alpha, \beta) H(\alpha, \beta) e^{j2\pi(\alpha x + \beta y)} \, d\alpha \, d\beta. \tag{5.26}$$

Because the peak-signal intensity is independent of the location of the signal, we design a filter which maximizes $|g(0, 0)|^2$. From Equations (5.21) and (5.25), we find that the noise power is

$$r_n(0, 0) = \iint_{-\infty}^{\infty} R_n(\alpha, \beta) |H(\alpha, \beta)|^2 \, d\alpha \, d\beta. \tag{5.27}$$

The ratio of peak-signal-to-noise power is obtained from Equations (5.26) and (5.27):

$$\text{SNR} = \frac{|g_s(0,0)|^2}{r_n(0,0)} = \frac{\left|\iint\limits_{-\infty}^{\infty} S(\alpha,\beta)H(\alpha,\beta)\,d\alpha\,d\beta\right|^2}{\iint\limits_{-\infty}^{\infty} R_n(\alpha,\beta)|H(\alpha,\beta)|^2\,d\alpha\,d\beta}. \tag{5.28}$$

The noise spectral density is always nonnegative, by definition, and is treated as the weighting function in the general form of the Schwartz inequality. The *Schwartz inequality* states that

$$\left|\int_{-\infty}^{\infty} f(x)g(x)\,dx\right|^2 \le \int_{-\infty}^{\infty} |f(x)|^2\,dx \int_{-\infty}^{\infty} |g(x)|^2\,dx. \tag{5.29}$$

If we let $f(x) = f_1(x)\sqrt{p(x)}$ and $g(x) = g_1(x)\sqrt{p(x)}$, with the real, nonnegative weighting function $p(x) \ge 0$, we get a more general form of the Schwartz inequality:

$$\left|\int_{-\infty}^{\infty} f_1(x)g_1(x)p(x)\,dx\right|^2 \le \int_{-\infty}^{\infty} |f_1(x)|^2 p(x)\,dx \int_{-\infty}^{\infty} |g_1(x)|^2 p(x)\,dx. \tag{5.30}$$

The equality holds if and only if $f_1(x) = cg_1^*(x)$, where c is some constant.

We rewrite Equation (5.28) in the form

$$\text{SNR} = \frac{\left|\iint\limits_{-\infty}^{\infty} \left[\frac{S(\alpha,\beta)}{R_n(\alpha,\beta)}\right] H(\alpha,\beta)R_n(\alpha,\beta)\,d\alpha\,d\beta\right|^2}{\iint\limits_{-\infty}^{\infty} R_n(\alpha,\beta)|H(\alpha,\beta)|^2\,d\alpha\,d\beta}. \tag{5.31}$$

We apply the generalized Schwartz inequality to the numerator, with $R_n(\alpha, \beta)$ as the nonnegative function, to find that

$$\text{SNR} \le \iint\limits_{-\infty}^{\infty} \frac{|S(\alpha,\beta)|^2}{R_n(\alpha,\beta)}\,d\alpha\,d\beta. \tag{5.32}$$

The equality holds if, and only if,

$$\boxed{H(\alpha,\beta) = \frac{S^*(\alpha,\beta)}{R_n(\alpha,\beta)}}. \tag{5.33}$$

The magnitude of the optimum filter is therefore proportional to the magnitude of the Fourier transform of the signal and inversely proportional to the noise spectral density. Because the phase of the optimum filter is conjugate, or *matched*, to the phase of the Fourier transform of the signal, $H(\alpha, \beta)$ is called a *matched filter*. As a special case, if the noise spectral density is uniform for all spatial frequencies, the filter becomes

$$H(\alpha, \beta) = S^*(\alpha, \beta), \tag{5.34}$$

the result most commonly obtained when dealing with time signals. We also refer to Equation (5.34) as the matched filter because it is a special form of Equation (5.33). A common misconception is that the matched filter is restricted to the white-noise solution as given by Equation (5.34); instead, this just happens to be the easiest form of the matched filter to derive when the causality constraint is imposed.

The output of the system, for the special case of Equation (5.34), is

$$g(x, y) = \iint_{-\infty}^{\infty} F(\alpha, \beta) S^*(\alpha, \beta) e^{j2\pi(\alpha x + \beta y)}\, d\alpha\, d\beta. \tag{5.35}$$

By using the convolution theorem given by Equation (5.6), we express Equation (5.35) as

$$g(x, y) = \iint_{-\infty}^{\infty} f(u, v) s^*(u + x, v + y)\, du\, dv, \tag{5.36}$$

which is a correlation operation. Systems that use matched filters to detect signals, such as radar waveforms in noise, are often called *correlation receivers*.

5.2.6. Inverse Filtering

From Equations (5.25) and (5.34) it is clear that when the noise spectral density is uniform, the signal part of the output is given by

$$g(x, y) = \iint_{-\infty}^{\infty} |S(\alpha, \beta)|^2 e^{j2\pi(\alpha x + \beta y)}\, d\alpha\, d\beta. \tag{5.37}$$

Thus the matched filter "corrects" the phase of the signal to produce a cophasal wavefront. But $g(0, 0)$ has its maximum value if, in addition to having a constant phase, the magnitude part of the integrand is a constant. The question arises whether improved detection is obtained if we use an *inverse filter* of the form

$$H(\alpha, \beta) = \frac{ce^{-j\phi(\alpha, \beta)}}{|S(\alpha, \beta)|}, \tag{5.38}$$

where c is a constant needed to keep the filter passive and $\phi(\alpha, \beta)$ is the phase of $S(\alpha, \beta)$. The inverse filter therefore not only corrects the phase of $S(\alpha, \beta)$ but it also makes the magnitude of the integrand equal to unity. Although the peak signal at the output is maximized, unfortunately the signal-to-noise ratio is not. In particular, if $S(\alpha, \beta)$ has any zeros, a large amount of noise passes through the system because the filter transmittance is highest at those frequencies where the signal-to-noise ratio is lowest. Hence, the signal-to-noise ratio is not as good as when the matched filter is used, leading to inferior performance.

A better use of inverse filtering is in equalizing a distorted communication channel. Suppose that a deterministic signal $s(x, y)$ passes through a system whose transfer function is $H(\alpha, \beta)$. The Fourier transform of the output signal is the product $H(\alpha, \beta)S(\alpha, \beta)$. The appropriate response of an *equalizing filter* is (43)

$$H_{\text{opt}}(\alpha, \beta) = \frac{[H(\alpha, \beta)]^{-1}}{1 + \left[R_n(\alpha, \beta)/R_s(\alpha, \beta)|H(\alpha, \beta)|^2\right]}, \tag{5.39}$$

where $R_n(\alpha, \beta)$ is the spectral density of the noise and $R_s(\alpha, \beta)$ is the spectral density of the data. This filter insures that the noise is not too highly amplified when the value of $|H(\alpha, \beta)|$ is small. On the other hand, if the spectral density of the distorted data $R_s(\alpha, \beta)|H(\alpha, \beta)|^2$ is much larger than $R_n(\alpha, \beta)$, the result from Equation (5.39) closely approximates that produced by the inverse filter $[H(\alpha, \beta)]^{-1}$. After the application of this filter, the spectrum of the output of the filter approximates $S(\alpha, \beta)$ as desired.

5.3. SPATIAL FILTERS

In Chapter 3 we showed that a coherently illuminated optical system produces the Fourier transform of a two-dimensional distribution of light under a wide range of geometries. We now study ways of modifying the spatial frequencies of the Fourier transform by constructing filters with arbitrary magnitude and phase responses so that optical systems can perform general processing operations (9).

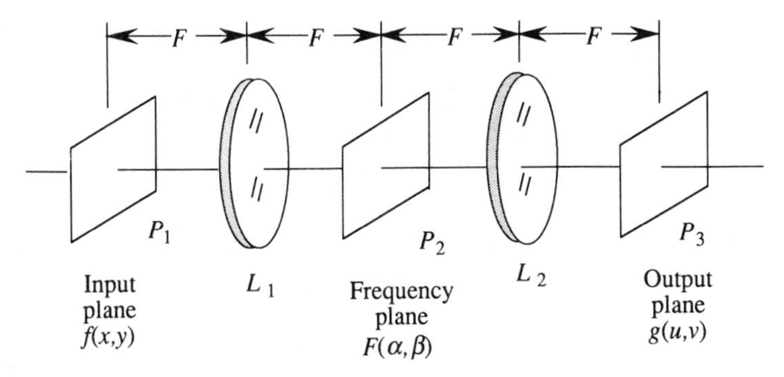

Figure 5.1. Basic spatial filtering system.

The optical system of Figure 5.1 coherently illuminates an object $f(x, y)$ stored on a spatial light modulator at plane P_1 and creates its Fourier transform $F(\alpha, \beta)$ at plane P_2. We place a transparency at plane P_2 whose complex-valued transmittance is

$$H(\alpha, \beta) = |H(\alpha, \beta)| e^{j\phi(\alpha, \beta)}, \tag{5.40}$$

where $|H(\alpha, \beta)|$ is the magnitude response and $\phi(\alpha, \beta)$ is the phase response of the transparency. Because the transmittance is a function of spatial frequencies, the transparency is called a *spatial filter*. The modified light distribution leaving plane P_2 is $G(\alpha, \beta) = F(\alpha, \beta)H(\alpha, \beta)$. Lens L_2 transforms the light to the output correlation plane P_3. The light distribution at plane P_3 is

$$g(u, v) = \iint_{-\infty}^{\infty} F(\alpha, \beta) H(\alpha, \beta) e^{j2\pi(\alpha u + \beta v)} \, d\alpha \, d\beta. \tag{5.41}$$

By using the convolution theorem, we express Equation (5.41) as

$$g(u, v) = \iint_{-\infty}^{\infty} f(x, y) h(u - x, v - y) \, dx \, dy. \tag{5.42}$$

Thus, an optical system performs a general linear integral operation as given by Equation (5.41) or Equation (5.42), provided that the spatial filter $H(\alpha, \beta)$ is physically implemented. These two relationships also imply that we can implement the operation by physically realizing $h(x, y)$ in a space plane or by realizing $H(\alpha, \beta)$ in a frequency plane. The choice between

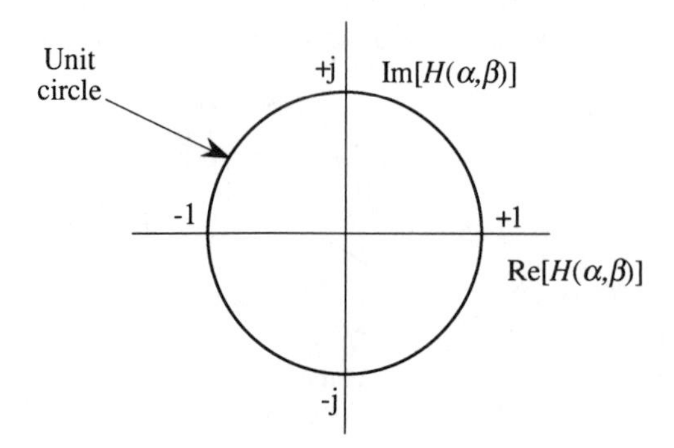

Figure 5.2. The spatial filter has values on or within the unit circle.

realizing $h(x, y)$ or $H(\alpha, \beta)$ rests on considerations discussed more fully in Section 5.16. For reasons which become clear later, we choose to concentrate on the physical realization of the spatial filter at the frequency plane. In electronic systems that operate in real time, of course, we have no choice but to implement the time-domain impulse response.

The spatial filter is most often implemented by storing a light distribution proportional to $H(\alpha, \beta)$ on a recording medium such as photographic film, if the filters are fixed in time, or on a spatial light modulator, if the filters are changed rapidly. In this chapter, we assume that the modulating medium has unlimited spatial frequency response, has unlimited dynamic range, and is linear. As spatial filters are passive elements that do not amplify light distributions, the magnitude of the spatial filter is less than or equal to one.

We describe the possible filter values at each spatial frequency by using the unit circle in the complex domain, as shown in Figure 5.2. The real part of the filter is confined to the interval $[-1, 1]$ on the real axis and the imaginary part to the interval $[-j, j]$ on the imaginary axis. The full potential of optical spatial filtering is realized only when filter values are on or within the unit circle; to operate at points outside the unit circle would require amplification of complex-valued light distributions. We discuss several important special cases in Sections 5.4–5.7 and more general cases in Sections 5.9 and 5.10. Additional information and other examples of basic filtering operations are given in the literature (44–48).

5.4. BINARY SPATIAL FILTERS

A *binary spatial filter* takes on a value of zero or one at any particular spatial frequency. The filter is physically implemented by using stops for

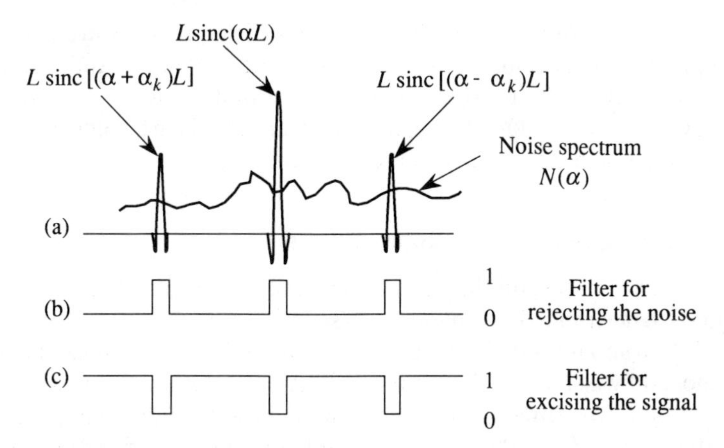

Figure 5.3. Binary filtering: (a) spectrum of signal plus noise, (b) signal acceptance filter, and (c) signal rejection filter.

zero transmittance and apertures for unity transmittance. For example, an elementary binary filter, occurring naturally in every optical system, is the aperture stop caused by the rim of a lens. Abbe, in his theory of coherent image formation in the microscope, showed that aperture stops in microscopes can alter the structure of an object considerably, as we discussed in Chapter 3, Section 3.5.3.

5.4.1. Binary Filters for Signal Detection or Excision

Binary filters are useful for detecting periodic signals in the presence of random noise. Suppose that the data is $f(x) = s(x) + n(x)$, where $s(x)$ is a sinusoid of the form $\text{rect}(x/L)[1 + \cos(2\pi\alpha_k x)]$ and $n(x)$ is noise whose spectral content is uniform. The Fourier transform $S(\alpha)$ of the signal $s(x)$ is shown in Figure 5.3(a), where the three sinc functions produced by $s(x)$ are combined with the noise spectrum $N(\alpha)$. The required filter passes the diffracted light at those positions in the frequency plane where the spectral orders of the signal occur (6), as shown in Figure 5.3(b). A significant improvement in the signal-to-noise ratio is thereby achieved (see Problem 5.3).

Binary filters are also used to remove or *excise* an unwanted periodic signal. An example is the excision of the 60-Hz frequency component present in a signal sensor or in an audio system. In this case, the appropriate filter is shown in Figure 5.3(c). It consists of *notches* at the positions in the frequency plane where energy from the periodic signal occurs. An interesting point is that the edges of the notches are extremely sharp, the notches are infinitely deep, and there is no amplitude ripple or phase distortion in the passband regions. The finite length L of the input data may, however, reduce the *apparent* steepness of the notch edges and the effective notch depth, as we discuss in Section 12.5.

5.4.2. Other Applications of Binary Filters

Binary filters are useful in processing seismic signals that contain, in addition to desirable information, noise caused by multiple reflections, diffraction from fault edges, or reverberations. When these noiselike terms are removed by a suitably chosen binary filter, the output signal is easier to interpret (49, 50). Binary filters are useful in spatial-domain processors when the reference masks are real-valued transmittance functions and in Schlieren methods for observing phase objects. They are useful for suppressing unwanted beam tracks in bubble-chamber photographs and for processing certain geophysical signals (51–53). Binary filtering techniques have been used by Kozma and Kelly and by Lohmann and his colleagues to generate filters that behave as complex-valued filters (54–56). Although this list of applications is not complete, it shows that the binary filter, in spite of its simplicity, has considerable utility.

5.5. MAGNITUDE SPATIAL FILTERS

Magnitude filters may have values anywhere on the real line between zero and one. When used alone, they are not that much more useful than binary filters. They are therefore most often used in combination with phase filters, in ways to be discussed in Section 5.8. Magnitude filters are sufficient, however, for implementing the *Weiner* filter, used for the optimal recovery of a random signal $s(x, y)$ in the presence of additive noise:

$$H(\alpha, \beta) = \frac{R_s(\alpha, \beta)}{R_s(\alpha, \beta) + R_n(\alpha, \beta)}, \tag{5.43}$$

where $R_s(\alpha, \beta)$ and $R_n(\alpha, \beta)$ are the spectral densities of the signal and noise, respectively. As both functions are nonnegative, the Weiner filter is also nonnegative and can be implemented by using a filter that has response in magnitude only.

Magnitude filters are sometimes used to increase the contrast of a photograph by controlling the low spatial frequencies in the spectrum.

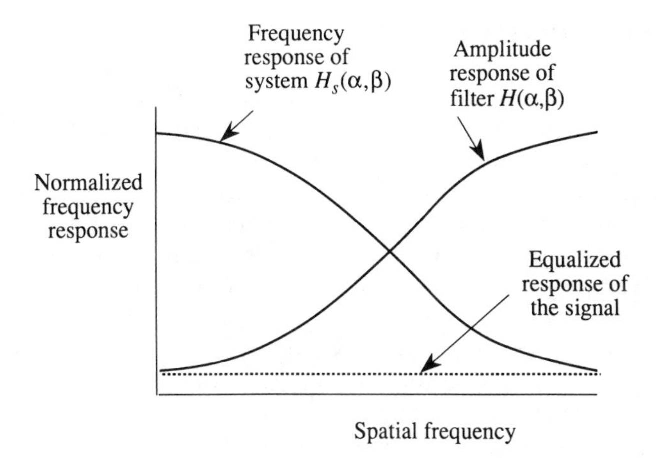

Figure 5.4. Magnitude filtering for equalization.

Although a binary filter can also be used for this purpose, the sharp transition from zero to unity transmittance in a binary filter may produce undesirable ringing and overshooting at the filtered output of the system. A smooth transition, obtained by using magnitude filters, minimizes these effects.

Magnitude filters are used to *equalize* the transfer function of collection systems whose response is nonnegative. Suppose that the input data $f_0(x, y)$ has a Fourier transform $F_0(\alpha, \beta)$. Suppose that the data-collection system has an overall frequency response $H_s(\alpha, \beta)$, which is the product of the frequency responses of the various components of the system: that is, if $H_L(\alpha, \beta)$ is the frequency response of the lens, $H_f(\alpha, \beta)$ is the frequency response of the film, and so forth, then

$$H_s(\alpha, \beta) = H_L(\alpha, \beta) H_f(\alpha, \beta) \cdots \tag{5.44}$$

is the overall system response as shown in Figure 5.4. The Fourier transform of the input data is therefore $F(\alpha, \beta) = H_s(\alpha, \beta)F_0(\alpha, \beta)$. To recover $f_0(x, y)$, we need a filter whose transmittance is

$$H(\alpha, \beta) = \frac{c}{H_s(\alpha, \beta)}, \tag{5.45}$$

where c is a constant needed to keep the filter passive. If all transfer functions are nonnegative, both $H_s(\alpha, \beta)$ and $H(\alpha, \beta)$ are nonnegative and we can realize $H(\alpha, \beta)$ with a magnitude filter whose response is shown in Figure 5.4. The equalized spectrum is shown by the dotted line.

5.6. PHASE SPATIAL FILTERS

A *phase filter* changes the optical path length at each spatial frequency position in the spectrum but has no effect on the magnitude of the spectrum. Such a filter is constructed by depositing a thin film on a supporting substrate, varying the refractive index of a thick film, or figuring a piece of glass. As an accurate, continuous-phase response is difficult to achieve by any of these methods, except when a lens provides the required phase response, phase filters are used mainly to advance or retard light by half a wavelength.

An excellent example of using continuous-phase response is the optical processing of synthetic-aperture radar signals. Measurements of the earth's reflectivity at microwave wavelengths by synthetic radar systems provide remote mapping of ground terrain, ocean waves, and other geological features. These systems are a significant departure from conventional radar systems in which large antennas are used to obtain high spatial resolution. For a large dish antenna, the angular beam resolution is obtained by a method similar to that used for optical telescopes; the resolution is Λ/D, where Λ is the radar wavelength and D is the aperture of the antenna. In a synthetic-aperture radar system, however, a small antenna provides better resolution than a large antenna. The increased resolution is made possible because the motion of the antenna, relative to the object, causes the radar return to be sampled sequentially in time instead of being collected in parallel as it is with a large dish antenna. Antennas with large effective apertures are thereby synthesized.

Figure 5.5(a) shows a small antenna, carried by an aircraft, taking on the sample positions of a much longer array as the aircraft moves. At each position, the antenna transmits a radar pulse that illuminates a portion of the ground in both range and azimuth directions. The smaller the antenna, the broader the transmitted beam pattern. This broader beam, in turn, increases the time that an object is illuminated and makes it possible for the radar system to receive signals over a longer flight path to generate a longer array. This sequence of events culminates in higher resolution because the space bandwidth product of the signal is increased relative to that of a single antenna.

To create a map of the radar reflectivity, we need a means for storing the radar returns as a two-dimensionally formatted signal and a means for processing the stored data, as shown in Figure 5.5(b). Because digital computers could not handle the computational load, new and more powerful techniques were required. In their classic paper on "Optical Data Processing and Filtering Systems," Cutrona, Leith, Palermo, and Porcello elegantly laid out the basic concepts for processing synthetic

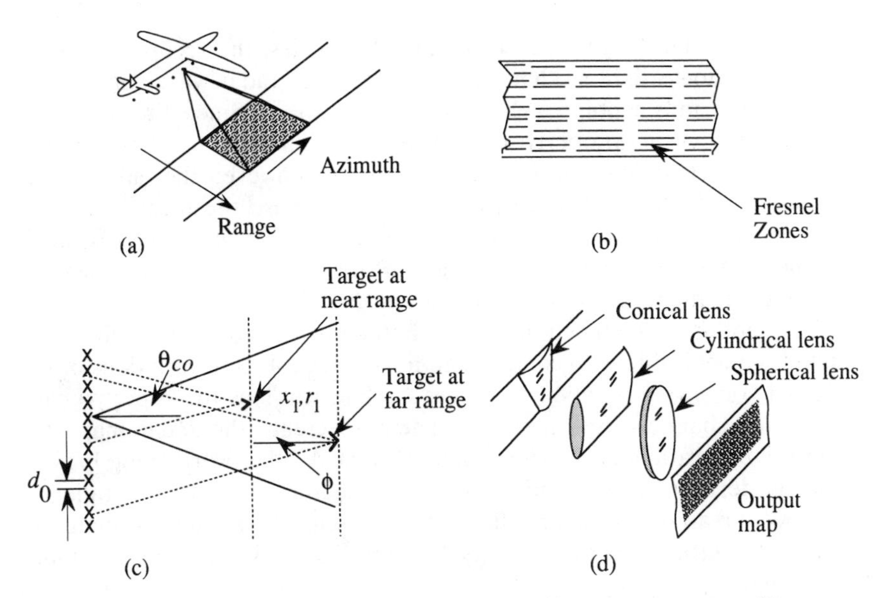

Figure 5.5. Synthetic-aperture radar data processing: (a) collection system, (b) storage format, (c) top view of collection system, and (d) optical processing system.

aperture radar signals (8). The reflectivity of the terrain obtained by a synthetic-aperture array is expressed as a sum of one-dimensional Fresnel zone patterns, whose focal lengths are a function of the orthogonal (range) direction. The antenna emits energy within a cone whose semiangle is θ_{co}, as shown in Figure 5.5(c); the small antenna samples the returned signals with sample spacing d_0. A typical return for a point at an azimuth position x_1 and a range position r_1 is written in the form

$$f(x,t) = A + \sigma(x_1, r_1)\cos\left[2\pi f_c t + \frac{2\pi}{\Lambda r_1}(x - x_1)^2\right], \quad (5.46)$$

where A is a bias, $\sigma(x_1, r_1)$ is the reflectivity of the object, f_c is an RF carrier frequency, and Λ is the radar wavelength. The reflected energy is confined mostly within a cone whose semiangle is ϕ, which is a function of the object size. As the airborne antenna moves, the sampled reflected energy is multiplied by a local oscillator to produce Fresnel zones, similar to those discussed in Chapter 3, whose parameters are a function of the variables given in Equation (5.46). These Fresnel zones are then recorded and stored on photographic film as shown in Figure 5.5(b).

To create the desired map of radar reflectivities, the recorded Fresnel zones must focus light so that a point scatterer in radar wavelength space is converted into a point in light wavelength space. Since Fresnel zones have a self-focusing property, free space is a suitable matched filter in the azimuth direction. Without further intervention, however, the radar map is reconstructed a long distance from the film because the focal lengths of the Fresnel zones are long. Furthermore, light focuses in a severely inclined plane because the focal lengths of the Fresnel zones are functions of the radar range, as shown by Equation (5.46).

The optimum matched filter, a continuous-valued phase-only filter, includes a conical lens to compensate for the focal-length variations of the Fresnel zones as a function of range, a cylindrical lens to properly image the range channels, and a spherical lens to create the radar map at a convenient location as shown in Figure 5.5(d). All the overlapping Fresnel zone patterns in range and azimuth are thereby sorted in parallel to produce a map of the radar reflectivity. The scale of the map is related to that of the ground by the ratio λ/Λ of the light wavelength to the radar wavelength.

This filtering technique for processing radar signals is ideal because the required phase response is a combination of lenses that are constructed with high accuracy and the matched filter does not attenuate the light. The synthetic-aperture radar configuration and the properties of the Fresnel transform, studied in Chapter 3, are nearly identical. In the radar case, a small source of electromagnetic radiation of size d_0 (the transmitting radar antenna length) produces a large angular illuminating beam θ_{co}, which covers large regions of the ground. The object reflects the electromagnetic radiation so that the same antenna samples the reflected field. Recall from Chapter 3 that the appropriate sampling interval in all planes containing a chirp function is d_0. As the sample sizes in all Fresnel domains are equal, the sample size at the ground is the same as that of the antenna, namely, d_0. A smaller antenna therefore leads to better resolution.

Creating radar maps was the first use of optical processing on a routine basis and was the first instance in which the equivalent of a matched spatial filter included a complicated phase function. Radar processing had a tremendous impact on the general field of optical signal processing and, in particular, on the development of holography.

5.7. REAL-VALUED SPATIAL FILTERS

When a binary phase filter, with a phase of either zero or π, is used in conjunction with a magnitude filter, the composite filter is a real-valued function with values between -1 and $+1$. An example is Zernike's phase contrast method for viewing objects that have no absorption but do have variations in thickness or refractive index; these objects are called *phase objects* (57). Zernike's filter advances or retards the light at the higher spatial frequencies, relative to the light at the low spatial frequencies, by $\gamma = \pm\pi/2$. The choice of γ determines the nature of the image: if $\gamma = +\pi/2$, the phase information is brighter than the average background intensity; if $\gamma = -\pi/2$, the phase information is darker than the average background intensity.

Real-valued filters have been used to correct aberrated images for certain forms of the aberration; this technique is generally called *image restoration*. Several ways for constructing the necessary filter are given by Tsujiuchi (58). Real-valued filters are also used to perform mathematical operations such as differentiation. For example, the filter required to differentiate a one-dimensional signal is $H(\alpha) = c\alpha$, where c is a constant that keeps the filter passive:

$$H(\alpha) = \begin{cases} c\alpha, & 0 \le c\alpha \le 1, \\ c|\alpha|e^{j\pi}, & -1 \le c\alpha < 0. \end{cases} \tag{5.47}$$

Therefore, we need a filter whose magnitude is proportional to the spatial frequency and whose phase is zero for $\alpha > 0$ and π for $\alpha < 0$.

5.8. EXPERIMENTAL EXAMPLES

Suppose we have a one-dimensional signal defined as $s(x) = \text{rect}(x/L)$. The required matched filter, when the noise spectral density is uniform, is $H(\alpha) = S^*(\alpha)$. Therefore, the matched filter has the form

$$H(\alpha) = \frac{1}{L}|S(\alpha)|e^{-j\phi(\alpha)}, \tag{5.48}$$

where $S(\alpha) = L\,\text{sinc}(\alpha L)$ and $\phi(\alpha) = 0$ or π according to whether $\text{sinc}(\alpha L)$ is positive or negative. The scaling factor $1/L$ in Equation (5.48) is introduced to keep the filter passive. Figure 5.6(a) shows a one-dimensional input pulse, and Figure 5.6(b) shows the pulse after it has passed through a phase-only filter with the phase response $\exp[-j\phi(\alpha)]$ defined by Equation (5.48). An interferogram of the phase-only filter is shown in Figure 5.7. The phase change is in the horizontal direction only; the phase in the vertical direction is constant, as indicated by the fixed period of the fringes, because the pulse is one dimensional. The fringes shift by a half-wavelength of light, as desired, at those spatial frequencies where $\text{sinc}(\alpha L)$ changes sign.

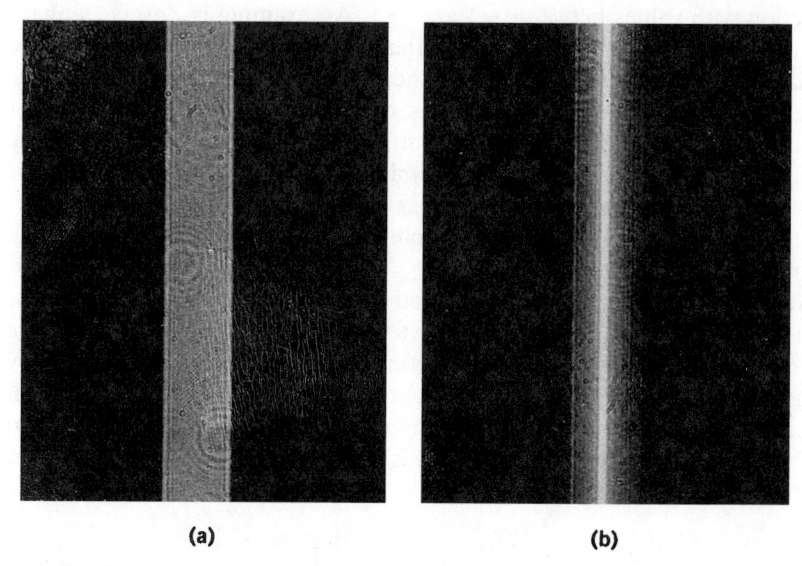

Figure 5.6. Example of phase-only filtering: (a) input pulse and (b) phase-only filtered pulse.

The peak intensity of the output pulse is

$$|g(0)|^2 = \left| \int_{-\alpha_{co}}^{\alpha_{co}} |\operatorname{sinc}(\alpha L)| \, d\alpha \right|^2, \tag{5.49}$$

where α_{co} is the spatial bandwidth of the system. Figure 5.8 shows the calculated value of $|g(x)|^2$ for $\alpha_{co} = 25$ Ab, superimposed on the original pulse for comparison purposes. Figure 5.9 shows a scan through $|g(x)|^2$

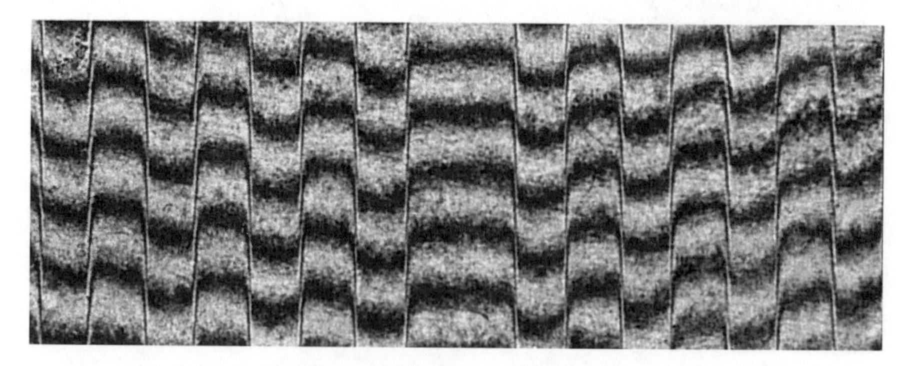

Figure 5.7. Interferogram of phase filter.

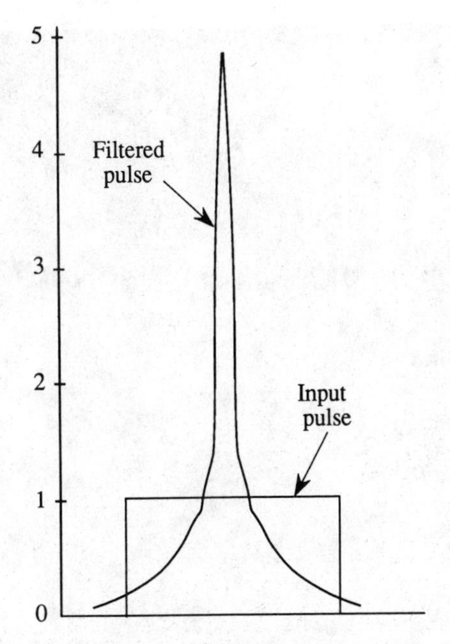

Figure 5.8. Calculated improvement in peak signal.

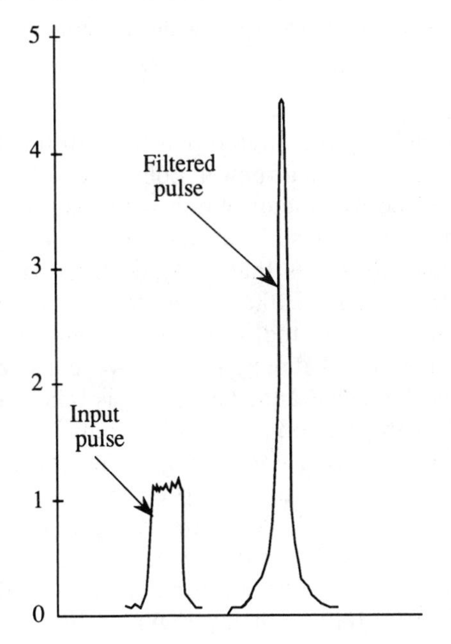

Figure 5.9. Experimental results.

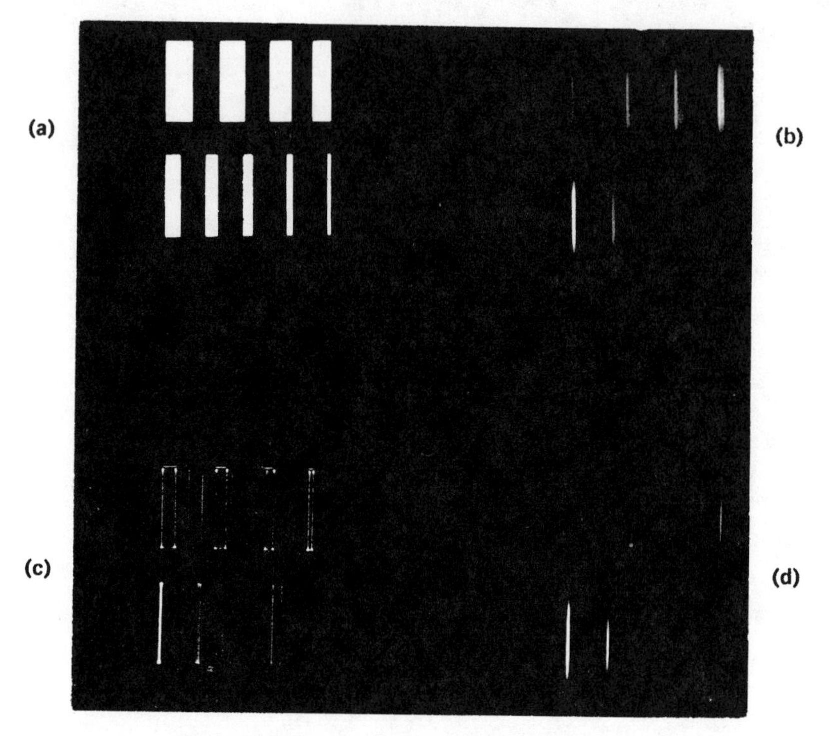

Figure 5.10. Results of various filtering techniques.

for both the unfiltered and the filtered pulses. With this filter, the pulse is spatially compressed at the half-power points by a factor of 13 : 1. The peak intensity increased by a factor of only 4.4 versus a predicted factor of 4.8 because the phase filter absorbs a small amount of light.

The next example illustrates that matched filtering provides the best signal-to-noise ratio. Figure 5.10(a) shows a set of nine pulses whose widths are in the ratio of the integers; all filtering operations are performed in the horizontal direction only. We consider the fifth pulse in the sequence as the signal and the others as false signals or "noise." Figure 5.10(b) shows the filtered output when only the correct *magnitude* response is used. Although the signal energy is not highly concentrated in the output, false signals are partially suppressed because of the magnitude response of the filter. Figure 5.10(c) shows the filtered output when only the correct *phase* response is used; the output signal is highly concentrated along a line and the light surrounding the signal peak resembles that shown in Figure 5.10(b). Finally, Figure 5.10(d) shows the filtered output when both the correct *phase* response and *magnitude* response are used; in this case, the filter is matched to the Fourier transform of the signal in both magnitude and phase. The magnitude part of the filter passes spatial frequencies according to whether the signal-to-noise ratio is high or low; the phase part contributes the important function of producing a plane wavefront so that the inverse transform lens concentrates the correlation peak in a small area. The two parts of the filter combine to maximize the signal-to-noise ratio.

In applications such as pattern recognition, several filters must be stored and recalled as needed. A phase filter requires less storage capacity than matched filters; in the extreme case where the signal is symmetric, binary phase values are sufficient. The requirements on the spatial light modulator required to implement the filters are also thereby reduced, and phase filters are more efficient because they have low attenuation factors. Although phase only filters do not maximize the signal-to-noise ratio under the conditions usually given, namely, that of maximizing the ratio of peak signal energy to average noise intensity, they may provide sufficient performance, while simplifying the filtering operations (59, 60).

5.9. THE SPATIAL CARRIER FREQUENCY FILTER

The full potential of optical spatial filtering is not achieved unless the filters are complex valued. As arbitrary phase responses are not obtainable by any of the methods described so far, we study a different technique for producing the desired phase response. The magnitude of the filter can be constructed if we place $h(x, y)$ in a two-dimensional Fourier-transforming system that produces $H(\alpha, \beta)$ at the output. The required phase response of the filter is constructed by combining the principles of Fourier transforms, interferometry, and communication theory. A complex-valued filter can be represented as a real-valued filter in which the magnitude and phase modulate a carrier frequency:

$$H(\alpha, \beta) = |S(\alpha, \beta)| \cos[2\pi x_0 \alpha + \phi(\alpha, \beta)], \tag{5.50}$$

where x_0 is the carrier frequency and $\phi(\alpha, \beta)$ is the phase of $S(\alpha, \beta)$. The result in Equation (5.50) has the same form as the real-valued signal produced by a communication system in which the magnitude and phase of the signal are simultaneously encoded on a temporal carrier frequency. We need to modify this technique by introducing a bias so that the filter is *nonnegative* and can therefore be stored on a spatial light modulator with magnitude-only response.

5.10. INTERFEROMETRIC METHODS FOR CONSTRUCTING FILTERS

We now describe interferometric systems for constructing spatial carrier frequency filters that have a prescribed magnitude and phase response (9). In the Mach-Zehnder interferometer shown in Figure 5.11, a point source of monochromatic light, collimated by lens L_c, is divided into two parts by the beamsplitter. The *signal beam* in the lower branch contains the signal $s(x, y)$ at plane P_1; lens L_1 produces the Fourier transform $S(\alpha, \beta)$ of the signal at plane P_2.

The upper branch of the interferometer is the *reference beam*, directed to plane P_2 by the second mirror and the beamcombiner. This beam is tilted through an angle θ with respect to the signal beam by rotating the beamcombiner. Although the reference beam can be tilted with respect to any axis, we generally assume that it is tilted with respect to the x axis for mathematical simplicity. The light distribution at plane P_2 is the sum of the contribution from the signal and reference branches:

$$H(\alpha, \beta) = A_1 e^{-j2\pi x_0 \alpha} + A_2 S(\alpha, \beta), \tag{5.51}$$

where A_1 and A_2 are constants and x_0 is equal to θF, where F is the focal length of the Fourier-transform lens. The intensity of the light distribution at plane P_2 is

$$\begin{aligned} I(\alpha, \beta) &= |H(\alpha, \beta)|^2 \\ &= A_1^2 + A_2^2 |S(\alpha, \beta)|^2 \\ &\quad + A_1 A_2 S(\alpha, \beta) e^{j2\pi x_0 \alpha} + A_1 A_2 S^*(\alpha, \beta) e^{-j2\pi x_0 \alpha}, \end{aligned} \tag{5.52}$$

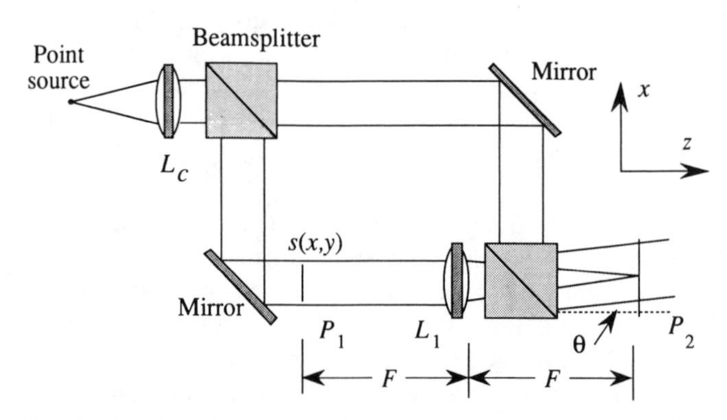

Figure 5.11. Mach-Zehnder interferometer for constructing spatial filters.

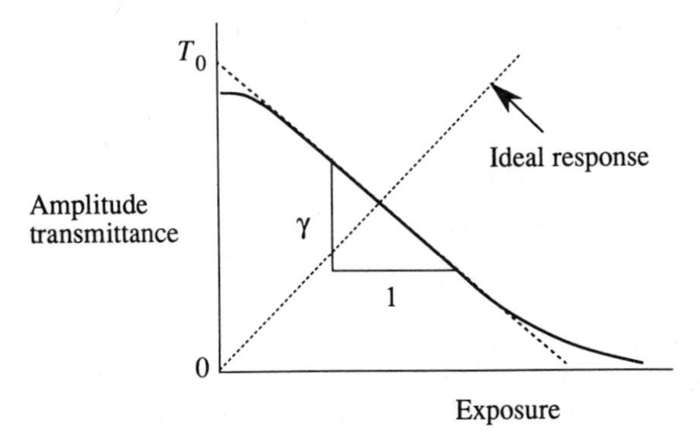

Figure 5.12. Transfer curve for photographic film.

which reduces to

$$I(\alpha,\beta) = A_1^2 + A_2^2|S(\alpha,\beta)|^2 + 2A_1A_2|S(\alpha,\beta)|\cos[2\pi x_0\alpha + \phi(\alpha,\beta)]. \tag{5.53}$$

The main difference between Equation (5.53) and a general signal in communication theory, as given by Equation (5.50), is that Equation (5.53) contains bias terms represented by a constant A_1^2 and by $|A_2S(\alpha,\beta)|^2$. The signal dependent bias term $|A_2S(\alpha,\beta)|^2$ ensures that the function is nonnegative, as required by the fact that $I(\alpha,\beta)$ is an intensity function.

Ideally, the amplitude transmittance of photographic film should be linearly related to the exposure. Most photographic films are negative working, however, in which an increase in the exposure leads to a decrease in the transmittance of the film, as shown in Figure 5.12 (61). The dynamic response of a negative working film is expressed as

$$T(\alpha,\beta) = T_0 - \gamma t_0 I(\alpha,\beta), \tag{5.54}$$

where $T(\alpha,\beta)$ is the amplitude transmittance of the film, T_0 is the amplitude intercept, γ is the slope in the linear region, t_0 is the exposure time, and $I(\alpha,\beta)$ is the intensity of the exposing light. By substituting Equation (5.52) into Equation (5.54), we find that the amplitude transmittance of the film is

$$\begin{aligned} T(\alpha\,\beta) = -\gamma t_0\Big[&\left(A_1^2 - T_0/\gamma t_0\right) + A_2^2|S(\alpha,\beta)|^2 \\ &+A_1A_2S(\alpha,\beta)e^{j2\pi x_0\alpha} + A_1A_2S^*(\alpha,\beta)e^{+j2\pi x_0\alpha}\Big]. \end{aligned} \tag{5.55}$$

We let $A_3^2 = (A_1^2 - T_0/\gamma t_0)$ and ignore the scaling factor $(-\gamma t_0)$ to obtain

$$T(\alpha, \beta) = A_3^2 + A_2^2 |S(\alpha, \beta)|^2 + A_1 A_2 S(\alpha, \beta) e^{j2\pi x_0 \alpha} + A_1 A_2 S^*(\alpha, \beta) e^{-j2\pi x_0 \alpha}. \tag{5.56}$$

Because Equation (5.56) has essentially the same form as Equation (5.52), we use either one to develop our results. We generally use the former result for simplicity, except when the relative magnitudes of the terms are important.

We emphasize the presence of the spatial carrier frequency by combining the last two terms of Equation (5.56) to find that

$$T(\alpha, \beta) = A_3^2 + A_2^2 |S(\alpha, \beta)|^2 + 2A_1 A_2 |S(\alpha, \beta)| \cos[2\pi x_0 \alpha + \phi(\alpha, \beta)]. \tag{5.57}$$

Equation (5.57) reveals that the spatial carrier frequency x_0 is modulated in phase according to $\phi(\alpha, \beta)$ and in magnitude according to $|S(\alpha, \beta)|$. The two bias terms ensure that $T(\alpha, \beta)$ is nonnegative. This filter, often called a *VanderLugt filter*, therefore encodes a complex-valued function as a nonnegative function.

Figure 5.13 shows the output of the interferometer when the signal is the symbol 3. The magnitude modulation, or envelope, of the filter is proportional to the magnitude of the Fourier transform of the signal. The diffraction in the vertical direction is caused by the horizontal stroke of the 3, the circular diffraction is caused by the loop of the 3, and the diagonal break is caused by the diagonal stroke of the 3.

The spatial carrier frequency is evident as the vertical interference fringe structure. Each fringe represents a contour of constant phase; for example, if the phase were constant over the entire frequency plane, all fringes would be straight. Thus, by comparing the deviation of the fringes with respect to straight lines, we can determine the phase of $S(\alpha, \beta)$ at each spatial frequency. Fortunately, such extensive and time-consuming measurements are not needed because the information has already been encoded so that the filter can be used directly in a signal-processing system which, at the output, decodes that information properly.

The decoding process is not immediately obvious, as Equation (5.56) contains four terms that overlap in the frequency domain. We want to use only the fourth term to perform correlation. To show how the optical

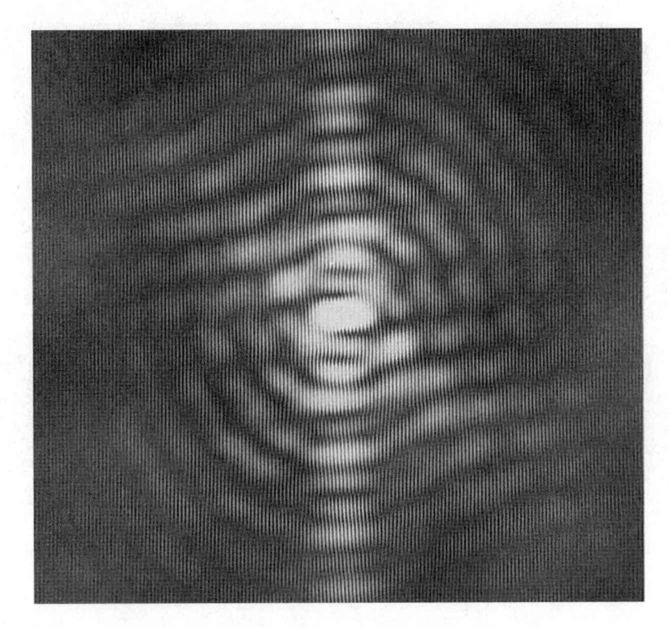

Figure 5.13. Spatial filter for the character 3.

processing system separates the effects of these terms at the output, we place the spatial carrier frequency filter at plane P_1 of the optical system shown in Figure 5.14 and calculate the impulse response of the filter given by Equation (5.57). We illuminate the filter $T(\alpha, \beta)$ at plane P_1 with a point source (the impulse function) provided by lens L_1. The Fourier transform produces the impulse response of the filter at plane P_2. The

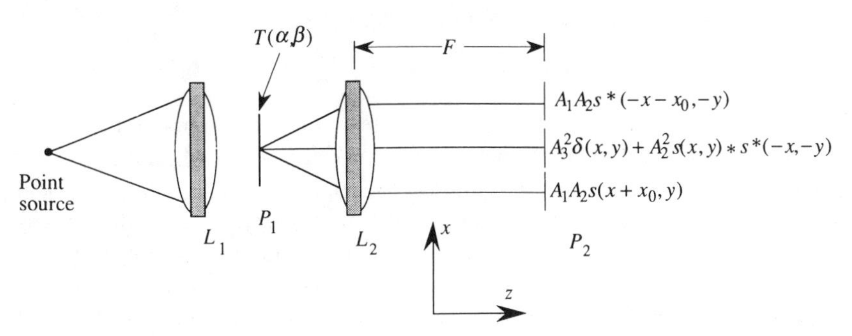

Figure 5.14. System for finding the impulse response of the filter.

Figure 5.15. Impulse response of a spatial filter (9) (copyright © IEEE, 1964).

spatial carrier frequency produces three distinct light distributions at the output whose positions are most simply found if we temporarily let $|S(\alpha, \beta)| = 1$ and $\phi(\alpha, \beta) = 0$. The transmittance of the filter is then

$$T(\alpha, \beta) = A_3^2 + A_2^2 + 2A_1A_2\cos(2\pi x_0\alpha), \tag{5.58}$$

so that the sinusoidal nature of the transmittance of the filter is obvious, as is the fact that x_0 is the spatial carrier frequency. The output in this case is simply three sinc functions, located at $x = 0$ and at $x = \pm x_0$.

For the more general case, we find the impulse response from Equation (5.56) as

$$\begin{aligned} t(x, y) = {} & A_3^2\delta(x, y) + A_2^2 s(x, y) * s^*(-x, -y) \\ & + A_1A_2 s(x + x_0, y) + A_1A_2 s^*(-x - x_0, -y). \end{aligned} \tag{5.59}$$

Figure 5.15 shows the impulse response when the input signal is the shape L. The sum of a delta function and the autocorrelation function of L, as given by the first two terms of Equation (5.59), are centered on the optical axis. The impulse response corresponding to the signal, as indicated by the third term of Equation (5.59), is located a distance x_0 from the axis. The complex conjugate of the signal occurs a distance x_0 in the opposite direction from the optical axis, as shown by the last term of Equation (5.59).

5.10.1. Limitations of the Mach-Zehnder Interferometer

The Mach-Zehnder interferometer used to construct the matched filter has several practical limitations. Unless the signal is placed in the front focal plane of lens L_1, its Fourier transform contains a spherical phase factor that interferes with the tilted reference wave, as discussed in Chapter 3. The result of this interference is an off-axis Fresnel zone pattern, recorded on the filter along with the desired information. The filter therefore has focal power so that the diffracted output images are located at planes axially displaced from the plane containing the undiffracted light. Unfortunately, the off-axis Fresnel zone behaves as a decentered thin lens whose aberrations increase rapidly as the field angle increases. Such filters are therefore useful only when the signal nearly fills the entire input processing aperture so that the correlation peak is always near the optical axis.

Another limitation of the Mach-Zehnder interferometer is that its two beams are nearly parallel in its normal mode of operation. Because large angles between the two beams are needed when the object is large, the interferometer must have large apertures. Finally, there is no convenient way to remove multiple reflections generated when the beamsplitter faces are not parallel.

5.10.2. The Rayleigh Interferometer

The Mach-Zehnder interferometer is useful as a guide for constructing more practical ones. If the tilted reference wave is projected back through lens L_1, we find it focused at a distance x_0 from the center of the signal at plane P_1. Therefore, if we provide a point source at that equivalent position, as shown in Figure 5.16, the beamsplitters and mirrors are not needed. Such a system, called a *modified Rayleigh interferometer*, operates by beam division rather than by magnitude division. Lens L_c collimates light from a source and lens L_r focuses a part of that light at plane P_1, a distance x_0 from the center of the signal. The signal at plane P_1 is expressed as

$$h(x, y) = s\left(x - \frac{x_0}{2}, y\right) + \delta\left(x + \frac{x_0}{2}, y\right), \tag{5.60}$$

where the source is represented by a delta function. As noted before, the delta function is equivalent to a $\mathrm{sinc}[(x - x_0/2)/d_0]$ light source produced by a lens. The sinc function and the delta function are equivalent if the sinc function produces a plane wave that covers the Fourier transform of the signal at plane P_2.

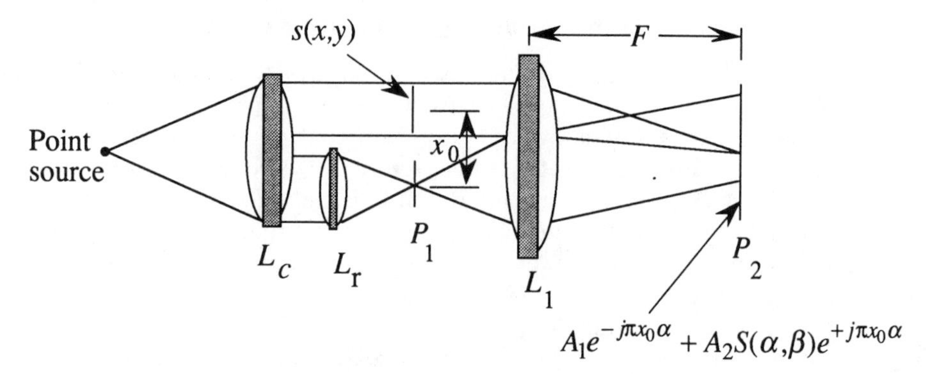

Figure 5.16. Modified Rayleigh interferometer.

Lens L_1 produces the Fourier transform of $h(x, y)$ at plane P_2:

$$H(\alpha, \beta) = \psi(\alpha, \beta)\left[A_1 e^{-j\pi x_0 \alpha} + A_2 S(\alpha, \beta) e^{j\pi x_0 \alpha}\right], \qquad (5.61)$$

where $\psi(\alpha, \beta)$ is a spherical phase factor. After $H(\alpha, \beta)$ is square-law detected, the recorded function is identical to that produced by the Mach-Zehnder interferometer. As the phase factor $\psi(\alpha, \beta)$ is common to both signals, it is not present in the intensity function that is recorded. Thus, even if plane P_1 is not the front focal plane of lens L_1, the recorded filters are always free of focal power because the spherical phase factor is common to both terms of $H(\alpha, \beta)$. Putting plane P_1 close to lens L_1 makes optimum use of the lens capacity; in fact, in some cases, as in the next interferometer described, plane P_1 is to the right of lens L_1.

The modified Rayleigh interferometer also has some drawbacks. The aperture of lens L_1 is large if a high spatial carrier frequency is needed, and the lens is used mostly near the edges of its aperture where aberrations are generally large. Finally, inefficient use is made of the available light when the signal has a small area. Nevertheless, the Rayleigh interferometer is often used because it is simple to construct, does not require beamsplitters or mirrors, is made from commonly available components, and is relatively insensitive to vibrations.

5.10.3. The Minimum-Aperture Interferometer

A more convenient and flexible interferometer, called a *minimum-aperture interferometer*, is developed by using the best features of those just described. As shown in Figure 5.17, coherent light is divided into two parts

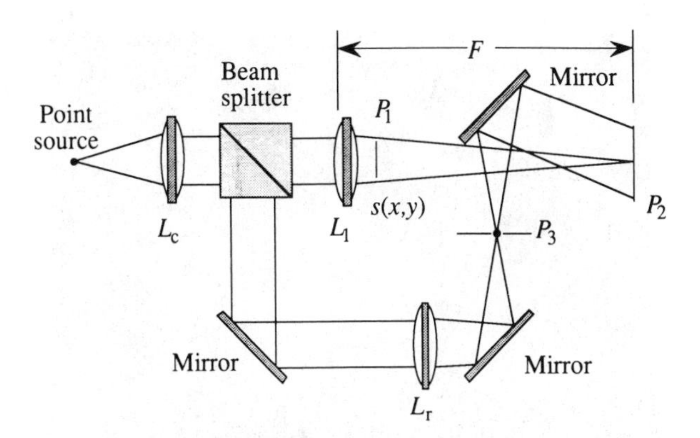

Figure 5.17. Minimum-aperture interferometer.

by the beamsplitter. The lower beam, passing through lens L_r, is focused at plane P_3, where a pinhole removes secondary reflections from the beamsplitter.

Two mirrors direct this beam to plane P_2 at an angle θ with respect to the optical axis. The direct beam passes through lens L_1 which produces the Fourier transform of the signal at plane P_2. To an observer at plane P_2, it appears that planes P_1 and P_3 are at the same axial distance and that the point source at plane P_3 is located a distance x_0 from the center of the signal. The amplitude at plane P_2 is then still given by Equation (5.51) and the recorded filter is the same as that generated by the other interferometers.

The minimum-aperture interferometer has several practical advantages. Each lens has a small relative aperture and works over small field angles, and the available light is efficiently used. Finally, the spatial carrier frequency can be increased by adjusting the angles of the last two mirrors without increasing the apertures of any of the lenses.

5.11. INFORMATION PROCESSING

We now show how the spatial carrier frequency filter is used to process information. We use the filter for the shape L as described in the last section. The data, for this example, is the set of geometric shapes shown in Figure 5.18. We want the output of the system to produce a strong correlation peak for the shape L when the filter is properly aligned in angle with each of the L's in the input. The output of the system, when

Figure 5.18. Input signal (9) (copyright © IEEE, 1964).

$f(x, y)$ is the input, is obtained from Equations (5.41) and (5.56):

$$g(u,v) = \iint_{-\infty}^{\infty} F(\alpha,\beta)T(\alpha,\beta)e^{j2\pi(\alpha u+\beta v)}\,d\alpha\,d\beta. \tag{5.62}$$

By expanding $T(\alpha, \beta)$ into its component terms, we find that

$$\begin{aligned} g(u,v) = &\iint_{-\infty}^{\infty} F(\alpha,\beta)\left[A_3^2 + A_2^2|S(\alpha,\beta)|^2\right]e^{j2\pi(\alpha u+\beta v)}\,d\alpha\,d\beta \\ &+ \iint_{-\infty}^{\infty} A_1A_2F(\alpha,\beta)S(\alpha,\beta)e^{j2\pi[\alpha(u+x_0)+\beta v]}\,d\alpha\,d\beta \\ &+ \iint_{-\infty}^{\infty} A_1A_2F(\alpha,\beta)S^*(\alpha,\beta)e^{j2\pi[\alpha(u-x_0)+\beta v]}\,d\alpha\,d\beta. \end{aligned} \tag{5.63}$$

The first term of Equation (5.63), centered on the optical axis, is of no particular interest in this discussion. Its structure and its size often resemble that of $f(x, y)$ provided that the delta function dominates the autocorrelation function of the signal.

The second term, centered at $u = -x_0$ and $v = 0$, is the *convolution* of data with the impulse response of the desired filter; this result is shown in Figure 5.19(a). The last term, centered at $u = x_0$ and $v = 0$, is the *correlation* of data with the impulse response of the desired filter; this result is shown in Figure 5.19(b). Except for their locations, the last two terms are equivalent to the convolution and correlation functions given by Equations (5.5) and (5.13). Thus, the optical system decodes the desired information and automatically produces the desired linear integral operation at the output. We point out that the central order term, which closely resembles the input, is not shown in Figure 5.19; it is physically centered between the convolution and correlations outputs shown there. The results shown in Figure 5.19 illustrate the significant difference between convolution and correlation in terms of peak-signal-to-noise ratio. As the two filters have identical magnitude responses, the noise levels at the output must be the same. It is clear, therefore, that *the phase response of the filter plays the most significant role in increasing the peak-signal level*.

To further explore the importance of the phase response, consider a signal whose Fourier transform has the form

$$S(\alpha, \beta) = e^{-j\pi a_1(\alpha^2+\beta^2)}, \tag{5.64}$$

where a_1 is a constant. The phase response shown in Figure 5.20(a) is quadratic in α and β, and the magnitude is constant. This phase response is represented as a wavefront curvature in the frequency plane as shown in Figure 5.20(b), similar to the way we represented aberrations in Chapter 2.

When the noise spectral density is uniform, the appropriate matched filter has the form

$$H(\alpha, \beta) = S^*(\alpha, \beta) = e^{j\pi a_1(\alpha^2+\beta^2)}. \tag{5.65}$$

In terms of the coordinates of the frequency plane, we find that Equation (5.65) becomes

$$H(\alpha, \beta) = e^{j[\pi a_1/(\lambda F)^2](\xi^2+\eta^2)}, \tag{5.66}$$

where F is the focal length of the Fourier transform lenses. This filter could, of course, be constructed using the interferometric technique developed in the last section. It is instructive to pursue another approach, however, and to recognize that Equation (5.66) essentially represents a phase function that we can implement with a lens. Recall that a positive

Figure 5.19. Result of spatial filtering: (a) convolution and (b) correlation (9) (copyright © IEEE, 1964).

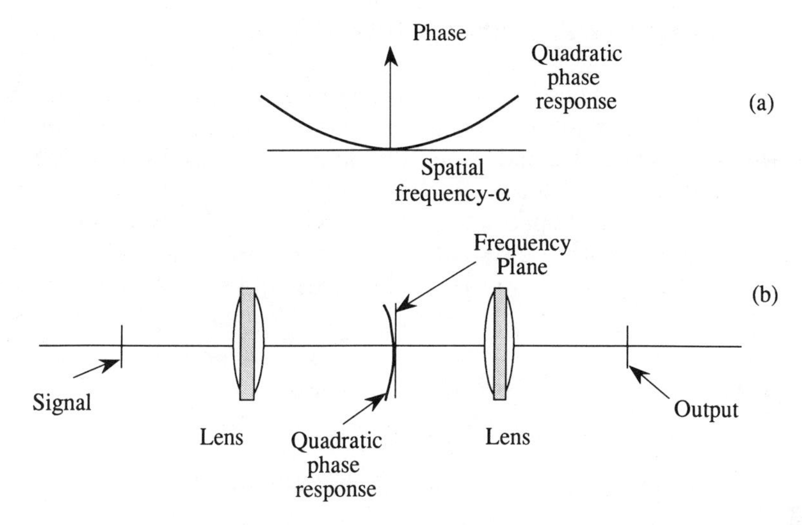

Figure 5.20. The importance of phase in pattern recognition.

spherical lens, of focal length F_1, is represented by the phase function $\exp[j(\pi/\lambda F_1)(\xi^2 + \eta^2)]$. The appropriate matched filter is therefore given by the solution to the equation

$$\frac{\pi a_1}{(\lambda F)^2} = \frac{\pi}{\lambda F_1}, \tag{5.67}$$

so that the focal length of the matched filter lens is

$$F_1 = \frac{\lambda F^2}{a_1}. \tag{5.68}$$

It is easy to visualize why the matched filter provides such a gain in the peak-signal level at the output of the system. In the absence of the filter, the image of the signal is a blur circle of light whose size is a function of the parameter a_1. The matched filter conjugates the phase of the Fourier transform of the signal, creating a plane wave just after the frequency plane so that the light is tightly focused into a diffraction-limited spot that occupies only one sample spacing d_0 at the output. Although the noise is spatially redistributed at the output plane, the noise spectral density is not changed by virtue of Equation (5.25).

5.12. ARBITRARY REFERENCE FUNCTION

We can also record filters by using an arbitrary signal in place of the point source. This technique can be used to form an *associative memory*, in which either signal responds when the other is used as a stimulus. When this technique is used for performing correlation, it is called *joint transform processing* (62). Suppose that $f(x, y)$ replaces the impulse function in one of the interferometric systems that we have discussed so far. As usual, we assign the signal $s(x, y)$ to the other sideband so that the total input amplitude is $f(x - x_0/2, y) + s(x + x_0/2)$. The intensity at the Fourier plane is

$$\begin{aligned} I(\alpha,\beta) &= |F(\alpha,\beta)e^{j\pi x_0\alpha} + S(\alpha,\beta)e^{-j\pi x_0\alpha}|^2 \\ &= |F(\alpha,\beta)|^2 + |S(\alpha,\beta)|^2 + F^*(\alpha,\beta)S(\alpha,\beta)e^{-j2\pi x_0\alpha} \\ &\quad + F(\alpha,\beta)S^*(\alpha,\beta)e^{j2\pi x_0\alpha}. \end{aligned} \tag{5.69}$$

We note that the product $F(\alpha, \beta)S^*(\alpha, \beta)$ is already formed in this intensity function. We can therefore record the intensity $I(\alpha, \beta)$ as a filter and produce the desired result by simply creating the impulse response of the filter. The results are then similar to those obtained from Equation (5.63). The first two terms of Equation (5.69) produce light distributions centered at the optical axis and are ignored, as is the third (convolution) term centered at x_0. The final term is centered at $-x_0$; by use of the convolution theorem we recognize this term as the correlation of $f(x, y)$ and $s^*(x, y)$. The joint-transform correlator therefore produces the same correlation output as when a matched filter is used, provided that the noise spectral density is uniform.

A positive feature of the joint transform is that it avoids the need to accurately register the matched filter with respect to the Fourier transform of the data (see Chapter 6, Section 6.8). A negative feature is that the joint transform does not incorporate the required noise spectral density when the noise is nonuniform. Its performance is therefore typically worse than that of a matched filter. Furthermore, a new filter must be made for every frame of data processed; in contrast, the matched filter can be left in the system to process a sequence of frames. Finally, parametric searches for the orientation or scale of the signal require a large number of joint-transform filters; these searches are reasonably easy to make using the matched filter, as we discuss in Chapter 6. The joint transform would be more attractive if an erasable, low-noise, nonlinear recording material were available. This material would perform the square-law detection required by Equation (5.69) to record the intensity, and the impulse response of the recorded joint transform could be created by a light source of a different wavelength. The joint transform would then be quickly erased and a new joint transform made from a new combination of signal and data.

5.13. BANDWIDTH CONSIDERATIONS

We find that the spatial carrier frequency x_0 prevents the output terms in Equation (5.63) from overlapping and plays a strong role in determining the bandwidth of the filter. We express the output given by Equation (5.63) in an alternative form as

$$\begin{aligned} g(u,v) &= A_3^2 f(u,v) + A_2^2[f(u,v) * s(u,v) * s^*(-u,-v)] \\ &\quad + A_1 A_2[f(u,v) * s(u+x_0,v)] \\ &\quad + A_1 A_2[f(u,v) * s^*(-u-x_0,-v)], \end{aligned} \tag{5.70}$$

where we have divided the first term of Equation (5.63) into two parts. The length of each diffracted order, as given by the last two terms of Equation (5.70), is $L + l$ because the signal of length l is convolved with data of length L. The length of the first term is L because it is simply an image of $f(x, y)$. The length of the second term is $L + 2l$ because it is the convolution of data with the autocorrelation of the signal whose length is $2l$. We consider the longer of these two terms when determining the minimum value of x_0; therefore, to prevent the three terms from overlapping in the output, we find that

$$\boxed{x_0 \geq L + \frac{3l}{2}.} \tag{5.71}$$

In some signal-processing applications the length of the signal is much less than the length of the data. The value of the carrier frequency is then determined chiefly by the length of the input signal and is at its minimum value. In other applications the length of the signal is equal to the data length; the value of the carrier frequency is then $x_0 \geq (5/2)L$, which is its maximum value. The maximum spatial frequency present in the filter is proportional to $L + 2l$, and the minimum spatial frequency is proportional to $L + l$. Therefore, the information on the filter is contained within a bandwidth proportional to l.

5.14. MULTIPLEXED FILTERS

The similarity of these techniques for constructing filters to modulation techniques used in communication theory suggests other useful concepts. One such concept is that of a multiplexed filter in which filters for several spatial signals are recorded on a single sheet of photographic film. Figure 5.21 is a sketch of the input plane of a filter generator when we record a multiplexed filter. We indicate the (i, j)th signal by $s_{ij}(x, y)$ so that the recorded filter is given by

$$I(\alpha, \beta) = \left| A_0 + \sum_{i=1}^{M} \sum_{j=1}^{M} A_1 S_{ij}(\alpha, \beta) e^{j2\pi(x_i\alpha + y_j\beta)} \right|^2, \tag{5.72}$$

where x_i is the horizontal distance from the reference-point source to the ith column and y_j is the vertical distance from the reference point to the jth row. The spectra of the signals are considered as having been translated onto a vector carrier frequency $\mathbf{r}_{ij}$, as shown in Figure 5.21.

For multiplexed filters, the differences in the values of the subcarriers must be sufficient to process the data, and are given by

$$\begin{aligned} |x_i - x_{i+1}| &\geq L + \tfrac{1}{2}l_{x,i} + \tfrac{1}{2}l_{x,i+1}, \\ |y_j - y_{j+1}| &\geq H + \tfrac{1}{2}l_{y,j} + \tfrac{1}{2}l_{y,j+1}, \end{aligned} \tag{5.73}$$

for all values of i and j, where L and H are the length and height of the

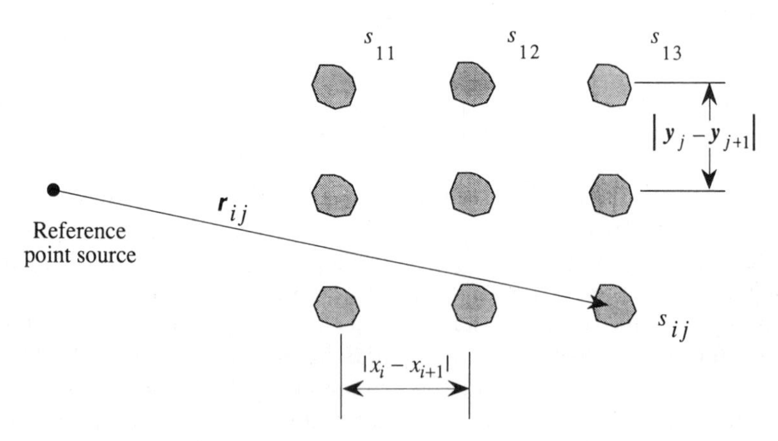

Figure 5.21. Multiplexed spatial filter input plane.

data we process. The physical interpretation of Equation (5.72) is that there is enough area between any four adjacent signals to accommodate the area of the data processed. A similar interpretation is given to the relationship given by Equation (5.71) for a single filter.

Figure 5.22 shows how a multiplexed filter is applied to word detection. In part (a) we see that the input to the filter generator consists of five key words, along with a point source that provides the reference beam, that are simultaneously recorded on a spatial filter. The five different vector subcarrier frequencies are indicated by the $\mathbf{r}_n$'s. This filter is placed in the Fourier plane of the processor shown in Figure 5.1, and one complete line of text is placed in the input plane, as shown in Figure 5.22(b). The output of the multiplexed filter then consists of five filtered versions of the line of text, each containing the correlation peaks corresponding to one of the key words, as shown in Figure 5.22(c). If we increase the distance between the key words in the vertical direction, we can search paragraphs of text, instead of just lines.

5.15. COMPUTER GENERATED FILTERS

Lohmann and Lee have shown how to use a computer to produce a binary filter with properties similar to those of a spatial carrier frequency filter constructed using interferometric techniques (55, 56, 63). Because a computer is used to calculate the filter, the impulse response does not have to exist physically; certain mathematical operations can be performed that are otherwise impossible to implement. The basic technique is to use a digital computer to find the magnitude and phase of a desired filter; this information drives a plotting device that modulates a sequence of pulses (corresponding to the spatial carrier frequency) on each line of a raster-scanned format. The width of the pulse is proportional to the required magnitude of the filter and the position of the pulse is proportional to the required phase. A filter constructed by pulse-width modulation and pulse-position modulation yields the same result in the first diffracted order as a spatial carrier frequency filter. A multiplicity of side orders are produced, of course, in the same way that a square-wave grating produces more side orders than does a sinusoidal grating.

5.16. REFERENCE FUNCTION OPTICAL PROCESSORS

Because correlation is expressed as a convolution of two space functions, as given by Equation (5.13), or as the multiplication of two frequency

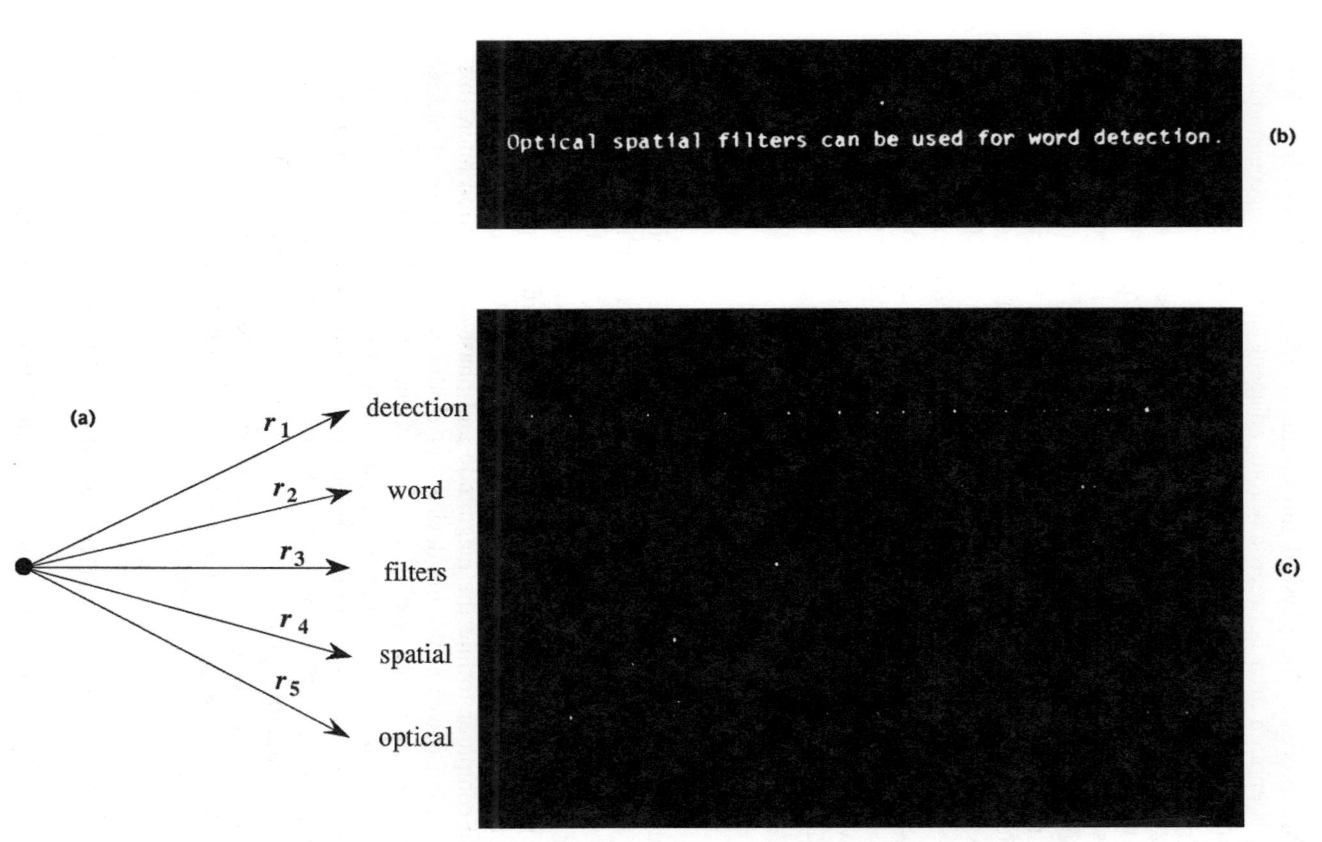

Figure 5.22. The use of a multiplexed spatial filter: (a) filter generating geometry to store five words simultaneously, (b) input line of text, and (c) output of the system showing the detection of the five words (45) (copyright © IEEE, 1974).

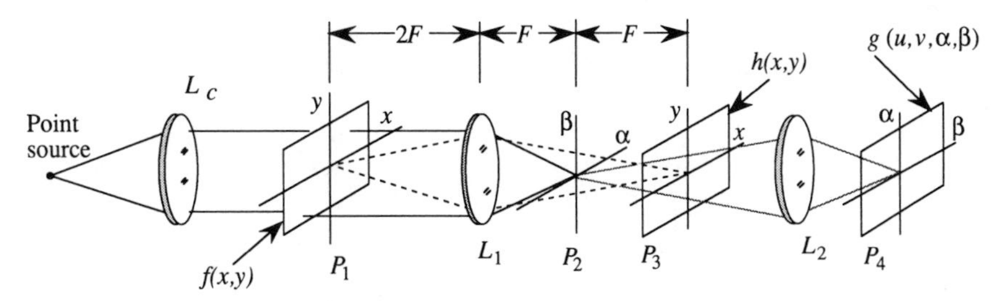

Figure 5.23. A multipurpose signal processor.

functions, as given by Equation (5.15), we can implement it optically by either method. In Figure 5.23 we show an optical system that operates in either of the two modes by proper adjustment of the parameters. As shown, lens L_c collimates a point source of monochromatic light that illuminates data $f(x, y)$ at plane P_1. Lens L_1 images $f(x, y)$ at plane P_3 and produces its Fourier transform $F(\alpha, \beta)$ at the intermediate plane P_2. Lens L_2 produces, at plane P_4, the Fourier transform of the light distribution at plane P_3.

If we choose to implement the correlation operation by realizing the impulse response $h(x, y)$ in a space domain, the system is called a *reference-function correlator* and $h(x, y)$ is called the *reference function*. Suppose that the reference function $h(x, y)$ is displaced a distance (u, v) in the x and y directions at plane P_3. The light distribution at plane P_4 is the Fourier transform of the product of $f(x, y)$ and the displaced reference function:

$$g(u, v, \alpha, \beta) = \iint_{-\infty}^{\infty} f(-x, -y)h(x + u, y + v)e^{j2\pi(\alpha x + \beta y)}\, dx\, dy, \quad (5.74)$$

where $h(x, y)$ is the Fourier transform of $H(\alpha, \beta)$ and we have ignored the rotation of the axes between planes P_1 and P_3. This function, when evaluated at $\alpha = \beta = 0$, becomes

$$g(u, v, 0, 0) = \iint_{-\infty}^{\infty} f(r, s)h(u - r, v - s)\, dr\, ds, \quad (5.75)$$

which is recognized as the convolution operation. Correlation is then implemented by setting the reference function $h(x, y)$ equal to the Fourier transform of either Equation (5.33) or Equation (5.34).

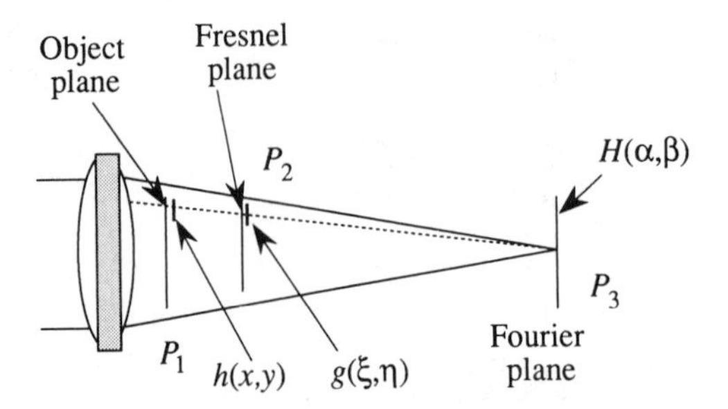

Figure 5.24. Reference-function scanning.

Alternatively, we can implement the correlation operation by placing a filter whose transmittance is proportional to either Equation (5.33) or Equation (5.34) at the Fourier plane P_2. Thus, the correlation operation can be implemented by realizing the impulse response $h(x, y)$ of the filter as a reference function in the space domain or by realizing the function $H(\alpha, \beta)$ as a filter in the frequency domain.

The display of the information is significantly different, however, in the two processors. As the displacement variables in the frequency-plane processor are coordinates of the output plane, *no scanning is needed* and the correlation peaks occur at the geometric image positions of the centers of the signals. Because the displacement variables in the reference function processor are generated as a function of time, scanning in both the x and y directions is needed because the correlation information occurs only at the point $\alpha = \beta = 0$.

To illustrate the need for scanning the reference function, consider the simplified optical system shown in Figure 5.24. Data is placed in a convergent illumination beam at plane P_1 so that the Fourier transform occurs at plane P_3. Suppose that the signal occurs with its center located on the dotted line at plane P_1. A correlation peak for a signal located at this position will not occur unless the reference function $h(x, y)$ is also centered at the same position. Hence, to detect the signal at all possible positions, the reference function must be scanned over the entire object.

Every plane between P_1 and P_3 is a Fresnel transform plane, and filtering operations are sometimes performed in these planes. The reference function now has the form

$$g(\xi, \eta) = \iint_{-\infty}^{\infty} h(x, y) e^{-j(\pi/\lambda D)[(\xi - x)^2 + (\eta - y)^2]} \, dx \, dy, \qquad (5.76)$$

where $g(\xi, \eta)$ is the Fresnel transform of $h(x, y)$ and D is the distance between planes P_1 and P_2. As before, the reference function must be scanned to obtain the required correlation function, but the scanning distance at plane P_2 is less than that required at plane P_1. The required scanning range continues to decrease as D increases until we reach plane P_3. Finally, no scanning is required when the mask, which now has become a spatial filter, is placed at plane P_3.

In some multichannel one-dimensional applications, such as synthetic-aperture radar signal processing, data is moving continuously through the system so that scanning is a natural phenomena; a reference-function processor then may be more useful than a Fourier-plane processor. As it is inconvenient to scan the functions in two dimensions, two-dimensional reference-function processors are seldom used, particularly as the required spatial filters are physically realizable using interferometric techniques.

PROBLEMS

5.1. An object consists of the pattern shown here:

$$\begin{array}{ccccc} +1 & & & & \\ & +1 & & & \\ & -1 & & & \\ & & +1 & & \\ & & & -1 & \end{array}$$

The samples all have the positive or negative unit *amplitudes* shown on the right. Sketch (a) the two-dimensional convolution of the object with itself and (b) the two-dimensional autocorrelation of the object. Annotate your answer to show the amplitude of each dot in the output for both cases. What is the ratio of the peak *intensities* for correlation versus convolution? To be more precise in your sketching and calculations, note that the signal consists of a set of samples whose centers are located at the points $(-1, +2)$, $(0, +1)$, $(0, 0)$, $(+1, -1)$, and $(+3, -2)$. Your answer should be a 9×9 matrix of values for both the correlation and convolution functions.

5.2. Parseval's theorem for Fourier transforms states that

$$\int_{-\infty}^{\infty} |f(x)|^2\, dx = \int_{-\infty}^{\infty} |F(\xi)|^2\, d\xi,$$

where $F(\xi)$ is the Fourier transform of $f(x)$. The result implies that the total signal energy is the same in both the space and the frequency domains. Prove that a similar result holds for the Fresnel transform:

$$\int_{-\infty}^{\infty} |f(x)|^2 \, dx = \int_{-\infty}^{\infty} |g(\xi)|^2 \, d\xi,$$

where $g(\xi)$ is the Fresnel transform of $f(x)$.

5.3. Show that the improvement in the signal-to-noise ratio of a sinusoidal signal embedded in white noise is approximately equal to $L\alpha_{co}/3$, where α_{co} is the noise bandwidth and L is the length of the input signal. Provide a sketch to support your answer.

5.4. Suppose that an object has a triangular spectrum equal to

$$S(\alpha) = S_0 \left[1 - \frac{|\alpha|}{100} \right].$$

Suppose that the system transfer function for an imperfect collection system is

$$H(\alpha) = \text{rect}(\alpha/100) + 0.1 \, \text{rect}(\alpha/200).$$

(a) Calculate the proper equalization filter for this system, and ensure that the filter is passive.

(b) Calculate the signal-to-noise ratio for the given system transfer function.

(c) Calculate the signal-to-noise ratio after the system transfer function has been equalized.

Assume that the noise spectral density is uniform at N_0 for all spatial frequencies.

5.5. Consider a signal whose spectral density is

$$R_s(\alpha) = \text{rect}[(\alpha - 20)/5) + \text{rect}[(\alpha + 20)/5) \text{ watts/Ab}.$$

Suppose that the noise spectral density is $R_n(\alpha) = 0.3$ watts/Ab. Sketch the spectral density of the signal plus noise. Suppose that you could implement only a binary filter. Calculate the proper representation of $H(\alpha)$ and sketch its response. Calculate the improvement in the signal-to-noise ratio if the system bandwidth is 50 Ab.

5.6. For the conditions of Problem 5.5, calculate and sketch the Wiener filter for recovering the signal with minimum mean-square error. Calculate the improvement in the signal-to-noise ratio if the system bandwidth is 50 Ab.

5.7. Suppose that a signal $s(x)$ has a Fourier transform given by $S(\alpha) = A_1 \exp[-c\alpha^2 - j2\pi\alpha^2 x_0^2]$, where c and x_0 are positive constants. Calculate and label the magnitude and the phase of the matched filter $H(\alpha)$. Discuss, by means of sketches, why the phase part of the filter helps significantly to produce a signal peak (i.e., illustrate, in a qualitative way, the shape of the output with and without the phase part of the filter) and discuss how the magnitude part improves the signal-to-noise ratio.

5.8. Suggest a good way to implement the phase part of the required matched filter from Problem 5.7. Describe, as much as you can, the parameters of this phase filter if $x_0 = 1$ mm, the focal length of the Fourier-transform lens is 100 mm, and the maximum spatial frequency is $\alpha_{co} = 25$ Ab.

5.9. A spatial carrier frequency filter is constructed using a modified Rayleigh interferometer. The focal length of the Fourier transform lens is 300 mm. The object has a cutoff frequency of $\alpha_{co} = 100$ Ab, a height of 10 mm, and a width of 6 mm. The point reference source is placed in the same plane as the object but it is displaced 50 mm away from the centroid of the object in the x direction. Calculate the spatial carrier frequency. Calculate the maximum and minimum spatial frequencies contained in the filter.

5.10. If a second clear aperture signal whose diameter is 7 mm were placed a distance of 100 mm from the point source, calculate and plot the range of spatial frequencies contained in the filter.

5.11. We construct a spatial filter of the function

$$
\begin{aligned}
S(\alpha) &= 1 - \alpha, \qquad 0 \le \alpha \le 4, \\
&= 0, \qquad \alpha < 0.
\end{aligned}
$$

Let the reference-beam amplitude at the Fourier plane be A_1 and let $A_2 = 1$. Sketch the intensity function

$$I(\alpha) = \left| A_1 e^{j2\pi b\alpha} + A_2 S(\alpha) \right|^2$$

for each of the following situations: (a) $A_1 = 1$ and (b) $A_1 = 0.1$. You may choose any convenient value of the carrier frequency b. Hint: First expand $I(\alpha)$ in terms of α, then calculate the envelope of $I(\alpha)$, and then fill in between the envelope with the chosen carrier frequency.

5.12. At what spatial frequency (or frequencies) is the percentage modulation the greatest in each of the cases given in Problem 5.11?

6

Spatial Filtering Systems

6.1. INTRODUCTION

In this chapter we discuss some practical aspects of optical spatial filtering systems. We begin with an overview of a general optical processing system and then discuss the basic modules of the optical processor in more detail. We treat the important problem of setting a threshold at the output of the optical processor to detect the signal. We discuss how to handle the usual case of nonuniform noise spectral densities and show how to process signals adaptively. Finally, we illustrate these techniques with various applications and discuss the effects induced by errors in the position of the matched filters in the Fourier plane.

6.2. OPTICAL SIGNAL PROCESSOR AND FILTER GENERATOR

Consider the basic elements needed in a combined optical signal processor and filter generator, as sketched in Figure 6.1. Light from the laser passes through a shutter and is divided into a reference branch and a processing branch by a beamsplitter. The reference and processing branches are derived from the same laser to ensure that a high degree of spatial and temporal coherence is produced at the filter plane. The transmitted light passes through a second shutter, reflects from a mirror, and expands to fill the Fourier-transform lens, which converges light toward the spatial filtering plane. A spatial light modulator, placed in the convergent beam, accepts data from an electronic source or from an incoherent optical source.

The Fourier transform of the information contained on the spatial light modulator occurs at the filter module plane. The filters might be recorded *in situ* on thin reusable materials such as thermoplastics, photoplastics, or elastomers. The filters might also be recorded on thick materials, such as one of the polymethylmethacrylates, or on nonlinear materials, such as strotium barium titanate. In any event, the filter module provides the bias voltages, charging stations, and erasing corona discharges required to

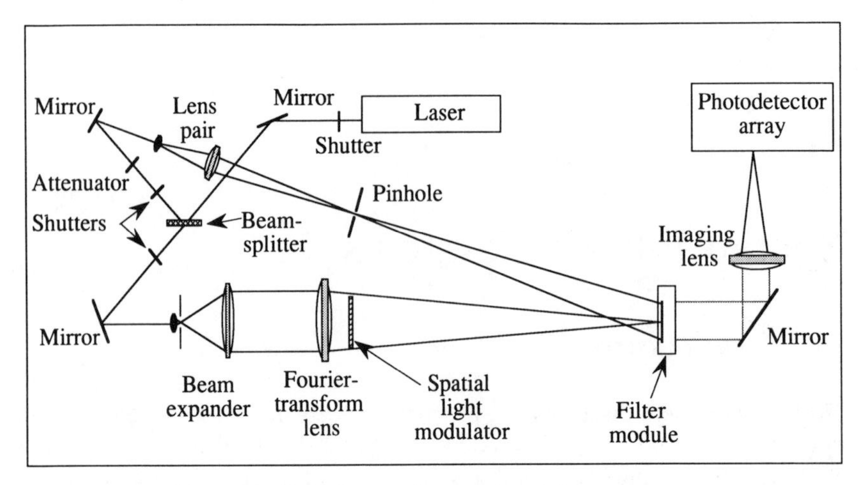

Figure 6.1. Optical processing system and filter generator.

support the recording material used. In some applications, the filters may be recorded on photographic media, either on separate glass plates or on sprocketed roll film. In the latter case, the filter module is a pin registration unit to accurately position the film both for recording filters and for data processing. The filter module may also include a rotating prism to keep the correlation output centered on the photodector array while an orientation search is performed.

After passing through the filter, the input data is imaged by a lens onto a photodetector array. A mirror, convenient for folding the system, deflects the light so that the path of the correlation output coincides with the optical axis of the lens. The photodetector-array module includes the necessary electronics for setting the required threshold level and for driving subsequent digital-processing systems.

The reference beam passes through a shutter and an attenuator, reflects from a mirror, and focuses, by means of a pair of lenses, through a pinhole. The pinhole, located at the signal plane so that the filters are recorded without focal power, removes the effects of minor magnitude imperfections in the light beam. The optical path lengths, as measured from the beamsplitter, are equal for the reference and processing branches so that the system has a high degree of temporal coherence.

To construct filters, we place the signal at the spatial light modulator plane and expose the recording material in the filter module with the intensity of the sum of the Fourier transforms of the signal and the reference beams:

$$E(\xi, \eta) = t_0 I(\xi, \eta) = t_0 \left| A_1 e^{j(2\pi/\lambda F)x_0 \xi} + A_2 S(\xi, \eta) \right|^2, \quad (6.1)$$

where $I(\xi, \eta)$ is the intensity of the light, t_0 is the exposure time, and x_0 is the distance from the pinhole to the center of the signal. The value of A_1 relative to A_2 is controlled by the attenuator placed in the reference branch, and the value of t_0 is set by the main shutter just after the laser. After the filter is recorded, the photodetector array is used to compare the filter impulse response with the original signal by toggling the shutters in the reference and processing branches. After satisfactory filters have been recorded, the reference branch shutter is closed and data to be processed replaces the signal in the spatial light modulator.

For a laboratory setup, the optical elements are selected according to the basic suggestions given in Chapters 2 and 5. A full optical, mechanical, and electrical design procedure is required, of course, before constructing an operational system. The information presented in the following sections is intended primarily for those who are beginning experimental work.

6.2.1. The Light Source

Either a gas laser or an injection laser diode is a good primary source of light because of its high degree of spatial and temporal coherence and high power levels. As the laser beam diameter is too small for direct use in processing large data formats, the beam must be expanded to illuminate data. Although a negative lens could be used to diverge the light, it generally induces large aberrations. A better method is to use two lenses in a telescopic configuration to create a real, instead of a virtual, point source of monochromatic light, as shown in Figure 6.2

Light from a gas laser has a Gaussian intensity profile, defined by the diameter at which the intensity has dropped to $1/e^2$ of the central intensity. As the laser beam typically has a diameter of a few millimeters or so, a reasonable first lens for the beam expander is a microscope objective. Microscope objectives are designed to operate at a 10:1 or larger conjugate ratio, which is near enough to the required infinite conjugate. A pinhole, placed at the focal plane of the microscope objective, removes light diffracted by dust particles in the laser or on the microscope objective, as well as light due to spontaneous emission from the laser. Although there is some latitude in selecting its size, the pinhole diameter is typically 3–5 times the diameter of the first dark ring of the

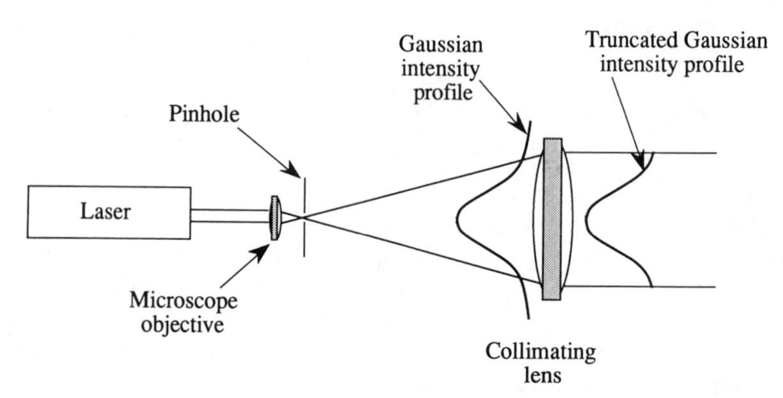

Figure 6.2. Beam-expanding module.

Airy disc pattern that would be formed at the focus if the microscope objective were uniformly illuminated. The pinhole diameter must be small enough to block the scattered light, but not so small that it seriously affects the magnitude weighting of the beam.

The beam emitted by the laser has a Gaussian intensity profile of the form $I(r) = I_0 \exp(-r^2/2\sigma^2)$, where I_0 is the central intensity, $r^2 = x^2 + y^2$ is the radial distance, and σ is the standard deviation of the Gaussian function. The amount of optical power collected within a collimating lens of radius R is

$$\begin{aligned} P_R &= I_0 \int_0^R \int_0^{2\pi} e^{-r^2/2\sigma^2} r\,dr\,d\theta \\ &= 2\pi\sigma^2 I_0 \left[1 - e^{-R^2/2\sigma^2}\right]. \end{aligned} \tag{6.2}$$

As $R \to \infty$, we capture all the available light, which is equal to $2\pi\sigma^2 I_0$. The fraction of light available to illuminate data by a beam of radius R (see Problem 6.1) is therefore $[1 - \exp(-R^2/2\sigma^2)]$. Thus, for a permissible variation in light intensity over the useful beam, we can quickly determine how much light is used. This uniformity constraint is severe if we try to produce a highly uniform light beam. For example, if the variation in the light intensity must be kept to 10%, we retain only 10% of the total available light; thus the percentage of energy lost equals the truncation intensity. Fortunately, correlation is not strongly affected by magnitude variations over small and isolated signals. The key problem caused by nonuniform illumination is the need to set the threshold properly at the output of the system, as we discuss in Section 6.3.

The collimating lens converts the divergent wavefront into a plane wave; its aperture must be at least as large as that of the data. As the collimator works over only a small field, it could be a single-element *aspherical* lens, designed to work slightly off axis to ease the centering tolerance. The relative aperture of this lens is not important with respect to its light-gathering properties; an $f/3$ collimator collects no more coherent light than does an $f/5$ collimator, provided that we fix the percentage variation in light over the aperture. We simply equate the numerical aperture of the microscope objective forming the point source to the numerical aperture of the collimator. To keep the length of the optical system to a minimum, we usually use the highest numerical aperture possible for the collimating lens, consistent with good aberration control.

6.2.2. The Spatial Light Modulator

A spatial light modulator, such as photographic film, may produce two kinds of phase errors: a thickness variation in the film substrate that is independent of the recorded function on the film and a thickness variation of the emulsion that is data dependent (64–67). Phase errors must be removed if the optical system is to operate on only the amplitude of the photographic film.

As the refractive index of the emulsion generally differs from that of the substrate, the question arises as to which error is more severe in a given application. The thickness of the substrate is usually a slowly varying spatial function with energy only at low spatial frequencies; in most processing applications these frequencies are suppressed because they do not help to recognize the signal. The variations of major concern, then, are those in the emulsion that are directly related to the recorded signal; these signal-dependent variations are a source of nonlinearities in the processing system. For small phase variations, the effective recorded transmittance is

$$f_1(x, y) = f(x, y)e^{jcf(x, y)}, \tag{6.3}$$

where c is a constant. This relationship shows that phase variations are proportional to the amplitude variations in the data. We expand $f_1(x, y)$ into a power series to find that

$$f_1(x, y) = f(x, y) + jcf^2(x, y) + \text{higher-order terms}, \tag{6.4}$$

and we want to remove all but the leading term in the series to avoid introducing nonlinearities.

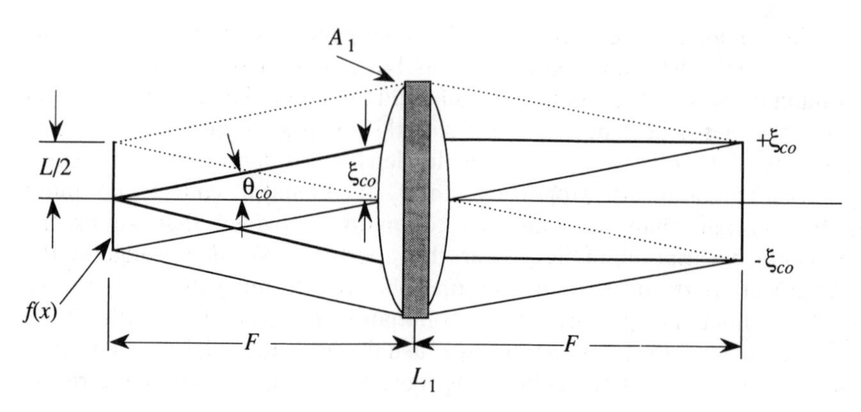

Figure 6.3. Aperture requirement on Fourier-transform lens.

Phase errors are removed by putting the film in a cell, formed by two optical windows, filled with a liquid whose refractive index matches that of the emulsion. For data recorded on a continuous film, a device similar to a liquid gate used in the motion-picture industry allows the phase errors to be removed as the film passes through a processing aperture.

Other spatial light modulators, such as liquid-crystal displays, may also have sources of phase errors. The techniques for controlling phase errors in these spatial light modulators are similar to those used with photographic film. In this case, optically flat windows are attached to each face of the spatial light modulator with an index-matching adhesive.

6.2.3. The Fourier-Transform Lens

The Fourier-transforming lens collects light diffracted by the object. The maximum object length is L and the highest spatial frequency in the object is α_{co}. Light diffracted by the highest spatial frequency is directed at a ray angle θ_{co} with respect to the optical axis, as shown in Figure 6.3, where

$$\theta_{co} = \lambda \alpha_{co}. \tag{6.5}$$

Diffracted light from the extreme upper sample of the data enters lens L_1 a distance $\xi_{co} + L/2$ from the optical axis, where

$$\xi_{co} = \lambda \alpha_{co} F, \tag{6.6}$$

and where F is the focal length of lens L_1. Therefore, the lens radius must be

$$A_1 = L/2 + \lambda \alpha_{co} F \tag{6.7}$$

to collect all the light.

In Chapter 3 we discussed alternative optical configurations that optimize the information capacity of a processing system. We showed that the lens apertures are minimized by using two lenses. The first lens, whose aperture is equal to L, illuminates the data with a convergent waveform. The second lens, whose aperture is equal to $2\xi_{co}$, produces the inverse transform of the filtered information. Lens design is therefore simplified because each lens has the minimum possible aperture.

The focal length of lens L_1 is chosen by balancing mechanical and optical considerations. We choose the scale of the Fourier transform so that it is expanded enough to relax the tolerance on the position of the filter (see Section 6.6). Too large a transform, however, causes difficulties in rapidly changing the more massive filters and increases the total optical path significantly. For a desired radius A_f of the filter, Equation (6.6) shows that

$$A_f = \lambda \alpha_{co} F. \tag{6.8}$$

Thus, after the radius of the filter is chosen, the focal length F of lens L_1 is determined by Equation (6.8), and the aperture is determined by Equation (6.7).

6.2.4. The Filter Plane

When the spatial filters are recorded on film, a second liquid cell is usually required at the frequency plane. The purpose of this cell is to remove the phase aberrations in the *substrate* of the film. Phase errors associated with a spatial carrier frequency function do not affect the form of the desired output; they simply introduce extraneous diffraction orders which are not, in general, bothersome.

The effect of variations in the substrate of the film is avoided by recording the filters on an emulsion that has been coated onto optically flat glass so that a liquid cell is not needed. This method sometimes has the added advantage of supplying a reference edge for positioning the filters. When the spatial filter is stored on other spatial light modulators, such as liquid crystal displays, the techniques for controlling phase errors in these spatial light modulators are similar to those discussed in Section 6.2.3.

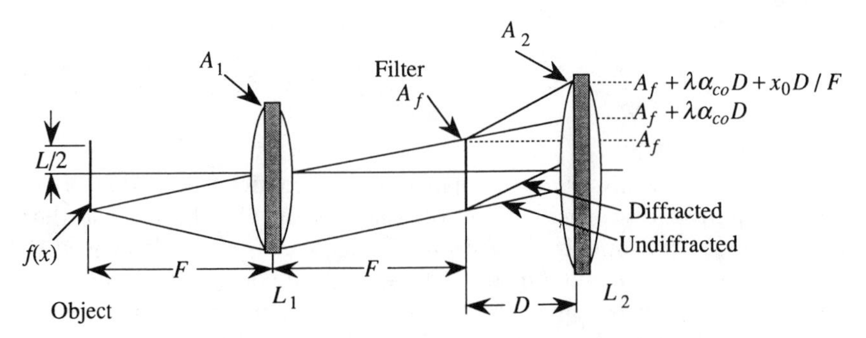

Figure 6.4. Aperture requirement on imaging lens.

6.2.5. The Imaging Lens

The imaging lens is a second Fourier-transform lens whose aperture must be at least as large as that of the filter. Suppose that the lens is placed a distance D from the frequency plane, as shown in Figure 6.4. The aperture of the lens must capture all rays from the edge samples of $f(x)$. Its aperture, in the absence of the spatial filter, is

$$A_2 = (D + F)\lambda\alpha_{\text{co}} = A_f + \lambda\alpha_{\text{co}}D. \tag{6.9}$$

When we use a spatial carrier frequency filter, the lens aperture must increase to capture all the rays caused by diffraction, as shown in Figure 6.4. The increased aperture size becomes

$$A_2 = A_f + \lambda\alpha_{\text{co}}D + \frac{x_0 D}{F}, \tag{6.10}$$

where x_0/F is the physical angle corresponding to the carrier frequency of the filter.

The total field over which lens L_2 operates, when all outputs from the filter are transformed, is $\theta_1 \approx \tan^{-1}[3L/F]$. A useful technique is to shift lens L_2 laterally by a distance x_0D/F, where x_0 is the carrier frequency, and to tilt the lens so that its axis is collinear with that of the diffracted order. The field angle is then reduced to $\theta_1 = \tan^{-1}(L/F)$ so that the optical invariant for lens L_2 is reduced by a factor of 3, leading to a reduction in aberrations. From Equation (6.10) we see that keeping the distance D small reduces the lens aperture and the aberrations. The focal length of lens L_2 is generally chosen to give a suitable magnification between the input and output planes.

6.3. THE READOUT MODULE

A photodetector module is needed to extract the desired information from the system. Recall that the matched filtering operation maximizes the ratio of peak signal intensity to the average noise intensity. As the matched filter produces a plane wave with some magnitude weighting, the correlation peak occupies a region on the order of a few sample diameters. The basic characteristics of the photodetector, such as sensitivity, number of elements, blooming, crosstalk, and dynamic range, are similar to those used in spectrum analysis applications, as discussed in Chapter 4, Section 4.2.

6.3.1. The Thresholding Operation

Figure 6.5(a) shows one scan line produced at the output of a photodetector array. The threshold level is determined from the probability distribution function of the output for signal plus noise and for noise alone, as shown in Figure 6.5(b). This probabilistic model assumes that the signal and noise are additive, a condition not usually met in pattern-recognition applications. However, the signal is usually *isolated* in a noise background as shown in Figure 6.6(a); examples of isolated signals are characters, words, or small geological phenomena in aerial photographs. As an isolated signal occupies only a small part of the total area of the data, the signal and noise are additive "almost everywhere." In these situations, the probabilistic model is helpful in setting the threshold level, given that the statistics shown in Figure 6.5(b) are experimentally measured. The threshold level, shown as a dotted line in Figure 6.5(b), does not need to be at the point where the two probability functions cross. Considerations

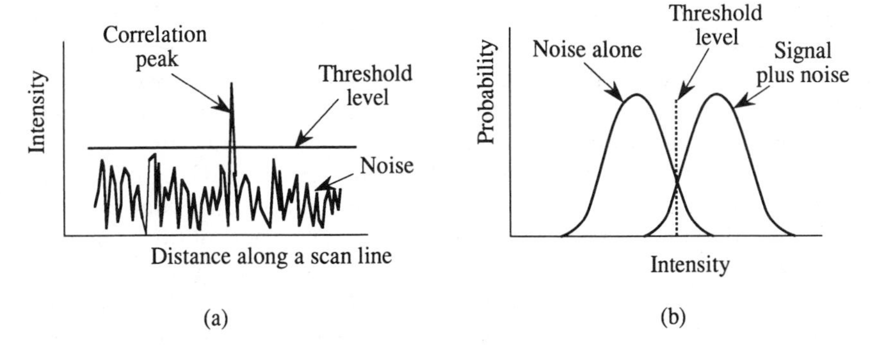

Figure 6.5. Thresholding operation: (a) space signal and (b) probability distributions.

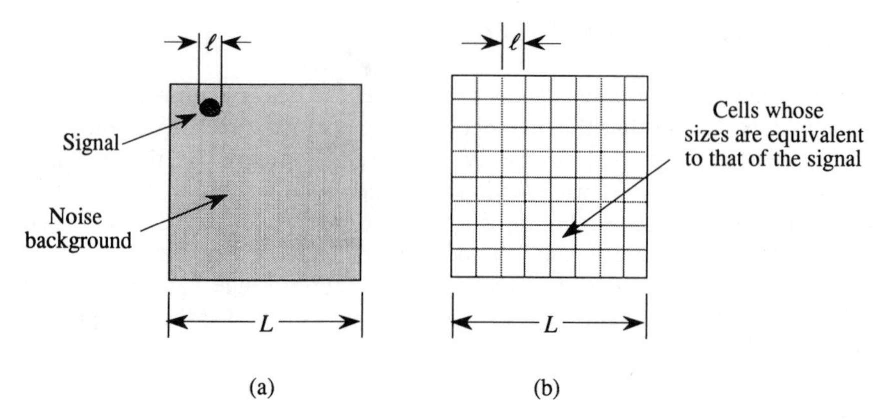

Figure 6.6. Data reduction: (a) nonoverlapping signal and noise and (b) equivalent cell size.

such as the cost of a missed signal, the cost of a false alarm, and the cost of a false signal determine the optimum threshold level.

6.3.2. The Importance of Nonoverlapping Signals

The readout device sets a threshold level. If the intensity exceeds threshold, we decide that a signal is present at that position in the data; if it does not, we decide that only noise is present. Information obtained after thresholding is either displayed for visual observation or, more likely, processed further digitally. The exact nature of the display or the subsequent processing is not of great significance here. The key issue is to determine the space bandwidth product required of the photodetector array. Because linear operations preserve the space bandwidth product throughout the system, the space bandwidth product of the output of the system is exactly equal to that of the input. Unfortunately, the space bandwidth product of an electric readout system, such as a CCD photodetector array, is generally much less than that of the optical system. Hence, it appears that the high data-handling capability of optical processing meets a space bandwidth constriction at the output of the system.

We can overcome the space bandwidth constriction of the photodetector array in those cases where we detect nonoverlapping signals. Either the signal or the noise is present at any position in the data, not the sum of the two; the signal and noise are therefore *mutually exclusive* with respect to the areas that they occupy. Figure 6.6(a) illustrates a condition in which the signal and noise do not occupy the same region. Suppose that the space bandwidth product of the input data is $2.5(10)^7$, based on a square-

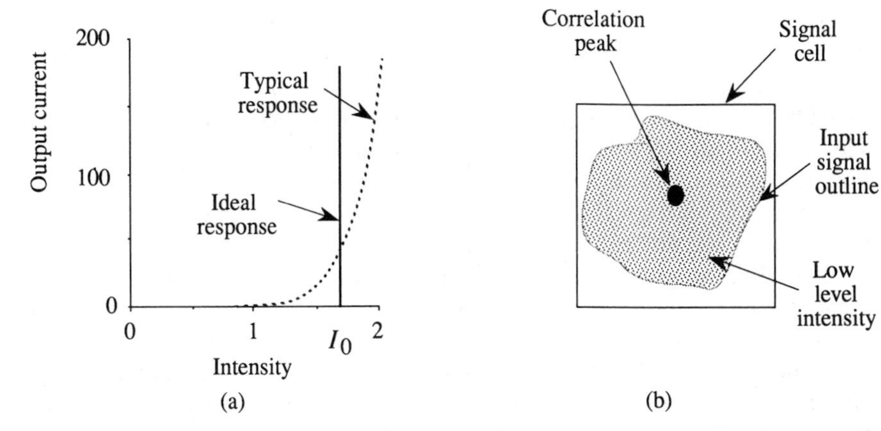

Figure 6.7. Desired detector characteristics: (a) nonlinear response of the photodetector and (b) correlation characteristics.

data format for which $L = H = 100$ mm and $\alpha_{co} = \beta_{co} = 50$ Ab. The isolated signal might have a space bandwidth product equal to 10^4, based on an area of 4 mm^2 and the same spatial bandwidth.

As the signals do not overlap, there is no danger that more than one correlation peak can occur within a cell shown in Figure 6.6(b) whose size and shape is similar to that of the signal. After thresholding the output, the region of uncertainty as to the location of the signal is increased to the cell size because it is unimportant where the correlation peak occurs within the cell. The space bandwidth product of the final display is therefore the quotient of the space bandwidth product of the data and the space bandwidth product of the signal. In our example, the required space bandwidth product of the display unit is reduced by four orders of magnitude from $2.5(10)^7$ to $2.5(10)^3$; such a space bandwidth product is available from fairly elementary CCD photodetector arrays.

This reduction in output information rate can be achieved *only* if the photodetector device has a sharp nonlinearity. Furthermore, the detector response must be dependent on the local *intensity* of the light at the output plane, not on the *integrated optical power*, which is the most common condition. Figure 6.7(a) shows the desired nonlinearity for an ideal detector element. A typical, nonideal form of the nonlinearity for a photodetector device is shown by the dotted line.

In comparing the performance of an ideal and a typical device, we see that the sharpness of the nonlinearity is important because the signal cell may contain from 10^3 to 10^4 samples and we must distinguish between situations when the magnitude of *only one sample* exceeds threshold and when the magnitude of all samples are *just below threshold*. Suppose that the size of each element of the detector matches that of a signal cell and we want to know if any one of the resolution elements within that cell exceeds the threshold. Suppose that we set the threshold level at I_0 as shown in Figure 6.7(a). If the light intensity of each of the 10^3–10^4 samples contained within the cell is just below threshold they may sum to give an output current that is greater than that from a single element just above threshold unless the device is sufficiently nonlinear. For the ideal nonlinearity, the output is zero except when the correlation peak occupying one sample space exceeds the threshold I_0.

No suitable threshold devices have been developed with the required predetection space bandwidth product, a sufficiently sharp nonlinearity, and a postdetection space bandwidth product of the order of 10^3. Those that are nonlinear tend to have low space bandwidth products, and those that have adequate space bandwidth products are not sufficiently nonlinear. The development of such a device is a pressing need for those optical signal-processing applications where the output data rate must be reduced.

6.3.3. On-Chip Processing

An alternative to using sharply nonlinear devices to reduce the output data rates is to use on-chip processing that includes nonlinear or decision operations. There are three possible methods for achieving the desired data rate reduction: (1) on-chip processing for arrays that output electronic signals, (2) methods for segmenting arrays, along with the possibility of on-chip processing, and (3) array types in which the output of the detector remains in an optical format for further processing.

We want to implement signal processing within the detector chips that significantly reduces the output data rates to the postprocessor. Some desired functions are

- Video amplitude compression, perhaps programmable
- Temporal change detection from frame to frame
- Spatial change detection along the elements of the array
- Dynamically programmable spatial convolutions for tasks such as centroiding, using cellular blocks of up to 7×7 elements
- Random access to any subregion of the array to isolate and dynamically track changes
- Threshold levels that are either globally or locally set and adjusted adaptively to achieve constant false-alarm rates

- Methods for synchronous detection to remove the strong bias terms or background signals that arise in some processing operations
- Analog first-in first-out (FIFO) or last-in first-out (LIFO) memory

One way to reduce the data rate is to segment the array of N elements into K subarrays of M elements each. Each subarray is then given the capability to process the output at a local level to reduce the amount of data it reports to the optical processor output. Processing operations such as those listed above are done in each subarray to simplify the output data. Processing at the subarray level can have a major impact on the processing rate needed at the postdetection stage of the system.

6.3.4. Constant False-Alarm Rate

Setting the threshold at the proper level is important for obtaining a constant false-alarm rate. In some applications the threshold is uniform over the output as suggested by the sketch in Figure 6.5. In other applications, however, we must vary the threshold to compensate for variations in ground illumination or other uncontrollable factors. A nonuniform threshold is also a useful alternative for compensating variations in the laser illumination. Figure 6.8 shows a noise background whose average value changes significantly as a function of x. If we set a uniform threshold to detect the correlation where the noise background is high, the correlation peak in the region where the noise is low is missed. If we lower the threshold to detect the correlation peak in the low-noise region, we generate a large number of false alarms.

The ideal threshold level is one that adapts to the local average noise level. We want to control the threshold over cells corresponding roughly to the size of signal. These cell areas are *local* with respect to the entire

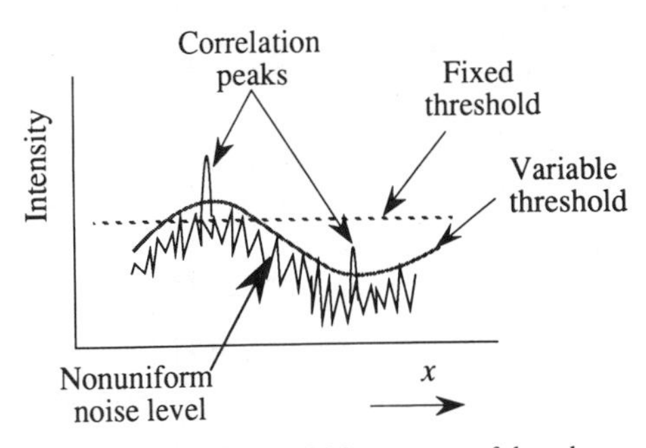

Figure 6.8. Variable threshold for constant false-alarm rate.

output but *global* with respect to an individual sample that represents the size of the correlation peak. The local threshold is therefore applied to 100 or so samples in each direction, depending on the signal characteristics.

In conventional radar systems, the threshold level is determined by averaging the received signal for an appropriate number of samples just before the current time. As we generally process data on a frame basis, we can implement the ideal adaptive threshold in which we average over samples both *before* and *after* the occurrence of the correlation peak. Correlation peaks located in a sharp transition from low to high noise levels are therefore not missed.

Recall from Equation (5.25) that the output noise spectral density is related to the input-noise spectral density by

$$R_g(\alpha, \beta) = |H(\alpha, \beta)|^2 R_f(\alpha, \beta). \tag{6.11}$$

As the filter for correlation is the same as that for convolution, except for a conjugate phase, we see that the value of the output spectral noise density is the same in the convolution output as in the correlation output. A method for providing this information to the readout device is shown in Figure 6.9. A low-resolution photodetector array averages the noise in the convolution output and subtracts the local average from the high-resolution correlation data so that the threshold can now be set at a fixed level over the entire output to achieve a *constant false-alarm rate*. Another method for obtaining a constant false-alarm rate is given in Section 6.6.2, in connection with transposed processing.

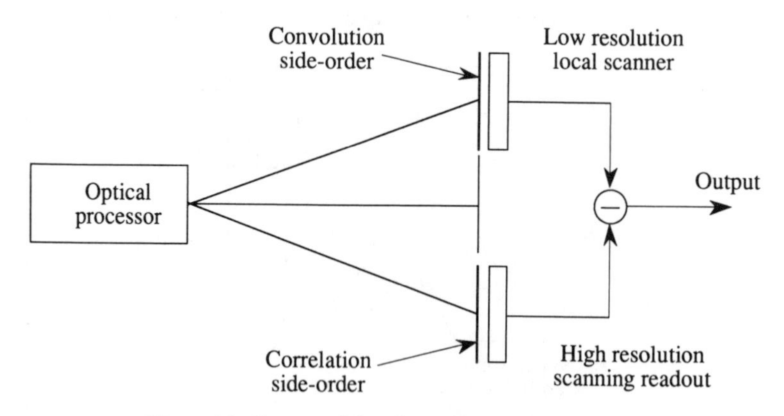

Figure 6.9. Constant false-alarm rate sensing scheme.

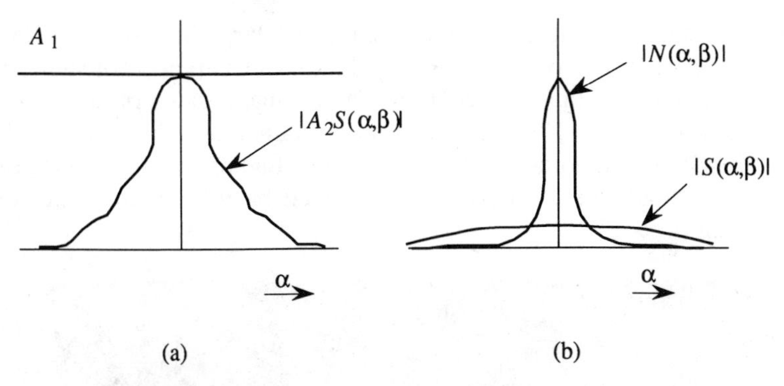

Figure 6.10. Reference-beam levels: (a) high and (b) low.

6.4. THE REFERENCE-TO-SIGNAL-BEAM RATIO

A fundamental issue in matched-filter construction is the level at which we set the reference-beam magnitude A_1 relative to the maximum value of the Fourier transform of the signal $A_2S(\alpha, \beta)$, as given in Equation (5.53). The theory of matched filtering requires that all frequencies in $S(\alpha, \beta)$ must be captured faithfully, if the noise is white. This means that A_1 must be set at a level equal to the peak value of $A_2S(\alpha, \beta)$, as shown in Figure 6.10(a).

The noise spectral density is not, however, generally white when processing imagery; as shown in Figure 6.10(b), the spectrum $N(\alpha, \beta)$ of the noise $n(x, y)$ associated with a frame of information is concentrated in the vicinity of $\alpha = \beta = 0$. Because the signal $s(x, y)$ generally occupies only a fraction of the total area of the frame $f(x, y)$, the total energy of $s(x, y)$ is often at least 40 dB down from that of $n(x, y)$. To prevent the filter from

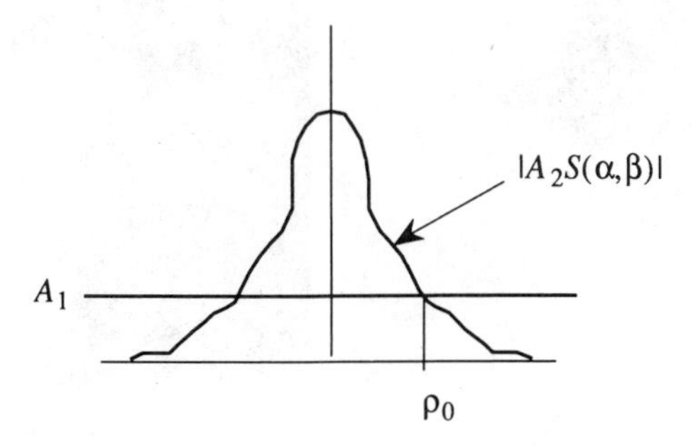

Figure 6.11. Reference-beam level for rejecting noise.

having a high transmittance where the noise is large, we adjust the level of the reference beam so that $A_1 \ll A_2 S(0,0)$. In Chapter 5 we showed that the maximum modulation occurs when the magnitudes of the reference beam and the signal Fourier transform are equal, as shown in Figure 6.11 at the spatial frequency ρ_0. All frequencies higher than ρ_0 are faithfully recorded; those less than ρ_0 are suppressed by the nonlinearities of the recording process.

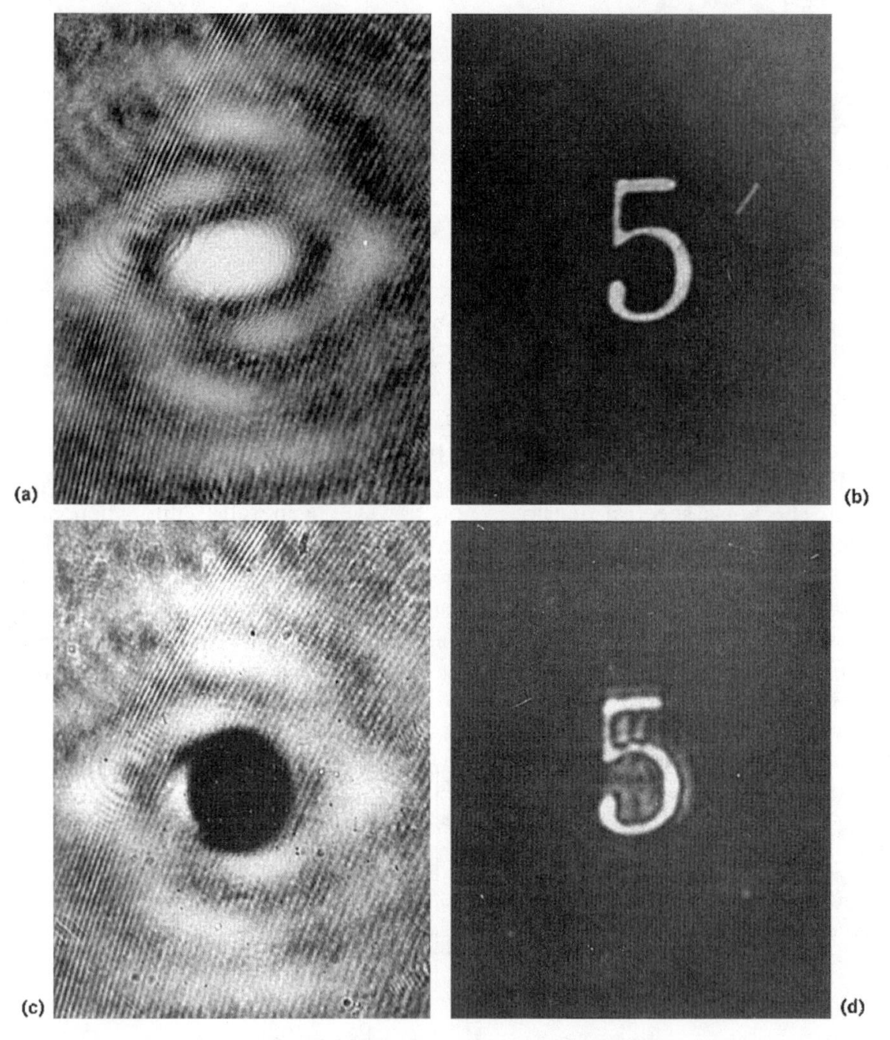

Figure 6.12. Examples of filtering: (a) broadband filter, (b) impulse response from broadband filter, (c) highpass filter, and (d) impulse response from highpass filter (68).

A secondary reason to set A_1 at a fairly low value relative to $A_2S(0,0)$ is that we often need to distinguish among closely related patterns in a set. For example, the difference between Q and O is the tail of Q. We increase the sensitivity of the filter to these subtle features by setting A_1 at a relatively low value so that the high spatial frequencies due to the diffraction from the edges are emphasized. In effect, setting the reference level at a low value is equivalent to not recording the low-frequency content of the signal (those frequencies of the order of the reciprocal of the character size). Some caution must be exercised, however, because a filter with high sensitivity may also produce a higher miss rate if the signal has variations in its shape. Hence the response in the low and high spatial frequency regions must be based on experimental results.

Figure 6.12 shows the impulse response for a broadband and a highpass spatial filter for the character 5. Figure 6.12(a) shows the *broadband* spatial filter when the reference beam is set at a fairly high level. The spatial modulation does not extend fully into the central lobe because the reference beam was not set exactly equal to $A_2S(0,0)$ and the chosen exposure led to saturation of the film at the low frequencies. The impulse response, nevertheless, is a reasonably faithful rendition of the character as we see in Figure 6.12(b). To simulate the low modulation level caused by a low reference-beam level, we physically blocked the central lobe of the spatial filter as shown in Figure 6.12(c). All frequencies less than the reciprocal of the character size are thereby removed. Figure 6.12(d) shows that the edges of the character are accentuated by this *highpass* filter (45, 68).

6.5. ORIENTATION AND SCALE-SEARCHING OPERATIONS

When we use frequency-plane filters the location of a two-dimensional signal is automatically determined without scanning. The signal may have orientation and scale parameters, however, that require searching operations. Both orientation and scale are sometimes unknown, although the scale is usually known within certain limits. These operations are not always required; for example, in character-recognition applications, signals have only one orientation and one scale for any given font type.

6.5.1. The Orientation Search

One alternative for performing the orientation search is to rotate the filter. Unfortunately, rotating the spatial carrier frequency filter causes the diffracted orders to rotate about the optical axis, as we deduce from

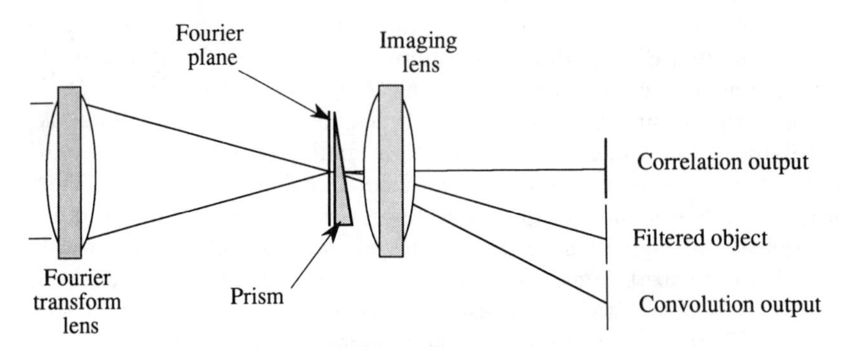

Figure 6.13. Prism/filter technique for orientation search.

the fundamental properties of rotating gratings. However, if we place a prism in contact with the filter, as shown in Figure 6.13, we stabilize the correlation output and force it to be centered on the optical axis, as the filter rotates. The apex angle γ of the prism is related to the spatial carrier frequency by the requirement that the deviation angle δ of the prism must be equal to the diffraction angle produced by the carrier frequency. We use the thin prism result from Equation (2.26) and the fact that the diffraction angle due to the carrier frequency is $\theta = x_0/F$ to find that

$$\gamma = \frac{x_0}{F(n_2 - 1)}, \tag{6.12}$$

where n_2 is the refractive index of the prism. Because the matched filter generates a plane wave, the prism does not add aberrations to the system if the Fourier-transform relationship is exact. A plane wave is not aberrated by a prism regardless of the angle at which it enters the prism.

Another alternative for performing the orientation search is to construct a sequence of filters for signals at different orientations. If the spatial carrier frequency has the same direction for each filter in the sequence, the prism is not needed. Another possibility, when we use an electronically driven spatial light modulator, such as a liquid-crystal display, is to rotate data electronically.

The choice among these methods for detecting the orientation of a signal rests largely on mechanical considerations. In the first alternative the filter/prism combination must be accurately rotated about its center but the filters are not often changed. In the second alternative the filters must be changed many times while searching through 180° and each filter in the sequence must be accurately positioned in the frequency plane (see Section 6.8). The third alternative is the best from a mechanical viewpoint

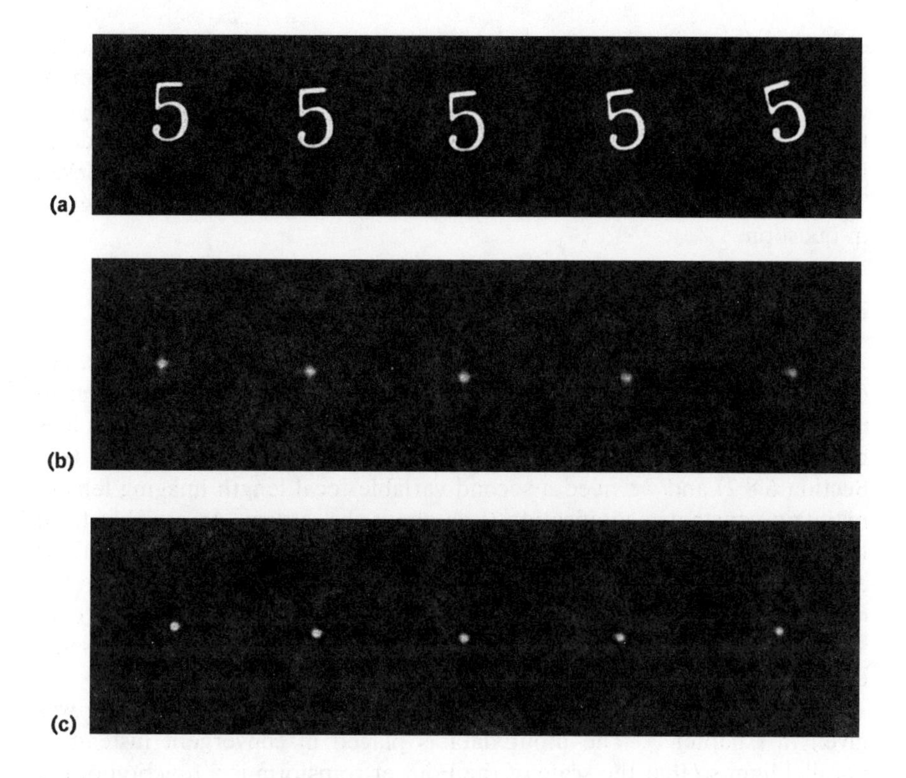

Figure 6.14. Orientation sensitivity: (a) input signals, (b) output from broadband filter, and (c) output from highpass filter (68).

because it requires no moving parts; the output of the system must, however, be synchronized to the input so that the locations of the signals in data are preserved.

The sensitivity of matched filtering to the orientation of the signal is a strong function of the shape of the signal. A rotationally symmetric signal is completely insensitive to orientation, but a signal with a high length-to-width ratio may exhibit a high degree of sensitivity to orientation. The character 5 is composed of some straight-line segments, with high length-to-width ratios, and a curved-line segment, nearly in the form of a circle. Figure 6.14 shows the output of a broadband filter and that of a highpass filter to a set of such characters that are rotated 2°, 4°, 8°, and 16° with respect to one that is vertically oriented. The peak correlation value declines somewhat more rapidly as a function of rotation for the highpass filter, relative to the rate of decline for the broadband filter. This result is expected because the highpass filter places more emphasis on the edge information in the signal; the overlap between the edges declines more quickly than does the overlap of the interior regions. For a typical signal, the sensitivity to orientation is such that measurements must be made at intervals of 5–10° to ensure that the correlation peak is still at a high level. The exact angular interval must be obtained experimentally for the signals in question.

6.5.2. The Scale Search

A search for the scale of the signal is made in one of several ways. An obvious method is to use a zoom lens instead of a fixed focal length lens to produce the Fourier transform. However, the back focal plane of this lens must not shift while zooming from the plane containing the filter (see Section 6.8.2) and we need a second variable focal length imaging lens to maintain constant magnification throughout the system. As variable focal length lenses with the quality demanded by most processing applications are costly, the scale search is better performed by a method similar to the one described for the orientation search; a sequence of filters is constructed with a fixed carrier frequency but with differing scales. The same comments about accurate positioning of the filters apply here.

An even simpler method for constructing a variable-scale correlator was given in Chapter 3. The input data is placed in convergent instead of parallel light so that the scale of the Fourier transform is a function of the distance from the input plane to the frequency plane. By mechanically coupling the input plane, the Fourier-transform lens, and the output plane, we search for the scale of the signal while moving these elements along the optical axis. The output is always a focused, fixed-magnification image of the filtered data. The search range is normally not greater than about $\pm 20\%$; if the size of the signal varies by more than $\pm 20\%$ relative to the filter, a new filter is generally used. Another possibility, when we use an electronically driven spatial light modulator such as a liquid-crystal display, is to change the size of the data electronically.

The sensitivity of matched filtering to the scale of the signal is independent of the shape of the signal. Figure 6.15 shows the output of a broadband filter and that of a highpass filter to a set of 5's whose scales are 7.5% and 15% both larger and smaller than that of the central character. The peak correlation value declines somewhat more rapidly as a function of scale for the highpass filter relative to the rate of decline for the broadband filter. Again, this result is expected because the highpass filter places more emphasis on the edge information in the signal, and the mutual overlap of these edges changes more rapidly as a function of scale

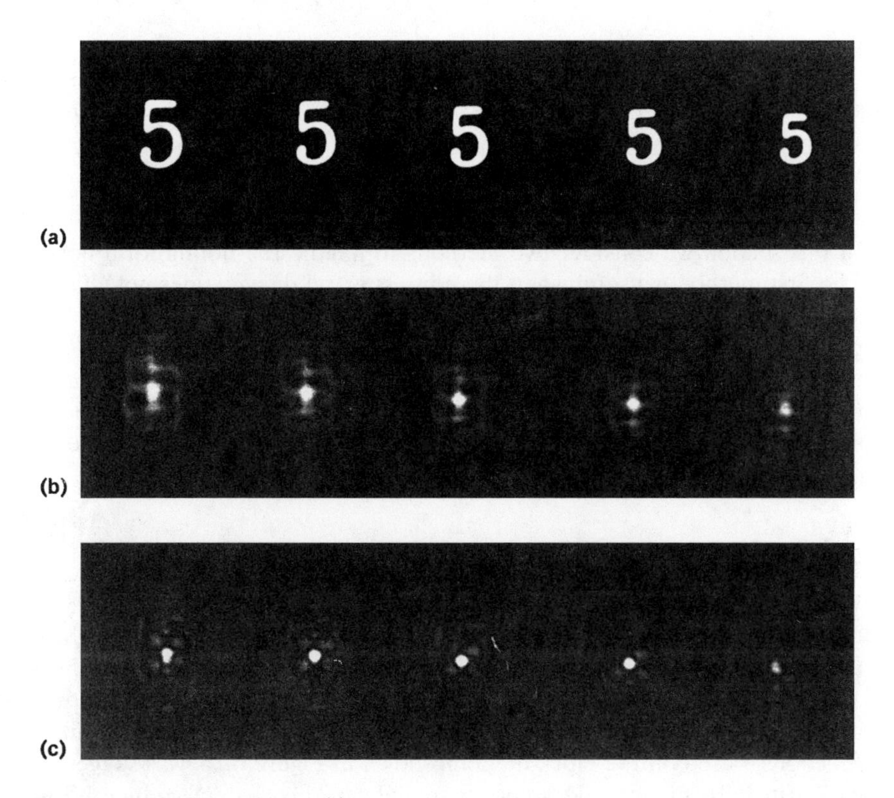

Figure 6.15. Scale sensitivity: (a) input signals, (b) output from broadband filter, and (c) output from highpass filter (68).

than does the overlap for the interior regions. For a typical signal, the sensitivity to scale requires measurements over a $\pm 20\%$ range to ensure that the correlation peak is still at a high level. The exact scale interval must be obtained experimentally for the signals in question.

6.6. METHODS FOR HANDLING NONUNIFORM NOISE SPECTRAL DENSITIES

The noise spectral density for images is rarely uniform, in contrast to time signals, for which this assumption is generally true. As a result, we typically need to implement filters that include the denominator of the matched filter as given by Equation (5.33):

$$H(\alpha,\beta) = \frac{S^*(\alpha,\beta)}{R_n(\alpha,\beta)}. \tag{6.13}$$

In this section we consider two methods to handle the nonuniform noise. The first method is useful when the noise spectral density does not change significantly from frame to frame so that we implement the noise-rejection part of the matched filter (the denominator) in a first frequency plane and the signal part of the matched filter in a second frequency plane of a *dual frequency-plane processor*. The second method is useful when the noise spectral density might vary *within* a frame of data; in this case we need an *adaptive filtering operation* to achieve optimum results.

6.6.1. Dual Frequency-Plane Processing

When the noise spectral density does not change significantly from frame to frame, we place the fixed noise-rejection part of the filter, which requires high positional accuracy, in one frequency plane and the changeable signal part of the filter, which requires relatively less positional accuracy, in the second frequency plane. A useful way to choose the filter functions, based on the block diagram shown in Figure 6.16, was suggested by Turin (69). Figure 6.17 shows a spatial filtering system configured to implement this solution. It has a first frequency plane P_2 which is imaged by lens L_2 into a second frequency plane P_3. Lens L_3 provides the required Fourier transform to produce the output $g(u,v)$.

A filter $H_1(\alpha,\beta)$ is placed at plane P_2, and a second filter $H_2(\alpha,\beta)$ is placed at plane P_3. The overall transfer function of the system is therefore

$$H(\alpha,\beta) = H_1(\alpha,\beta)H_2(\alpha,\beta). \tag{6.14}$$

Suppose that the data has noise spectral density $R_n(\alpha,\beta)$ and that it

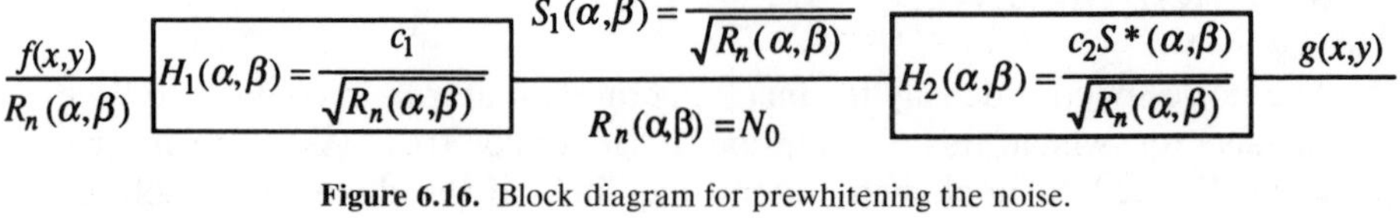

$\dfrac{f(x,y)}{R_n(\alpha,\beta)}$ → $H_1(\alpha,\beta) = \dfrac{c_1}{\sqrt{R_n(\alpha,\beta)}}$ → $S_1(\alpha,\beta) = \dfrac{c_1 S(\alpha,\beta)}{\sqrt{R_n(\alpha,\beta)}}$, $R_n(\alpha,\beta) = N_0$ → $H_2(\alpha,\beta) = \dfrac{c_2 S^*(\alpha,\beta)}{\sqrt{R_n(\alpha,\beta)}}$ → $g(x,y)$

Figure 6.16. Block diagram for prewhitening the noise.

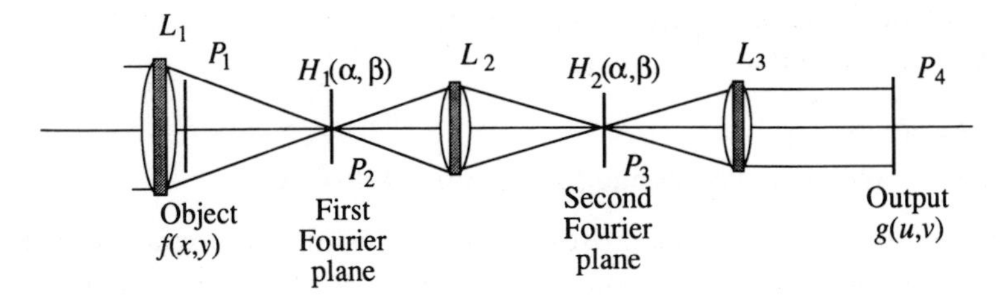

Figure 6.17. Dual frequency-plane processor.

passes through a filter

$$H_1(\alpha, \beta) = \frac{c_1}{\sqrt{R_n(\alpha, \beta)}}, \tag{6.15}$$

as shown in Figure 6.16. The square-root operation on $R_n(\alpha, \beta)$ is permissible because it is a nonnegative function. The noise spectrum at the output of the first filter will then be uniform but the signal Fourier transform will be distorted, becoming

$$S_1(\alpha, \beta) = \frac{c_1 S(\alpha, \beta)}{\sqrt{R_n(\alpha, \beta)}}. \tag{6.16}$$

As the noise spectral density is now uniform, the basic theory of matched filtering states that the second filter should be matched to the Fourier transform of the distorted signal:

$$H_2(\alpha, \beta) = \frac{c_2 S^*(\alpha, \beta)}{\sqrt{R_n(\alpha, \beta)}}. \tag{6.17}$$

From Equations (6.15) and (6.17), we see that the overall filtering operation is

$$\boxed{H(\alpha, \beta) = \frac{c_1}{\sqrt{R_n(\alpha, \beta)}} \frac{c_2 S^*(\alpha, \beta)}{\sqrt{R_n(\alpha, \beta)}} = \frac{c S^*(\alpha, \beta)}{R_n(\alpha, \beta)},} \tag{6.18}$$

where $c_1 c_2 = c$, as required by Equation (5.33). This development confirms that the matched filter is the optimum process for maximizing the ratio of peak signal to average noise intensity even if the noise spectral density is not white.

6.6.2. Transposed Processing for Adaptive Filtering

Nonstationary noise often arises when processing photographs such as those generated in earth resource monitoring or in surveillance activities. A particular frame may contain terrain that is predominantly forests, lakes, deserts, or plains. Estimates based on the sample data (a particular frame of imagery) may be quite different from the statistical properties of the random process (all such frames). Batch processing of data provides new opportunities for signal processing. If we consider only one frame at a time, the total energy of the sample is finite; $f(x, y)$ is then a deterministic function rather than a random process. As the data is Fourier transformable, we can obtain a complete knowledge of the "noise"; that is, we do not need to rely on its statistical properties for constructing the filter. Images are naturally in a batch form and we can therefore learn about the noise before processing the data.

One way to take advantage of this process is to *interchange the role of data and its transform, as far as the space and Fourier domains are concerned*. We make a filter $F(\alpha, \beta)$ from $f(x, y)$ and use the conjugate signal $s^*(-x, -y)$ as the input function. This technique is called *transposed processing*; its success rests on adaptively realizing, in an unusual way, the optimum denominator for the matched filter (70). The analogy is that we can think of the input data $f(x, y)$ as a broadband IF filter of bandwidth L, followed by a hard limiter in the form of the photographic film and a narrowband matched filter given by $s^*(-x, -y)$, the bandwidth of which is l (71).

We want to construct a filter from $f(x, y)$ that also contains the proper noise-rejection response. Because the signals are isolated, the exposure is equivalent to $|N(\alpha, \beta)|^2$ in the low-frequency region. If we set the reference level low, relative to the magnitude at $\alpha = \beta = 0$, the effect is equivalent to constructing a filter with a prewhitening component that is adapted to the specific data being processed; the average value of the data is now zero, and the frequency content is nearly uniform. If we use $s(x, y)$ as the signal, we find that the light leaving the Fourier plane is (71)

$$G(\alpha, \beta) = \frac{S(\alpha, \beta) F^*(\alpha, \beta)}{|N(\alpha, \beta)|^2}, \tag{6.19}$$

which is simply the complex conjugate of what we wish to have. Since the output is square-law detected, the conjugation is of little consequence. The two cases are compared as follows:

$$G(\alpha,\beta) = \overbrace{\underbrace{\frac{S^*(\alpha,\beta)}{|N(\alpha,\beta)|^2}}_{\text{filter}}\underbrace{F(\alpha,\beta)}_{\text{from input}}}^{\text{Normal processing}} = \overbrace{\underbrace{\frac{F^*(\alpha,\beta)}{|N(\alpha,\beta)|^2}}_{\text{filter}}\underbrace{S(\alpha,\beta)}_{\text{from input}}}^{\text{Transposed processing}}. \tag{6.20}$$

To use the full potential of transposed processing, we need an optical processing system that constructs filters on a frame-by-frame basis in near real time. Thermoplastic recording materials are useful for recording filters, and recent work on nonlinear materials used for phase conjugation and real-time holography make transposed processing attractive.

An example of transposed processing, which illustrates both this technique and that of multiplexing, is character recognition. Figure 6.18(a) shows a character set from a type face, such as HELVETICA, that is distinguished by having a high occurrence of vertical and horizontal strokes. These common strokes are not particularly helpful in recognizing a given character from the set. A transposed filter, in which we consider the entire alphabet the signal, is recorded with the reference-beam level set so that the common frequency components due to the vertical and horizontal strokes are automatically suppressed. Figure 6.18(b) shows the impulse response of the transposed filter; we see that the vertical strokes are most strongly suppressed, the horizontal ones less so, and the others still less. To a first approximation, the various strokes are attenuated proportionally to their frequency of concurrence. For example, the intensity of the horizontal and vertical strokes, relative to the diagonal stroke of the letter X, are attenuated by 4 and 9 dB, respectively. The value of transposed processing, then, is that the common features among a set of signals are adaptively suppressed in favor of the features that best distinguish among the various signals. This characteristic of transposed processing is important when hypothesis testing among many possible received signals. It is particularly useful for character recognition where the crucial issue is to distinguish one character from all the others; the other characters represent the noise.

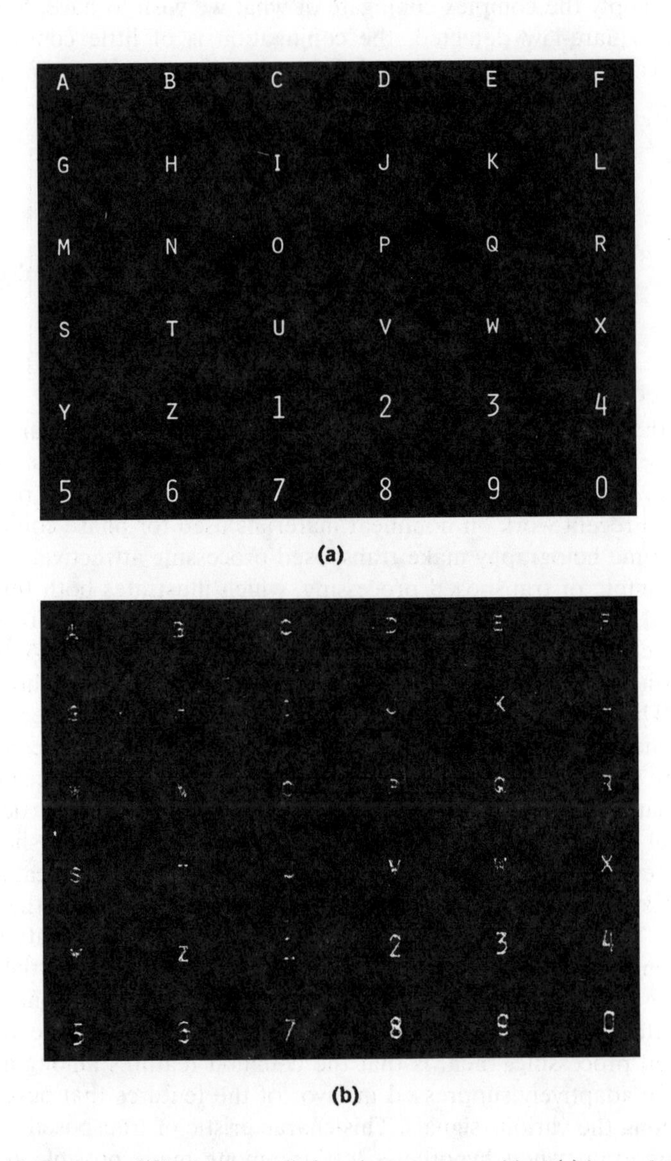

Figure 6.18. Transposed filtering as applied to character recognition: (a) input character set for the multiplexed filter, (b) impulse response of the nonlinearly recorded transposed filter (70).

6.7. OTHER APPLICATIONS FOR OPTICAL SPATIAL FILTERING

In the following paragraphs, we discuss several applications of optical spatial filtering to illustrate the theory as developed to this point. The applications we describe in Sections 6.7.2 and 6.7.3 are interesting examples in which the output plane coordinates are *differential displacement variables* instead of *space variables*.

6.7.1. Target Recognition

Most of the examples given so far use binary signals, such as characters, to illustrate various features of optical signal processing. We now show that these techniques are equally applicable to detecting analog signals in an arbitrary noise background. For example, Figure 6.19(a) shows an aerial photograph of some terrain that includes an automobile on a road. A spatial filter constructed for this car produced the output shown in Figure 6.19(b). The signal-to-noise ratio for the correlation peak is clearly sufficient to provide for reliable detection.

6.7.2. Motion Analysis

An offshoot of transposed processing is a method for tracking the motion of signals, such as clouds, icebergs, ships, or road traffic in a sequence of photographs separated by time delays T_f. As an example, we apply

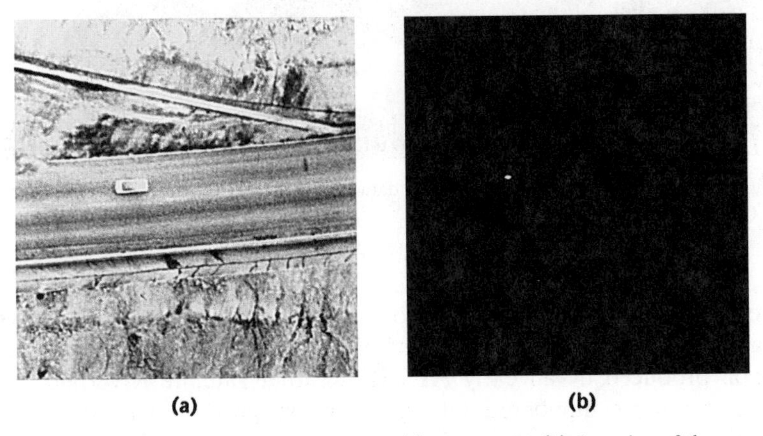

Figure 6.19. Detection of an analog signal: (a) input scene, (b) detection of the car.

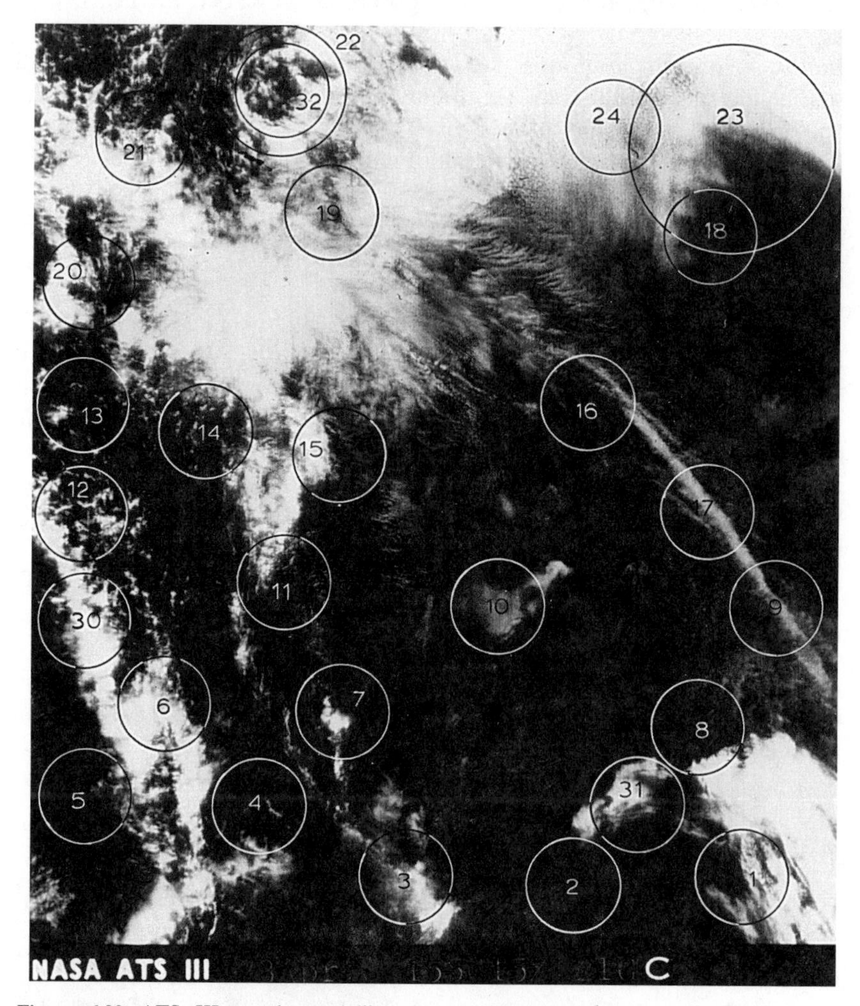

Figure 6.20. ATS III weather satellite data with overlay (courtesy F. B. Rotz and M. O. Greer) (72).

filtering techniques to cloud-motion analysis (72). Figure 6.20 shows an image of the cloud coverage in the Baja California and Central Mexico region produced by an early ATS III weather satellite. Overlaid on the image are some numbered subregions whose purpose will become clear shortly. From a sequence of such photographs the displacements of various cloud patterns can be plotted and estimates made of the velocity of the cloud masses. We begin by constructing a transposed filter from the first frame of imagery and storing it at the frequency plane. We calibrate the processing system by autocorrelating the first image so that the zero displacement position is found at the output plane.

After T_f seconds, we introduce the second frame of imagery and correlate subregions of the second frame relative to the entirety of the first frame. If all cloud patterns move with equal vector velocities, we expect to find the correlation peak displaced by a distance and in a direction consistent with the overall cloud mass velocity. Because cloud patterns often move in different directions and at different velocities, we generally find several correlation peaks present at the output. The issue is to relate the correlation peak positions to the various cloud masses.

The trick is to isolate cloud patterns by illuminating subregions of the second frame of data, sequentially, in a raster-scanning fashion. This scanning technique is similar to the telescopic scanner, described in Section 4.4.1 in connection with the spectrum analysis technique that uses a ring/wedge detector for feature analysis. In this case, the position of the correlation peak is a function of the subregion being currently illuminated. Suppose that the raster scanning proceeds from the upper-left corner of the data at a scanning velocity v; after the first line is scanned, the illumination is indexed by an amount Δy in the vertical direction.

The correlation at the output of the system for the nth scan can be expressed as

$$g_n(r, s, t) = \iint_{-\infty}^{\infty} a(vt, n\Delta y) f_2(x, y) f_1^*(x + r, y + s)\, dx\, dy, \quad (6.21)$$

where r and s are the coordinates of the output correlation plane and $a(vt, n\Delta y)$ is the aperture function that determines the subregion of data that is being illuminated. Thus, at some instant in time t_0, the scanning illumination is $a(vt_0, n\Delta y)$ and we sample the value of $g_n(r, s, t)$ at this instant in time.

The interval between samples, as the illumination scans a given line, is dependent on the application. We usually measure the correlation function whenever the illumination has moved to a new area. The illumination size and shape are generally selected to be roughly equal to the size and geometry of the desired cloud patterns. The set of coordinates r and s at which the correlation peaks occur indicate the relative cloud motion between frames.

After the relative motions of all cloud masses between frames one and two have been measured and recorded, we erase the filter, construct a new

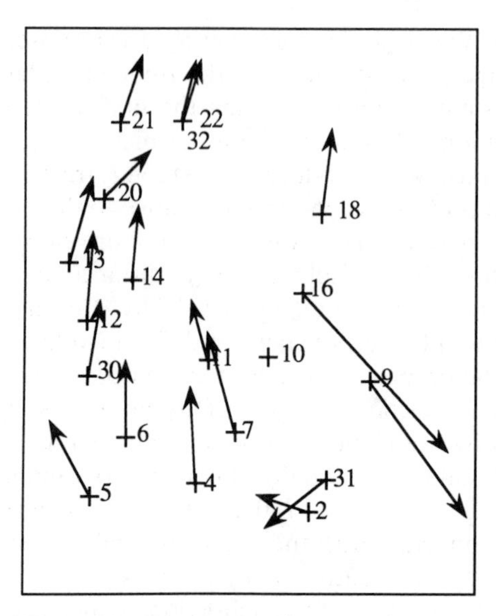

Figure 6.21. Displacement (or velocity) vectors.

transposed filter for frame two, introduce frame three, and repeat the process. To illustrate the results, we superimposed several circles on Figure 6.20 that indicate selected cloud masses. We measured the correlation-peak displacement for each of these regions and converted the displacement information to a velocity vector. Figure 6.21 shows velocity vectors for one-hour time intervals between frames of data. Cloud mass No. 10 is a stationary pattern on the western coastline of Central Mexico. The vector velocities for cloud masses No. 22 and No. 32 are in close agreement, indicating that the two cloud masses, which are made up of several smaller clouds, are moving with nearly the same velocity and that the dispersion is not great in this region.

By using this technique, clouds are tracked over large distances, even though they change from frame to frame. The adaptive tracking is achieved by virtue of the fact that the transposed filter is constantly being updated with new versions of the cloud mass as time progresses so that the correlation peaks remain well formed (72). Cloud mass No. 16 is associated with the jet stream; these cloud patterns change so rapidly that some inaccuracies in the velocity vectors result from uncertainties as to the position of the correlation peak.

6.7.3. Frame Alignment and Stereo Compilation

When frames of imagery are collected from an airborne platform, it is necessary to orient and align successive frames to make a useful montage. Transposed processing, similar to that described in Section 6.7.2, is also useful in this application. We generate a filter for the first image and place it in the frequency plane. The autocorrelation function for this image establishes the datum from which all future measurements are made. The second image is placed into the system and subregions common to the two images are correlated as described in Section 6.7.2. The displacement measurements indicate where the second image should be placed relative to the first image when building the montage. The processing operation generally includes an orientation search because the flight path is not exactly straight. This search is done using the techniques described in Section 6.5.1. Furthermore, a scale search as described in Section 6.5.2 may be necessary if the terrain changes elevation abruptly or the aircraft changes altitude significantly between images.

Figure 6.22 illustrates the image alignment process. Figure 6.22(a) and Figure 6.22(b) show two successive images from a set of images; the subregion shown in Figure 6.22(c) is common to both images. Figure 6.22(d) is the correlation of this subregion with the image from Figure 6.22(a); the position of this correlation peak establishes the datum from which subsequent displacements are measured. Figure 6.22(e) is the correlation of the selected subregion with the image from Figure 6.22(b), the second image in the set. The correlation peak has moved upward and to the right, indicating the displacement of the second frame relative to the first. As the displacement is essentially the same for all subregions, we can average the results over several such regions to improve the accuracy of the overlay. An experimental program showed that the overlaying function, as illustrated in Figure 6.23, can be performed to an accuracy of 1/600th of an image dimension in both directions at a speed of 50 images per minute.

The displacements of subregions may, however, change as a function of its elevation. This change suggests that elevation profiles for generating contour maps can be made from stereo pairs. From the displacement information, the height of the small patch of ground currently being illuminated is calculated relative to a datum plane. The fundamental difference between the image-alignment technique and parallax measurements for stereo compilation is that the size of the subregion is much smaller in the latter case, the exact size being dependent on the desired resolution of the elevation contours.

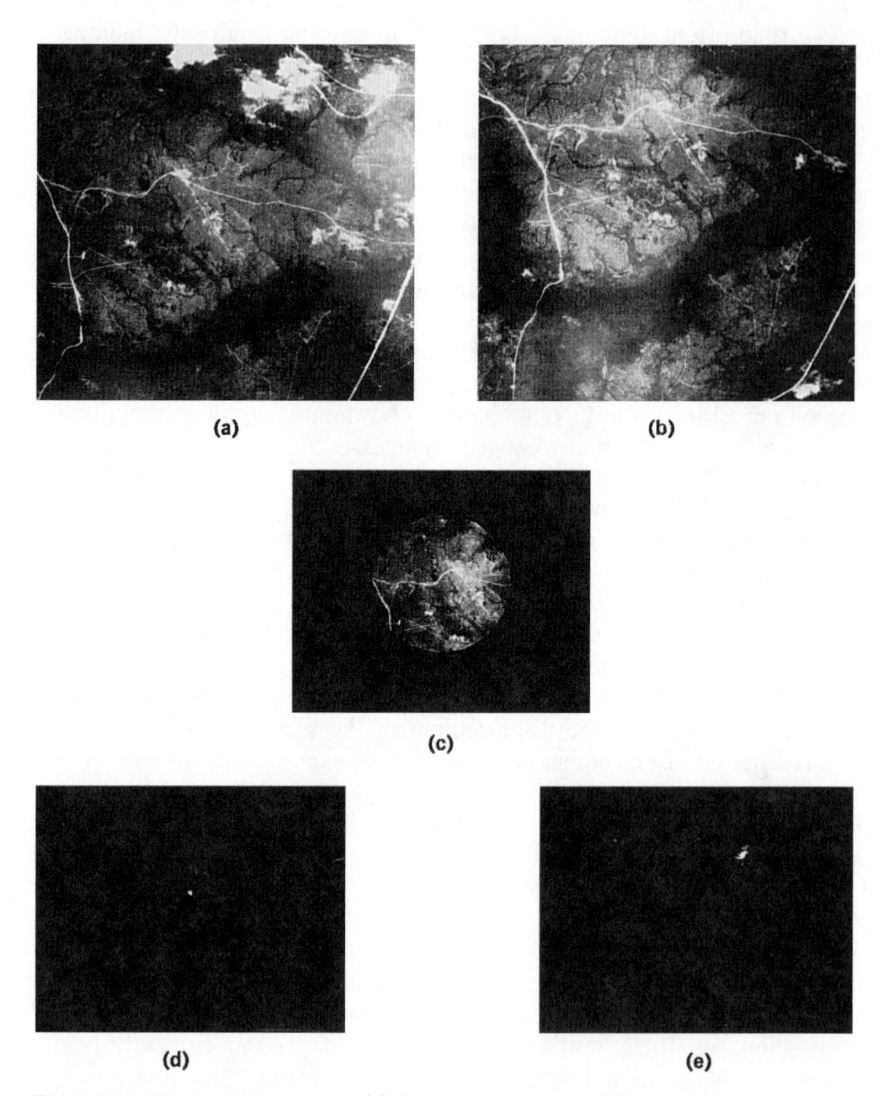

Figure 6.22. Photographic mapping: (a) first frame of imagery, (b) second frame of imagery, (c) region common to both frames, (d) autocorrelation to establish datum, and (e) cross correlation to determine displacement (45) (copyright © IEEE, 1974).

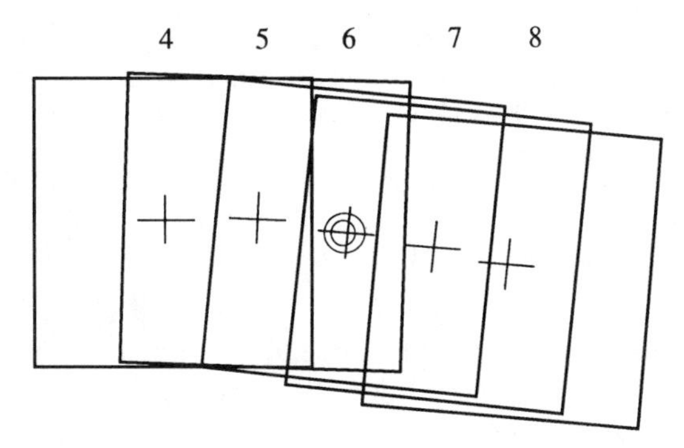

Figure 6.23. A typical frame sequence.

6.8. THE EFFECTS OF SMALL DISPLACEMENTS OF SPATIAL FILTERS

When spatial filters are changed, they may become displaced from their optimum position in the frequency plane. Sometimes the filter is rotated to detect randomly oriented signals in two-dimensional data; an error in centering the filter causes a displacement from its optimum position. Vibrations may cause time-varying displacement errors in the position of the filter. These displacements occur normal to the optical axis or parallel to it.

Displacements of spatial filters affect the performance of a matched filtering system. As the relative performance of the system does not depend on whether a spatial carrier frequency filter or a baseband complex-valued filter is used, we use the latter to simplify the analysis. We calculate the effect of both lateral and longitudinal displacement errors for both uniform and nonuniform noise spectral densities; the analysis presented here is a condensation of a more complete analysis (73). We use one-dimensional notation, thus simplifying the results without losing any important information.

6.8.1. Lateral Displacements

The first displacement errors we consider are those normal to the optical axis. We first examine the situation in which the noise spectral density is uniform. The output, when the filter is displaced by an amount $\Delta\alpha$, is

$$g(x, \Delta\alpha) = \int_{-\infty}^{\infty} F(\alpha)H(\alpha + \Delta\alpha)e^{j2\pi\alpha x}\, d\alpha, \qquad (6.22)$$

where $H(\alpha)$ is given by Equation (5.34) and $F(\alpha)$ is the Fourier transform of the signal $f(x)$. The signal-to-noise ratio is given by Equation (5.32), using the equality sign. By using the convolution theorem, we express the signal part of the output as

$$g(x, \Delta\alpha) = \int_{-\infty}^{\infty} s(u) s^*(u + x) e^{j2\pi\Delta\alpha(u+x)}\, du. \tag{6.23}$$

The average noise intensity is, of course, independent of the displacement of the filter if the noise spectral density is uniform; the noise is therefore ignored in this calculation.

The performance P, as a function of $\Delta\alpha$, is defined as the peak value of Equation (6.23), normalized to the peak signal level when the displacement is equal to zero:

$$P = \frac{|g(0, \Delta\alpha)|^2}{|g(0,0)|^2} = \frac{\left|\int_{-\infty}^{\infty} |s(u)|^2 e^{j2\pi\Delta\alpha u}\, du\right|^2}{\left|\int_{-\infty}^{\infty} |s(u)|^2\, du\right|^2}. \tag{6.24}$$

The derivative of P with respect to the displacement variable $\Delta\alpha$ is determined from the numerator of Equation (6.24) as

$$\frac{\partial P}{\partial(\Delta\alpha)} = 2\int_{-\infty}^{\infty} j2\pi u |s(u)|^2 e^{j2\pi\Delta\alpha u}\, du, \tag{6.25}$$

which is maximized, for small values of $\Delta\alpha$ when the integral I is maximized, where

$$I = \int_{-\infty}^{\infty} u |s(u)|^2\, du. \tag{6.26}$$

To make some calculations of the performance of the system, we choose a signal $s(u)$ that leads to the most rapid loss of performance for small values of $\Delta\alpha$. The signal must have a finite extent and a magnitude not exceeding unity. The integral from Equation (6.26) is maximized under the given constraints if

$$|s(u)|^2 = \text{rect}(x/l), \tag{6.27}$$

where l is the length of the signal. An intuitive way to obtain this result is

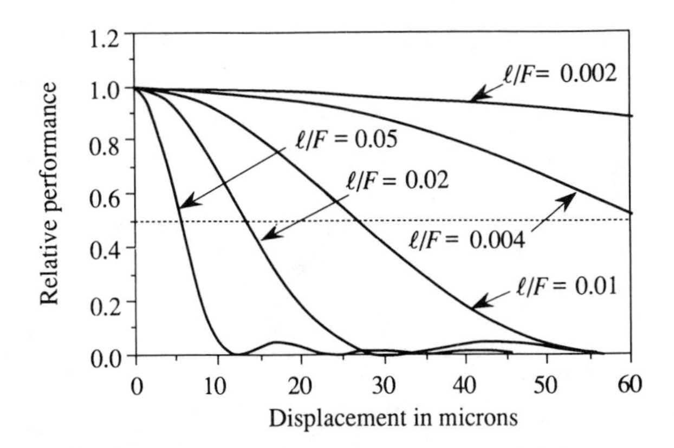

Figure 6.24. Correlation-peak intensity for various l/F ratios (white noise).

to note that if we interchange our concept of the input and frequency planes, then the rect function, as viewed from the frequency plane, has the highest possible bandwidth of any function in the input plane.

By substituting $s(u)$ into Equation (6.24) and by using the well-known Fourier-transform pair relationship, we express the performance of the system as

$$\boxed{P = \text{sinc}^2(\Delta\alpha l).} \tag{6.28}$$

Several curves for the performance of the system are shown in Figure 6.24, where we have used the relationship that $\Delta\xi = \Delta\alpha\lambda F$, where F is the focal length of a Fourier-transforming lens such as that shown in Figure 6.3 and $\lambda = 632.8$ nm. The parameter l/F is the relative aperture of the detected signal and the physical displacement of the filter from the optical axis is $\Delta\xi$.

From Equation (6.28) and Figure 6.24 we see, for example, that the performance of the system is reduced by 3 dB, for $l/F = 4(10)^{-3}$, when $\Delta\xi \approx 63\ \mu$. For $l/F = 10(10)^{-3}$, $\Delta\xi$ must be held to less than 26 μ for the same loss in performance. Therefore, the placement of the filter becomes more critical as the length of the signal increases, which is hardly a surprising result, because this dimension is directly related to the space bandwidth product of the signal; the space bandwidth product, in turn, generally specifies the quality level required of any optical system.

Obtaining comparable performance data for a nonuniform noise spectral density is more difficult because the spectral density is highly depen-

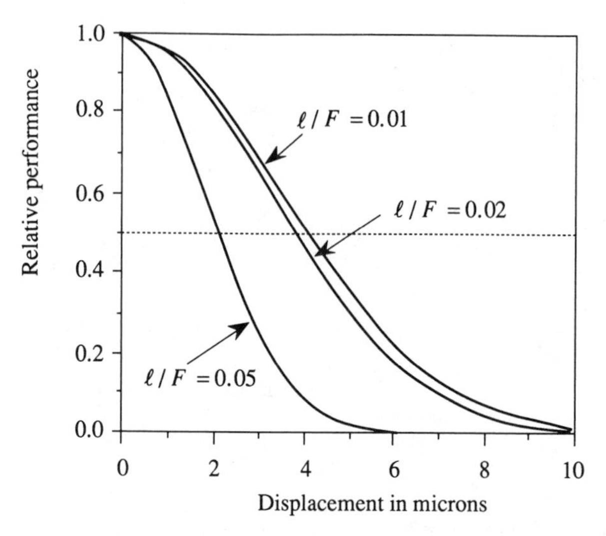

Figure 6.25. Correlation-peak intensity for various l/F ratios (nonuniform noise).

dent on the type of data being processed. As nonuniform spectral densities often occur in practice, we illustrate the procedure based on the noise spectral density for a collection of aerial photographs of various terrains. The measured spectral density for this collection is of the form (73)

$$\log[R_n(\alpha)] = e^{-|\alpha|/2} - 1. \tag{6.29}$$

The performance results for small lateral displacements are shown in Figure 6.25. The general trends are the same as for uniform noise, with the important difference that the permissible displacement for a given loss of performance is much less. For example, for uniform noise we found a 3-dB loss of performance, when $\Delta\xi = 26\ \mu$, for $l/F = 10(10)^{-3}$; for nonuniform noise we find that $\Delta\xi$ must be held to less than 3.3 μ for the same level of performance. This difference in positioning tolerance is nearly an order of magnitude; the difference increases greatly as l/F decreases.

The performance of the system does not vary appreciably for $l/F > 10(10)^{-3}$ when the noise spectral density is nonuniform. Because the spectra of such small signals are spread over a large area of the frequency plane relative to the spread in the noise, the performance depends primarily on the form of the spectral density $R_n(\alpha)$. For example, when $l/F = 4(10)^{-3}$, the permissible displacement for nonuniform noise is still 3.3 μ, whereas for uniform noise the displacement tolerance is relaxed to 63 μ.

6.8.2. Longitudinal Displacements

The second kind of displacement errors are those parallel to the optical axis. Their effects are calculated much the same as those of lateral displacements; the main difference is that the system is now space variant because the filter is in a Fresnel, rather than the Fourier, diffraction plane.

We showed that an error in lateral positioning does not affect the average noise intensity at the output when the noise spectral density is uniform; a similar argument is used here. Furthermore, because the length of the signal is usually much less than the focal length of the transforming lens, the Fourier transform is a slowly varying function of the longitudinal displacement ΔZ. We analyze the effects of longitudinal errors by referring to Figure 6.26 in which the displacement error is highly exaggerated. Normally this error is less than the depth of focus of the signal; the first-order effect is equivalent to that of a lateral displacement. We set L as the aperture of the input plane, x_0 as the distance that the signal is displaced from the optical axis, and ΔZ as the distance the filter is displaced from plane P_2.

We do not show the Fourier-transforming lens in this diagram because it has no significant impact on the result. Because the center of the Fourier transform of the signal passes through plane P_2 at the optical axis, it is displaced by a distance $\Delta\xi$ when it reaches the filter plane, where $\Delta\xi = x_0 \Delta Z / F$. Using the relationship between $\Delta\xi$ and $\Delta\alpha$, we have

$$\Delta\alpha = \frac{x_0 \Delta Z}{\lambda F^2}, \tag{6.30}$$

which is substituted into Equation (6.28) to give the performance of the

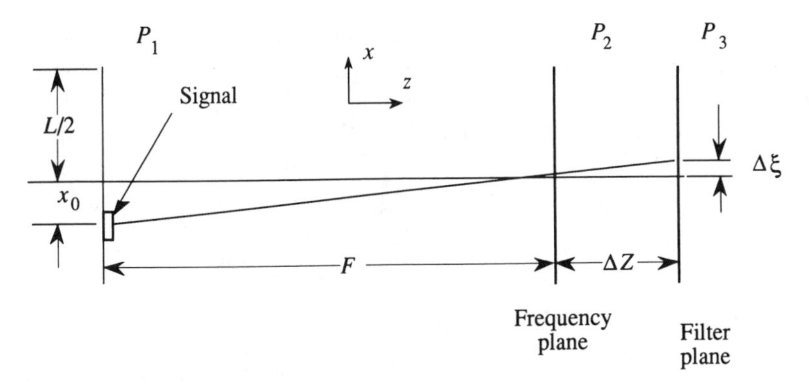

Figure 6.26. Sketch to find the effects of longitudinal filter displacements.

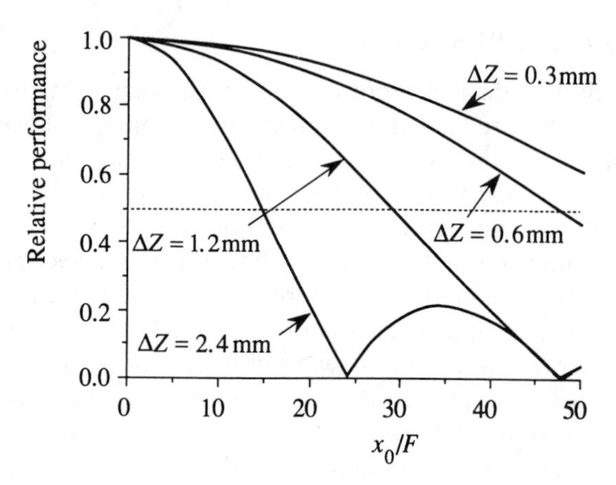

Figure 6.27. Correlation-peak intensity for various values of displacement; plotted for $l/F = 0.01$.

system as a function of both ΔZ and x_0. For fixed values of ΔZ and l/F, we can determine the performance P of the system as a function of x_0/F. Some results are shown in Figure 6.27 for $l/F = 10(10)^{-3}$; the space-variant nature of the system is readily apparent because the performance is a strong function of x_0/F. From Equations (6.30) and (6.28) we find that a 3-dB loss of performance at the edge of the aperture does not occur until $\Delta Z \approx 400\ \mu$. By comparison, for the same loss in performance and for the same value of l/F, the permissible value of $\Delta\xi$ is only 26 μ for lateral displacements. A longitudinal displacement, then, affects the performance of the system much less than an equivalent lateral displacement. Clearly, we must be concerned primarily with accurate lateral placement of the filters.

The dual frequency-plane processor described in Section 6.6.1 overcomes some of the sensitivity to filter displacement. Because the first filter does not have to be changed, its position can be accurately controlled. As the noise spectral density is uniform at the second frequency plane, we then can expect that the performance of the filter, under various conditions of displacement, will follow that of the white noise case as given in Sections 6.8.1.

6.8.3. Random Motion of the Filter

The effects of random motions of the filter are important if we use a motion picture projector or similar mechanisms for changing filters recorded on sprocketed roll film. The frames may not always be centered with respect to the center of the spectrum when the filters are made or when they are used. A partial analysis of this problem is given here; a more complete analysis has been given (73). We assume that the noise spectral density is uniform. The performance of the system is given by

$$P = \frac{\left| \int_{-\infty}^{\infty} |s(u)|^2 M_m(u)\, du \right|^2}{\left| \int_{-\infty}^{\infty} |s(u)|^2\, du \right|^2}, \tag{6.31}$$

where $M_m(u)$ is the characteristic function of the probability density $P(m)$ of the random motion (74). As an example, suppose that the probability density function $P(m)$ of a motion-picture film drive is a Gaussian distribution with zero mean and variance σ^2:

$$P(m) = \frac{1}{\sqrt{2\pi}\,\sigma} e^{-m^2/2\sigma^2}. \tag{6.32}$$

The corresponding characteristic function is the Fourier transform of $P(m)$ and is given by

$$M_m(u) = e^{-2\pi^2\sigma^2 u^2/\lambda^2 F^2}. \tag{6.33}$$

By substituting Equation (6.33) into Equation (6.31) and by using the form of the signal given by Equation (6.27), we find that the performance of the system is

$$P = \frac{\lambda^2 F^2}{8\pi\sigma^2 L^2} \operatorname{erf}\left(\sqrt{2\pi}\sigma L/\lambda F\right), \tag{6.34}$$

where

$$\operatorname{erf}(z) = \sqrt{\frac{4}{\pi}} \int_0^z e^{-t^2}\, dt. \tag{6.35}$$

A curve of the performance of the system as a function of $\sigma l/F$ is shown in Figure 6.28. For $l/F = 5(10)^{-3}$, a 3-dB loss in performance occurs at $\sigma = 20\ \mu$. If two such mechanisms are used, one to record the filters and one to change them, and if the motion processes are independent, then $\sigma^2 = \sigma_1^2 + \sigma_2^2$. The standard deviation for each unit must be less than

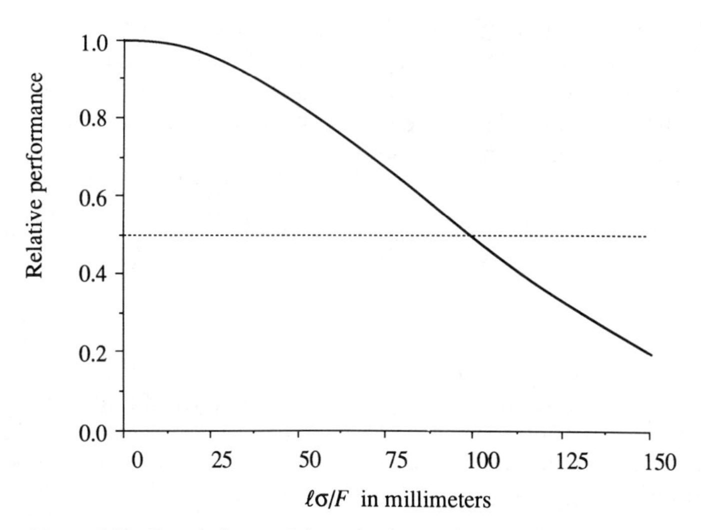

Figure 6.28. Correlation-peak intensity for random motion of the filter.

approximately 14 μ, which is not an unreasonable performance to expect from conventional motion-picture projectors and is well exceeded by film pin registration units that are used to accurately register film overlays.

PROBLEMS

6.1. Prove that the fraction of useful light power in a Gaussian beam of standard deviation σ is, for a light beam radius R, equal to $[1 - \exp(-R^2/2\sigma^2)]$.

6.2. For a Gaussian laser beam, calculate the fraction of usable light power if we require that the *amplitude* of the light must not vary by more than 8% from the peak value.

6.3. Consider a Gaussian laser beam whose amplitude is characterized by

$$a(x, y) = e^{-(2A/L^2)(x^2+y^2)},$$

where A is a scaling factor and $x = L/2$ is the position at which the Gaussian beam reaches its $1/e^A$ value in intensity. Suppose that you integrate the amplitude of the light in the y direction. Calculate and comment on the magnitude of the resulting amplitude function $a(x)$.

Note: This integration operation is similar to using a cylindrical lens to focus a two-dimensional Gaussian beam into a one-dimensional line.

6.4. We process data whose length is $L = 25$ mm and whose height is $H = 20$ mm. The cutoff frequency is $\alpha_{co} = \beta_{co} = 30$ Ab. Suppose that the filter for the target has uniform magnitude over the entire data bandwidth and some unspecified phase. Calculate the expected size of the correlation peak in the x and y directions and determine how many elements are required of the photodetector array.

6.5. For the conditions of Problem 6.4, how would the results change if $\beta_{co} = 15$ Ab and α_{co} remained unchanged?

6.6. For a target that has a 10 : 1 length-to-width ratio, calculate the approximate angle at which the correlation peak value decreases by 3 dB from its peak value. Assume that all spatial frequencies are faithfully recorded on the filter. Hint: This problem is most easily solved in the space domain.

6.7. Calculate the maximum allowable lateral displacement of a spatial filter if the correlation peak must remain within 3 dB of its maximum value for a system with the following parameters: target length $l = 7.5$ mm, focal length of transform lens $F = 600$ mm, $\lambda = 0.5\ \mu$, and uniform noise.

7

Acousto-Optic Devices

7.1. INTRODUCTION

Spatial light modulators form the interfaces between electrical and optical systems. In Chapter 4 we described several two-dimensional spatial light modulators whose inputs are either incoherently illuminated objects or raster-scanned electrical signals. None of those modulators, however, are able to accept signals with hundreds-of-megahertz bandwidths. We therefore concentrate in this chapter on acousto-optic spatial light modulators that help to implement a wide range of processing operations on wide-bandwidth signals. These devices are key to the signal-processing architectures discussed in the remainder of this book.

7.2. ACOUSTO-OPTIC CELL SPATIAL LIGHT MODULATORS

The advantages of optical systems, based on the use of acousto-optic cells for processing either analog or digital signals, may be summarized as a combination of high throughput, a small volume relative to competing rf systems, and low power consumption. Optical systems offer the potential for a large number of parallel channels with complete connectivity, and the high carrier frequencies ($\approx 10^{14}$ Hz) allow very high channel bandwidths with little crosstalk of the type present in electronic processors. Also, optical channels have comparatively smaller power requirements, as the dissipative losses associated with electrical transmission are not present; the losses in typical optical transmission media, such as air and glass, are low.

Brillouin (75) predicted in 1922 how light and sound would interact, and early experimental results were obtained by Debye and Sears (76) and Lucas and Biquard (77). Raman and Nath (78) put the interaction phenomena on a solid mathematical foundation in 1935. It was not until the 1960's, however, that devices with large bandwidths and good optical quality were developed. Acousto-optic cells have bandwidths up to 2 GHz,

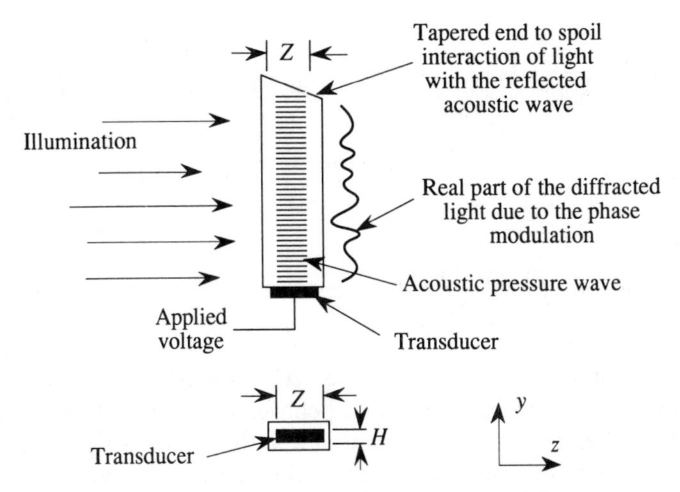

Figure 7.1. Acousto-optic cell spatial light modulator.

frame times of about 1 μsec, diffraction efficiencies of up to 90%, time bandwidth products of 1000–3000, good phase responses, and reasonable dynamic ranges.

An acousto-optic cell consists of an *interaction material*, such as water, glass, or an exotic crystal, to which a piezoelectric *transducer* is bonded, as shown in Figure 7.1. These one-dimensional devices are driven by an electrical signal connected to the transducer. The transducer launches either a compression or a shear acoustic wave into the x direction of the material which, in turn, creates strain waves. The strain waves lead to density changes in the interaction medium and, consequently, to index of refraction changes. The net result is that light passing through the acousto-optic cell in the z direction is modulated in phase according to changes in the optical path. The end of the acousto-optic cell is generally angled so that the reflected acoustic wave does not interact with the incident illumination.

As most signal-processing operations require many acoustic cycles in the cell to support sophisticated signals, the signal to be processed is translated to a center frequency f_c. Suppose that $s(t)$ is a baseband signal with highest frequency $W/2$. We mix this signal with $\cos(2\pi f_c t)$, as shown in Figure 7.2(a), to produce a double sideband modulated signal $f(t)$ whose spectral bandwidth is $W = f_2 - f_1$, centered at $\pm f_c$, as shown in Figure 7.2(b). In those applications where the rf signal spectrum falls naturally between f_1 and f_2, the signal can be fed directly into the acousto-optic cell without further preprocessing.

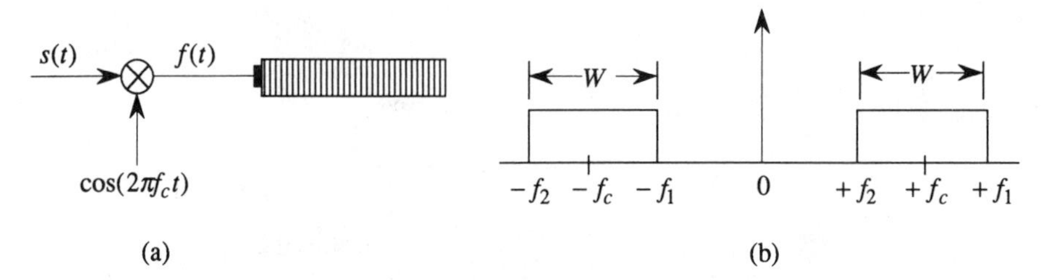

Figure 7.2. Acousto-optic cell: (a) electrical connection and (b) spectrum of drive signal.

When the interaction width Z of the acousto-optic cell is short relative to an acoustic wavelength, the device behaves as a thin diffracting material, as studied extensively by Raman and Nath in their 1935 papers (78). At the other extreme, the axial width of the cell may be large relative to the acoustic wavelength. We then speak of the device as operating in the Bragg mode, resulting in effects similar to those produced by x-ray diffraction in three-dimensional crystals.

7.2.1. Raman-Nath Mode

In Chapter 3, Section 3.1 we showed that a light wave is phase and amplitude modulated as it passes through an element whose response is $|a(x)|\exp[j\phi(x)]$; the wave then has the form $|a(x)|\cos[2\pi f_l t + \phi(x)]$, where f_l is the frequency of light. In a similar fashion, light passing through the acousto-optic cell, when driven by a pure sinusoidal frequency f_j, is phase modulated so that

$$A(x,t) = A_0 \cos\left[2\pi f_l t + \frac{2\pi Z}{\lambda}\left\{n_0 + \Delta n \cos\left[2\pi f_j\left(t - \frac{T}{2} - \frac{x}{v}\right)\right]\right\}\right], \tag{7.1}$$

where $A(x,t)$ is the amplitude of the output wave in space and time, A_0 is the amplitude of the incident light wave, and Δn is the change in the index of refraction within the interaction medium induced by the traveling strain wave. In Equation (7.1), the argument $t - T/2 - x/v$ shows that $A(x,t)$ is a wave traveling in the positive x direction with velocity v and that the cell has a transit time of $T = L/v$, where L is the length of the acousto-optic cell and v is the velocity of sound in the interaction medium.

The strain wave within the cell is proportional to the amplitude of the acoustic wave. As the modulated optical wave given by Equation (7.1) is a function of both space and time, a sinusoidal input signal causes the cell to

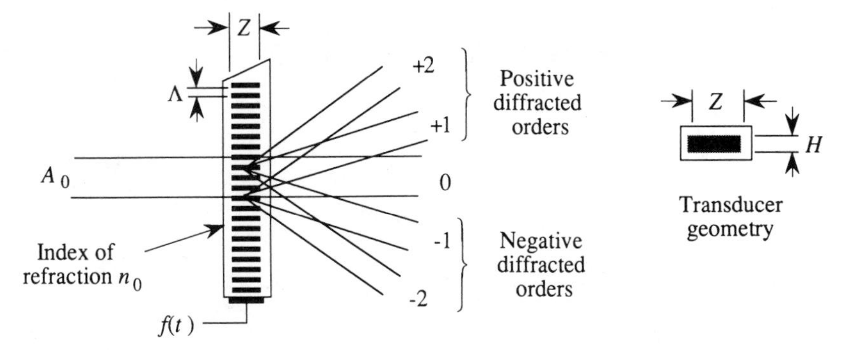

Figure 7.3. Diffracted orders in the Raman-Nath mode.

behave as a phase diffraction grating traveling in the x direction. Light is diffracted by the phase grating to produce several positive and negative diffracted orders. Figure 7.3 shows the rays associated with the incident illumination on the acousto-optic cell, as well as the first two of many diffracted waves. A rigorous analysis of the operation of an acousto-optic cell yields the Raman-Nath equation for the amplitude of the ith diffracted order (79):

$$|A_i| = |A_0 J_i(\gamma)|, \tag{7.2}$$

where A_0 is the amplitude of the incident light, J_i is the ith-order Bessel function, and γ is the phase shift of the light induced by the refractive index change. The normalized diffracted amplitude of the ith wave is indicated by m_i (79):

$$m_i = \frac{A_i}{A_0} = (-j)^i J_i\left[\frac{2\pi Z \Delta n}{\lambda}\right], \tag{7.3}$$

where $J_i(\cdot)$ is an ith-order Bessel function of the first kind. We define m_i as the *modulation index* for the ith order; it is defined as the ratio of the diffracted light amplitude to the incident light amplitude. The amplitude of the diffracted light, as a function of the phase shift of the Bessel function, is shown in Figure 7.4 for the undiffracted light and for the first two diffracted orders. The phase of each order, relative to its neighbors, is shifted by 90° as indicated by the $(-j)^i$ factor in Equation (7.3). For example, we find that the positive and negative orders are 180° out of phase. This result is also found from the fact that $J_{-n}(x) = (-1)^n J_n(x)$

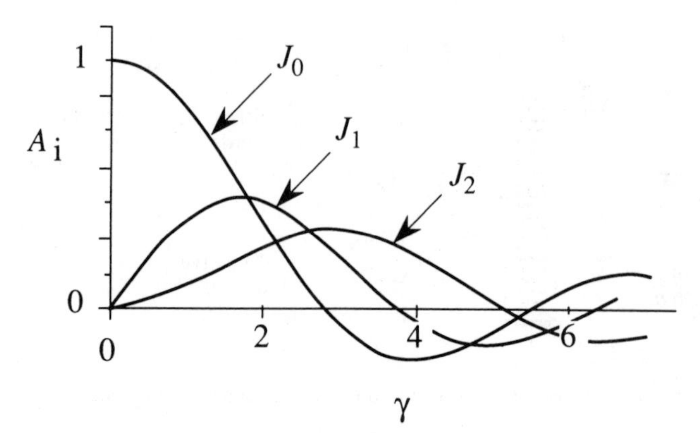

Figure 7.4. Bessel functions of order 0, 1, and 2.

and that

$$J_1(x) = \left(\tfrac{1}{2}x\right)\left[1 - \frac{\tfrac{1}{4}x^2}{2!} + \frac{\left(\tfrac{1}{4}x^2\right)^2}{2!3!} - \frac{\left(\tfrac{1}{4}x^2\right)^3}{3!4!} + \cdots\right]. \qquad (7.4)$$

As $J_1(x)$ is odd, the sign reversal between the two orders is apparent.

7.2.2. The Bragg Mode

From Equation (7.3) we see that large values of Δn and Z lead to high diffraction efficiencies. As light is wasted when multiple orders are generated, we generally increase the interaction width Z until the Bragg mode of operation is reached. We characterize the transition from the Raman-Nath mode to the Bragg mode by defining a Q factor:

$$Q \equiv \frac{2\pi}{n_0} \frac{\lambda Z}{\Lambda^2}, \qquad (7.5)$$

where n_0 is the index of refraction of the interaction material and Λ is the acoustic wavelength. The acoustic wavelength is related to the applied drive frequency f by $\Lambda = v/f$, where v is the velocity of the acoustic wave. A $Q \approx 2\pi$ establishes a boundary between the two modes. If $Q < 2\pi$, the acousto-optic cell is operating in the *Raman-Nath mode*; if $Q > 2\pi$, it is operating in the *Bragg mode*. In optical signal processing, the acousto-optic cell is primarily used in the Bragg mode because more of

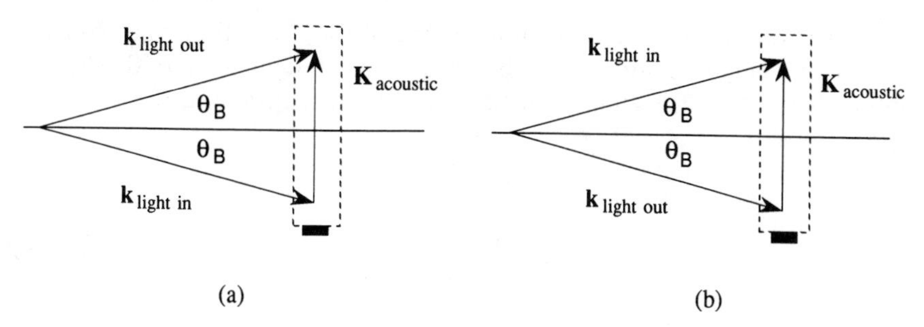

Figure 7.5. Bragg diffraction: (a) upshift mode and (b) downshift mode.

the optical power is coupled into a single diffracted order. In the strong Bragg region, only two diffracted orders are present, either the zero and one positive diffracted order or the zero and one negative diffracted order.

The essential properties of acousto-optic diffraction are explained with the aid of a model showing the collision between photons and phonons. The momenta of the interacting particles are given by $\hbar\mathbf{k}$ and $\hbar\mathbf{K}$, where $\hbar$ is Planck's constant and $\mathbf{k}$ and $\mathbf{K}$ are the wave vectors of light and sound. From Figure 7.5(a), we see that the optimum illumination angle for wave matching occurs when $\mathbf{k}_{\text{out}} = \mathbf{k}_{\text{in}} + \mathbf{K}$ so that

$$\sin\theta_B = \frac{|\mathbf{K}|}{2|\mathbf{k}|}. \tag{7.6}$$

The magnitude of the wave vectors are inversely proportional to the wavelengths (e.g., $|\mathbf{k}| = 2\pi/\lambda$), so that

$$\boxed{\sin\theta_B = \frac{\lambda}{2\Lambda},} \tag{7.7}$$

where Λ is the wavelength of the acoustic signal in the medium. In general, $\Lambda \gg \lambda$, so that

$$\theta_B \approx \sin\theta_B = \frac{\lambda}{2\Lambda}. \tag{7.8}$$

The optimum illumination for the Bragg mode is therefore at the off-axis *Bragg angle* θ_B, whereas the illumination is normal to the surface of the acousto-optic cell in the Raman-Nath mode.

In addition to the geometrical matching of the wave directions, conservation of energy requires that the frequency of light is shifted when it interacts with the traveling acoustic wave. We therefore find that

$$\left.\begin{aligned} \omega_+ &= \omega + \Omega \\ \omega_- &= \omega - \Omega \end{aligned}\right\}, \tag{7.9}$$

where ω_+ and ω_- refer to the radian frequency of the diffracted light in the positive and negative orders and where ω and Ω refer to the frequencies of the incident light and the sound wave. These relationships predict a frequency shift in the light, visualized by considering the motion of the sound wave. If the sound wave is moving toward the incident light, as shown in Figure 7.5(a), it shortens the wavelength of the diffracted light; the diffracted light is Doppler shifted upward by an amount equal to the frequency of the sound wave, as shown by ω_+ in Equation (7.9). The frequency ω_+ refers to the *upshifted* condition associated with the *positive diffracted order*. If we reverse the angle of the incident light on the acousto-optic cell, as shown in Figure 7.5(b), we see that $\mathbf{k}_{\text{out}}$ and $\mathbf{k}_{\text{in}}$ exchange positions to produce the negative diffracted order; the frequency of light is then downshifted. In this case, ω_- refers to the *downshifted* condition associated with the *negative diffracted order*.

Although we almost always use the Bragg mode in practice to produce the highest diffraction efficiency, there are compelling reasons to use the Raman-Nath mode in explaining basic optical signal-processing technology. In the Raman-Nath mode there is a nice degree of symmetry in the results and having the choice of which diffracted order to use is often convenient for developing the proper processing architectures, as we see in subsequent chapters. We generally confine our attention to the undiffracted light and the first positive and negative diffracted orders. The physical angles between the diffracted orders are typically of the order of milliradians; we exaggerate them in our diagrams for clarity.

7.2.3. Diffraction Angles, Spatial Frequencies, and Temporal Frequencies

Figure 7.6 shows the connections among diffraction angles, spatial frequencies, temporal frequencies, and acoustic wavelengths. For a given drive frequency f, we find that the acoustic wavelength is $\Lambda = v/f$. By definition, the spatial frequency is therefore $\alpha = 1/\Lambda$, which gives an

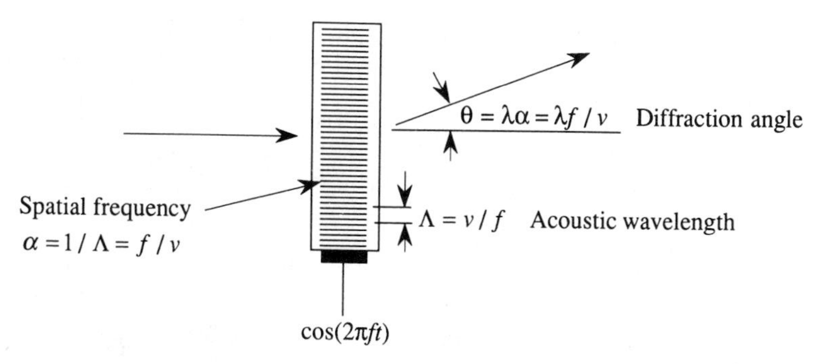

Figure 7.6. Relationships among wavelength, spatial frequency, and diffraction angle.

important relationship between spatial frequencies and temporal frequencies:

$$\boxed{\alpha \equiv \frac{1}{\Lambda} = \frac{f}{v}.} \tag{7.10}$$

Thus, there is a unique spatial frequency associated with every temporal frequency present in the acousto-optic cell. Furthermore, the diffracted ray angle is connected to the spatial and temporal frequencies by

$$\boxed{\theta = \lambda \alpha = \frac{\lambda f}{v},} \tag{7.11}$$

which nicely ties together all the important parameters.

When the drive signal is an arbitrary sum of cw components, light is diffracted over a large set of angles simultaneously, with angles and amplitudes determined by the frequencies and amplitudes of the cw components. In particular, suppose that the drive signal is

$$f(t) = \sum_{n=-\infty}^{\infty} a_n e^{j2\pi n f_0 t} \operatorname{rect}\left[\frac{nf_0 - f_c}{W}\right], \tag{7.12}$$

where f_0 is the smallest resolvable frequency. The rect function shows that the spectrum of $f(t)$ is centered at f_c and that it has bandwidth $W = f_2 - f_1 = K_2 f_0 - K_1 f_0$. This signal contains $M = K_2 - K_1 + 1$ discrete frequency components, each a multiple of f_0, beginning at frequency f_1 and ending at f_2. The signal $f(t)$ may be generated directly by an rf signal or by a baseband signal

$$s(t) = \sum_{n=0}^{\infty} a_n \cos(2\pi n f_0 t) \operatorname{rect}\left[\frac{n f_0}{W}\right], \tag{7.13}$$

that is multiplied by $\cos(2\pi f_c t)$, where $f_c = (f_2 + f_1)/2$, to put its spectrum at the center of the passband of the acousto-optic cell. Thus, the signal within the acousto-optic cell can be represented by M discrete temporal/spatial frequencies, leading to M discrete diffraction angles.

7.2.4. The Time Bandwidth Product

An important parameter in signal processing is the *time bandwidth product*, which is the product of the bandwidth and the time duration of the processed signal. The time bandwidth product tells us, in general, the degree of complexity of the signal or of an optical system. We begin the derivation of the time bandwidth product for an acousto-optic cell by relating the total angular deflection range $\Delta\theta$ to the bandwidth Δf:

$$\Delta\theta = \frac{\lambda}{v}\Delta f, \tag{7.14}$$

so that the deflection angle is linear in applied frequency. The number M of resolvable angles produced by the cell is

$$M = \frac{\text{angular range}}{\text{angular resolution}} = \frac{\Delta\theta}{\lambda/L}, \tag{7.15}$$

where L is the length of the acousto-optic cell and λ/L is the intrinsic angular resolution of any physical system, as discussed in Chapter 3, Section 3.5.2. We use Equation (7.14) in Equation (7.15) to find that

$$\boxed{M = \frac{\lambda \Delta f/v}{\lambda/L} = \frac{L}{v}\Delta f = T\Delta f = TW,} \tag{7.16}$$

where we made use of the fact that $L = vT$ and that the total bandwidth of the cell is $\Delta f = W$. Depending on the application, T is called the *transit time* of the cell, the *time delay* of the cell, the *access time* of the cell, or the *time duration* of the signal within the cell.

Equation (7.16) shows that the number of resolvable angles M is equal to the time bandwidth product of the cell. The angular resolution λ/L is predicated on illuminating the cell with a uniform-magnitude plane wave of light. If the incident wave is amplitude weighted in the x direction, as in most spectrum analysis applications, the angular resolution decreases; the value of M is therefore reduced accordingly. Time bandwidth products for acousto-optic cells are generally in the 1000–3000 range. The time bandwidth product is limited by the basic tradeoff between time and bandwidth and involves the physical limitations of important material properties such as attenuation and available crystal sizes. The reader is referred to the literature for more detailed discussions of the design relationships that govern acousto-optic cells (80, 81).

7.3. DYNAMIC TRANSFER RELATIONSHIPS

In Chapter 3, Section 3.8, we noted that the input/output relationship for an optical system is linear in amplitude when the system is coherently illuminated, is linear in intensity when the system is incoherently illuminated, or is linear in neither amplitude nor intensity when the system is partially coherently illuminated. Acousto-optic systems also provide linearity in any of these quantities depending on how they are illuminated and on the nature of the drive signal.

7.3.1. Diffraction Efficiency

A key performance parameter associated with acousto-optic cells is the amount of light diffracted into the first order. Suppose that the input electrical signal has voltage V_s and that the incident light has intensity I_0. The *diffraction efficiency* of the acousto-optic cell is defined as (80)

$$\eta \equiv m^2 \equiv \frac{I_d}{I_0} = \sin^2\left[\frac{\pi^2 P_s}{2\lambda^2}\frac{Z}{H}M_2\right]^{1/2}, \tag{7.17}$$

where I_d is the intensity of the diffracted light, I_0 is the intensity of the incident light, P_s is the acoustic power within the material, Z and H are the transducer dimensions as shown in Figure 7.3, and M_2 is a figure of merit used to evaluate acousto-optic configurations. The *figure of merit* is defined as

$$M_2 = \frac{n_0^6 p^2}{\rho v^3}, \tag{7.18}$$

where n_0 is the refractive index of the material, p is a strain-optic coefficient, ρ is the density of the material, and v is the sound velocity. From Equation (7.17) we note that the diffraction efficiency increases for large values of M_2; from Equation (7.18) we see that the diffraction efficiency, in turn, increases for high indices of refraction and low acoustic-wave velocities. Equation (7.17) shows that a large value for the interaction width Z and a small transducer height help to achieve high diffraction efficiencies.

A high diffraction efficiency must be balanced against increased attenuation as the sound propagates through the cell and against a reduction in the cell bandwidth. As the drive frequency f_j changes, a mismatch of the wave vectors occurs within the acousto-optic cell and the diffraction efficiency suffers. We express the 3 dB bandwidth Δf as (80)

$$\Delta f = \frac{2v^2}{\lambda f_c Z}, \tag{7.19}$$

where f_c is the center frequency of the cell. From Equations (7.17) and (7.19) we note a conflict among bandwidth, velocity, and interaction width. For example, acousto-optic cells that have low acoustic velocities have high diffraction efficiencies, as we see from Equation (7.17), but also tend to have narrow bandwidths, as we see from Equation (7.19).

An important parameter of acoustic devices is the attenuation of the acoustic wave as it propagates in the acousto-optic cell. Attenuation reduces the effective diffraction efficiency, reduces the effective time aperture of the cell, and reduces the frequency response. High attenuation also results in heating of the acoustic material, which can lead to effects such as a change in the acoustic velocity and defocusing of the optical beam. Some work has been done to obtain higher frequency responses and better time bandwidth products by cooling the acoustic material (82), taking advantage of the decrease in attenuation with temperature below 30 K (83).

The material must also have a high homogeneous optical transmission and low defect levels to minimize optical loss and scatter. Relatively few materials are sufficiently developed to provide high quality routinely. Single-crystal $LiNbO_3$ and TeO_2 are popular largely due to their availability, in spite of their less than optimum values for M_2 and attenuation; GaP is particularly well suited for operation at the wavelengths available from injection laser diodes.

7.3.2. Input / Output Relationships

From Equation (7.17) we see that the diffraction efficiency of the acousto-optic cell is given by

$$\eta = \sin^2\left(\sqrt{BP_s}\right), \tag{7.20}$$

where the input acoustic power P_s is proportional to the square of the input signal voltage V_s and B is a constant that accounts for the parameters associated with the interaction medium. The value of B is

$$B = \frac{\pi^2}{2\lambda^2}\frac{Z}{H}M_2, \tag{7.21}$$

where Z is the interaction width and H is the height of the transducer. The ratio of the amplitude of the diffracted light to the amplitude of the incident light is the *modulation index* as given by Equation (7.3) so that

$$m = \sqrt{\eta} = \sin\left(\sqrt{BP_s}\right). \tag{7.22}$$

For $\sqrt{BP_s} \ll \pi/2$, we can replace the sine function by its argument so that

$$m = \sqrt{BP_s} = B'V_s, \tag{7.23}$$

where B' is a new constant. As the modulation index is linearly proportional to the input voltage, we find that *the amplitude of the diffracted light is linearly proportional to the input voltage*. Figure 7.7(a) shows the operating condition for achieving linearity in amplitude; the signal is double-sideband modulated with a low modulation level so that the signal envelope stays on the linear part of the sine function.

We sometimes make the diffraction efficiency η proportional to the signal voltage V_s to operate the system so that it is linear in intensity. We first add a bias voltage V_b to V_s, so that the new drive signal $V_b + V_s$ is nonnegative. The diffraction efficiency is then

$$\eta = \sin^2\{B''[V_b + V_s]\}, \tag{7.24}$$

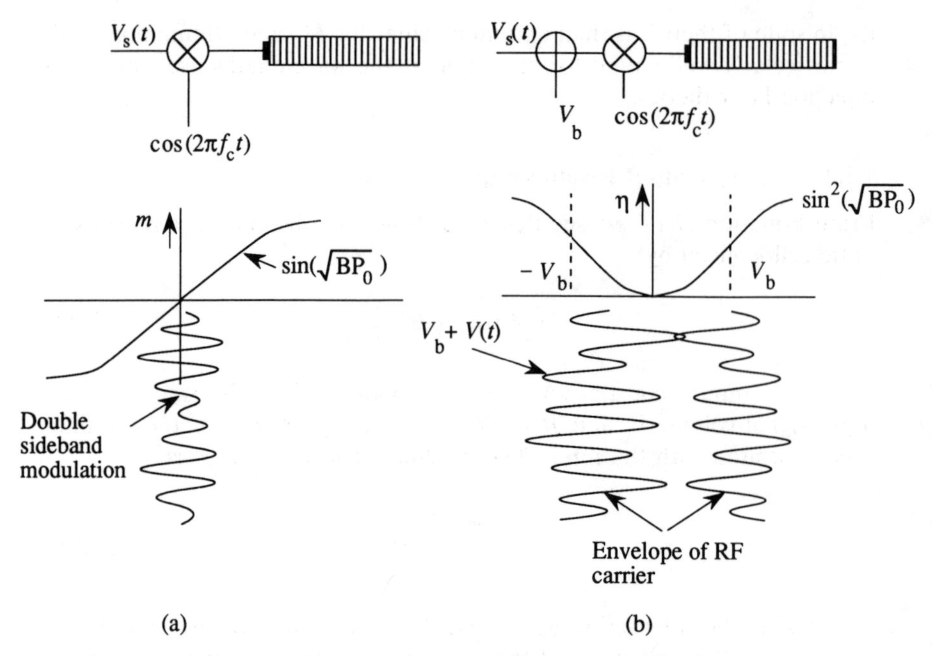

Figure 7.7. Operating conditions: (a) linearity in amplitude or (b) linearity in intensity.

where B'' is a constant. We use the fact that $\sin^2 x = (1 - \cos 2x)/2$ and that $\cos(x + y) = [\cos x \cos y - \sin x \sin y]$ to find that

$$\eta = \tfrac{1}{2}[1 - \cos(2B''V_b)\cos(2B''V_s) + \sin(2B''V_b)\sin(2B''V_s)]. \quad (7.25)$$

When we choose the bias so that $B''V_b = \pi/4$, the second term of Equation (7.25) vanishes and we achieve the highest degree of linearity. The diffraction efficiency then becomes

$$\eta = \tfrac{1}{2}\{1 + \sin[2B''V_s]\}. \quad (7.26)$$

If the value of the input voltage is small, we find that

$$\eta = \tfrac{1}{2}\{1 + 2B''V_s\}. \quad (7.27)$$

From Equation (7.27) we see that, for biased signals, *the intensity of the diffracted light is linearly proportional to the input signal voltage*. Figure 7.7(b) shows the operating conditions required for achieving linearity in intensity. Here we arrange for the signal envelope to be centered on the most linear region of the square of the sine function.

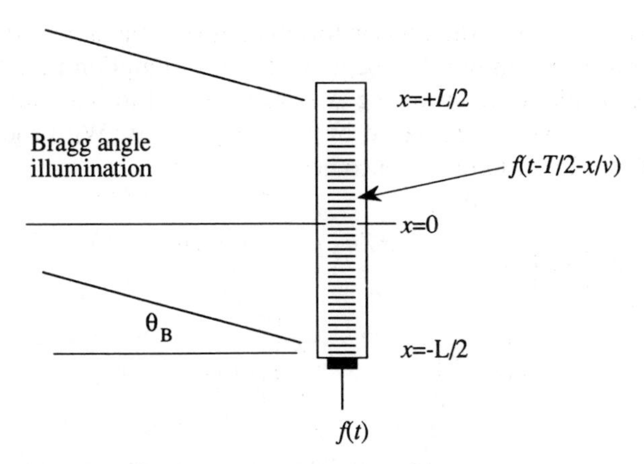

Figure 7.8. Traveling wave geometry.

Finally, the input voltage generally varies as a function of time. To represent the time dependence, we let $V_s \to f(t)$ represent the input signal, we let $m \to A(x, t)$ represent the output amplitude signal, and we let $\eta \to I(x, t)$ represent the output intensity signal.

7.4. TIME DELAYS AND NOTATION

We regard the acousto-optic cell as a delay line that is tapped optically, or as a device that stores a certain time history of the signal. The time signal propagates through the cell with velocity v so that it has characteristics of a traveling wave of the form $f(t - x/v)$, as shown in Figure 7.8, where x is a coordinate along the acousto-optic cell. As $x = 0$ represents the midpoint of the acousto-optic cell, we introduce a delay $T/2$ so that the signal becomes $f(t - T/2 - x/v)$. We use this notation so that when we evaluate the space/time function at the transducer, for which $x = -L/2$, we obtain $f(t)$ as expected. At the midpoint of the acousto-optic cell where $x = 0$, we obtain $f(t - T/2)$, which is the input signal delayed $T/2$ seconds. Finally, at the far end of the cell, where $x = L/2$, we obtain $f(t - T)$, which represents the eldest signal value and the maximum time delay.

7.5. PHASE-MODULATION NOTATION

For mathematical convenience, we want to represent complex-valued functions in phasor form instead of in trigonometric form. The transition from Equation (7.1) to the phasor form is not obvious unless the intermediate steps are spelled out. We begin with the assumption that Δn is small so that the cosine terms involving Δn are set equal to one and the sine of terms involving Δn are replaced by their arguments. With these conventions in place, Equation (7.1) becomes

$$A(x,t) = A_0\left\{\cos\left(2\pi f_l t + \frac{2\pi Z n_0}{\lambda}\right) - \frac{2\pi Z \Delta n}{\lambda} \times \cos\left[2\pi f_c\left(t - \frac{T}{2} - \frac{x}{v}\right)\right]\sin\left(2\pi f_l t + \frac{2\pi Z n_0}{\lambda}\right)\right\}, \quad (7.28)$$

where $2\pi Z n_0/\lambda$ is the phase associated with the optical path through the acousto-optic cell and f_c is the carrier frequency of the baseband signal that produces the index variations Δn. By some straightforward expansions using the relationships that $\sin\gamma = \cos(\gamma - \pi/2)$ and that $j = \exp(j\pi/2)$, we find

$$A(x,t) = A_0 \operatorname{Re}\left[e^{j(2\pi f_l t + 2\pi Z n_0/\lambda)}\left\{1 + jms\left(t - \frac{T}{2} - \frac{x}{v}\right) \times \cos\left[2\pi f_c\left(t - \frac{T}{2} - \frac{x}{v}\right)\right]\right\}\right], \quad (7.29)$$

where we have replaced the change in refractive index Δn by the corresponding time signal $ms(t - T/2 - x/v)$.

In earlier chapters we generally ignored the frequency of light in deriving the results because the light distributions with which we dealt were generally functions of space only. When we treat optical signal-processing systems using acousto-optic cells, the light distributions are functions of both space and time. It is therefore important to recognize that the light amplitude leaving the acousto-optic cell is given by Equation (7.29). For mathematical convenience, however, we often ignore the exponential terms in derivations and express the light distribution leaving the acousto-optic cell as

$$A(x,t) = A_0\left[1 + jms\left(t - \frac{T}{2} - \frac{x}{v}\right)\cos\left\{2\pi f_c\left(t - \frac{T}{2} - \frac{x}{v}\right)\right\}\right], \quad (7.30)$$

with the understanding that Equation (7.29) provides the complete result.

7.6. SIGN NOTATION

In optical signal-processing applications we generally use only one of the diffracted orders. We therefore further refine our description of the transmitted signal as just the positive or just the negative diffracted order. When we propagate the acoustic wave in the positive x direction, the light leaving the acousto-optic cell in the first positive diffracted order is expressed in phasor form by the equivalent function

$$f_+(x,t) = jma(x)s\left(t - \frac{T}{2} - \frac{x}{v}\right)e^{j2\pi f_c(t - T/2 - x/v)}. \tag{7.31}$$

In this expression, $f_+(x,t)$ indicates the amplitude of the light for the positive order just outside the acousto-optic cell. The *amplitude weighting function* $a(x)$ includes the illumination function, attenuation factors, the laser power level, and truncation effects due to the acousto-optic cell itself or other optical elements. The next term is the real-valued baseband signal $s(t)$, traveling in the positive x direction and time delayed by $T/2$. The final term implies that $s(t)$ has been multiplied by $\cos(2\pi f_c t)$ to translate it to the center frequency f_c of the acousto-optic cell. The positive sign associated with the overall argument of the exponential shows that the positive diffracted order is selected.

There are other combinations of illumination direction, acoustic-wave direction, and choice of diffracted order. The most general way to express the equivalent amplitude function of the acousto-optic cell at its exit face is

$$\boxed{f_\pm(x,t) = jma(x)s\left(t - \frac{T}{2} \pm \frac{x}{v}\right)e^{\pm j2\pi f_c(t - T/2 \pm x/v)}.} \tag{7.32}$$

In Equation (7.32) the $\pm$ signs combine to produce four possibilities. We first determine which diffracted order we need to use; this decision determines the $f_+(x,t)$ or $f_-(x,t)$ notation for the positive and negative orders, respectively. The sign of the exponential is the same as that of the diffracted order when the acoustic wave is propagating in the positive x direction. The frequency of light is therefore upshifted in the positive order and downshifted in the negative order. If the acoustic wave is propagating in the negative x direction, the sign associated with the exponential is opposite to that of the diffracted order; the sign associated with x/v in the argument of the signal envelope is then positive.

Recall that a basic assumption in the development of diffraction theory was that $\exp[j(\omega t - kr)]$ is a satisfactory eigenfunction for the wave equation. As the term that led to a positive kernel function for the spatial Fourier transform is $\exp(-jkr)$, we associate a positive exponential $\exp(j2\pi f_l t)$ with the temporal frequency of light. The last term in $f_{\pm}(x, t)$ is of the form $\exp(j2\pi f_c t)$ which shows that, relative to $\exp(j2\pi f_l t)$, the frequency of light is upshifted or downshifted because f_c is added to or subtracted from f_l.

The four possible configurations of the acousto-optic cell are shown schematically in Figure 7.9. These schematics are easy to construct and remember, plus they tell us pictorially the spatial direction of the diffracted light as well as its temporal shift. Note that both of these characteristics come from the last term of $f_{\pm}(x, t)$:

$$e^{\pm j2\pi f_c(t - T/2 \pm x/v)} = e^{\pm j2\pi f_c(t - T/2)} e^{\pm j2\pi f_c x/v}. \tag{7.33}$$

After the direction of the acoustic wave is chosen, we know whether the + or − sign is used for the x/v term; we then choose the + or − sign in the temporal frequency exponential to shift the frequency in the correct direction.

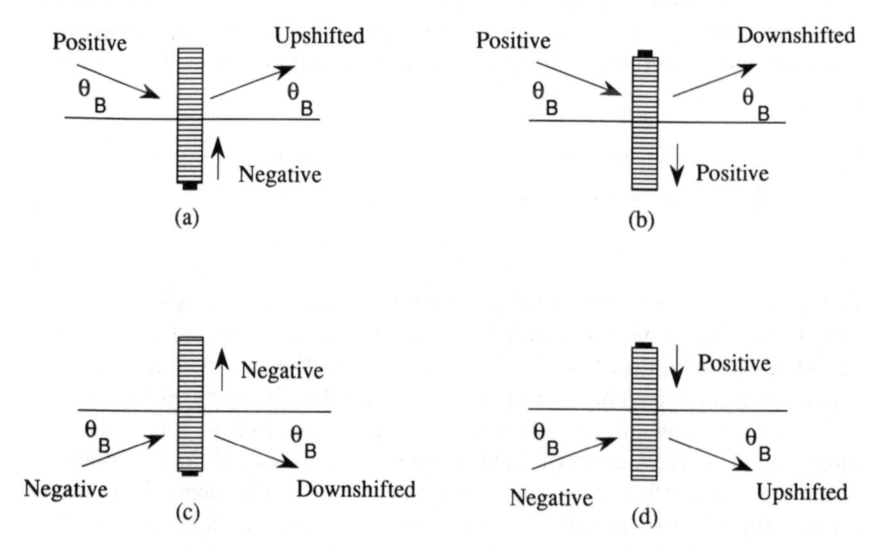

Figure 7.9. Various combinations of illumination and propagation directions.

7.7. CONJUGATE RELATIONSHIPS

The conjugate relationships between the positive and negative orders are obtained if we begin with the representation of the real-valued drive signal:

$$\begin{aligned}\text{Drive signal} &= |s(t)|\cos[2\pi f_c t + \phi(t)] \\ &= |s(t)|\text{Re}[e^{j[2\pi f_c t + \phi(t)]}],\end{aligned} \tag{7.34}$$

where $\phi(t)$ is the phase of the complex-valued signal $s(t)$. From Equation (7.33) we see that the terms in x/v have the same sign, as they are due solely to the direction of acoustic-wave propagation. Suppose that we propagate the acoustic wave into the positive x direction so that this sign is negative. If we arrange the illumination direction so that the light is downshifted in frequency, the signal becomes

$$\begin{aligned} & jma(x)\left|s\left(t - \frac{T}{2} - \frac{x}{v}\right)\right| e^{-j[2\pi f_c(t-T/2-x/v) - \phi(t-T/2-x/v)]} \\ &\qquad = jma(x)s^*\left(t - \frac{T}{2} - \frac{x}{v}\right)e^{-j2\pi f_c(t-T/2-x/v)}. \end{aligned} \tag{7.35}$$

Thus we see that the *downshifted signal is the conjugate of the upshifted signal*. When we want to obtain a "true" signal spectrum, we use the upshift mode of operation; when we want to obtain the conjugate of the spectrum of a signal, we simply use the downshifted mode of operation. Nice, and useful, too.

7.8. VISUALIZATION OF THE ACOUSTO-OPTIC INTERACTION

The observed light intensity leaving an acousto-optic cell operating in the Raman-Nath mode is the same as the illuminating beam because the effect of the acoustic interaction is pure phase modulation. In Figure 7.10(a) we use a Schlieren method (see Chapter 4, Section 4.3.1) to help visualize the interaction of light and sound for a single-channel acousto-optic cell driven by a pure rf signal at 400 MHz (84). The acoustic waves spread in the vertical direction as they propagate away from the transducer, which is at the left end of the cell. In the region near the transducer, we see some detailed structure caused by acoustic interference; this structure is equivalent to the near-field diffraction pattern or the Fresnel diffraction of a slit,

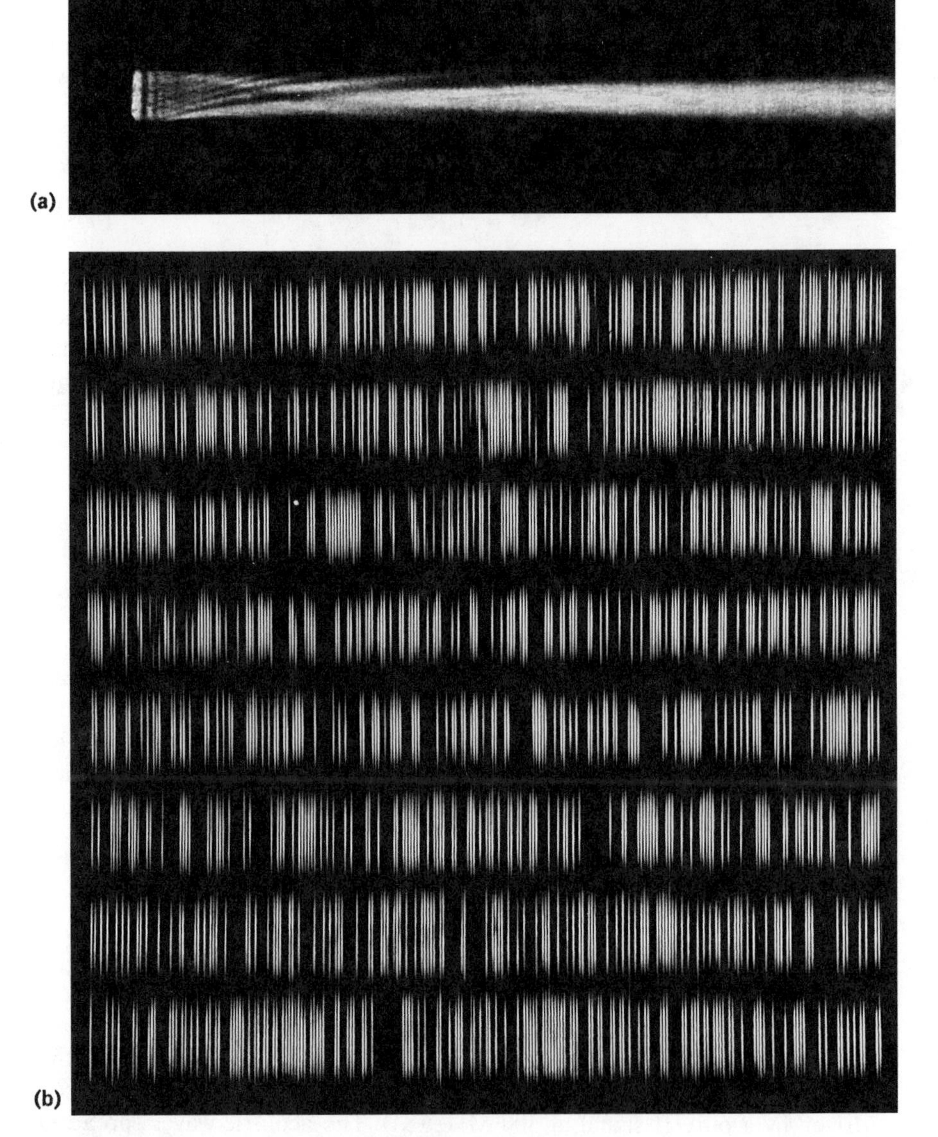

Figure 7.10. Schlieren images of acoustic waves within Bragg cells: (a) single channel cell and (b) multichannel cell (courtesy H. N. Roberts et al.) (85).

as we showed in Chapter 3, Section 3.2.5. In fact, the acoustic diffraction pattern is a scaled version of the diffraction patterns that occur at light wavelengths; the scaling factor is Λ/λ.

Figure 7.10(b) shows a Schlieren image of an eight-channel acousto-optic cell, wherein the center frequency f_c is modulated by eight different pseudorandom sequences (85). The acoustic spreading is not as pronounced in this example because the time bandwidth product of the cell is lower. Multichannel acousto-optic cells with up to 128 active channels have been built, although 32 channel cells are more commonly available.

7.9. APPLICATIONS OF ACOUSTO-OPTIC DEVICES

Acousto-optic cells are configured in different ways, such as with single-channel or multichannel transducers. They perform different functions, such as modulating light temporally, deflecting light, or serving as a delay line. In this section we review some of these functions.

7.9.1. Acousto-Optic Modulation

For optical signal processing we are primarily interested in using acousto-optic cells as delay lines or as short-term memories. They are sometimes useful, however, as *modulators* to impart a temporal variation onto the entire optical beam. In our model of these applications, we reduce the length of acousto-optic cell by letting $L \to 0$ so that the space/time signal $f(t - T/2 - x/v)$ degenerates to the purely temporal signal $f(t)$, as illustrated in Figure 7.11. The illumination must be a focused beam because the rise time associated with the modulation bandwidth is proportional to the time τ required for the acoustic beam to travel across the optical beam. As the acoustic transit time decreases, the rise time decreases so that the corresponding bandwidth increases.

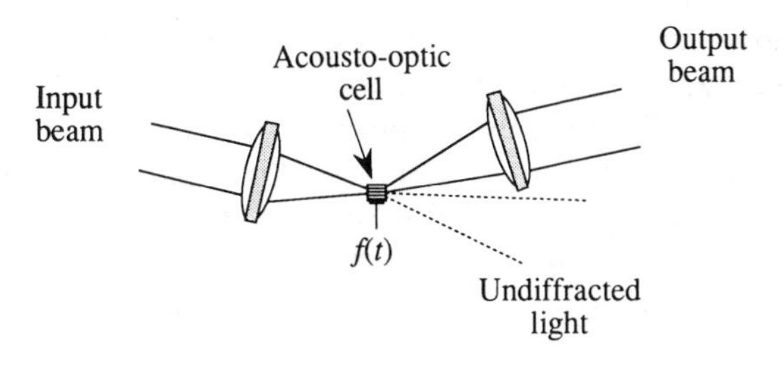

Figure 7.11. Acousto-optic temporal modulator.

As the bandwidth of the acousto-optic cell increases, however, the output cone of light becomes elliptical because the optical and acoustic waves do not match over the full range of input ray angles (80). The momentum-conservation law is given by $\mathbf{k}_{\text{out}} = \mathbf{k}_{\text{in}} + \mathbf{K}$, as shown in Section 7.2.2. In a modulator, the incident light beam has a range of $\mathbf{k}_{\text{in}}$ vectors of constant magnitude, distributed over an angular range $\delta\theta_0$ because the light beam is convergent. To satisfy the vector relationship, the acoustic wave must have a corresponding range of angular directions $\delta\theta$. When $\delta\theta < \delta\theta_0$, momentum cannot be conserved for all optical wave components; this loss of momentum produces an elliptical output beam cross section. If $\delta\theta > \delta\theta_0$, the bandwidth is reduced.

The 10–90% rise time t_r for an acousto-optic modulator is a function of the transit time T. For a Gaussian input beam profile, truncated at the $1/e^2$ points in intensity, we find that $t_r \approx T/1.5$ and that the frequency response rolloff β, defined in decibels, is (80)

$$\beta = 10 \log\left[e^{-\pi^2 f^2 T^2/8}\right], \tag{7.36}$$

where f is the frequency. We solve for the cutoff frequency f_{co} at which the response has rolled off to the value β:

$$f_{\text{co}} = \frac{c\sqrt{\beta}}{\pi T}, \tag{7.37}$$

where $c = \sqrt{0.8 \ln 10} \approx 1.4$. The relationship between rise time and modulation bandwidth is, therefore, that

$$\Delta f = \frac{0.29\sqrt{\beta}}{t_r}. \tag{7.38}$$

Note that Δf is the *modulation bandwidth of a baseband signal*; the rf bandwidth is twice Δf because we retain both the upper and lower sidebands of the signal about the carrier frequency.

Acousto-optic cells also have application as Q switches, mode lockers, cavity dumpers, and other devices associated with lasers for controlling light (79–81). Table 7.1 gives the properties of selected interaction materials as used in specific acousto-optic configurations (80). The attenuation of these devices is indicated by Γ and is stated as the number of decibels of attenuation per μsec of cell length per GHz^2 of the applied frequency. For example, a 3-μsec cell made from $LiNbO_3$ and operating at a frequency of 750 MHz has an attenuation of $0.098 \times 3 \times (0.75)^2 = 0.17$ dB at the end of the cell.

Table 7.1 Properties of Selected Acousto-Optic Interactions

Material	Velocity v (10^3 m/s)	Index n	Attenuation Γ dB/(μsec · Ghz^2)	Figure of Merit M_2 (10^{-15} sec^3/kg)
$LiTaO_2$	6.19	2.18	0.062	1.37
$LiNbO_3$	6.57	2.20	0.098	7.00
TiO_2	8.03	2.584	0.566	3.93
$Sr_{0.75}Ba_{0.25}Nb_2O_6$	5.50	2.299	2.20	38.6
GaP	6.32	3.31	3.80	44.6
TeO_2 (longitudinal)	4.20	2.26	6.30	34.6
TeO_2 (slow shear)	0.617	2.26	17.6	1200

The data given in Table 7.1 is representative of the interaction parameters. The specific values are dependent on factors such as the strain mode (longitudinal or shear), the polarization and direction of the incident light, and the acoustic **K** vector direction with respect to the crystal axes. Furthermore, the acoustic attenuation dependence on frequency is sometimes proportional to the 1.5 power instead of the square of the frequency. In a similar fashion, the figure of merit M_2 is not a pure material constant but is dependent on the factors cited above.

7.9.2. Acousto-Optic Beam Deflectors

The acousto-optic cell, when driven by a cw frequency, behaves as a random access beam deflector which addresses a specific position at the focal plane of a lens. Figure 7.12 shows an acousto-optic cell at plane P_1 with an acoustic velocity v and a length $L = vT$. We drive the cell with a signal $f(t) = \cos(2\pi f_k t)$ to access the kth spot position in the scan line. This signal produces a positive diffracted order whose Fourier transform is

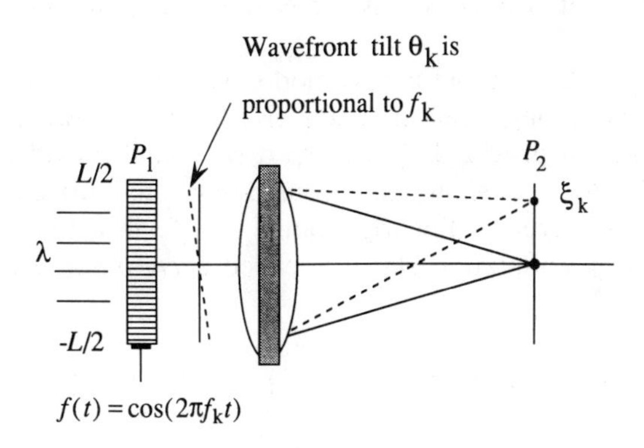

Figure 7.12. Acousto-optic scanner.

generated by a lens with focal length F:

$$F_+(\xi, t) = \int_{-\infty}^{\infty} f_+(x, t) e^{j(2\pi/\lambda F)\xi x}\, dx, \tag{7.39}$$

where $f_+(x, t) = \text{rect}(x/L)\exp[j2\pi f_k(t - T/2 - x/v)]$. The Fourier transform of $f_+(x, t)$ is

$$\begin{aligned} F_+(\xi, t) &= \int_{-L/2}^{L/2} e^{j2\pi f_k(t - T/2 - x/v)} e^{j(2\pi/\lambda F)\xi x}\, dx \\ &= e^{j2\pi f_k(t - T/2)} \operatorname{sinc}\left[\left(\frac{\xi}{\lambda F} - \frac{f_k}{v}\right)L\right], \end{aligned} \tag{7.40}$$

when we ignore amplitude scaling factors. From Equation (7.40), we find that the lens focuses light at the spatial position

$$\xi_k = \frac{\lambda F}{v} f_k \tag{7.41}$$

at plane P_2. This result shows that the spot position is linearly proportional to the applied frequency. The first zero of the sinc function occurs at

$$\xi_0 = \frac{\lambda F}{L} = d_0, \tag{7.42}$$

in accordance with basic Fourier-transform theory. We use d_0 both as a measure of the spot size as well as the Nyquist sampling interval at plane P_2.

In addition to the random-access mode, we can use cw frequencies to scan a light beam along a line in a stepwise fashion. However, a continuous scanning action provided by a chirp drive signal provides higher line scan rates. Figure 7.13 shows an acousto-optic cell driven by a signal whose frequency increases linearly from f_1 to f_2 in a time duration T_c. Such a frequency-modulation signal is called a *chirp signal* as characterized by

$$c(t) = \cos(2\pi f_1 t + \pi a t^2); \qquad 0 \le t \le T_c, \tag{7.43}$$

where a is the *chirp rate*, expressed in Hz/sec and T_c is the *chirp*

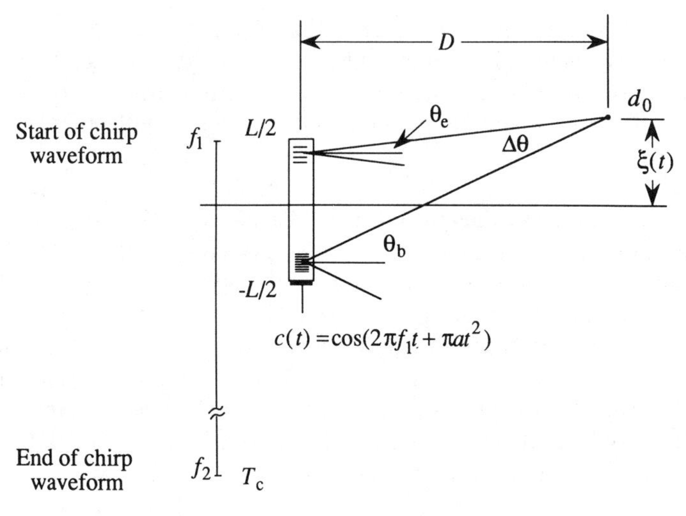

Figure 7.13. Linear scanning with chirp waveform.

duration. From Equation (7.43), we see that the *instantaneous frequency* f_i is

$$\begin{aligned} f_i &= \frac{1}{2\pi}\frac{\partial}{\partial t}\left(2\pi f_1 t + \pi a t^2\right) \\ &= f_1 + at; \qquad 0 \leq t \leq T_c. \end{aligned} \tag{7.44}$$

The instantaneous frequency of the chirp sweeps over the bandwidth $W = f_2 - f_1$ of the acousto-optic cell in the chirp duration T_c so that the chirp rate a is

$$\boxed{a = \frac{W}{T_c}.} \tag{7.45}$$

Because the instantaneous frequency at the end of the acousto-optic cell is f_e, the frequency at the beginning of the cell must be

$$f_b = f_e + aT = f_e + \frac{WT}{T_c}. \tag{7.46}$$

In the example shown, the chirp frequency is increasing in time, generally called the *upchirp* condition; the chirp frequency may also decrease in time, called the *downchirp* condition.

The behavior of the scanning action produced by the cell can be explained by using elementary diffraction theory and geometrical ray tracing or by using a diffraction integral. Each method provides useful insights into the scanning phenomena; we begin with the ray tracing approach.

7.9.2.1. Linear Scanning with Chirp Waveforms: The Ray-Tracing Approach. Consider a ray trace for a stationary chirp segment that has just filled the acousto-optic cell, as shown in Figure 7.13. Basic diffraction theory shows that the instantaneous frequency in the small region near the end of the cell produces an undiffracted waveform, indicated by a ray traveling parallel to the optical axis, along with positive and negative diffracted waveforms, indicated by rays that each make an angle θ_e with respect to the undiffracted light. The diffraction angle is related to the spatial frequency α_e and the temporal frequency f_e by

$$\theta_e = \lambda\alpha_e = \frac{\lambda f_e}{v}. \tag{7.47}$$

A similar relationship holds for the region near the beginning of the cell:

$$\theta_b = \lambda\alpha_b = \frac{\lambda f_b}{v} = \frac{\lambda(f_e + WT/T_c)}{v}. \tag{7.48}$$

When we trace the rays associated with the positive diffracted orders of each subregion within the cell, we find that they intersect a distance D from the cell to form a spot whose size is d_0. For the small diffraction angles produced by the acousto-optic cell, the included angle between the extreme rays is

$$\Delta\theta = \theta_b - \theta_e = \frac{\lambda WT}{vT_c}, \tag{7.49}$$

so that the distance to the plane of focus is

$$D = \frac{L}{\Delta\theta} = \frac{vLT_c}{\lambda WT} = \frac{v^2T_c}{\lambda W}. \tag{7.50}$$

From Equation (7.50) we find a useful relationship between the chirp rate a and the radius of curvature D of the chirp wavefront within the cell. By rearranging the factors, we find that

$$\boxed{\frac{v^2}{\lambda D} = \frac{W}{T_c} = a.} \tag{7.51}$$

The key geometrical parameters of the acoustic signal, such as v, λ, and D, are on the left of this relationship while the key drive signal parameters, such as W and T_c, are on the right.

The spot size, following the Rayleigh criterion as given in Chapter 3, Section 3.5.2, is $d_0 = \lambda/\Delta\theta$, where $\Delta\theta$ is the angle between the two rays. For the configuration of Figure 7.13, we find that

$$d_0 = \frac{\lambda}{\Delta\theta} = \frac{vT_c}{TW}. \tag{7.52}$$

For a given chirp duration, the spot size is inversely proportional to the time bandwidth product of the cell.

The scanning velocity is most easily calculated by noting that the spot position, as a function of time, is

$$\begin{aligned} \xi(t) &= -\frac{L}{2} + D\theta_b(t) \\ &= -\frac{L}{2} + D\frac{\lambda(f_e + Wt/T_c)}{v}, \end{aligned} \tag{7.53}$$

where we used the general form of Equation (7.48) to produce Equation (7.53). The scanning velocity v_s is then

$$v_s = \frac{\partial}{\partial t}\xi(t) = D\frac{\lambda W}{vT_c}. \tag{7.54}$$

We now use the value of D from Equation (7.50) in Equation (7.54) to find that

$$v_s = \frac{v^2 T_c}{\lambda W}\frac{\lambda W}{vT_c} = v. \tag{7.55}$$

The scanning velocity is therefore always equal to the acoustic velocity and cannot be controlled by any of the system parameters.

The length of the scan line is equal to the product of the scan velocity and the *active* scanning time. Scanning begins at $t = T$ and continues until $t = T_c$ so that the active scan time interval is $(T_c - T)$. The length of the scan line is therefore

$$L_s = v_s(T_c - T) = v(T_c - T) = \left[\frac{T_c}{T} - 1\right]L, \tag{7.56}$$

so that the scan line is longer than the length of the acousto-optic cell. The number of samples in a scan line is

$$\boxed{M = \frac{L_s}{d_0} = \frac{v(T_c - T)}{vT_c/TW} = \left[1 - \frac{T}{T_c}\right]TW,} \tag{7.57}$$

so that the number of samples in a scan line approaches the time bandwidth product of the cell, if $T_c \gg T$.

7.9.2.2. Linear Scanning with Chirp Waveforms: The Diffraction Approach. To control the scanning velocity, we must introduce a lens to the right of the acousto-optic cell. To analyze this condition, we use the diffraction method exclusively; in the process, we develop some new analytical tools and provide other useful insights. Furthermore, we can now more fully address the effects produced by the temporal characteristics of the chirp waveform.

Figure 7.14 shows a condition for which the acousto-optic cell aperture is small compared to the chirp duration. The chirp duration T_c is the time between the lowest and highest frequencies of the chirp, the difference

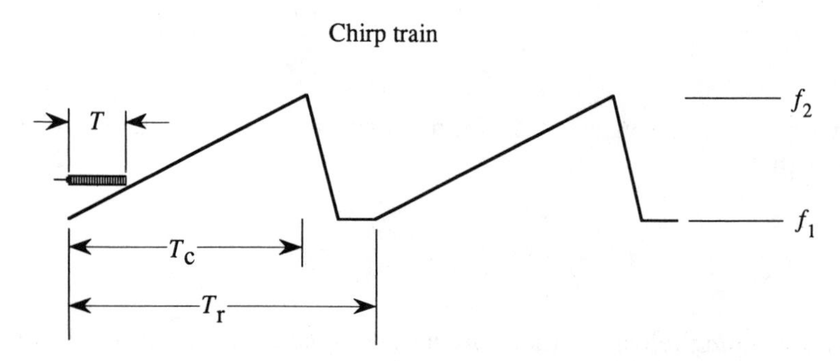

Figure 7.14. Frequency/time relationship for a chirp train.

being $W = f_2 - f_1$. The *repetition period* T_r is the time interval between a given point on one chirp segment and a similar point on the next chirp segment, for example, the time between the highest frequencies of two adjacent segments. The repetitive nature of the *chirp train* between the highest frequencies of two adjacent segments. The repetitive nature of the *chirp train*, shown in Figure 7.14, is expressed by a time convolution of the chirp signal with an impulse train:

$$\begin{aligned} f(t) &= c(t) * \sum_{n=-\infty}^{\infty} \delta(t - nT_r) \\ &= \cos(2\pi f_1 t + \pi a t^2) * \sum_{n=-\infty}^{\infty} \delta(t - nT_r), \end{aligned} \tag{7.58}$$

where T_r is the repetition period of the chirp train.

We consider the general case in which T_c may be larger than, comparable to, or even less than, the aperture time T of the acousto-optic cell. We classify scanners according to two criteria: the active aperture time and the active scan time. If $T_c \geq T$, the active aperture time is governed by the length of the acousto-optic cell; we refer to this condition as the *long-chirp scanner*. If $T_c < T$, the active aperture time is governed by the length of the chirp, we refer to this condition as the *short-chirp scanner*. For the long-chirp scanner, the active scan time is $T_s = T_c - T$, as noted in Section 7.9.2.1. For the short-chirp scanner, the active scan time is $T_s = T - T_c$. These two scan times can be combined to give a single active scan time of $T_s = |T - T_c|$.

When we use a voltage-controlled oscillator to generate the chirp signal, T_r must be greater than T_c because the signal does not return instantaneously from f_2 to f_1. A part of the chirp train is therefore not available for active scanning. We can, however, arrange for the chirp waveforms to overlap to an arbitrary extent by impulsing a surface acoustic wave device that produces a chirp waveform at arbitrary repetition intervals. The active scan time is then either T_s or T_r, whichever is shorter. When the active scan time is T_s, the system is *aperture limited*. When the active scan time is T_r, the system is *repetition rate limited*.

7.9.2.2.1. A Long-Chirp, Aperture-Limited Scanner. There are four basic scanner configurations: a long or short chirp scanner; each is either aperture or repetition rate limited. We begin our diffraction analysis for a long-chirp scanner that is aperture limited. The optical arrangement is essentially the same as that shown in Figure 7.12, except that the acousto-optic cell is now driven by a chirp signal represented by Equation (7.58).

For this exercise, we select the negative diffracted order, whose amplitude just to the right of the acousto-optic cell is

$$f_{-}(x,t) = \text{rect}(x/L)e^{-j[2\pi f_1(t-T/2-x/v)+\pi a(t-T/2-x/v)^2]}; \qquad T \le t \le T_c, \tag{7.59}$$

where we have dropped scaling factors and used uniform illumination. The scan time starts at $t = T$ and finishes when the end of the chirp segment arrives at the transducer. For the moment we assume that the lens is in contact with the acousto-optic cell; we show how to handle a finite separation later in this section. A positive lens is represented by the phase response

$$h(x) = e^{j(\pi/\lambda F)x^2}, \tag{7.60}$$

so that the light distribution to the right of the lens is

$$f_{-}(x,t) = \text{rect}(x/L)e^{-j[2\pi f_1(t-T/2-x/v)+\pi a(t-T/2-x/v)^2]}e^{j(\pi/\lambda F)x^2}; \qquad T \le t \le T_c. \tag{7.61}$$

The light distribution at any plane a distance D_f to the right of the lens is given by the Fresnel transform of $r_{-}(x,t)$:

$$F(\xi,t) = \int_{-\infty}^{\infty} f_{-}(x,t)e^{-j(\pi/\lambda D_f)(\xi-x)^2}\,dx. \tag{7.62}$$

We substitute Equation (7.61) into Equation (7.62), to find that

$$F(\xi,t) = \int_{-\infty}^{\infty} \text{rect}(x/L)e^{-j[2\pi f_1(t-T/2-x/v)+\pi a(t-T/2-x/v)^2]} e^{j(\pi/\lambda F)x^2}e^{-j(\pi/\lambda D_f)(\xi-x)^2}\,dx; \qquad T \le t \le T_c. \tag{7.63}$$

We use Equation (7.51) in Equation (7.63) to find that

$$\begin{aligned} F(\xi,t) &= e^{-j2\pi f_1(t-T/2)}\int_{-\infty}^{\infty} \text{rect}(x/L)e^{j2\pi f_1 x/v}e^{-j[(\pi v^2/\lambda D)(t-T/2-x/v)^2]} \\ &\qquad \times e^{j(\pi/\lambda F)x^2}e^{-j(\pi/\lambda D_f)(\xi-x)^2}\,dx \\ &= e^{j\phi}\int_{-L/2}^{L/2} e^{j(\pi x^2/\lambda)[1/F-1/D-1/D_f]} \\ &\qquad \times e^{j(2\pi x/\lambda)[v(t-T/2)/D+\xi/D_f+\lambda f_1/v]}\,dx; \qquad T \le t \le (T_c - T), \end{aligned} \tag{7.64}$$

where we collect all phase factors that are not functions of x into the term ϕ. Note that the chirp rate $a = W/T_c$ is positive when we use the upchirp mode, as we do here, and negative when we use the downchirp mode of modulation.

The focal position occurs when the integral has its maximum value so that the light intensity is highest. The integral in Equation (7.64) has its maximum value when the integrand is set equal to one. Let us begin, however, by setting just the value of the exponential that is quadratic in x equal to one. The first condition necessary to obtain focus is therefore that

$$\frac{1}{F} - \frac{1}{D} - \frac{1}{D_f} = 0, \tag{7.65}$$

or that

$$D_f = \frac{DF}{D - F}. \tag{7.66}$$

When Equation (7.66) is satisfied, Equation (7.64) produces the spatial light distribution at the focal point:

$$\begin{aligned} F(\xi, t) &= \int_{-L/2}^{L/2} e^{j(2\pi x/\lambda)[v(t-T/2)/D + \xi/D_f + \lambda f_1/v]}\, dx \\ &= L \operatorname{sinc}\left[\frac{v(t - T/2)L}{\lambda D} + \frac{\xi(D - F)L}{\lambda FD} + \frac{f_1 L}{v}\right]; \qquad T \le t \le T_c. \end{aligned} \tag{7.67}$$

The position of the scanning spot at any instant in time is found by setting the argument of the sinc function equal to zero, equivalent to setting the value of the exponential in Equation (7.64) that is linear in x equal to 1:

$$\xi = -\frac{\lambda f_1 DF}{v(D - F)} - \frac{v(t - T/2)F}{D - F}, \qquad T \le t \le T_c. \tag{7.68}$$

The spot position at the beginning of scan when $t = T$ is

$$\xi_b = -\frac{\lambda f_1 DF}{v(D - F)} - \frac{v(T/2)F}{D - F}, \tag{7.69}$$

and the spot position at the end of scan when $t = T_c$ is

$$\xi_e = -\frac{\lambda f_1 DF}{v(D-F)} - \frac{v(T_c - T/2)F}{D-F}, \tag{7.70}$$

so that the length of scan is

$$L_s = |\xi_e - \xi_b| = \left|\frac{v(T_c - T)F}{D-F}\right|. \tag{7.71}$$

The scanning velocity is readily obtained from Equation (7.68) as

$$\boxed{v_s = \frac{\partial \xi}{\partial t} = -\frac{vF}{D-F}.} \tag{7.72}$$

The scanning velocity v_s has the same or opposite direction as v depending on the value D of the wavefront radius of curvature. When we use a configuration in which the acoustic wave is traveling in the positive x direction, the rules are that:

1. When D is positive and greater than F, as for the case analyzed here, v_s is negative so that the spot moves in the negative x direction. In this case, the chirp signal in the acousto-optic cell is equivalent to a negative lens whose focal length is longer than that of the positive lens. The light therefore focuses at some plane to the right of the lens because the distance D_f is positive, as we see from Equation (7.66).
2. When D is negative, the scanning spot moves in the positive x direction. In this case, the focal length of the chirp is equivalent to a positive lens and the net result, as confirmed by Equation (7.66), is that of two positive lenses working together.
3. When D is positive and less than F, the scanning velocity is positive so that the spot moves in the same direction as v, but the light does not focus anywhere to the right of the lens. In this case, the focal length of the chirp is equivalent to a negative lens whose focal length is shorter than that of the positive lens, and Equation (7.66) confirms that D_f is negative.

These rules are illustrated in Figure 7.15. The negative diffracted order satisfies the first rule. The value of D is positive, equivalent to stating that

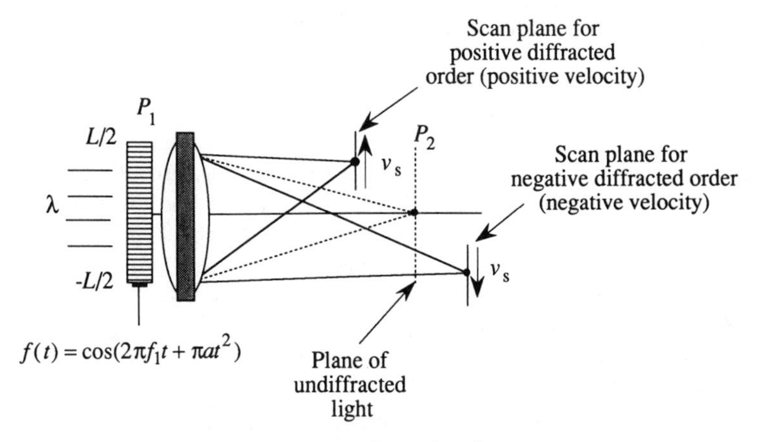

Figure 7.15. Scanning action diagram.

the focal power of the chirp is negative so that the scan plane lies to the right of the plane at which the undiffracted light is focused. As the chirp signal flows through the acousto-optic cell, the ray angles increase in the negative direction, leading to a negative scan velocity. The start-of-scan position, as seen from Equation (7.69), is negative as is the end-of-scan position, as we see from Equation (7.70).

The positive diffracted order satisfies the second rule. The value of D is negative, equivalent to stating that the focal power of the chirp is positive, so that the scan plane lies to the left of the plane at which the undiffracted light is focused. As the chirp signal flows through the acousto-optic cell, the ray angles increase in the positive direction, leading to a positive scan velocity. The start-of-scan position, as seen from Equation (7.69), is positive; the end-of-scan position, as we see from Equation (7.70), is also positive. The positive diffracted order exists for all values of D because the equivalent focal length of two lenses with positive powers must be positive.

Rule three applies to a special case for the negative diffracted order and states that the light may not focus at any plane to the right of the lens for certain values of D. For example, as the value of D approaches F, the negative power due to the chirp signal subtracts from the positive power of the lens; the focal plane for the negative diffracted order therefore recedes to infinity. When $D = F$, the two focal powers exactly cancel and the focal plane is at infinity. As stated in the third rule, the negative diffracted order does not focus at any plane to the right of the lens if D is positive and less than F; it generates a virtual scan plane.

When we drive the acousto-optic cell with a downchirp signal of the form

$$f(t) = \cos(2\pi f_2 t - \pi a t^2), \tag{7.73}$$

instead of with the upchirp signal, the same general results apply except that we interchange f_1 and f_2 to account for the different starting frequency and replace D by $-D$ to account for the negative chirp rate. The roles of the two scan planes shown in Figure 7.16 are then interchanged so that the negative diffracted order focuses to the left of the positive diffracted order. As expected, the scan velocities also have opposite signs so that the spots scan toward the optical axis instead of away from the optical axis.

Equation (7.72) shows how to control the scanning velocity by selecting the value of the focal length of the lens. For a desired scan velocity, the required focal length of the lens is

$$F = \frac{D}{1 - v/v_s}. \tag{7.74}$$

The signs of D and v_s can combine, according to the rules, only to cause the focal length of the lens to be positive.

The size of the scanning spot is obtained from Equations (7.52) and (7.66):

$$d_0 = \frac{\lambda}{L/D_f} = \frac{\lambda DF}{(D - F)L}. \tag{7.75}$$

As the chirp radius of curvature $D \to \infty$, the spot size tends to a value of $d_0 = \lambda F/L$, as expected, because then the chirp waveform contributes no power to the system; the lens alone acts on the diffracted light.

The number of samples in the scan line is

$$M = \frac{L_s}{d_0} = \left| \frac{v(T_c - T)L}{\lambda D} \right|. \tag{7.76}$$

We use Equation (7.51) in Equation (7.76) to find that

$$M = \left[1 - \frac{T}{T_c}\right] TW, \tag{7.77}$$

just as we found from the geometrical analysis.

The *scan duty cycle* is defined as the ratio of the active scan time divided by the repetition period:

$$U = \frac{\min(T_s, T_r)}{T_r} = \frac{T_c - T}{T_r}. \tag{7.78}$$

The *sample rate* at which samples are recorded is given by the ratio of the scan velocity to the spot size:

$$R_s = \frac{|v_s|}{d_0} = \frac{T}{T_c} W. \tag{7.79}$$

The throughput rate is the average number of samples recorded per unit time and is the product of the sample rate and the scan duty cycle;

$$R_t = U R_s = \left[1 - \frac{T}{T_c}\right] \frac{T}{T_r} W, \tag{7.80}$$

where we have used Equations (7.78) and (7.79) to produce Equation (7.80).

If the acousto-optic cell and the lens are not in contact, as shown in Figure 7.15, we can use the thin-lens formula to find the equivalent position of the plane at which the light is focused. If the separation between two thin lenses with powers $K_1 = 1/F_1$ and $K_2 = 1/F_2$ is z_{12}, the equivalent power of the combination, as given in Chapter 2, Section 2.5.8, is

$$K_{eq} = K_1 + K_2 - z_{12} K_1 K_2. \tag{7.81}$$

We associate the power of the chirp signal in the acousto-optic cell with K_1 so that $K_1 = -1/D$ and associate the lens power with K_2. The net power of the combination gives the distance to the scan plane from the acousto-optic cell: $D_f = 1/K_{eq}$.

7.9.2.2.2. A Short-Chirp, Aperture-Limited Scanner. In the short-chirp scanner, the active aperture time is limited by the chirp duration T_c. Suppose that one of the chirp segments from the chirp train is completely within the acousto-optic cell, as shown in Figure 7.16. The relationships given from Equation (7.67) onward are modified for application to the short-chirp, aperture-limited scanner. For example, the start-of-scan time is T_c and the end-of-scan time is $T - T_c$, and the integrations are over a

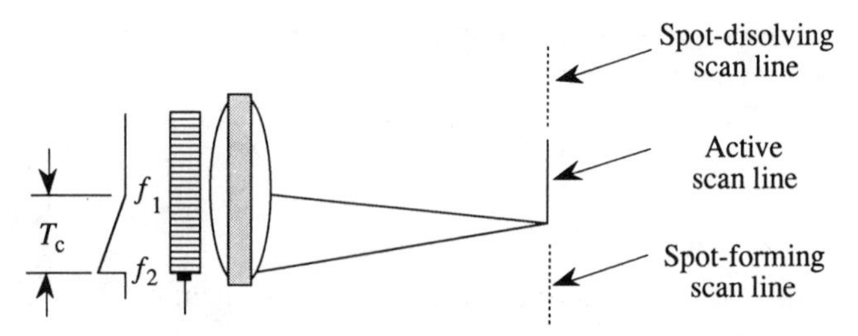

Figure 7.16. Short-chirp scanner.

spatial range $L_c = vT_c$. We study the diffraction phenomenon as the chirp transitions into and out of the cell at the end of this section.

The new form of Equation (7.67) becomes

$$F(\xi, t) = \int_{-L/2}^{(L-L_c)/2} e^{j(2\pi x/\lambda)[v(t-T/2)/D+\xi/D_f+\lambda f_1/v]}\, dx$$

$$= L_c \operatorname{sinc}\left[\frac{v(t - T/2)L_c}{\lambda D} + \frac{\xi(D - F)L_c}{\lambda FD} + \frac{f_1 L_c}{v}\right];$$

$$T_c \leq t \leq (T - T_c), \quad (7.82)$$

where we ignore unessential magnitude and phase factors. As before, the position of the scanning spot at any instant in time is found by setting the argument of the sinc function equal to zero, from which we find that

$$\xi = -\frac{\lambda f_1 DF}{v(D - F)} - \frac{v(t - T/2)F}{D - F}, \qquad T_c \leq t \leq (T - T_c). \quad (7.83)$$

The spot position of the beginning of scan when $t = T_c$ is

$$\xi_b = -\frac{\lambda f_1 DF}{v(D - F)} - \frac{v(T_c/2)F}{D - F}, \quad (7.84)$$

and the spot position at the end of scan when $t = T - T_c$ is

$$\xi_e = -\frac{\lambda f_1 DF}{v(D - F)} - \frac{v(T/2 - T_c)F}{D - F}, \quad (7.85)$$

so that the length of scan is

$$L_s = |\xi_e - \xi_b| = \left| \frac{v(T - T_c)F}{D - F} \right|. \tag{7.86}$$

The scanning velocity is still given by Equation (7.72), but the spot size is slightly different:

$$d_0 = \frac{\lambda}{L_c/D_f} = \frac{\lambda DF}{(D - F)L_c}, \tag{7.87}$$

which is similar to Equation (7.75), except that L is replaced by L_c because the spot size is now determined by the length of the chirp, not by the length of the acousto-optic cell. The number of samples in the scan line is

$$M = \frac{L_s}{d_0} = \left| \frac{vF(T - T_c)}{(D - F)} \right| \frac{(D - F)L_c}{\lambda DF} = \left| \frac{v(T - T_c)L_c}{\lambda D} \right|. \tag{7.88}$$

We now use Equation (7.51) in Equation (7.88) to find that the number of samples in the scan line is

$$M = \left[1 - \frac{T_c}{T} \right] TW. \tag{7.89}$$

Because the chirp duration is less than the cell duration, the time bandwidth product of the acousto-optic cell is not fully utilized.

The scan duty cycle for this configuration is

$$U = \frac{T - T_c}{T_r}, \tag{7.90}$$

and the sample rate is

$$R_s = \frac{|v_s|}{d_0} = W, \tag{7.91}$$

which is the maximum achievable sample rate. The throughput rate is

$$R = U\frac{|v_s|}{d_0} = \frac{T - T_c}{T_r}W, \tag{7.92}$$

obtained in a fashion similar to that used to produce Equation (7.79).

Figure 7.16 shows the situation when at least one period of a chirp signal is fully in the acousto-optic cell and the scanning spots are therefore well formed. We now examine the scanning spot shape and position as the chirp segments enter and leave the cell. The light from these transition times is located in regions just before and just after the scan line. To account for the spot-forming condition, we modify the limits of integration in Equation (7.64):

$$F(\xi, t) = \int_{-L/2}^{-L/2+vt} e^{j(\pi x^2/\lambda)[1/F-1/D-1/D_f]} \times e^{j(2\pi x/\lambda)[v(t-T/2)/D+\xi/D_f+\lambda f_1/v]}\, dx; \qquad 0 \leq t \leq T_c, \tag{7.93}$$

which is applicable for a chirp waveform as it just enters the cell. The limits of integration show that the integral is over a small spatial region when t is small and that the region of integration increases linearly for $0 \leq t \leq T_c$. As before, we set the value of D_f so that the quadratic term in x is equal to unity, leaving the integral

$$\begin{aligned} F(\xi, t) &= \int_{-L/2}^{-L/2+vt} e^{j(2\pi x/\lambda)[v(t-T/2)/D+\xi/D_f+\lambda f_1/v]}\, dx \\ &= vt \operatorname{sinc}\left[\left\{\frac{v(t-T/2)}{D} + \frac{\xi}{D_f} + \frac{\lambda f_1}{v}\right\} vt/\lambda\right]; \qquad 0 \leq t \leq T_c, \end{aligned} \tag{7.94}$$

where we ignore unimportant scale factors.

The behavior of this sinc function, whose argument is quadratic in the time variable, has some interesting features that are exhibited in the dotted line region of Figure 7.16 where the spot is first formed:

1. The magnitude of the sinc is small for small values of t, as we expect from a consideration of the region of integration, and reaches a limit that is proportional to vT_c when the chirp in the cell is fully illuminated.

2. The centroid of the spot as a function of time is located at

$$\xi = -\frac{\lambda f_1 DF}{v(D-F)} - \frac{v(t-T/2)F}{D-F}, \qquad 0 \le t \le T_c, \quad (7.95)$$

just as in Equation (7.83) but with a slightly different time interval of validity. The spot position when the chirp just enters the cell is

$$\xi_b = -\frac{\lambda f_1 DF}{v(D-F)} - \frac{v(-T/2)F}{D-F}, \qquad (7.96)$$

and its position when it is fully in the cell is

$$\xi_e = -\frac{\lambda f_1 DF}{v(D-F)} - \frac{v(T/2)F}{D-F}. \qquad (7.97)$$

By comparing Equations (7.96) and (7.97) with Equations (7.84) and (7.85), we see that the end positions of the scanning spot produced by the chirp as it enters the cell are displaced a distance L_s below that of the active scan line.

3. The most interesting feature of the sinc function is that the spot changes its size continuously as the chirp enters the cell. The spot size is determined by finding the position of the first zero of the sinc function relative to its centroid. This distance is

$$\Delta\xi = d_0 = \frac{\lambda D_f}{vt}; \qquad 0 \le t \le T_c. \qquad (7.98)$$

From Equation (7.94) we see that the sinc function is infinitely broad when $t = 0$, but its magnitude is zero. As time increases, the spot moves towards the active scanning region and its size decreases while its magnitude increases. The rate at which the spot size decreases, as the centroid moves closer to the beginning of the active scan position, is just sufficient to keep the light from spilling into the active scanning region prematurely. When the chirp has fully entered the cell at $t = T_c$, the spot has full resolution and the active scanning begins as the chirp travels through the remainder of the acousto-optic cell.

As the chirp segment leaves the cell, the spot dissolves in an order that is a reversal of its evolution. The spot gradually loses intensity as it broadens, until it reaches the end of the spot dissolving scan line shown in Figure 7.16.

7.9.2.2.3. A Long-Chirp, Repetition-Rate-Limited Scanner. To achieve a high throughput rate, we need a high scan duty cycle; Equation (7.78) shows that we want the active scan time to equal the repetition period of the chirp train. Suppose that we use a surface acoustic device to generate a chirp segment whenever it is driven by an impulse function. By controlling the timing of the impulses, we produce chirp segments with any desired repetition period T_r. Depending on the ratio of the repetition period to the chirp duration, one or more overlapping chirp segments may be in the cell at the same time. If the response of the cell is linear, the chirp signals do not interfere and the only effect of the overlapping chirps is to lower the diffraction efficiency. As the chirp signal is on a carrier frequency, a nonlinear response from the cell produces higher-order terms that are easily eliminated by spatial filters.

In this section, we assume that the chirp segments overlap so that $T_r \leq T_c$, and that $T_r > T$. The system parameters that are changed are the scan length which, through a line of analysis similar to that given in Section 7.9.2.2.1, is now

$$L_s = |\xi_e - \xi_b| = \left| \frac{vT_r F}{D - F} \right|. \tag{7.99}$$

The spot size is still determined by the cell aperture because $T_r > T$:

$$d_0 = \frac{\lambda DF}{(D - F)L}, \tag{7.100}$$

so that the number of samples in a scan line is

$$M = \frac{T_r}{T_c} TW. \tag{7.101}$$

The scan duty cycle for this scanner configuration is

$$U = \frac{\min(T_s, T_r)}{T_r} = \frac{T_r}{T_r} = 1, \tag{7.102}$$

as expected. The sample rate and the throughput rate are equal in this configuration at

$$R_t = UR_s = \frac{|v_s|}{d_0} = \frac{T}{T_c} W. \tag{7.103}$$

As before, we see that the throughput rate is maximized only when $T_c = T$. To achieve this condition, we consider the final of the four basic configurations.

7.9.2.2.4. A Short-Chirp, Repetition-Rate-Limited Scanner. In this configuration, the chirp segments also overlap so that $T_r \leq T_c$, and we assume that $T_r \leq T$. The scan length, found through a line of analysis similar to that given in Section 7.9.2.2.1, is

$$L_s = |\xi_e - \xi_b| = \left| \frac{vT_rF}{D - F} \right|. \tag{7.104}$$

The spot size is now determined by the active scan aperture because $L_c = vT_c$ so that

$$d_0 = \frac{\lambda DF}{(D - F)L_c}, \tag{7.105}$$

and the number of samples in a scan line is

$$M = T_rW, \tag{7.106}$$

which achieves its maximum value when $T_r = T$. The scan duty cycle for this configuration is

$$U = \frac{\min(T_s, T_r)}{T_r} = \frac{T_r}{T_r} = 1, \tag{7.107}$$

as expected. The sample rate and the throughput rate are equal in this configuration at

$$R_t = UR_s = \frac{|v_s|}{d_0} = W. \tag{7.108}$$

In this configuration, the throughput rate is maximized independently of the values of T_r or T_c, provided that the constraints necessary to implement the short-chirp, repetition-rate-limited scanner are observed.

7.9.2.3. Summary of Scanner Performance Criteria. Table 7.2 gives a summary of the important performance parameters of the four basic scanning configurations and serves as a useful aid in beginning a design. For example, some applications require a high throughput rate R_t. The

Table 7.2 Acousto-Optic Scanner Parameters

Type of Chirp	Long Chirp $T_c \geq T$		Short Chirp $T_c < T$	
Scan Time	Aperture Limited $T_c - T \leq T_r$	Repetition-Rate Limited $T_c - T > T_r$	Aperture Limited $T_c + T_r \geq T$	Repetition-Rate Limited $T_c + T_r < T$
Active scan time = $\min(T_s, T_r)$	$T_c - T$	$T_c - T$	$T - T_c$	$T - T_c$
Active aperture time = $\min(T, T_c)$	T	T	T_c	T_c
Scan length = L_s	$\left\lvert \frac{v(T_c - T)F}{D - F} \right\rvert$	$\left\lvert \frac{vT_rF}{D - F} \right\rvert$	$\left\lvert \frac{v(T - T_c)F}{D - F} \right\rvert$	$\left\lvert \frac{vT_rF}{D - F} \right\rvert$
Spot size = d_0	$\frac{\lambda DF}{(D - F)L}$	$\frac{\lambda DF}{(D - F)L}$	$\frac{\lambda DF}{(D - F)L_c}$	$\frac{\lambda DF}{(D - F)L_c}$
Number of samples per line = M	$\left[1 - \frac{T}{T_c}\right]TW$	$\frac{T_r}{T_c}TW$	$\left[1 - \frac{T_c}{T}\right]TW$	$\frac{T_r}{T}TW$
Scan duty cycle = U $\min(T_s, T_r)/T_r$	$\frac{T_c - T}{T_r}$	1	$\frac{T - T_c}{T_r}$	1
Sample rate = R_s	$\frac{T}{T_c}W$	$\frac{T}{T_c}W$	W	W
Throughput rate = R_t	$\left[1 - \frac{T}{T_c}\right]\frac{T}{T_r}W$	$\frac{T}{T_c}W$	$\frac{T - T_c}{T_r}W$	W

maximum rate of W samples per second can be achieved with a short-chirp scanner that is repetition-rate limited. The number of samples per line, however, is always less than TW because the highest useful ratio for T_r/T is $\frac{1}{2}$. On the other hand, we can achieve nearly TW spots per scan line with either of the long-chirp scanners, but only with a reduction in the throughput rate.

To more fully appreciate the relationships among the design parameters, we use two key graphic representations. The first graphic, shown in Figure 7.17(a), illustrates the number of samples M per scan line, normalized to its maximum value of TW, as a function of the ratios T/T_c and T_r/T_c. The vertical line passing through $T/T_c = 1$ is the dividing line between the long- and short-chirp configurations. The horizontal line for which $T_r = T_c$ is the boundary between those configurations in which the chirp segments do or do not overlap. The diagonal lines passing through the points (0, 1), (1, 0), and (2, 1) represent the boundaries between the full scan duty cycle configurations (below the diagonal lines) and the partial scan duty cycles conditions (above the diagonal lines). The loci of constant number of samples per scan line are shown for each of the four basic configurations. We note that the largest number is obtained by using a long-chirp scanner for which the ratio T/T_c is small; the scanner may be either aperture- or repetition-rate limited. When $T/T_c = 1$, the number of samples reaches its minimum value because the active scan time is at its minimum value so that only one spot can be formed in each scan line. For aperture-limited short-chirp scanners, the number of samples per scan line is reciprocally related to the ratio T/T_c, while the lines for repetition-rate-limited short-chirp scanners have slopes whose values are equal to the normalized values themselves.

The second graphic, shown in Figure 7.17(b), illustrates the throughput rate R_t, normalized to its maximum value of W samples per second, as a function of the ratios T/T_c and T_r/T_c. The normalized throughput rate follows parabolic curves when the scanner is aperture limited. Aperture-limited short-chirp scanners have throughput rates that follow straight-line segments, and the normalized throughput rate is fixed at unity for all repetition-rate limited short-chirp scanners. The throughput rate is not a function of the ratio T/T_c for repetition-rate-limited long-chirp scanners.

7.9.2.4. Examples of an Acousto-Optic Recording System. In this section, we provide some brief design guidelines for using the results summarized in Table 7.2 and in Figure 7.17.

Example 1. Suppose that we design a relatively low-performance system, such as a facsimile scanner or recorder. In this case, a large number of

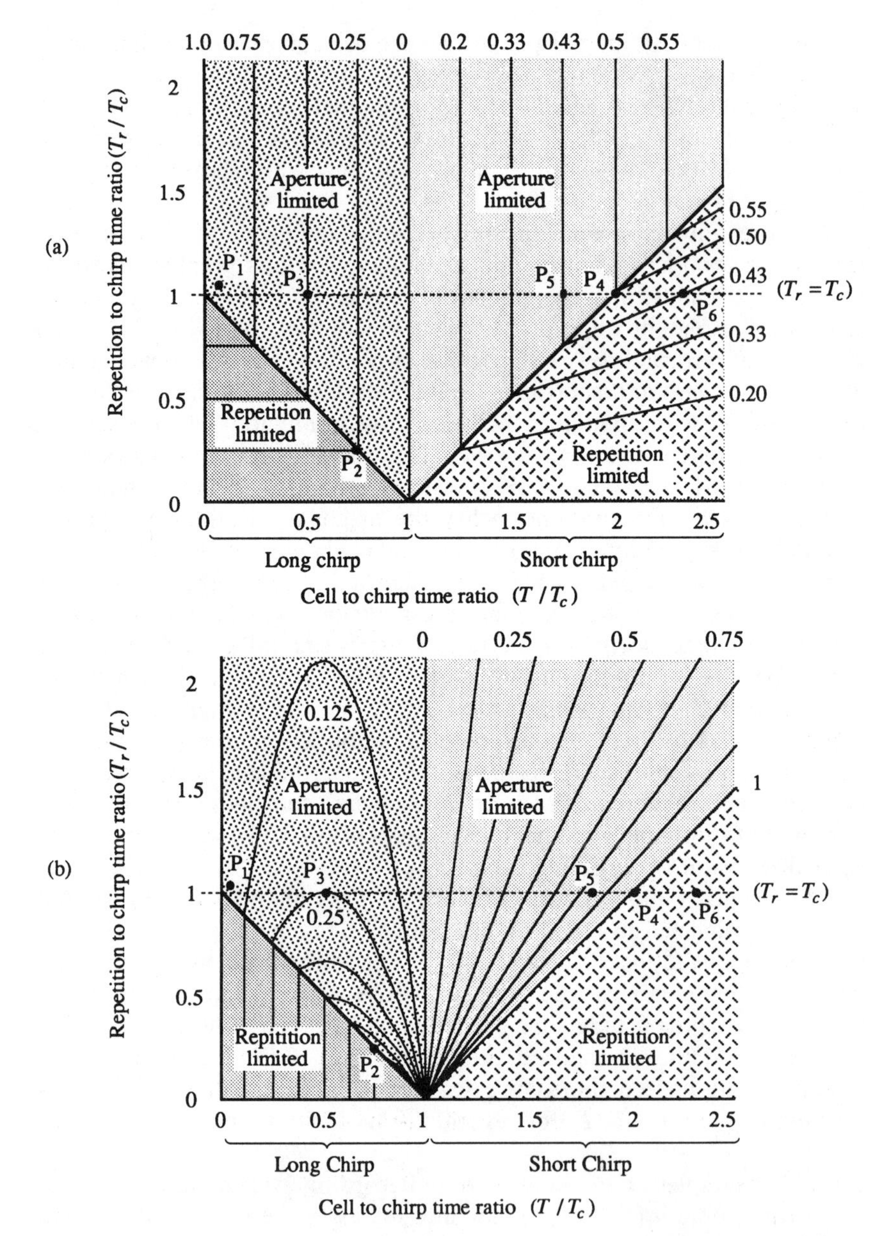

Figure 7.17. Normalized plots: (a) number of samples per scan and (b) throughput rate.

samples per scan line is typically more important than a high throughput rate. We therefore select an acousto-optic cell, such as one made from slow shear-wave tellurium dioxide material with a large time bandwidth product (for example, $T = 50$ μs and $W = 40$ MHz so that $TW = 2000$). Because the required throughput rate for a typical facsimile is well under 1 MHz, the normalized throughput rate is much less than 0.025; as the repetition period is nearly equal to the chirp length ($T_r = T_c$), only a small data buffer is needed. This scanner/recorder configuration is represented by the point P_1 in the graphics of Figure 7.17.

Example 2. Suppose that the requirements are the same as in the first example, but we need to operate at a much higher throughput rate of $30(10^6)$ samples per second. If we use the same acousto-optic cell as before, the normalized throughput rate is 0.75. For a long-chirp scanner, Figure 7.17(b) shows that the operating point is at P_2. Unfortunately, Figure 7.17(a) shows that the normalized number of samples per scan line is only 0.25 for this arrangement and a significant amount of high-speed buffering is needed because the scan duty cycle is low $[(T_c - T)/T_c < 1]$. To achieve better performance, we might consider using an 8.3 μs, 120 MHz acousto-optic cell, operating as an aperture-limited, long-chirp scanner with a normalized throughput rate of 0.25 to provide a normalized number of samples per scan line of 0.5, operating at point P_3. Although the actual number of samples per line are the same in the two alternatives [(0.25)(50 μs)(40 MHz) = 500 as compared to (0.5)(8.3 μs)(120 MHz) = 500], the requirements on the data buffer are not as severe in the second instance. An even better solution for these requirements may be to use just 30 MHz of the 40 MHz bandwidth of the slow shear-wave cell and to operate a short-chirp scanner at point P_4, where the scan duty cycle is 100% and where the normalized number of samples is 0.48, to provide (0.48)(50 μs)(30 MHz) = 720 samples per scan line.

The graphs of Figure 7.17, coupled with the data from Table 7.2, provide the information needed to quickly sort through the possible scanner solutions for a particular problem. For example, the two scanner configurations shown by P_5 and P_6 in Figure 7.17 provide the same number of samples per scan line, but the solution at P_5 has a normalized throughput rate of only 0.8, as compared to a normalized rate of one for the solution at P_6.

7.9.2.5. Other Considerations. In the analyses given so far, we have assumed uniform illumination of the acousto-optic cell so that the design relationships can be clearly stated in closed form. In practice, the

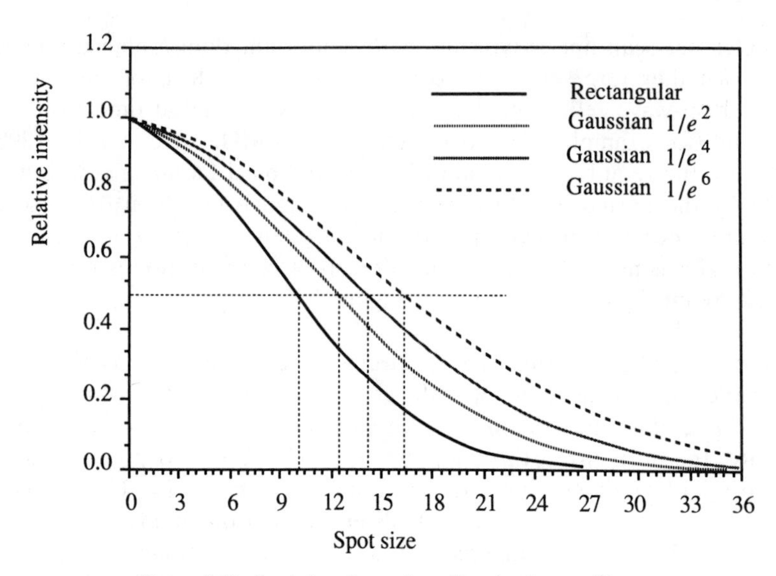

Figure 7.18. Spot sizes for various illumination profiles.

acousto-optic cell is usually illuminated by a laser beam with a Gaussian intensity weighting so that the spot size, for a given aperture, is greater than that for a uniform illuminating beam. Figure 7.18 shows the spot sizes for a uniform illumination and for Gaussian illuminations in which the intensity at the edges of the cell drops to $1/e^2$, $1/e^4$, and $1/e^6$ of the central value. If we use the half-power response of the spot distribution as a convenient measure of the spot size, the spot sizes for these Gaussian illuminations have increased by a factor G, where G is equal to 1.15 for the $1/e^2$ illumination, to 1.36 for the $1/e^4$ illumination, and to 1.58 for the $1/e^6$ illumination. All the relationships developed in previous sections are still valid, except that the spot size d_0 must be multiplied by G, while the number of samples per scan line M, the sample rate R_s, and the throughput rate R_t must all be divided by G.

PROBLEMS

7.1. A spatial signal contains a maximum frequency $\alpha_{co} = 150$ Ab. What is the required sample spacing d_0 to satisfy the Nyquist criterion? If the signal is illuminated with light of wavelength $\lambda = 0.5\ \mu$, calculate the maximum physical angle θ_{co} that the geometric rays can have as the wavefront diverges from any sample point of the signal.

7.2. We have an acousto-optic cell constructed from gallium phosphide material. The index of refraction is 3.31 and the velocity of sound is $v = 6320$ m/sec. The crystal is $L = 12$ mm long and we use light of wavelength $\lambda = 0.5\ \mu$. For a center frequency $f_c = 500$ MHz and a total bandwidth of $W = 200$ MHz, calculate

(a) the acoustic wavelength at the center frequency

(b) the Bragg angle needed for optimum illumination, and

(c) the time bandwidth product.

7.3. For the parameters given in Problem 7.2, calculate

(a) the angular spread $\Delta\theta$, the maximum diffracted angle, and the minimum diffracted angle, and

(b) the distance occupied by the spectrum of the signal in the Fourier plane if we use a 100-mm focal length lens. Sketch and label the regions where the spectrum lies if we operate in the Raman-Nath mode.

7.4. Suppose that you have a scanner of the long-chirp, aperture-limited type, using an acousto-optic cell made of TeO_2 operated in the longitudinal mode. Further, suppose that $W = 1000$ MHz, $T = 1\ \mu$sec, $T_c = 10\ \mu$sec, $\lambda = 0.5\ \mu$, $f_1 = 900$ MHz, and you use a lens having a 50-mm focal length. Calculate

(a) the distance from the lens to the scan plane,

(b) the position for the beginning of scan,

(c) the position of the end of scan,

(d) the length of the scan line,

(e) the spot size,

(f) the scan velocity, and

(g) the number of spots in the scan line.

Be sure to include a sketch of the system that clearly shows where the scan interval lies relative to the focal plane of the lens. Hint: Use a consistent set of relationships to solve this problem and then use an independent set, where possible, as a sanity check.

7.5. A periodic chirp signal has a chirp rate CR $= 100(10^{12})$ Hz/sec, with a chirp length of $T_c = 1\ \mu$sec. Suppose that exactly two periods of the chirp can fit within an acousto-optic cell made of GaP material.

(a) What is the scanner type?

(b) What is the required cell bandwidth?

(c) Calculate the distance from the acousto-optic cell to the plane where the chirp is focused.

(d) What is the scanning spot size d_0? Note that T_c sets the time bandwidth product of the signal in this example.

(e) What is the scanning spot velocity?

7.6. We want to scan a focused beam over a distance $L_s = 200$ mm at a rate of 100,000 scans per second. We want a spot size of $d_0 = 0.4$ mm at the plane of focus and we need a 100% duty cycle so that no data buffers are needed. Design an acousto-optic scanning system to achieve these goals. What is the best type of scanner to use? Select a suitable interaction material and determine values for T, W, L, v, chirp rate, center frequency, focal lengths, magnifications, etc. Sketch and label the drive signal with time and frequency. Provide a sketch of all the basic optical elements needed to make the scanner work (use a top and side view sketch). Note: Your design rational should lead you to use a TeO_2 slow shear-wave acousto-optic cell (be sure to support this conclusion).

7.7. From a crumpled and torn spec sheet, you note that a manufacturer has an acousto-optic scanner made of TeO_2, operating in the slow shear mode, with a line scan rate of 40,000 scans/second. They claim that, when used with a 1000-mm focal length lens, the scanning spot velocity is 5,000 m/sec, in a direction opposite to that of the acoustic velocity. Furthermore, they claim that the system has a 100% duty cycle and that the sample rate is equal to the bandwidth when the time duration of the chirp segment is just equal to the repetition period. Unfortunately, the information about the bandwidth of the cell cannot be read from the spec sheet. Calculate it from the data given. Also, calculate the required time bandwidth product. Be careful with the signs!

8

Acousto-Optic Power Spectrum Analyzers

8.1. INTRODUCTION

In Chapters 3 and 4, we showed how coherently illuminated optical systems naturally display the Fourier transform of a signal and how to use this information in spectrum-analysis applications. Optical spectrum analyzers are divided into several major architectural classes based on the variable of integration for the Fourier-transform operation. One architecture, described as *space integrating*, performs a Fourier transform with respect to a space variable. This is the natural mode of operation because lenses collect or integrate light over a given area. The second architecture, described as *time integrating*, performs a Fourier transform with respect to a time variable. The integration is achieved by collecting light on a photodetector array for a given time period. In some cases the two types are combined to form *hybrid* architectures. Both one- and two-dimensional Fourier transforms exist for all types of architectures.

In this chapter we concentrate our attention on one-dimensional power spectrum analyzers of the space integrating type. These systems are often called *instantaneous power spectrum analyzers* because the Fourier transform is computed for the signal history resident in the cell at every instant in time. Because the calculations are completed as soon as light has propagated to the Fourier-transform plane, generally in a few nanoseconds, the computation is essentially instantaneous. A photodetector array in the Fourier-transform plane measures the intensity of the light, which is directly proportional to the rf power of the temporal frequencies contained in the input signal.

It was not until the mid-1950's that the connection between spectrum analysis and diffraction in a coherently illuminated optical system was fully appreciated. The application of acousto-optic cells in spectrum analysis began with the work of Rosenthal (86), Wilmotte (87), and Lambert (88). The interaction medium, in these early systems, was often a liquid such as water. As liquids support only low-frequency signals, the cells were generally used in the Raman-Nath mode. In the 1960's, better interaction materials became available that could handle wideband signals and retain the wavefront quality needed for accurate spectrum analysis.

8.2. A BASIC SPECTRUM ANALYZER

One way to electronically analyze the spectrum of a signal is to measure its energy in a narrow frequency band, over the frequency range of interest, with a superheterodyne receiver. Unfortunately, when the measurement bandwidth is made small to increase the sensitivity of the receiver, the scan rate is reduced and the time required to cover a given band of frequencies is increased. Such a receiver does not monitor all frequencies at all times and the probability that a given signal is intercepted may not be 100%. When the electromagnetic environment is densely filled with exotic emitters, such as frequency-hopped radios, a probability of intercept less than 100% means that some of the signals are missed at least some of the time.

To produce a high probability of intercept, electronic spectrum analyzers must process the signal by implementing a large number of narrow bandpass filters, as shown in Figure 8.1(a). The outputs of the bandpass filters drive nonlinear devices, such as square-law detectors, to provide an output proportional to the signal spectrum. The spectrum is integrated over a time interval consistent with the channel bandwidth. The integrator outputs are multiplexed and sampled at a sufficiently high rate so that signals are not missed. Because the bandpass filter in each channel has a different center frequency, filter design and implementation is complicated. Furthermore, as the wideband signals are typically in the rf region, the component values are often awkward; high-order filters are needed to suppress adjacent signals. Finally, these systems become bulky and power hungry as the number of channels increase. As a result, the number of channels in an electronic spectrum analyzer of this type is typically limited to well under 100.

Consider the relatively simple and compact acousto-optic spectrum analyzer shown in Figure 8.1(b). A source of monochromatic light at wavelength λ is collimated by lens L_1 and illuminates the acousto-optic cell. A wideband signal $f(t)$ drives the acousto-optic cell so that light is diffracted at angles that are linearly related to the temporal frequencies of the signal, as explained in Chapter 7, Section 7.2.3. Lens L_2 converts the angular spectrum into a spatial spectrum at its back focal plane where it is detected by a linear photodetector array. The photodetector array in-

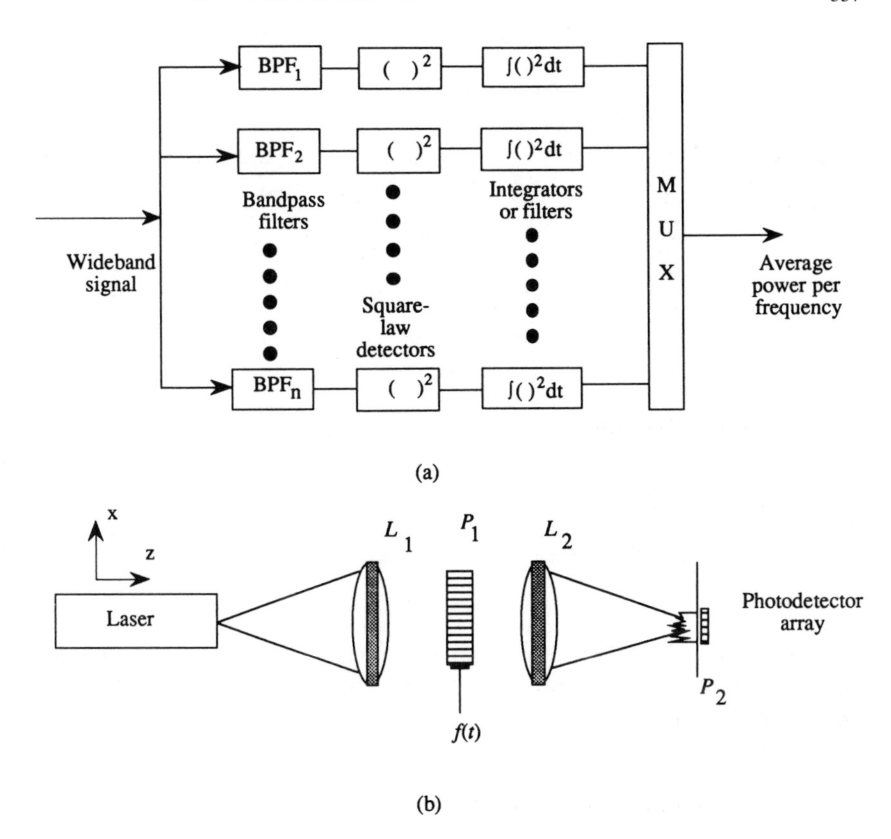

Figure 8.1. Spectrum analyzers: (a) electronic spectrum analyzer and (b) optical spectrum analyzer.

cludes a charge-coupled device that multiplexes the detected spectrum and delivers it to a postprocessor as a sampled video signal.

As we showed in Chapter 4, Section 4.5.4, we generally require three photodetector elements to properly detect and resolve distinct spatial frequencies. As arrays containing up to 4096 elements are readily available, acousto-optic spectrum analyzers can resolve well over 1000 frequencies. The highly parallel structure of the optical system provides a tremendous gain in processing capability, the most important advantage being a high probability of intercept for signals anywhere in the bandwidth of the spectrum analyzer. The acousto-optic spectrum analyzer provides the equivalent of thousands of narrowband filters, as we shall show shortly, followed by integrators to measure the signal power. The entire subsystem

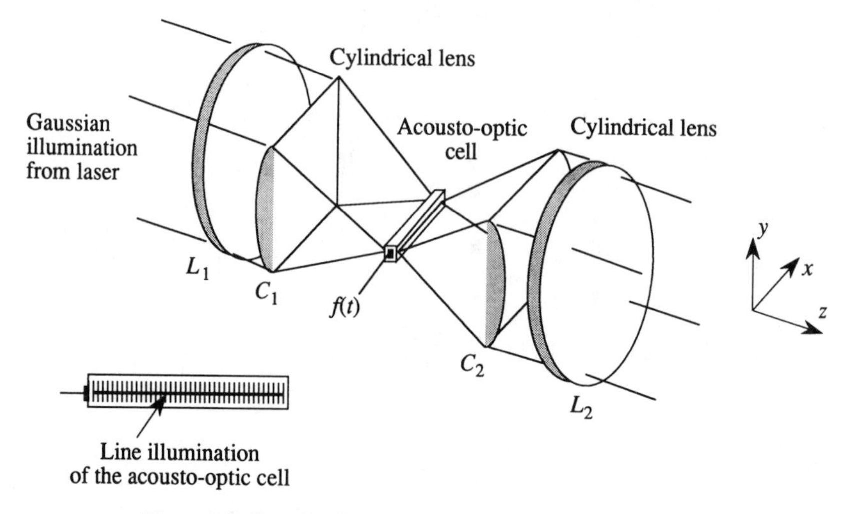

Figure 8.2. Gaussian illumination of a Bragg cell in a line focus.

consists of just three active components: the laser, the acousto-optic cell, and the photodetector array. The optical package is small and the power requirements are low.

8.2.1. The Illumination Subsystem

The illumination subsystem in Figure 8.1(b) is idealized in the sense that a slightly more complicated lens system is needed to efficiently illuminate the cell and to produce the spectrum. To efficiently illuminate the acousto-optic cell in the vertical direction, we focus light to a line by means of cylindrical lens C_1, as shown in Figure 8.2. The combination of L_1 and C_1 produces a *line illumination* pattern at the cell. The illumination along the direction of acoustic propagation must be collimated to obtain a high diffraction efficiency from a cell operating in the Bragg mode, as noted in Chapter 7. As cylindrical lens C_1 has no power in the direction of acoustic propagation, light remains collimated until it reaches the acousto-optic cell. Lens C_2 collimates the line source of illumination and lens L_2 is the final Fourier-transform lens.

Illumination from a coherent laser source has the Gaussian form $\exp[-4A(x/L)^2]$, where L is the length of the cell and the parameter A defines the amplitude at the edges of the cell. The truncation of the Gaussian illumination by the ends of the acousto-optic cell or by other optical elements in the system is represented by $\text{rect}(x/L)$. Finally, the signal is modified by a frequency-dependent attenuation factor of the form $\exp[-c(x + L/2)f^2]$, where c is an attenuation constant and $x = 0$ defines the center of the cell. In this development the attenuation is expressed in dB/(mm · GHz^2) so that $c = 0.23\Gamma/v$, where Γ is the attenuation constant from Table 7.1 in Chapter 7, and where v is the acoustic velocity in the medium. Although the attenuation is not a part of the illumination function, we find it convenient to include its effect in the illumination function.

We can combine the Gaussian, exponential, and rectangular functions to form the *aperture weighting function* for the cell. For example, we note that

$$\begin{aligned} a(x) &= \left[e^{-(c/2)(x+L/2)f^2} e^{-2Ax^2/L^2}\right]\text{rect}(x/L) \\ &= \left[e^{-(cLf^2/4)(1-cLf^2/8A)} e^{-2A/L^2(x+cL^2f^2/8A)^2}\right]\text{rect}(x/L) \\ &= a_0 e^{-2A[(x-x_0)/L]^2}\,\text{rect}(x/L), \end{aligned} \tag{8.1}$$

so that the combination remains Gaussian, but with a shifted central value, as obtained from Equation (8.1) and illustrated in Figure 8.3. The new Gaussian function has its center at

$$x_0 = -\frac{cL^2f^2}{8A}, \tag{8.2}$$

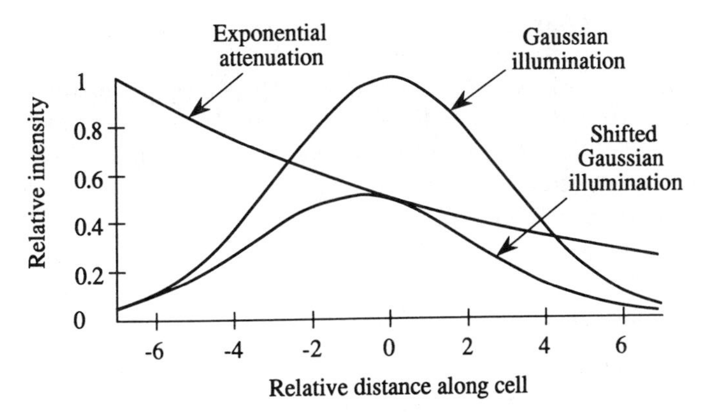

Figure 8.3. Effect of exponential attenuation and Gaussian illumination.

and is attenuated by a factor

$$a_0 = e^{-(cLf^2/4)(1 - cLf^2/8A)} \tag{8.3}$$

by the exponential attenuation.

As an example, consider the GaP cell cited in Table 7.1 of Chapter 7, operating at a center frequency of 500 MHz. For this cell material, $\Gamma = 3.80\ \text{dB}/(\mu\text{sec} \cdot \text{GHz}^2)$ so that, for $v = 6.32$ km/s, we calculate $c = 0.138\ \text{dB}/(\text{mm} \cdot \text{GHz}^2)$. Suppose that $A = 4$, which is equivalent to truncating the Gaussian illumination at the $1/e^4$ points in intensity at the ends of the cell. From Equation (8.2) we find that the center of the new Gaussian function, for a cell of length $L = 12.64$ mm, has moved to $x_0 = -0.172$ mm and that the overall attenuation factor is $a_0 = 0.9$.

The temporal and spatial coherence properties of injection laser diodes are adequate, in general, for use in power spectral analyzers. As the emitting surface is typically rectangular, the angle of divergence of the output intensity pattern is different in the two directions. The focal lengths of C_1, C_2, L_1, and L_2 are therefore chosen to match the characteristics of the source to the illumination requirements of the cell. The major purpose of the second cylindrical lens is to image the line illumination of the cell onto the photodetector array so that light is efficiently collected. As noted in Chapter 2, Section 2.4.3, prisms are sometimes used as an alternative method to equalize the beam dimensions in orthogonal directions.

In the chapters to follow we typically do not sketch the optical system in the direction orthogonal to that of the acoustic propagation unless such sketches are necessary to fully understand the operation of the system. The figures portray, with reasonable accuracy, the key elements needed for a particular processing operation and clearly show whether we are dealing with space domains or Fourier domains. To simplify the diagrams, we usually omit lenses that do not have power in the plane of the diagram. The reader is expected to anticipate where such lenses, or other elements such as mirrors, are needed to construct a practical system.

8.2.2. A Raman-Nath-Mode Spectrum Analyzer

When the spectrum analyzer is operated in the Raman-Nath mode, the illumination is normal to the entrance face of the acousto-optic cell, as shown in Figure 8.4. The cell is driven by the signal $f(t) = s(t)\cos(2\pi f_c t)$,

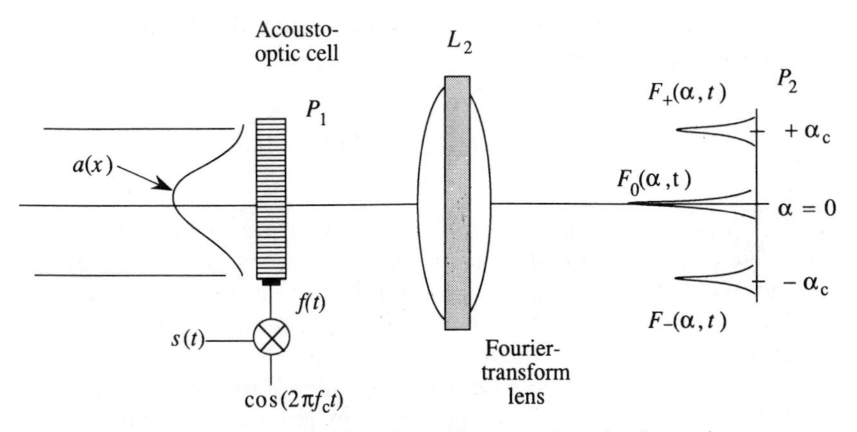

Figure 8.4. Spectrum analyzer in the Raman-Nath mode of operation.

and lens L_2 produces the Fourier transform $F(\alpha, t)$ at plane P_2, where

$$F(\alpha, t) = \int_{-\infty}^{\infty} f(x, t) e^{j2\pi\alpha x} \, dx, \tag{8.4}$$

and $f(x, t)$ is the light distribution at the exit face of the acousto-optic cell. As the integration in Equation (8.4) is with respect to the spatial coordinate of the input plane, this spectrum analyzer clearly belongs to the class of *space integrating* architectures.

The expression for $f(x, t)$ is obtained by a slight modification of Equation (7.30):

$$f(x, t) = a(x)\left[1 + jms\left(t - \frac{T}{2} - \frac{x}{v}\right)\cos\left\{2\pi f_c\left(t - \frac{T}{2} - \frac{x}{v}\right)\right\}\right], \tag{8.5}$$

where $a(x)$ is the aperture function. The Fourier transform of $f(x, t)$ at plane P_2 is therefore

$$F(\alpha) = \int_{-\infty}^{\infty} a(x)\left[1 + jms\left(t - \frac{T}{2} - \frac{x}{v}\right)\right.$$
$$\left.\times \cos\left\{2\pi f_c\left(t - \frac{T}{2} - \frac{x}{v}\right)\right\}\right] e^{j2\pi\alpha x} \, dx. \tag{8.6}$$

The Fourier transform of the optical wave, at the exit face of the acousto-optic cell, consists of undiffracted light, indicated by $F_0(\alpha, t)$, and the positive and negative diffracted orders, indicated by $F_+(\alpha, t)$ and

$F_{-}(\alpha, t)$. Consider first the constant term inside the brackets in the integrand of Equation (8.6):

$$F_0(\alpha, t) = \int_{-\infty}^{\infty} a(x) e^{j2\pi\alpha x}\, dx = A(\alpha), \tag{8.7}$$

where $A(\alpha)$ is the Fourier transform of the aperture function $a(x)$. The undiffracted light therefore simply produces a response $A(\alpha)$, centered at $\alpha = 0$ at plane P_2, as shown in Figure 8.4. Note that $F_0(\alpha, t)$ has no time dependence, other than that implied by the light frequency f_l; it is purely a function of the spatial frequency α.

Consider next the simple case where $s(t) = c_0$, so that the baseband signal $s(t)$ contains only a dc component of magnitude c_0. For this signal, we expand the cosine carrier term in Equation (8.6), using the Euler formula, to find that the positive diffracted term becomes

$$F_{+}(\alpha, t) = \int_{-\infty}^{\infty} jmc_0 a(x) e^{j2\pi f_c(t - T/2 - x/v)} e^{j2\pi\alpha x}\, dx, \tag{8.8}$$

where we ignore unessential constants. The negative sign associated with x/v in the integrand of Equation (8.8) indicates that the acoustic wave is traveling in the positive x direction. The diffracted light is therefore upshifted, as confirmed by the positive sign of the exponential involving time.

We move the temporal term outside the integral because it is not a function of the variable of integration. We also recall from Chapter 7 that $\alpha_c = f_c/v$ is the connection between spatial and temporal frequencies. We then combine the exponential terms in x to produce

$$\begin{aligned} F_{+}(\alpha, t) &= jmc_0 e^{j2\pi f_c(t - T/2)} \int_{-\infty}^{\infty} a(x) e^{j2\pi(\alpha - \alpha_c)x}\, dx, \\ &= jmc_0 e^{j(2\pi f_c t - \phi)} A(\alpha - \alpha_c). \end{aligned} \tag{8.9}$$

This result reveals that the spectrum has a magnitude proportional to the modulation index m and to c_0, the magnitude of the baseband signal $s(t)$. The fixed phase from Equation (8.9) is $\phi = f_c T/2$ and is generally of little interest, except in some applications where the phase of the Fourier transform is important. The last factor of $F_{+}(\alpha, t)$ is the Fourier transform of the aperture function, shifted by the carrier frequency f_c so that it is centered at α_c, as shown in Figure 8.4.

The exponential factor in Equation (8.9) shows that the light amplitude function $A(\alpha - \alpha_c)$ is unshifted relative to the frequency of light by f_c, the carrier frequency of the drive signal. As the sidelobes of $A(\alpha - \alpha_c)$ due to the aperture function may be spread over a large spatial frequency range and oscillate at f_c, we see that there is not an exact coupling between spatial and temporal frequencies. This observation contradicts the assertion made in Chapter 7 that spatial and temporal frequencies are coupled by the relationship that $\alpha = f/v$.

An exact coupling occurs *only in the limit* as $L \to \infty$ so that $A(\alpha - \alpha_c)$ collapses to $\delta(\alpha - \alpha_c)$. A key point to remember is that the spatial diffraction $A(\alpha - \alpha_c)$ is due to integration over a finite range in the space plane. The diffraction due to the aperture function therefore has the same temporal frequency as the underlying cw signal. This phenomenon also occurs in conventional spectrum analyzers if the time history of the cw signal being analyzed is truncated to a time duration of T seconds. We discuss these spatial and temporal frequency relationships in greater detail in Chapter 10, Section 10.3.

The Fourier transform $F_{-}(\alpha, t)$ is the negative diffracted order:

$$F_{-}(\alpha, t) = \int_{-\infty}^{\infty} jmc_0 a(x) e^{-j2\pi f_c(t - T/2 - x/v)} e^{j2\pi\alpha x}\, dx. \qquad (8.10)$$

As before, we move the temporal term outside the integral and combine all the exponential terms in x to produce

$$\begin{aligned} F_{-}(\alpha, t) &= jmc_0 e^{-j2\pi f_c(t - T/2)} \int_{-\infty}^{\infty} a(x) e^{j2\pi(\alpha + \alpha_c)x}\, dx \\ &= jmc_0 e^{-j(2\pi f_c t - \phi)} A(\alpha + \alpha_c). \end{aligned} \qquad (8.11)$$

This result is similar to that given by Equation (8.9), except that it is centered at $-\alpha_c$ and its temporal frequency is downshifted relative to the frequency of light. The j factors show that both $F_{+}(\alpha, t)$ and $F_{-}(\alpha, t)$ are 90° out of phase with respect to the undiffracted light. We ignore this factor in the rest of this chapter; its importance surfaces in subsequent chapters.

So far we have disregarded the frequency of light in our description of the spectrum. Recall from Equation (7.29) that the complete form of the light distribution should be

$$\boxed{\begin{aligned} F_{+}(\alpha, t) &= \mathrm{Re}\left[e^{j(2\pi f_l t - \phi)} F_{+}(\alpha, t)\right] \\ &= mc_0 |A(\alpha - \alpha_c)| \cos[2\pi(f_l + f_c)t - \phi + \theta(\alpha, t)], \end{aligned}} \qquad (8.12)$$

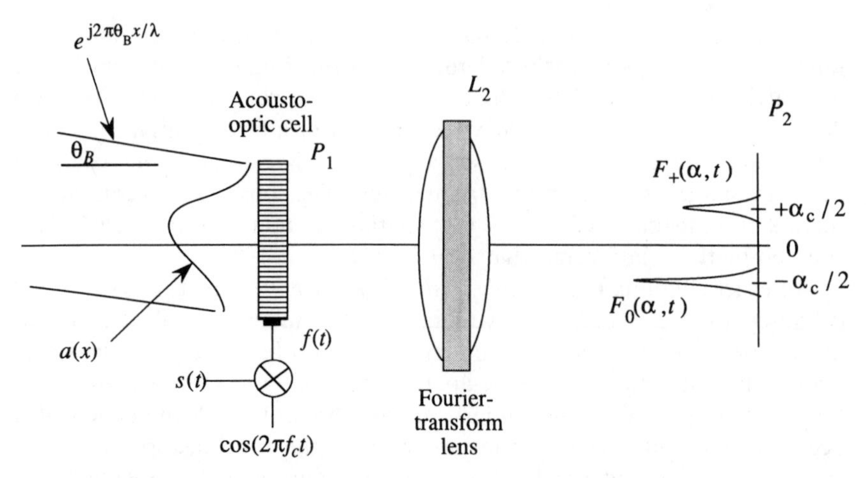

Figure 8.5. Spectrum analyzer in Bragg mode of operation.

where ϕ represents all the fixed phase factors and $\theta(\alpha, t)$ is the phase of $A(\alpha - \alpha_c)$. This result shows that light is transported as a real-valued function, at a carrier frequency $f_l + f_c$ that is magnitude and phase modulated by $A(\alpha - \alpha_c)$. All these salient points are evident in Equations (8.9) and (8.11), but it is worthwhile to remember that they are just a shorthand way to describe the complete waveform that actually propagates in space and time.

8.2.3. A Bragg-Mode Spectrum Analyzer

The details of the Bragg mode of operation are slightly more complicated than those for the Raman-Nath mode of operation. The acousto-optic cell is now illuminated by a plane wave traveling at the Bragg angle θ_B with respect to the optical axis, as shown in Figure 8.5. As in the Raman-Nath mode of operation, lens L_2 produces the Fourier transform $F(\alpha, t)$ at plane P_2. The optical wave $f(x, t)$ at the exit face of the cell is modified to include the Bragg illumination angle θ_B:

$$f(x,t) = a(x)e^{j2\pi\theta_B x/\lambda}\left\{1 + jma(x)s\left(t - \frac{T}{2} - \frac{x}{v}\right)\right.$$
$$\left.\times\cos\left[2\pi f_c\left(t - \frac{T}{2} - \frac{x}{v}\right)\right]\right\}, \quad (8.13)$$

where $\exp[j2\pi\theta_B x/\lambda]$ represents the Bragg illumination. The positive sign in this exponential shows that the illumination is directed downward and to the right at the Bragg angle θ_B. The negative sign associated with x/v indicates that the acoustic wave is traveling in the positive x direction.

The Fourier transform at plane P_2 of $f(x, t)$ is

$$F(\alpha, t) = \int_{-\infty}^{\infty} a(x) e^{j2\pi\alpha_B x}\left[1 + jms\left(t - \frac{T}{2} - \frac{x}{v}\right) \times \cos\left\{2\pi f_c\left(t - \frac{T}{2} - \frac{x}{v}\right)\right\}\right] e^{j2\pi\alpha x}\, dx, \quad (8.14)$$

where we have used the fact that $\alpha_B = \theta_B/\lambda$. Consider first the constant term inside the brackets in the integrand of Equation (8.14):

$$F_0(\alpha, t) = \int_{-\infty}^{\infty} a(x) e^{j2\pi\alpha_B x} e^{j2\pi\alpha x}\, dx = A(\alpha + \alpha_B), \quad (8.15)$$

where $A(\alpha)$ is the Fourier transform of the aperture function $a(x)$. As we discussed in Chapter 7, Section 7.2, the Bragg angle is $\theta_B = \lambda\alpha_B = \lambda/2\Lambda = \lambda f_c/2v$ so that $\theta_B = \lambda\alpha_c/2$. The undiffracted light therefore produces a response $A(\alpha + \alpha_c/2)$ at plane P_2, as shown in Figure 8.5. We recognize that $-\alpha_c/2$ is the position where the light would focus in the Fourier plane if the cell were removed from the system. As in the Raman-Nath mode, $F_0(\alpha, t)$ has no time dependence. When $s(t) = c_0$ so that $s(t)$ contains only a dc component of magnitude c_0, the positive diffracted order becomes

$$F_+(\alpha, t) = \int_{-\infty}^{\infty} jmc_0 a(x) e^{j2\pi\alpha_B x} e^{j2\pi f_c(t - T/2 - x/v)} e^{j2\pi\alpha x}\, dx. \quad (8.16)$$

We combine all the exponential terms in x to produce

$$\begin{aligned} F_+(\alpha, t) &= jmc_0 e^{j2\pi f_c(t - T/2)} \int_{-\infty}^{\infty} a(x) e^{j2\pi(\alpha + \alpha_B - \alpha_c)x}\, dx \\ &= jmc_0 e^{j(2\pi f_c - \phi)} A(\alpha - \alpha_c/2). \end{aligned} \quad (8.17)$$

This result is similar to that given by Equation (8.9), except that it is centered at $\alpha_c/2$ as shown in Figure 8.5. We can see why $A(\alpha)$ is centered at $\alpha_c/2$ by the following sequence of events. The Fourier transform of $a(x)$, in the absence of the acousto-optic cell, is $A(\alpha)$, which is centered at $\alpha = 0$ if the illumination is parallel to the optical axis. For the Bragg mode under consideration, the illumination angle is not parallel to the optical axis; the Bragg angle shifts $A(\alpha)$ to $\alpha_B = \alpha_c/2$, as indicated by $A(\alpha + \alpha_B)$

in Equation (8.15). The carrier frequency f_c then shifts $A(\alpha)$ to its final position, as indicated $A(\alpha + \alpha_B - \alpha_c) = A(\alpha - \alpha_c/2)$; the undiffracted and diffracted beams are therefore located at a distance corresponding to $\alpha_c/2$ on either side of the optical axis and are separated by a spatial frequency interval of α_c, just as in the Raman-Nath mode of operation.

Although the Bragg mode of operation provides better performance because the theoretical diffraction efficiency is much higher than that of the Raman-Nath mode (100% maximum versus 33.8% maximum), it provides no additional insights relative to the Raman-Nath mode from an analysis viewpoint. As the Raman-Nath mode has several attractive features, chief among them are a simplified notation and a symmetrical spectrum about the optical axis, we use it for analytical purposes in the remainder of this book unless otherwise stated.

8.2.4. The Generalization to Arbitrary Signals

So far we have discussed the Fourier transform of a simple signal for the Raman-Nath and Bragg modes of operation. The results for an arbitrary signal $s(t)$ can be expressed as the sum of such simple signals:

$$s(t) = \sum_{k=K_1}^{K_2} c_k \cos(2\pi k f_0 t + \phi_k), \tag{8.18}$$

where $M = K_2 - K_1 + 1$ is the number of frequencies that the system can resolve; that is, M is equal to the time bandwidth product of the signal. When $K_1 = 0$, $s(t)$ is a baseband signal and the Fourier-transform component $F_+(\alpha, t)$ is obtained by using Equation (8.18) in Equation (8.8):

$$\begin{aligned}
F_+(\alpha, t) &= \int_{-\infty}^{\infty} m a(x) \sum_{k=0}^{K_2} c_k \cos\left[2\pi k f_0\left(t - \frac{T}{2} - \frac{x}{v}\right) + \phi_k\right] \\
&\quad \times e^{j2\pi f_c(t - T/2 - x/v)} e^{j2\pi\alpha x}\, dx, \\
&= \int_{-\infty}^{\infty} \frac{m}{2} a(x) \sum_{k=0}^{K_2} c_k e^{j[2\pi k f_0(t - T/2 - x/v) + \phi_k]} \\
&\quad \times e^{j2\pi f_c(t - T/2 - x/v)} e^{j2\pi\alpha x}\, dx \\
&\quad + \int_{-\infty}^{\infty} \frac{m}{2} a(x) \sum_{k=0}^{K_2} c_k e^{-j[2\pi k f_0(t - T/2 - x/v) + \phi_k]} \\
&\quad \times e^{j2\pi f_c(t - T/2 - x/v)} e^{j2\pi\alpha x}\, dx.
\end{aligned} \tag{8.19}$$

We now apply the integration over the space variable to find that

$$F_{+}(\alpha, t) = \frac{m}{2} \sum_{k=0}^{K_2} e^{j[2\pi(f_c + kf_0)t + \phi_k]} c_k A(\alpha - \alpha_c + k\alpha_0)$$

$$+ \frac{m}{2} \sum_{k=0}^{K_2} e^{j[2\pi(f_c - kf_0)t - \phi_k]} c_k A(\alpha - \alpha_c - k\alpha_0). \quad (8.20)$$

From Equation (8.20) we see that the spectrum of $s(t)$ is centered at α_c and consists of $2M + 1$ versions of $A(\alpha)$, weighted by the magnitudes c_k, and distributed about α_c at intervals of α_0. Equation (8.20) shows that the dc component of the signal, given by $k = 0$, is located at α_c in accordance with the result given by Equation (8.9). Further, each spectral component carries with it a temporal frequency revealed by the exponential factor; we have ignored unimportant phase terms in Equation (8.20).

As $s(t)$ is a baseband signal, each spectral component produces a pair of responses, displaced by an amount $\pm k\alpha_0$ from α_c and weighted according to the magnitude c_k of the spectral component. The spectrum is therefore redundant about α_c so that only half the spectrum needs to be detected. On the other hand, the spectrum for a bandpass signal is not redundant about α_c; instead, it contains completely independent spectral components. In both cases, of course, there is redundancy about $\alpha = 0$; that is, $F_{-}(\alpha, t)$ is always the mirror image of $F_{+}(\alpha, t)$ when $s(t)$ is real valued, with $\alpha = 0$ being the point of symmetry.

8.3. APERTURE WEIGHTING FOR SIDELOBE CONTROL

The transform $A(\alpha)$ of the aperture function $a(x)$ plays an important role in spectrum analysis. The aperture function is often referred to as the *window function* when we discuss spectrum analysis using DFT's or FFT's and is used primarily to control the sidelobe levels of strong signals so that weak signals are also detectable. We discussed the role of the aperture function in detail in Chapter 4. As all the key results are applicable here, the reader may wish to review Chapter 4, Section 4.5, on how to choose the appropriate aperture function to meet the system specifications. We briefly review the major points here for reference.

The magnitude of the sidelobes must be controlled so that weak signals can be detected in the presence of nearby strong signals. When the aperture function is a rect function, the intensity of its Fourier transform

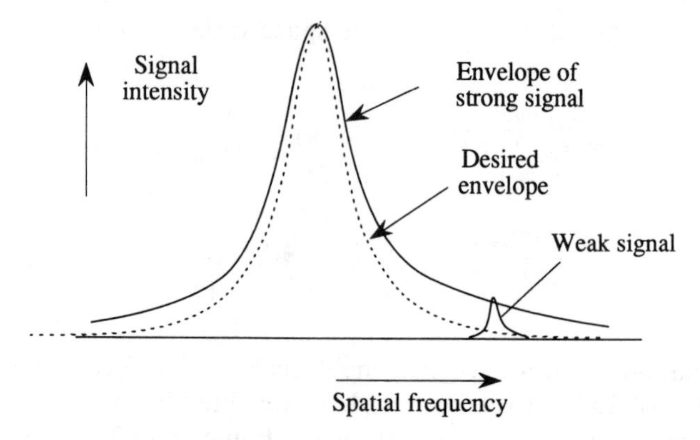

Figure 8.6. Weak signal in strong signal sidelobes.

is a sinc^2 function. The sidelobes of the sinc^2 function decrease as $1/\xi^2$, where ξ is the distance from the main lobe. The first sidelobe is only 13 dB down from the mainlobe and the rate of decrease is only 6 dB per octave. This low rate of decrease means that weak signals may be lost in the sidelobe energy from a nearby strong signal, as illustrated in Figure 8.6. The objective, then, is to use an aperture weighting function to suppress the sidelobes.

In Chapter 4, Section 4.5 we discussed in detail how to choose an aperture function whose Fourier transform has a sidelobe envelope below the specified levels, and we stated that a typical criterion is that the envelope of the sidelobes must be 50 dB below the mainlobe level at five resolvable frequencies away from the strong signal. We compared alternative forms of window functions such as Bartlett, Hanning, Hamming, and Kaiser, each with its own resolution limit and sidelobe level. We argued that a Gaussian weighting function is the logical choice because it provides adequate performance and is obtained naturally from a laser.

8.4. RESOLUTION

Sidelobe suppression is gained at the expense of some reduction in frequency resolution because the width of the mainlobe is broadened. We must therefore increase the distance between resolvable frequencies in the Fourier plane to achieve the required dip between frequencies. This increased distance, in turn, reduces the number of frequencies that can be resolved by the system. Recall from Chapter 4, Section 4.5 that frequency resolution is based on achieving a given dip in the intensity between two

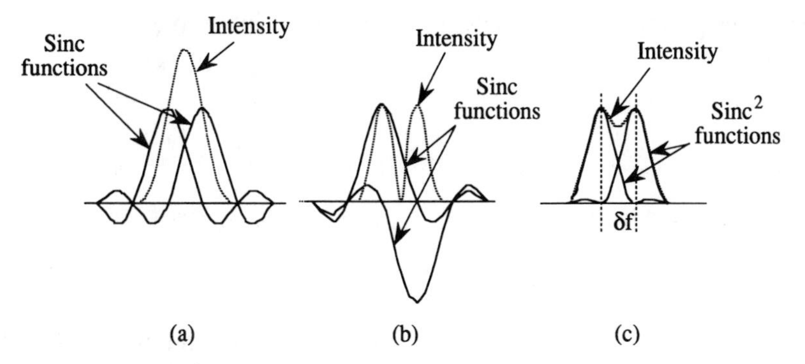

Figure 8.7. Coherent spectrum analyzers: (a) signals in phase, (b) signals out of phase, and (c) signals with different temporal frequencies.

frequencies. This criterion assumes that the system is incoherently illuminated and has its roots in the Rayleigh criterion for the angular resolution of a telescope. We revisit the question of resolution to determine the minimum frequency difference that is detected at the Fourier plane.

In Chapter 4, we discussed input signals recorded on spatial light modulators, such as photographic film, that were functions of space only. When the spectral components are in phase, the dip between the spatial frequencies may disappear completely, as shown in Figure 8.7(a); the spatial frequencies are therefore not resolved. However, when the spectral components are out of phase, as shown in Figure 8.7(b), the dip is complete; the spatial frequencies are well resolved. The spatial frequency resolution is therefore a strong function of the phase relationships of the adjacent frequencies and the only recourse to ensure good resolution for all phase relationships is to increase significantly the distance between the spatial frequencies.

In an acousto-optic cell spectrum analyzer, however, the signals within the cell are functions of both space and time. Consider the resolution criterion for an rf signal at f_k and an adjacent signal at f_j, as shown in Figure 8.7(c), where $\delta f = f_k - f_j$ is the minimum resolvable frequency. The two spectral components are

$$F_+(\alpha_k, t) = mc_k e^{j(2\pi f_k t - \phi_k)} A(\alpha - \alpha_k), \tag{8.21}$$

and

$$F_+(\alpha_j, t) = mc_j e^{j(2\pi f_j t - \phi_j)} A(\alpha - \alpha_j). \tag{8.22}$$

When these two signals are added, the resultant intensity is

$$\begin{aligned} I(\alpha, t) &= \left| F_{+}(\alpha_k, t) + F_{+}(\alpha_j, t) \right|^2 \\ &= m^2 c_k^2 \left| A(\alpha - \alpha_k) \right|^2 + m^2 c_j^2 \left| A(\alpha - \alpha_k) \right|^2 \\ &\quad + 2\left\{ \mathrm{Re}\left[m^2 c_k c_j A(\alpha - \alpha_k) A^*(\alpha - \alpha_k) e^{j[2\pi(f_k - f_j)t - \phi_k + \phi_j]} \right] \right\}. \end{aligned} \tag{8.23}$$

If the bandpass of the photodetector element is low relative to δf, the time average of the last term is zero, independently of the phases. As we show shortly, the frequency resolution of the acousto-optic cell spectrum analyzer is of the order of $\delta f = 1/T$, where T is the transit time of the cell. The integration time of the photodetector array in most power spectrum analyzers is much longer than T seconds; thus, the detected spectrum does not contain a contribution from the third term of $I(\alpha, t)$. The first two terms of Equation (8.23) are analogous to the situation of incoherent illumination of a telescope, for which the Rayleigh criterion is applicable.

The Rayleigh-resolution criterion requires a dip between adjacent frequencies of 0.8 ($\approx$ 1 dB), which is adequate for visual resolution. A more frequently specified dip in spectrum analyzers that use photodetectors at the output is 2–3 dB. The corresponding intensity dips are 63% and 50%, and the corresponding increases in separation at the Fourier plane are factors of 1.11 and 1.21 over that established by the Rayleigh limit for a rectangular aperture function (see Problem 8.6). The resolution limit also changes when we modify the shape of the function $A(\alpha)$ to control the sidelobe levels, as we showed in Figure 4.13.

8.5. DYNAMIC RANGE AND SIGNAL-TO-NOISE RATIO

In Chapter 4, Section 4.6 we discussed the dynamic range of an optical spectrum analyzer; the reader is encouraged to review that material because most of it also applies to acousto-optic spectrum analyzers. Here we consider some aspects of the dynamic range unique to the use of acousto-optic cells in the spectrum analyzer. The maximum signal level at the output of the system is determined by the required spur-free dynamic range, by the allowable degree of signal compression, or by the required degree of linearity as discussed in subsequent sections. The minimum signal level at the output is determined by the signal-to-noise ratio available at the output of the system, as discussed in Chapter 4, Section 4.6.

In a well-designed spectrum analyzer, the principal source of noise at the output is the photodetector and its associated circuitry. In Chapter 4, Section 4.6 we modeled the circuit for a photodetector element and showed that the signal-to-noise ratio at the kth frequency is

$$\mathrm{SNR} = \frac{\langle i_k^2 \rangle R_L}{2eB(i_d + \bar{i}_k)R_L + 4kTB}, \tag{8.24}$$

where $i_k = SI_k\, \delta A$ is the photocurrent, S is the responsivity of the photodetector, I_k is the light intensity, and δA is the area of the photodetector element. Also, $e = 1.6(10)^{-19}$ Coulomb is the charge for an electron, B is the postdetection-noise bandwidth, i_d is the dark current, $\bar{i}_k$ is the average signal current, $k = 1.38(10)^{-23}$ J/K is Boltzmann's constant, and T is the temperature in degrees Kelvin. The first term in the denominator of Equation (8.24) is the shot noise, and the second term is the thermal noise.

Equation (8.24) suggests that we can increase the signal-to-noise ratio arbitrarily by increasing the load resistance until we are shot-noise limited; this increase may, however, lead to inadequate bandwidth. We must ensure that we have enough photodetector bandwidth to pass the highest detected frequency. The relationship among the load resistance, capacitance, and cutoff frequency f_{co} is obtained from basic circuit theory as

$$f_{\mathrm{co}} = \frac{1}{2\pi c_d R_L}. \tag{8.25}$$

We solve Equation (8.25) for R_L and use it in Equation (8.24) to generate a more useful form of the signal-to-noise ratio equation:

$$\boxed{\mathrm{SNR} = \frac{\langle i_k^2 \rangle}{2eB(i_d + \bar{i}_k) + 8\pi kTBf_{\mathrm{co}}c_d}.} \tag{8.26}$$

In subsequent discussions we still refer to the two terms in the denominator of Equation (8.26) as "shot" and "thermal" noise even though they represent the square of the noise current instead of the noise power as they normally do.

The general procedure for calculating the dynamic range is to first calculate the minimum signal current when the SNR = 1 (or x dB of your choice). The *dynamic range* for a single spatial frequency in the system is then given by

$$\boxed{\mathrm{DR} = 10 \log\left[\frac{c_{k\,\max}^2}{c_{k\,\min}^2}\right],} \tag{8.27}$$

where $c_{k\,\max}^2$ is the maximum input signal power and $c_{k\,\min}^2$ is the minimum input signal power.

As in Chapter 4, we need to relate the photodetector current to the signal magnitude. The spectrum analyzer shown in Figure 8.1(b) contains the key components of the system, even though it does not show them in detail. Suppose that the laser has power P_0, the optical system, aside from the acousto-optic cell, has an overall efficiency ε, and the acousto-optic cell has an aperture of length L. For convenience, we reference the overall system efficiency to the input plane so that the effective intensity in the x direction is $I_0 = P_0\varepsilon/L$ W/mm and the amplitude is $A_0 = \sqrt{P_0\varepsilon/L}$. We do not consider the intensity in the vertical direction because, based on the optical system shown in Figure 8.2, all optical power in that direction is focused onto the photodetector.

For the calculation of dynamic range, consider just one cw signal $f(t) = c_k \cos(2\pi f_k t)$, whose frequency is f_k and whose magnitude is c_k. The light amplitude at the exit face of the acousto-optic cell is

$$f(x,t) = A_0 a(x)\left[\tfrac{1}{2} + \tfrac{1}{2}jmc_k \cos\left(t - \frac{T}{2} - \frac{x}{v}\right)\right], \tag{8.28}$$

where m is the modulation index, $a(x)$ is the normalized aperture function required to control the sidelobe levels, and the factor of $\frac{1}{2}$ is needed to keep the spatial light modulator passive. As we must calculate the optical power falling on the photodetector, we use the exact Fourier-transform relationship, including all the relevant constants, and express the result in terms of the spatial frequency coordinate ξ. The positive diffracted order then becomes

$$F_+(\xi,t) = A_0\sqrt{\frac{j}{\lambda F_2}}\int_{-\infty}^{\infty} \tfrac{1}{2}ma(x)\frac{c_k}{2}e^{j2\pi f_k(t-T/2-x/v)}e^{j2\pi\xi x/\lambda F}\,dx, \tag{8.29}$$

where F is the focal length of the spherical lens that creates the Fourier transform, as shown in Figure 8.4, and the second factor of $\frac{1}{2}$ in the integrand is due to the Euler expansion of the cosine term in $f(t)$. As we ultimately calculate the intensity at the output plane, we ignore the j factor in the radical of Equation (8.29).

To obtain closed-form solutions, we set the aperture weighting function to $a(x) = \text{rect}(x/L)$. We perform the integration in Equation (8.29) to find that

$$F_+(\xi, t) = \sqrt{\frac{P_0\varepsilon}{16\lambda LF}}\, mc_k e^{j2\pi f_k(t-T/2)} L \,\text{sinc}\left[(\xi - \xi_k)\frac{L}{\lambda F}\right]. \tag{8.30}$$

A photodetector element in the Fourier plane integrates the light intensity $I(\xi) = |F_+(\xi, t)|^2$ to produce the optical power $P_k = I(\xi)\,\delta\xi$, where $\delta\xi$ is the length of the photodetector element. As noted above, we have already accounted for the height of the photodetector element in our assumption that all the light in the vertical direction is concentrated on the photodetector element by the cylindrical lens C_2. The remaining integration is therefore only over the variable ξ. The optical power collected by a photodetector element whose width is $\delta\xi$, centered at ξ_k, is

$$P_k = \frac{P_0\varepsilon m^2 c_k^2}{16\lambda LF}\int_{-\delta\xi/2}^{\delta\xi/2} L^2 \,\text{sinc}^2\left[(\xi - \xi_k)\frac{L}{\lambda F}\right] d\xi. \tag{8.31}$$

The width of the photodetector element, based on the criterion of three photodetector elements per resolvable frequency, is $\delta\xi = \lambda F/3L$. In Chapter 4, Section 4.6 we showed that the integral of a sinc^2 function is linear for small photodetector elements so that power collected by the photodetector is

$$P_k = \frac{P_0\varepsilon m^2 c_k^2}{16\lambda LF} L^2 \frac{\lambda F}{3L} = 0.02 P_0 \varepsilon m^2 c_k^2. \tag{8.32}$$

Finally, the photocurrent i_k is the product of the photodetector responsivity S and the optical power:

$$i_k = 0.02 P_0 \varepsilon S \eta_f c_k^2, \tag{8.33}$$

where we use the fact that $n_f = m^2$ is the diffraction efficiency per frequency. As noted in Chapter 4, this result is interesting for three reasons: (1) it shows that the detected power is not a function of λ, L, or

F, (2) it gives the connection between the photocurrent and the signal power, and (3) it shows that the photocurrent is proportional to the *power* of the input signal.

We use Equation (8.33) in Equation (8.26) to find that

$$c_{k\,\min}^2 = \frac{\sqrt{2eBi_d + 8\pi kTBf_{co}c_d}}{0.02P_0\varepsilon S\eta_f} \tag{8.34}$$

is the minimum input signal level. We substitute Equation (8.34) into Equation (8.27) and find that

$$\boxed{\mathrm{DR} = 10\log\left[\frac{0.02P_0\varepsilon S\eta_f}{\sqrt{2eBi_d + 8\pi kTBf_{co}c_d}}\right],} \tag{8.35}$$

which clearly reveals how the various system parameters affect the dynamic range. Recall that $c_{k\,\max}^2 = 1$ because the modulation index m accounts for the conversion from electrons to photons at the acousto-optic cell. As expected, the dynamic range is logarithmically proportional to the laser power, the system efficiency, the photodetector responsivity, and the diffraction efficiency per frequency. The dynamic range is logarithmically proportional to the reciprocal square root of the system bandwidth.

So far, the signal-to-noise ratios and dynamic range have been calculated on the basis of using discrete photodetector elements. Although discrete elements are always used in heterodyne spectrum analyzers, as we discuss in Chapter 10, and are sometimes used in instantaneous spectrum analyzers, such as radar threat warning receivers, a more commonly used device in power spectrum analyzers is a photodetector array in which the elements are integrated on a silicon chip. In this device, light falling on an array element creates a charge that is accumulated at the photodetector site until the array element reaches saturation. The charge pattern is then transferred to a CCD structure and read out onto one or more video lines as discussed in Chapter 4, Section 4.4. In this case, the dynamic range is set by the ratio of the saturation level of the photodetector array to the minimum signal level. The dynamic range can be extended by the use of nonlinearities in the photodetector response, as discussed in Chapter 4, Section 4.4.

8.6. SPUR-FREE DYNAMIC RANGE

In this section we concentrate on the limitations imposed by the acousto-optic cell on the spur-free dynamic range. From Chapter 7, Section 7.3.2, we find that the dynamic transfer curve for an acousto-optic device is

$$\eta = \sin^2\left(\sqrt{BP_s}\right), \tag{8.36}$$

where η is the diffraction efficiency, P_s is the rf drive power of the applied signal, and B is a constant involving the dimensions of the acoustic beam and the figure of merit of the interaction material. The diffraction efficiency is defined as the ratio of the intensity of the diffracted beam to the intensity of the incident beam. When we deal with coherently illuminated systems, we use the modulation index $m = \sqrt{\eta}$. We then have that

$$m = \sin\sqrt{BP_0}. \tag{8.37}$$

When $\sqrt{BP_s} \ll \pi/2$, we can replace the sine function by its argument:

$$m \approx \sqrt{BP_s} = B'V_s, \tag{8.38}$$

where B' is a new constant and V_s is the voltage of the rf signal. The amplitude of the light, as measured by m, is therefore linearly related to the voltage of the applied rf signal.

When the signal level exceeds the linearity constraint, the nonlinear transfer curve produces harmonic distortion. This is not a problem with a single cw signal because the generated harmonics are out of band. That is, if f_2 and f_1 are the highest and lowest frequencies of the acousto-optic cell and if we design the system so that $f_2 = 2f_1$, the harmonics of any pure cw signal are outside the passband of the acousto-optic cell. We are mostly concerned about *spurious signals* generated by two- or three-tone signals that may produce in-band frequency components due to acousto-optic nonlinearities.

8.6.1. Intermodulation Products Due to Acousto-Optic Cells

In Chapter 7, Section 7.2 we described the basic mechanism of how an acousto-optic cell generates diffracted beams according to the frequency of the applied signal. Suppose that we have two equal-magnitude cw signals of frequencies f_1 and f_2. If the response of the acousto-optic cell were completely linear, we would expect just two diffracted beams at

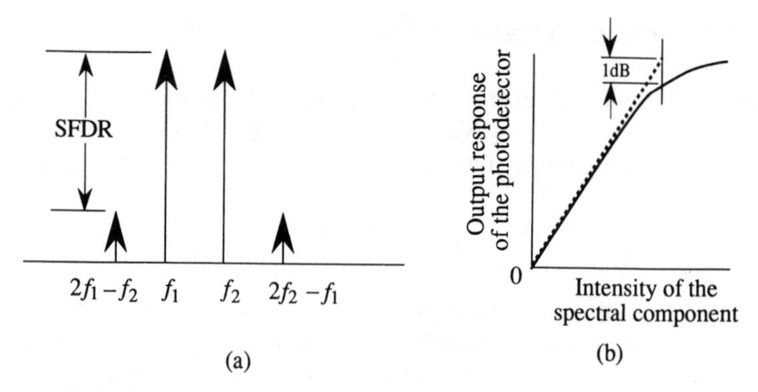

Figure 8.8. Performance measures: (a) spur-free dynamic range and (b) compression.

angles θ_1 and θ_2 with respect to the undiffracted beam. When the response is nonlinear, additional beams called *cross-coupling* or *cross-modulation* terms occur at the sum and difference frequencies. These components are out of band and are eliminated by the bandpass nature of the acousto-optic cell or simply ignored by the photodetector array. Of greater significance are the *intermodulation products* generated at frequencies $2f_1 - f_2$ or $2f_2 - f_1$; these are called *two-tone, third-order intermodulation products.*

Consider the case when the input consists of two frequencies f_1 and f_2 that produce frequencies at $2f_1 - f_2$ and at $2f_2 - f_1$ within the passband of the system. Suppose that the signal $f(t) = V_1 \cos(2\pi f_1 t)$ produces an intensity I_1 and an associated diffraction efficiency $\eta_1 = m_1^2 = I_1/I_0$, where I_0 is the incident intensity. In a similar fashion, suppose that the signal $f(t) = V_2 \cos(2\pi f_2 t)$ produces an intensity I_2 and an associated diffraction efficiency $\eta_2 = m_2^2 = I_2/I_0$. A typical intermodulation product which falls within the passband, when both tones are present, produces an intensity I_3 at each of the spurious signal locations.

The *spur-free dynamic range* (SFDR) is the ratio of the signal intensity at the true frequency locations f_1 or f_2 to the intensity of the intermodulation products at $2f_1 - f_2$ and $2f_2 - f_1$, as shown in Figure 8.8(a). The spur-free dynamic range is given by

$$\mathrm{SFDR} = \frac{I_1}{I_3}. \tag{8.39}$$

Hecht gave an extensive analysis of the intermodulation products produced by an acousto-optic cell and found that the spur-free dynamic range can be expressed in terms of the diffraction efficiency of each of the two frequencies with the result that SFDR = $36/\eta_f^2$, where η_f is the *diffraction efficiency per frequency* and the SFDR is expressed in linear terms (89). We solve for the maximum diffraction efficiency for each signal that still meets the SFDR specification:

$$\boxed{\eta_f = \frac{6}{\sqrt{\text{SFDR}}}.} \tag{8.40}$$

For example, if we require a spur-free dynamic range of 10^5 (50 dB), we find that $I_1/I_3 = 10^5$ and we conclude from Equation (8.40) that $\eta_f \leq 0.019$ for any single frequency.

Hecht also discussed the third-order, three-tone case in which the allowable diffraction efficiency per frequency is reduced by a factor of 2 and, for completeness, calculated the allowable diffraction efficiency for the Raman-Nath mode of operation as well (89).

A second source of nonlinear behavior in acousto-optic cells arises from the *nonlinear elastic* response of the crystalline lattice structure. This source of nonlinearities is most noticed when we use acousto-optic cells designed to operate with wide bandwidths or with long time apertures. Whereas the dynamic nonlinearities are constant over the aperture of the cell, the magnitude of the elastic nonlinearity depends on the propagation distance from the transducer, being small near the transducer and increasing monotonically with propagation distance of the acoustic wave. At some point within the cell, the elastic nonlinearities may exceed those of the dynamic nonlinearities described earlier; the integrated effect over the entire aperture of the cell causes the spur-free dynamic range to be less than that calculated from Equation (8.40) due purely to dynamic nonlinearities.

The difference between the dynamic and elastic intermodulation product levels is proportional to the *sixth power* of the interaction length Z of the transducer. A long interaction length is therefore highly desirable; a byproduct of a long transducer is an increase in the diffraction efficiency as shown by Equation (7.17). A strong disadvantage of a long transducer is lower signal bandwidth, as shown by Equation (7.19). A phased-array transducer has been used to improve the cell bandwidth; recent experiments show that such a transducer can also reduce the elastic intermodulation products by as much as 15 dB so that the dynamic nonlinearities dominate and the spur-free dynamic range is at its maximum value (90). The phased-array transducer has the property that the mainlobe of the acoustic wave tracks the Bragg angle over a wide range of operating frequencies, while achieving optimum Bragg matching at two frequencies (91). The diffraction efficiency therefore increases and, of greater importance for the elastic nonlinearities, the acoustic power density within the cell decreases.

8.6.2. Signal Compression

If we have a spectrum analyzer with $TW = 500$, we cannot drive all frequencies at a diffraction efficiency permitted by the spur-free dynamic-range test because the total diffraction efficiency cannot exceed 100%. Well before we reach a 100% overall diffraction efficiency level, however, we experience a phenomenon called signal compression. Suppose that there are several cw signals present in the spectrum, and suppose that we introduce yet another one that we call the probe signal. As the voltage of the probe signal is increased, we find that the response to this signal eventually deviates from linearity. This departure from linearity is commonly called *signal compression* and is most noted when the acousto-optic cell is loaded with many signals from a dense electromagnetic environment.

When many signals share the available diffraction efficiency, the acousto-optic cell becomes *source depleted*, which means that the total amount of light diffracted is significant relative to the undiffracted light. Suppose that there are Q equal-strength signals, each with diffraction efficiency η_f. The source is depleted according to the relationship (89)

$$\eta_e = \eta_f \left[1 - \frac{2Q - 1}{3} \eta_f \right], \tag{8.41}$$

where η_e is the *effective* diffraction efficiency per frequency. As a result, when many frequencies are competing for the same laser power, the diffraction efficiency per frequency is somewhat less than expected. The small signal approximations used to develop this result are valid provided that the source depletion does not exceed 50% [i.e., $\{(2Q - 1)/3\}\eta_f \leq 0.5$].

When the depletion is too great, the measured output in a spectrum analyzer is compressed, as shown in Figure 8.8(b). We indicate the relative degree of compression by C with the requirement that $C \leq \eta_e/\eta_f$. We find a relationship for η_f in terms of Q and C from Equation (8.41):

$$\eta_f = \frac{3(1 - C)}{2Q - 1}. \tag{8.42}$$

We now have two criteria for setting η_f; we find that either Equation (8.40) or Equation (8.42) may set the limit on the per frequency diffraction efficiency. If the degree of signal compression C is close to one, and the number Q of frequencies at saturation is high, Equation (8.42) may set the limit. If the required spur-free dynamic range is large, Equation (8.40) may set the limit.

As an example, suppose that we want a spur-free dynamic range of 60 dB for two-tone, third-order signals, a time bandwidth product of $TW = 100$, signal compression of 1 dB or less, and the number of saturated frequencies to be $Q \leq 20$. From Equation (8.40) we find $\eta_f \leq 6(10^{-3})$, and from Equation (8.42) we find $\eta_f \leq 15(10^{-3})$. Thus we see that Equation (8.40) is the more restrictive of the two limits for these performance parameters. We now find either Q or C from Equation (8.42), using the constraint that $\eta_f \leq 6(10^{-3})$. If $Q = 20$, we find that $C = 0.4$ dB, which exceeds the specification; if $C = 1$ dB, we find that $Q = 53$, which also exceeds the specification. Thus, given a dynamic range of 60 dB, we can exceed the compression specification or the number of frequencies that can go to saturation, or both.

When η_f as calculated from Equation (8.42) is less than that calculated from Equation (8.40), the system has an excess spur-free dynamic range or an excess value for Q; the system is overdesigned. A consistent specification results when the diffraction efficiencies per frequency, as calculated from Equations (8.40) and (8.42), are equal:

$$\frac{2}{\sqrt{\mathrm{SFDR}}} = \frac{1 - C}{2Q - 1}, \tag{8.43}$$

which relates the relative linearity required and the number of saturated frequencies to the spur-free dynamic range as set by the two-tone criterion. Two of the three parameters can be set independently; the third is subject to the constraint given by Equation (8.43). The drive amplifiers must have compatible compression specification, of course, to achieve the required intermodulation product performance level.

8.6.3. Scattered Light

Light scattered from optical components may also affect the available dynamic range if it raises the noise floor (92). Because the overall diffraction efficiency of the acousto-optic cell, as set by the intermodulation product specification, is of the order of a few percent, most of the light is transmitted in the undiffracted beam and may be scattered by other optical elements into the spatial frequency band of interest. Lenses with imperfections such as scratches, bubbles, dirt, dust, smudges, and films are prime culprits. Such elements must be properly specified during the design phase of a system and kept clean when the system is in operation.

A high value for the center frequency of the acousto-optic cell helps to move the passband further from the region where the undiffracted light is focused. This approach, however, is generally undesirable because it complicates the transducer design. A better approach is to trap the undiffracted light as early as possible to minimize scattering.

Power spectrum analyzers are sensitive to scattered light because they respond only to the intensity of light. As we show in Chapter 10, scattering is not as great a problem with heterodyne detection because most of the scattered light is not at the proper temporal frequency for detection.

8.7. PHOTODETECTOR GEOMETRIC CONSIDERATIONS

After the operational requirements have been met, the final step in the design of a spectrum analyzer is to select the discrete photodetector elements or the photodetector array. Arrays are commercially available with various numbers and sizes of elements. As a custom-designed array generally implies a significant nonrecurring cost, it is desirable to use standard arrays whenever possible.

We begin by selecting a photodetector array with the proper number of elements for the application. This number is usually three times the number of resolvable frequencies of the overall system. We want a high spatial duty cycle for the elements to efficiently collect light. If d' is the width of the active area of an element, as shown in Figure 4.9, and d is the spacing, we want the duty cycle $c = d'/d$ as close to one as possible.

The value of d from the photodetector array establishes the requisite value of the physical distance $\delta\xi$ between the resolvable frequencies at the output of the system. This distance is a function of λ, L, and F. Because L is determined by the required aperture weighting function and λ is a restricted choice, the only free parameter is F, the focal length that determines the scale of the Fourier transform. If the required value of F leads to a high relative aperture so that aberrations are a problem, we can sometimes split the lens into two or three elements to control the aberrations. Generally this is not a problem in optical spectrum analyzers, since typical values of d are of the order of 13 μ or so for a 2048-element array. If we allow approximately 3 elements per frequency, we find that $\delta\xi \approx 39\ \mu$. This spacing between frequencies is conveniently obtained using a lens with a modest $f/\#$ and with a reasonable focal length.

As we do not need to image the acousto-optic cell in the horizontal direction and as we can tolerate a spherical phase factor in the Fourier domain for a power spectrum analyzer, we can relieve the constraint that lenses L_1 and L_2 are exactly one focal length from the acousto-optic cell, provided that there is no mechanical interference with the cylindrical lens. We can therefore reduce the required lens aperture and reduce the distance from plane P_1 to plane P_2.

8.8. EXAMPLE

From Equation (8.33), recall that a signal of the form $f(t) = c_k \cos(2\pi f_k t)$ leads to a signal current $i_k = 0.02\varepsilon S\eta_f P_0 c_k^2$. Suppose that the system parameters are such that $P_0 = 10$ mW, $i_d = 10$ nA, $c_d = 4$ pF, $\varepsilon = 0.5$, $S = 0.4$ A/W, $T = 300$ K, $B = 50$ kHz, and $f_{co} = 50$ kHz. In this example, we determine the value of the cutoff frequency f_{co} for which the shot-noise and thermal-noise terms are equal, the required load resistor value, the value of η_f required to achieve a spur-free dynamic range of 55 dB under the two-tone criterion, and the dynamic range.

(a) The value of the cutoff frequency f_{co} for which the shot-noise and thermal-noise terms are equal, when the signal current $i_k \ll i_d$, is

$$f_{co} = \frac{2eBi_d}{8\pi kTBc_d} = \frac{ei_d}{4\pi kTc_d} = 7{,}688 \text{ Hz}. \tag{8.44}$$

As this value is well below the requirement, the system is thermal-noise limited. This thermal-noise-limited condition is due solely to the parameters of the photodetector; no amount of additional laser power can change this condition.

(b) The required load resistor for the parameters given is

$$R_L = \frac{1}{2\pi f_{co} c_d} = 796 \text{ k}\Omega. \tag{8.45}$$

(c) We confirm that the system is thermal-noise limited because

$$\begin{aligned} 2eBi_d &= 1.6(10^{-22}) \text{ A}^2, \\ 8\pi kTBf_{co}c_d &= 1(10^{-21}) \text{ A}^2. \end{aligned} \tag{8.46}$$

(d) The value of η_f required to achieve a spur-free dynamic range of 55 dB, under the two-tone criterion, is

$$\eta_f = \frac{6}{\sqrt{\text{SFDR}}} = \frac{6}{\sqrt{10^{5.5}}} = 0.01. \tag{8.47}$$

(e) The dynamic range is obtained by calculating the value of $c_{k\,\min}^2$ that produces a signal-to-noise ratio of 1:

$$\text{SNR} = 1 = \frac{\langle i_k^2 \rangle}{2eBi_d + 8\pi kTBf_{co}c_d}, \tag{8.48}$$

which implies that $c_{k\,\min}^2 = 8.1(10^{-5})$. The dynamic range is obtained from Equation (8.27) with the underlying assumption that $c_{k\,\max}^2 = 1$; for the parameters given, the dynamic range is 40.9 dB. As the dynamic range is much less than the spur-free dynamic range, the latter is overspecified. In general, the spur-free dynamic range specification is not as severe as that for the dynamic range. An interesting problem is to find the conditions for which the spur-free dynamic range is just equal to the dynamic range (see Problem 8.4).

We can use an avalanche photodiode, as discussed in Chapter 4, Section 4.6, to improve the performance of this system somewhat. For the parameters cited above, an avalanche photodiode gain of $G = 2.26$ is optimum in the sense that it equalizes the shot noise and thermal noise; the dynamic range then increases to 43.2 dB. In practice, the additional complexity of the avalanche photodiode circuitry is generally not worth the 2.3-dB improvement in dynamic range.

In some rf systems, the dynamic range is specified as the ratio of the maximum input signal level of a single frequency, as determined by a linearity specification, to the minimum input signal level as determined by the signal-to-noise ratio test. The linearity specification limits the input signal level to that value for which the output signal departs from linearity by a given amount, often given as 1 dB. At first glance, this method for determining the maximum signal level seems similar to the signal compression specification given in Section 8.6.2. In the linearity test, however, there is only one frequency present in the input signal. As a discrete photodetector element with its associated circuitry is generally linear up to 100 dB, we could increase the diffraction efficiency per frequency from the value given by either Equation (8.40) or (8.42) to a value of $\eta_f = 1$ for this test. From Equations (8.35) and (8.47), we see that the dynamic range of this spectrum analyzer would increase by 20 dB to 60.9 dB by this less-restrictive test.

8.9. THE SIGNAL-TO-NOISE RATIO

In a spectrum analyzer, the dynamic range is far more important than the signal-to-noise ratio. Nevertheless, we make a few comments illustrating

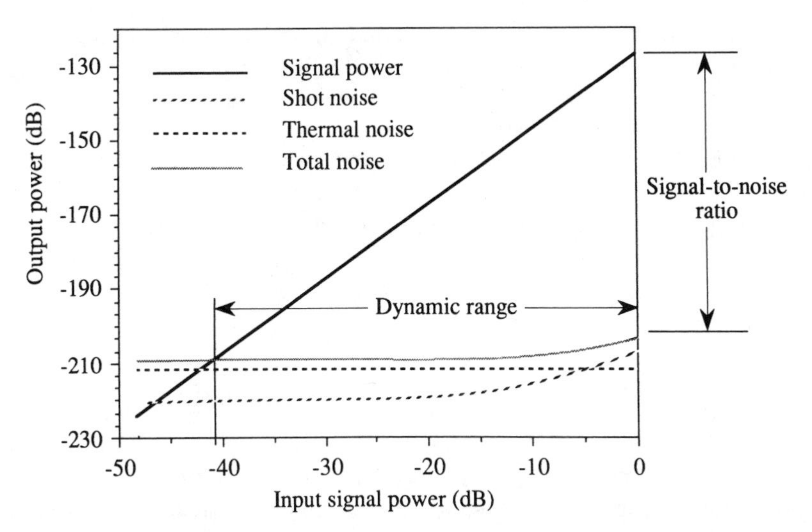

Figure 8.9. Signal and noise levels at the output of the spectrum analyzer.

the connection between the two measures of performance. In the example just given, the system is thermal-noise limited because the bandwidth is high. As long as the system is thermal-noise limited, the signal-to-noise ratio increases linearly as the input signal level increases from its minimum value. When the input signal rises above a certain level, however, shot noise due to the average current $\bar{i}_k$ may equal or exceed the thermal-noise level, thus affecting the signal-to-noise ratio. The total noise floor is the sum of the thermal and shot noise and is therefore a function of the signal level.

In Figure 8.9 we plot the detector output signal power on the vertical axis against the relative rf input power on the horizontal axis. The slope of the straight line is two, as we expect, because i_k is proportional to c_k^2. The line intercepts the thermal noise floor at -40.9 dB; this intercept gives the dynamic range. When the input signal is maximized, the SNR, given as the difference between the output power and the combined noise, is 73.6 dB. The signal-to-noise ratio varies, of course, as the input signal power varies. At the intercept point with the thermal-noise line, SNR $= 1$ by definition; it is this condition that establishes the dynamic range.

8.10. RADIOMETERS

So far we have assumed that the signal-to-noise ratio of the signals we wish to detect is high, on an instantaneous basis, within the channel of interest. Sometimes, however, signals may not have a positive SNR on an instantaneous basis. Suppose that σ_s^2 is the signal power and σ_W^2 is the noise power in the received band W. The spectrum analyzer divides the total signal bandwidth W into TW frequency bins so that the noise power in a bandwidth of $1/T$ is σ_W^2/TW. The SNR at the output of the spectrum analyzer within a frequency bin is therefore a factor TW greater than that at the input, referenced to the total bandwidth:

$$\mathrm{SNR}_0 = \frac{\sigma_s^2}{\sigma_W^2/TW} = TW\,(\mathrm{SNR}_i). \tag{8.49}$$

If SNR_0 approaches 0 dB, we need more processing gain to detect the signal. But the amount of time history T available is limited by the length of the acousto-optic cell crystal, acoustic attenuation, or other constraints.

A way to overcome a limited signal-to-noise ratio is to integrate light on the array for a longer period of time which, in effect, narrows the filter bandwidth so that the signal can be detected. Integrating for long periods of time, either on the array or with postdetection integration, is referred to as *radiometry*, a technique developed by radio astronomers for detecting weak signals. The demarcation between instantaneous spectrum analyzers and radiometers is not always sharply drawn. We characterize the former by integration times up to T, the argument being that the spectrum does not change significantly over periods of time shorter than T. Radiometry is characterized by integration times longer than T.

As noted before, we often violate this characterization and integrate light on a CCD array for $T_p > T$ seconds even in an instantaneous spectrum analyzer. Perhaps a better characterization of a radiometer is that the integration time is much greater than T_p. For example, the integration time for most wideband (> 80-MHz) spectrum analyzers is of the order of a few microseconds. Radiometers have integration times ranging from a few milliseconds to seconds. Thus, the radiometer tends to integrate over time periods that are factors of 10^3–10^5 longer than the integration time T_p of an "instantaneous" power spectrum analyzer.

Suppose that the total integration time is kT_p, where $k \gg 1$, and that the photodetector array is read out every T_p seconds. If the signal is coherent over the time kT_p, it grows according to k. The noise, however, being random, fluctuates over this time period. We can think of the noise taking on different values for each time segment T_p. These samples are incoherently summed so that the noise grows according to $\sqrt{k}$. The output

SNR_0 therefore grows as $k/\sqrt{k} = \sqrt{k}$. Sometimes we cannot integrate for the desired time period on the array because it saturates. In this case, we read out the array just before saturation occurs, sample and quantize the contents of the CCD, and accumulate the values in a digital memory.

Digital postprocessing is also used to correct systematic errors. For example, we calibrate the system initially to correct variations in photodetector responsivity as a function of spatial frequency position. We may periodically insert pilot tones to ensure that the Fourier transform has not shifted relative to the array due to thermal changes. We test for fixed pattern noise due to clock signals or other switching noise sources. We measure the scattered light contribution as a function of spatial frequency. These and other factors that are not functions of time or, at best, are slowly varying, are then removed before the signal is accumulated.

8.11. SUMMARY OF THE MAIN DESIGN CONCEPTS

The major steps in the design of a spectrum analyzer are similar to those given in Chapter 4, in connection with the design of a power spectrum analyzer. There are a few minor differences, however, and the main results are summarized in the following notes:

1. From the required frequency resolution δf and total bandwidth W to be covered, calculate the required time bandwidth product $M = W/\delta f = TW$. This number gives a quick assessment of the difficulty of the design task and an initial cut at the type of acousto-optic cell required. It also gives a preliminary estimate of the time duration T of the input cell.
2. From the specification on the sidelobe level needed to detect weak signals in the presence of strong ones, determine the required value of A for the Gaussian illumination; calculate the Fourier transform $A(\alpha)$ for the chosen truncated Gaussian function.
3. For the chosen Gaussian aperture weighting, calculate the relative loss in resolution and increase the required time bandwidth product by this factor. In particular, the revised length of signal is now increased to reflect the loss in resolution needed to gain the desired sidelobe control.
4. Determine, from the required dynamic range considerations, what type of photodetector subsystem is needed (discrete detectors, a photodetector array, etc.)

5. Using three photodetector elements per resolved frequency, calculate the convolution of the aperture response $A(\alpha)$ and a single detector width.
6. Using the convolved aperture response, determine the value of $\delta\alpha = \delta f/v$ needed to satisfy the dip criterion for frequency resolution. Several iterations may be required to achieve the desired result for the worst case condition.
7. From the value of $\delta\alpha$ and the photodetector array spacing, calculate the focal length of the Fourier transform lens. This calculation matches the scale of the displayed spectrum to that of the detector array.
8. From the required dynamic range, determine the required laser power and what type of laser is needed. Find its Gaussian beam illumination parameter.
9. Design an illumination subsystem to magnify the laser beam to the plane of the signal such that the proper truncation takes place at the edges of the revised signal length. Calculate the exponential weighting due to the absorption of the acoustic signal within the cell. Use this information to determine the central location of the Gaussian illumination so that the combination of the Gaussian and exponential weightings produce a Gaussian weighting that is centered on the acousto-optic cell and is symmetric about its midpoint.
10. From the spur-free dynamic range specification, determine the value of the diffraction efficiency per frequency.

PROBLEMS

8.1. In a power spectrum analyzer we have $W = 500$ MHz and $T = 2$ μsec. We want the highest sidelobe at five resolution spacings from a strong signal to be at least 35 dB down from the mainlobe. How would you truncate the Gaussian illumination to achieve this specification? Estimate the number of frequencies that can be resolved if we require a 3-dB dip between frequencies. How does this number compare with that of $M = TW$ for a uniformly illuminated aperture? Sketch and label the graphs used.

8.2. We design a power spectrum analyzer using direct detection of the light in the Fourier domain. Assume that a signal of the form $f(t) = \frac{1}{2}[1 + c_k \cos(2\pi f_k t)]$ leads to a signal current

$i_k = 0.02\varepsilon S\eta_f P_0 c_k^2$. The photodetector parameters are

$$\begin{aligned} P_0 &= 30 \text{ mW}, \\ i_d &= 10 \text{ nA}, \\ c_d &= 2 \text{ pF}, \\ \varepsilon &= 0.5, \\ S &= 0.4 \text{ A/W}, \\ T &= 300 \text{ K}, \\ B &= 40 \text{ kHz, and,} \\ f_{co} &= 40 \text{ kHz.} \end{aligned}$$

(a) Calculate the value of the frequency f_{co} that makes the shot-noise and thermal-noise terms equal when the average signal current $\bar{i}_k$ is very much less than the shot-noise current i_d.

(b) Calculate the required load resistor for the parameters given.

(c) Is the system shot-noise or thermal-noise limited? Calculate both $2eBi_d$ and $8\pi kTBf_{co}c_d$ to support your answer.

8.3. For the parameters given in Problem 8.2 (a) calculate the value of the diffraction efficiency η_f required to achieve a spur-free dynamic range of 55 dB using the two-tone criterion, and (b) calculate the dynamic range. Does it make sense to have the specified spur free dynamic range? Why, or why not?

8.4. Derive a general formula that equates the spur-free dynamic range to the dynamic range for the two-tone case (in this case the spurs are just at the noise floor). For the parameters of Problem 8.2, calculate the spur-free dynamic range and the value of η_f that will produce the equality. Be careful here! Remember that the dynamic range is the ratio of $c_{k\,\max}^2/c_{k\,\min}^2$ (you probably don't want to work in decibels) and that the spur-free dynamic range is a number greater than one (you probably want to work with its inverse to get it on the same footing as the dynamic range). Check your result by using the value of η_f to see if you do indeed have the correct answer for the dynamic range.

8.5. One of the photodetector arrays in the Reticon D linear series has 2048 elements that are on 13-μ center spacings. What is the maximum number of frequencies that can be detected? Suppose that you need a frequency resolution of 1 MHz and decide to use a cell made from TeO_2 used in the longitudinal mode. Calculate the focal length of the spherical lens between the acousto-optic cell and the detector. Assume that $\lambda = 0.5\ \mu$. If you use all the elements in the array, calculate the length L of the cell and the total bandwidth W of the system.

8.6. The Rayleigh-resolution criterion requires a dip between adjacent frequencies of 0.8 (≈ 1 dB). A more frequently specified dip in spectrum analyzers is 2–3 dB. The corresponding intensity dips are 0.631 and 0.50, and the corresponding increases in separation at the Fourier plane are factors of 1.13 and 1.19 over that established by the Rayleigh limit for a rectangular aperture function. Show, by analysis, that these separation factors are correct.

8.7. For GaP the sound velocity is given as $v = 6.32$ km/sec and $\Gamma = 3.8\ \text{dB}/\mu\text{sec}/\text{GHz}^2$ (from Table 7.1 of Chapter 7). For $f_c = 750$ MHz, $T = 2\ \mu$sec, and $A = 4$, calculate the attenuation of the center frequency at the end of the cell and calculate the amount and direction of the shift. Also calculate the ratio of the peak intensity of the effective illumination to that of the original illumination. Be careful regarding amplitudes vs intensities.

8.8. If five cw signals of different frequencies are injected into an acousto-optic cell at a power level of 100 mW each, what is the diffraction efficiency per frequency if the parameter $B = 0.0156\pi/\text{mW}$? Calculate the compression.

8.9. A radar warning system requires that a spectrum analyzer have a SFDR of 45 dB and that the dynamic range be no less than 10 dB greater than the SFDR. Calculate the minimum laser power required to implement the system for the parameters given in Problem 8.2.

9

Heterodyne Systems

9.1. INTRODUCTION

In Chapter 8 we discussed acousto-optic power spectrum analyzers in which we use *direct detection* of the intensity of the light at the Fourier-transform plane. Detecting light intensities is sometimes restrictive because both the phase and the temporal frequency of the signal are lost. In this chapter we show how the range of signal processing operations can be expanded considerably by using *heterodyne detection* in which we add a reference wave, sometimes called the *local oscillator*, to the light distribution to be detected. The interference between the signal and reference waves produces an output signal that is linearly proportional to the input signal voltage so that magnitude, frequency, and phase information are preserved. More sophisticated signal-processing operations, based on heterodyne detection, are discussed in subsequent chapters.

Heterodyne detection is also used in holography, matched filtering, and synthetic aperture radar processing. In the first two instances the signals are functions of two or three *spatial* dimensions while, in the last instance, we perform heterodyne detection on the *temporal* radar returns, which are then recorded on film as a raster-scanned, two-dimensional spatial function. Leith and Upatnieks recognized that the angle between the interfering waves in the holographic process must be large enough to separate the desired terms from all others upon reconstruction. They applied the principles of communication theory to the problem and recognized that the holographic fringe structure is similar to a temporal carrier frequency that is modulated in both magnitude and phase (93). If the carrier frequency is at least twice the signal bandwidth, the information can be completely recovered. In Chapter 5, we showed how these ideas, suitably modified, are used for constructing matched filters.

Heterodyne detection in either the spatial or frequency domain dates to the early work on spectrum analysis (94) based on even earlier work on correlation (86, 95). The basic ideas were brought together in an interesting series of papers related to probing coherent light fields by means of heterodyne techniques (96–98). In the study of heterodyne systems we sometimes encounter surprising results that do not, at first, seem consistent with our intuition. Further exploration of these concepts, however, reveals a satisfying richness of information and new arrangements for visualizing the fundamentals of optical signal processing.

9.2. THE INTERFERENCE BETWEEN TWO WAVES

As the basic heterodyne process is caused by the interference between two waves of light, we begin with a summary of the key results from Chapter 3 associated with spatially modulated signals. We then introduce a temporal modulation on one of the signals to illustrate the results of both spatial and temporal interference.

9.2.1. Spatial Interference

Consider the spatial interference caused by two plane waves traveling in directions θ_1 and θ_2 with respect to the optical axis and with magnitudes A_1 and A_2 as shown in Figure 9.1. Recall that the relationship between the angles and the spatial frequencies is $\alpha = \theta/\lambda$ so that the amplitude at plane P_2 is

$$A(x) = A_1 e^{-j2\pi\alpha_1 x} + A_2 e^{-j(2\pi\alpha_2 x + \phi_0)}, \tag{9.1}$$

where ϕ_0 is the relative phase between the two waves and where we have suppressed the time-dependent factor due to the frequency of light. The physical meaning of the phase is that one wave has advanced, at some instant in time, a distance $\lambda\phi_0/2\pi$ relative to the other wave.

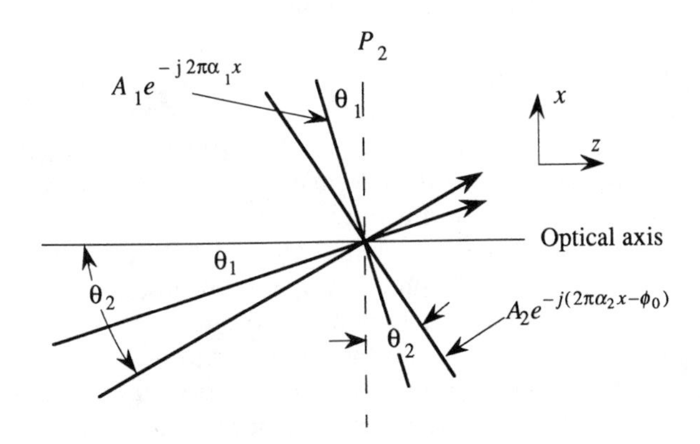

Figure 9.1. Interference between two plane waves.

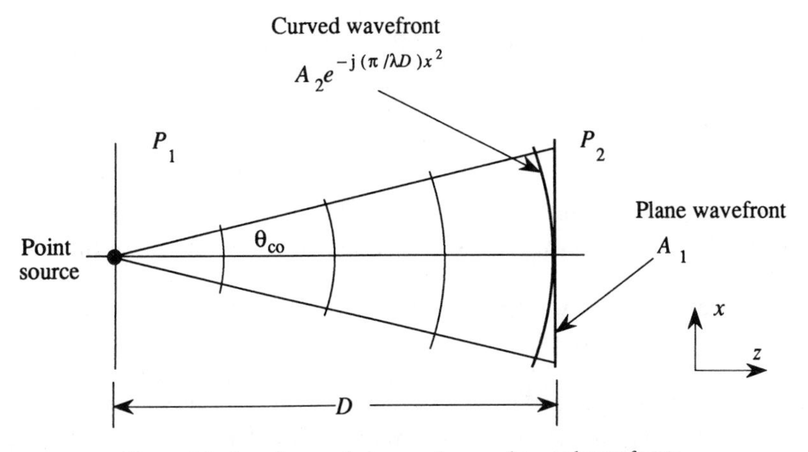

Figure 9.2. Interference between plane and curved wavefronts.

The intensity is the product of the amplitude and its complex conjugate:

$$\begin{aligned} I(x) &= \left| A_1 e^{-j2\pi\alpha_1 x} + A_2 e^{-j(2\pi\alpha_2 x + \phi_0)} \right|^2 \\ &= A_1^2 + A_2^2 + 2A_1A_2\cos[2\pi(\alpha_1 - \alpha_2)x - \phi_0]. \end{aligned} \tag{9.2}$$

From Equation (9.2) we learn that the spatial frequency of the resultant intensity is proportional to the angle between the two waves; that is, $\alpha_1 - \alpha_2 = (\theta_1 - \theta_2)/\lambda$. Because the phase accumulates more rapidly as the included angle increases, the greater the included angle, the higher the spatial frequency. The fringe pattern produced by two plane waves is called a *linear, one-dimensional fringe pattern* because the fringes are equally spaced in the x direction and do not vary in the y direction.

Spatial fringes are also produced by the interference of a plane wave and a cylindrically diverging wave from a point source, as shown in Figure 9.2. The plane wave has amplitude A_1, which is the limiting form of $A_1 \exp(-j\pi x^2/\lambda D)$ as $D \to \infty$. The intensity of the Fresnel zone pattern at plane P_2 is

$$I(x) = \left| A_1 + A_2 e^{-j(\pi x^2/\lambda D)} \right|^2 = A_1^2 + A_2^2 + 2A_1A_2\cos\left(\frac{\pi x^2}{\lambda D}\right). \tag{9.3}$$

The spatial frequency of the interference pattern, at plane P_2, is

$$\alpha = \frac{1}{2\pi}\frac{\partial}{\partial x}\left[\frac{\pi x^2}{\lambda D}\right] = \frac{x}{\lambda D}, \tag{9.4}$$

so that the spatial frequency is a linear function of the position variable x at plane P_2. We associate the instantaneous spatial frequency of the fringe pattern at any value of x to an associated ray angle; in particular, the cutoff spatial frequency α_{co} is associated with the highest ray angle θ_{co}. Once again, we see that the spatial frequency at any point in plane P_2 is proportional to the angle between the interfering waves at that point. The fringe frequency is *quadratic* in the x direction and is called a *chirp signal*.

As the interference phenomena reviewed so far are due to waves that have the same temporal frequency, we have suppressed the temporal frequency for mathematical simplicity. When we deal with heterodyne detection in acousto-optic signal-processing systems, however, we generally encounter the interference between waves with different temporal frequencies.

9.2.2. Temporal and Spatial Interference

Figure 9.3 shows a Mach-Zehnder interferometer in which the upper branch contains no spatial or temporal modulators, whereas the lower branch contains an acousto-optic cell driven by a cw signal at a frequency f_k. Suppose that the light amplitude at plane P_2 from the upper branch is

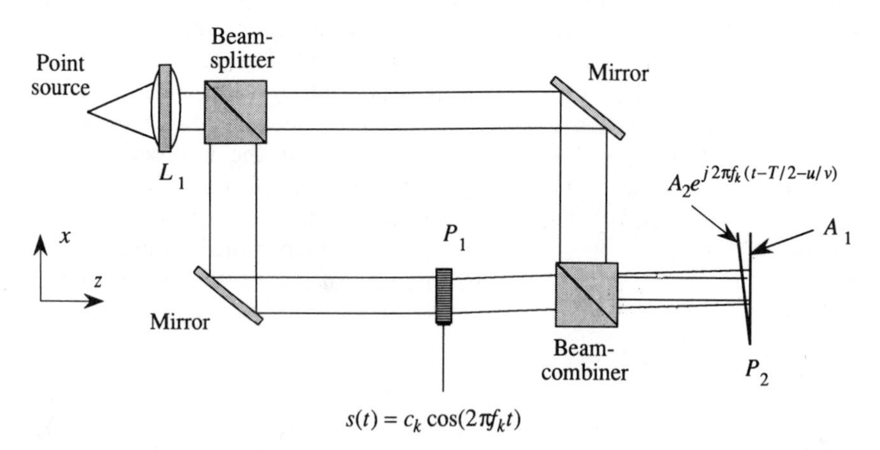

Figure 9.3. Interference between wave of different temporal and spatial frequencies.

represented by $A_1 \exp(j0)$, that is, a plane wave traveling parallel to the optical axis. For the moment, consider only the positive diffracted order from the acousto-optic cell. The intensity at plane P_2, due to these two plane waves, is

$$I(u,t) = \left| A_1 + A_2 e^{j2\pi f_k(t-T/2-u/v)} \right|^2, \tag{9.5}$$

where u is the coordinate at plane P_2. As usual, we have dropped the explicit dependence of these two waves on the frequency f_l of light. The intensity from Equation (9.5) is

$$\boxed{I(u,t) = A_0^2 + A_1^2 + 2A_1A_1 \cos[2\pi f_k(t - T/2 - u/v)],} \tag{9.6}$$

and we see that the linear interference pattern is now a function of both space and time. If we freeze the pattern at some time t_0, the intensity is

$$I(u,t_0) = A_0^2 + A_1^2 + 2A_0A_1 \cos[2\pi\alpha_k - \phi_0], \tag{9.7}$$

where $\alpha_k = f_k/v$ and $\phi_0 = 2\pi f_k(t_0 - T/2)$ is a fixed phase that is independent of the space variable. This intensity pattern is similar to that given by Equation (9.2) due to two plane waves with the same temporal frequency. On the other hand, if we focus our attention at the point u_0, we find that the intensity, as a function of time, is

$$I(u_0,t) = A_1^2 + A_2^2 + 2A_1A_2 \cos[2\pi f_k t - \phi_1], \tag{9.8}$$

where $\phi_1 = 2\pi f_k(T/2 + u_0/v)$ is a fixed phase. Here we note that the intensity pattern oscillates in time according to the temporal frequency f_k, which confirms our notion that the temporal frequency content of an applied signal is retained when we invoke heterodyne detection.

We visualize Equation (9.6) as a spatial fringe pattern that is traveling in the positive u direction with velocity v. The connection between the spatial and temporal frequencies then becomes clear: a photodetector, placed at some point u_0, senses a moving fringe structure whose spatial frequency is α_k and generates a temporal frequency $f_k = v\alpha_k$. The contrast, visibility, or modulation of the detected signal is, however, a function of the photodetector size; we now turn our attention to this question of the optimum photodetector size.

9.3. OVERLAPPING WAVES AND PHOTODETECTOR SIZE

In Chapters 4 and 8 we developed the design guidelines for determining the photodetector size required to achieve a specified dip between frequencies in a spectrum analyzer. A similar issue arises with heterodyne detection, but calculating the required photodetector size is more subtle. In a direct detection system, it is more or less a matter of "what you see is what you get." Light falling on the photodetector surface contributes to the induced photocurrent, more or less independently of its direction of arrival or temporal frequency. Light also contributes to the photocurrent in heterodyne detection, but not necessarily to the *cross-product* term, which is the third term of the intensity given, for example, by Equation (9.8). The cross-product term is separated from the bias terms by a bandpass filter centered at f_k. We therefore retain only the temporally oscillating part from Equation (9.8), the bias $A_1^2 + A_2^2$ being rejected by the filter.

Both the signal and reference beams must *overlap* to achieve heterodyne detection. In heterodyne detection we often call the reference beam the *probe* that allows us to detect both the magnitude and phase of a light distribution at some position in the optical system; in this sense, its purpose is similar to that of an oscilloscope probe used to determine the voltage waveform at a particular point in an electronic circuit.

Consider the plane-wave signal beam, represented in Figure 9.4 by solid rays and by a solid plane wavefront, and the reference beam, represented by dotted rays and by a dotted curved wavefront. These waves can be created by the interferometer shown in Figure 9.3 by placing a lens in the upper branch of the system. A photodetector placed anywhere between planes A and C will provide the same total current because the two

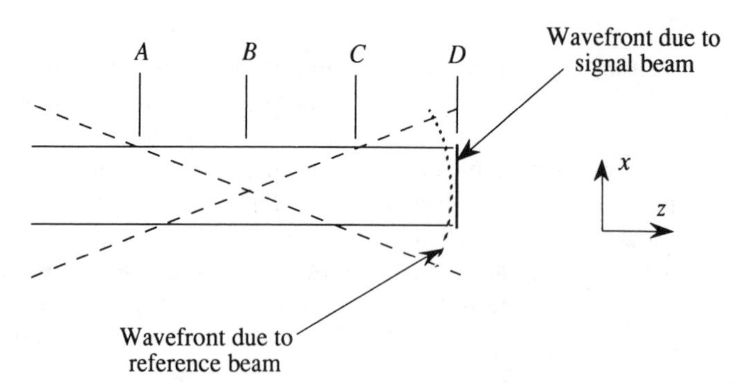

Figure 9.4. Reference- and signal-beam geometries.

beams overlap completely in this region. What is the situation at plane D? Here the reference beam extends beyond the signal beam, and we might expect that the amplitude of the cross-product signal will be reduced because the two beams do not completely overlap. As we show in the next section, a surprising result is that the cross-product output has the same value at all the planes shown in Figure 9.4. This nonintuitive result is due to a second key principle of heterodyne detection; namely, wavefronts must both overlap *and be nearly parallel*. How parallel need they be?

9.3.1. Optimum Photodetector Size for Plane-Wave Interference

To determine the required degree of parallelism, consider the simple case of a photodetector, a plane-wave signal beam, and a plane-wave reference probe as shown in Figure 9.5. The angle between the signal and reference beams is θ_k. The reference beam is represented by a plane wave of magnitude A_1 with zero spatial and temporal frequencies. The signal beam is represented as a plane wave with magnitude A_2 and temporal frequency f_k so that it is a function of both space and time. The photodetector current is the integral over the photodetector surface of the intensity of the sum of the reference and signal beam amplitudes:

$$\begin{aligned} g(t) &= S\int_{-\infty}^{\infty} I(x)\operatorname{rect}(x/h)\,dx \\ &= S\int_{-\infty}^{\infty} \left|A_1 + A_2 e^{j(2\pi f_k t + 2\pi\alpha_k x)}\right|^2 \operatorname{rect}(x/h)\,dx \\ &= S\int_{-h/2}^{h/2} \left[A_1^2 + A_2^2 + 2A_1A_2\operatorname{Re}\{e^{j(2\pi f_k t - 2\pi\alpha_k x)}\}\right] dx, \end{aligned} \tag{9.9}$$

where $\alpha_k = \theta_k/\lambda$, S is the responsivity of the photodetector, and the rect function shows that the photodetector has a total width h. When we expand the integrand, we find three contributions to the photodetector current. We see by inspection that $g_1(t) = A_1^2 hS$ and $g_2(t) = A_2^2 hS$ are signal components that are not functions of time; their temporal spectra are therefore centered at zero frequency. The third, or cross-product term, is a bandpass signal centered at f_k:

$$\begin{aligned} g_3(t) &= 2\operatorname{Re}\left[SA_1A_2e^{j2\pi f_k t}\int_{-h/2}^{h/2} e^{j2\pi\alpha_k x}\,dx\right] \\ &= 2hSA_1A_2\operatorname{sinc}(\alpha_k h)\cos(2\pi f_k t). \end{aligned} \tag{9.10}$$

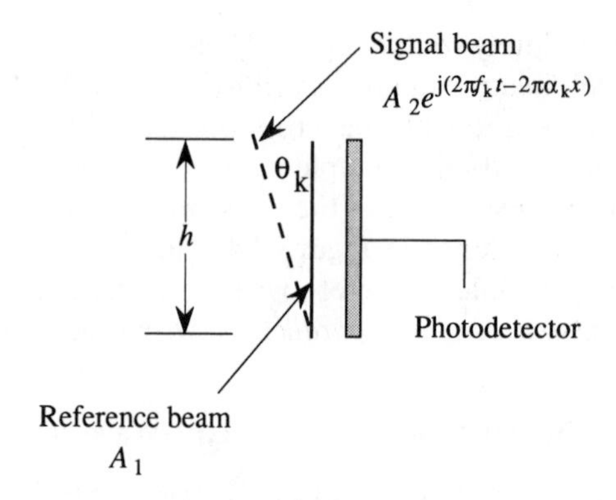

Figure 9.5. Plane-wave and photodetector geometry.

As expected, we find that $g_3(t)$ is proportional to the magnitudes of the signal and reference beams. This result also reveals, however, a new key factor. The magnitude of the output signal is controlled by a sinc function whose argument is a function of $\alpha_k = \theta_k/\lambda$, where θ_k is the angle between the signal and reference beams; the argument is also a function of h, the photodetector size. This sinc function is, in effect, a modulation transfer function that determines the magnitude of the cross-product temporal signal.

The condition for maximizing the output is found by expanding the sinc function:

$$\boxed{\begin{aligned} g_3(t) &= 2hSA_1A_2 \operatorname{sinc}(\alpha_k h)\cos(2\pi f_k t) \\ &= 2hSA_1A_2 \frac{\sin(\pi\alpha_k h)}{(\pi\alpha_k h)} \cos(2\pi f_k t) \\ &= \frac{2SA_1A_2}{\pi\alpha_k} \sin(\pi\alpha_k h)\cos(2\pi f_k t). \end{aligned}} \tag{9.11}$$

From Equation (9.11) we see that the output is small when the photodetector size h is small, as expected. The output increases as h increases, according to the sine function, until it reaches its maximum value when the argument of the sine function is $\pi/2$. This result shows that the optimum photodetector size is

$$h = \frac{1}{2\alpha_k}. \tag{9.12}$$

From Equation (9.12) we discover that the optimum photodetector size is the same as the optimum sample spacing d_0 for a spatial frequency α_k. As $\alpha_k = f_k/\lambda$, we find that the maximum allowable angle between the two waves for a photodetector of size h is

$$\theta_k = \frac{\lambda}{2h}. \tag{9.13}$$

If the photodetector size increases, the heterodyned signal is reduced, as shown by Equation (9.10), and reaches zero when $\theta_k h = \lambda$. This means that, over the physical aperture h of the photodetector, the phase change between the two waves is equal to exactly one-half wavelength of light.

One way to visualize this result is to note that if $\theta_k \gg \lambda/h$, there are several spatial interference fringes over the aperture, as shown in Figure 9.6(a). The spatial integral of the oscillating part of the interference determines the magnitude of the sinc function, and the value of the cross-product term is small in this case. As θ_k decreases, so too does the spatial frequency produced by the cross-product term until $\theta_k = \lambda/2h$ so that we have one-half cycle over the aperture, as shown in Figure 9.6(b).

From Equation (9.10) we also see that if θ_k is large, we need a small photodetector to keep the modulation transfer function at a high level. To keep the signal level within at least 3 dB of the maximum, we require that $\text{sinc}(\alpha_k h)$, as contained in Equation (9.10), has a value of 0.5 or greater.

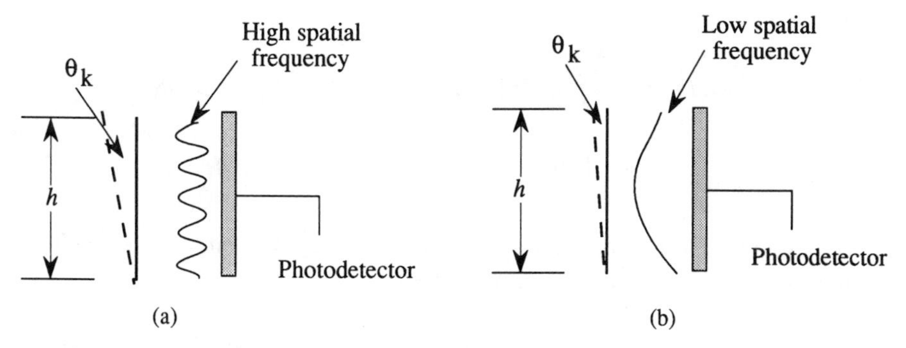

Figure 9.6. Interference fringe period and photodetector geometry. (a) high spatial frequency and (b) low spatial frequency.

This occurs whenever the argument of the sinc function is greater than 0.6 so that we require $\theta_k \le 0.6\lambda/h$. We have, therefore, established a criterion for how parallel plane waves must be to produce a contribution to the cross-product term at the output of the system.

9.3.2. Optimum Photodetector Size for a Two-Dimensional Chirp

Consider the spatial/temporal interference produced by a spherically diverging wave at frequency f_k and a plane wave. The plane-wave reference beam is represented by A_1 and the spherically diverging signal beam is represented by

$$s(\rho, t) = A_2 e^{j(2\pi f_k t - \pi\rho^2/\lambda D)}, \tag{9.14}$$

where ρ is a polar coordinate at plane P_2. The intensity is the square of the sum of the reference and signal light distributions:

$$\begin{aligned} I(\rho, t) &= \left| A_1 + A_2 e^{j(2\pi f_k t - \pi\rho^2/\lambda D)} \right|^2 \\ &= A_1^2 + A_2^2 + 2\,\mathrm{Re}\left\{ A_1 A_2 e^{j(2\pi f_k t - \pi\rho^2/\lambda D)} \right\}. \end{aligned} \tag{9.15}$$

The general form of the output after ignoring unessential constants is

$$\begin{aligned} g(t) &= \int_0^{2\pi} \int_0^R I(\rho, t)\rho\, d\rho\, d\theta, \\ &= g_1(t) + g_2(t) + g_3(t), \end{aligned} \tag{9.16}$$

where R is the radius of the photodetector. The first two terms are simply the constants $g_1(t) = \pi A_1^2 R^2$ and $g_2(t) = \pi A_2^2 R^2$. The cross-product term is

$$\begin{aligned} g_3(t) &= \int_0^{2\pi} \int_0^R 2\,\mathrm{Re}\left\{ A_1 A_2 e^{j(2\pi f_k t - \pi\rho^2/\lambda D)} \right\} \rho\, d\rho\, d\theta, \\ &= 4\pi A_1 A_2\, \mathrm{Re}\left\{ e^{j(2\pi f_k t)} \int_0^R e^{-j(\pi\rho^2/\lambda D)} \rho\, d\rho \right\}. \end{aligned} \tag{9.17}$$

To integrate this function, we let $\pi\rho^2/\lambda D = z^2$ and supply the factors needed for a perfect differential to produce

$$g_3(t) = 4\pi A_1 A_2 \left[\frac{\lambda D}{-j2\pi} \right] \mathrm{Re}\left\{ e^{j(2\pi f_k t)} \left[e^{-j(\pi R^2/\lambda D)} - 1 \right] \right\}. \tag{9.18}$$

The condition for maximizing the output is found by rearranging the terms in Equation (9.18) to produce

$$g_3(t) = 2\pi A_1 A_2 R^2 \,\mathrm{Re}\left\{ e^{j(2\pi f_k t)} e^{-j(\pi R^2/2\lambda D)} \left[\frac{e^{-j(\pi R^2/2\lambda D)} - e^{+j(\pi R^2/2\lambda D)}}{-j(\pi R^2/\lambda D)} \right] \right\}$$
$$= 2\pi A_1 A_2 R^2 \,\mathrm{sinc}\left[\frac{R^2}{2\lambda D} \right] \cos\left(2\pi f_k t - \frac{\pi R^2}{2\lambda D} \right). \tag{9.19}$$

We see that, once again, the magnitude of the output is determined by a sinc function that plays the role of a modulation transfer function with R as a parameter. We maximize Equation (9.19) with respect to the photodetector radius R by noting that

$$\pi R^2 \,\mathrm{sinc}\left[R^2/2\lambda D \right] = 2\lambda D \sin\left(\pi R^2/2\lambda D \right), \tag{9.20}$$

which reaches its maximum value, for a given value of D, when

$$R = \sqrt{\lambda D}\,. \tag{9.21}$$

The optimum photodetector size, for this case, is therefore simply a function of its distance from the source and of the wavelength of light.

An unanticipated and interesting result from Equation (9.19) is that the phase of the cosine carrier is a function of both R and D when the output is maximized. What does this mean physically? The phase reveals the shape of the *spatial* chirp pattern at that moment in time when $g_3(t)$ is at its maximum value. If we include the phase term from Equation (9.19) in Equation (9.15), we find that the spatial intensity pattern for the maximum output condition becomes

$$I(\rho, t) = \left| A_1 + A_2 e^{j(2\pi f_k t - \pi\rho^2/\lambda D - \pi R^2/2\lambda D)} \right|^2$$
$$= A_1^2 + A_2^2 + 2A_1 A_2 \cos\left(2\pi f_k t - \frac{\pi\rho^2}{\lambda D} - \frac{\pi R^2}{2\lambda D} \right). \tag{9.22}$$

We can associate the phase with either the spatial or the temporal part of Equation (9.22). If we select the spatial part, the chirp function changes from the cophasal function, shown in Figure 9.7, to the one shifted by 90°.

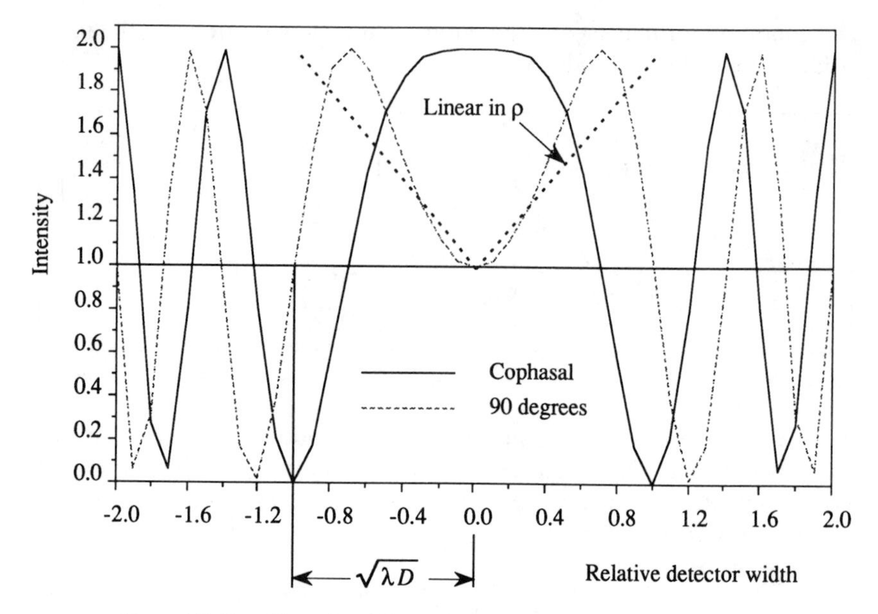

Figure 9.7. Two-dimensional Fresnel zones with different initial phases.

It is clear that there is significantly more positive contribution to the integral from the phase-shifted chirp than from the cophasal chirp for a photodetector radius of $\sqrt{\lambda D}$. The key to this difference is that the integrand in Equation (9.16) contains a term linear in ρ that multiplies the intensity $I(\rho, t)$ before the integration is carried out; this linear term is shown in Figure 9.7 as two straight dotted lines. In effect, the phase term serves to push more of the energy toward the larger values of ρ so that the integral is maximized. Also note that when $R = \sqrt{\lambda D}$ the phase-shifted chirp is just crossing the bias level so that any larger photodetector will produce a smaller output for the cross-product term.

9.3.3. Optimum Photodetector Size for a One-Dimensional Chirp

The optimum photodetector size for the one-dimensional chirp case is obtained by an analysis similar to that used in Section 9.3.2. We use the same general definitions for the reference and signal beams to find the intensity

$$I(x, t) = A_1^2 + A_2^2 + 2\,\mathrm{Re}\left\{A_1 A_2 e^{j(2\pi f_k t - \pi x^2/\lambda D)}\right\}, \qquad (9.23)$$

where x is the spatial variable at the detector plane. The cross-product output is

$$g_3(t) = 2A_1A_2 \operatorname{Re}\left\{ e^{j2\pi f_k t} \int_{-h/2}^{h/2} e^{-j\pi x^2/\lambda D}\, dx \right\}, \tag{9.24}$$

where h is the photodetector width. In this case the integral plays the role of the modulation transfer function. We recognize the integral as a Fresnel integral, which we cannot evaluate in closed form. We put the integral into its standard form by a change of variables in which $\pi x^2/\lambda D = \pi z^2/2$

$$g_3(t) = \sqrt{2\lambda D}\, A_1A_2 \operatorname{Re}\left\{ e^{j2\pi f_k t} \int_{-\sqrt{2/\lambda D}\, h/2}^{\sqrt{2/\lambda D}\, h/2} e^{-j\pi z^2/2}\, dz \right\}. \tag{9.25}$$

As we showed in Chapter 3, Section 3.2.5, the maximum value of the Fresnel integral occurs when

$$\sqrt{\frac{2}{\lambda D}}\frac{h}{2} = 1.21, \tag{9.26}$$

so that the optimum value of h is $1.72\sqrt{\lambda D}$. At this value of h, the integral is equal to $1.8\exp(j\phi)$, where $\phi = 42.3°$ so that the maximum value of $g_3(t)$ becomes

$$\boxed{g_3(t) = 1.8\sqrt{2\lambda D}\, A_1A_2 \cos(2\pi f_k t + \phi).} \tag{9.27}$$

As before, we relate the optimum photodetector size to the spatial interference patterns as shown in Figure 9.8. The phase shift of 42.3° maximizes the integral of the function from zero to $1.72\sqrt{\lambda D}$. If the phase shift were greater, the dip near $x = 0$ would become deeper and would more than offset any gain from an increased photodetector size.

9.3.4. Optimum Photodetector Size for a General Signal

Would the optimum detector size change if, in Figure 9.4, we were to place at plane B a signal with length L and high spatial frequency content? From the sketch in Figure 9.9 we have two ways to proceed; each provides additional insights into the detection process.

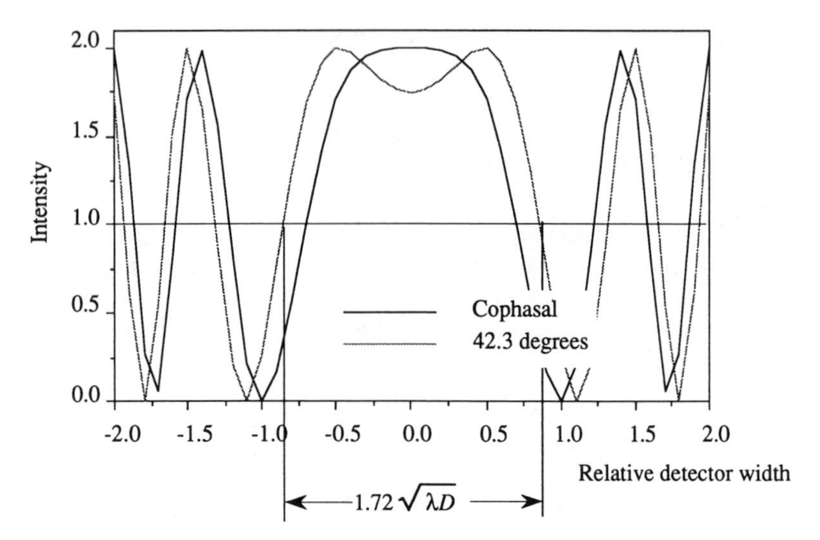

Figure 9.8. One-dimensional Fresnel zones with different initial phases.

The first method is to represent the signal by M plane waves whose incremental angles change by $\theta_0 = \lambda/L$, where L is the total length of the signal at plane B. To capture the undiffracted light from the signal, we must have some photodetector surface available at position $x = 0$, as we just explained. To capture the highest positive frequency, we need some photodetector surface at $+M\theta_0 D$, where D is the distance from plane B to plane D. Note that the latter position is where the Mth plane wave from the signal is *tangent* to the diverging wavefront produced by the probe. It is only near this point that the signal and probe wavefronts are

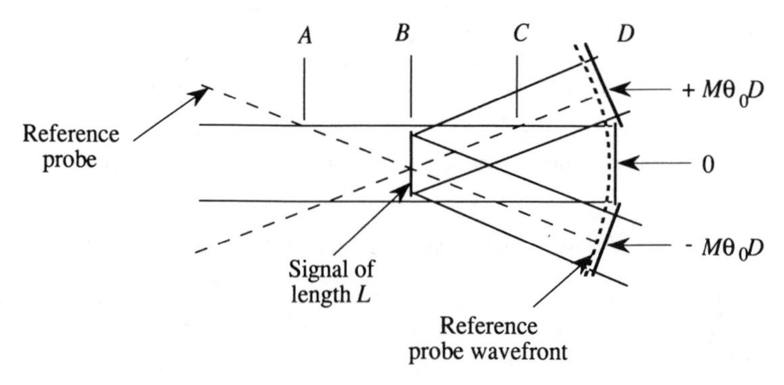

Figure 9.9. Heterodyne action with plane-wave signal decomposition.

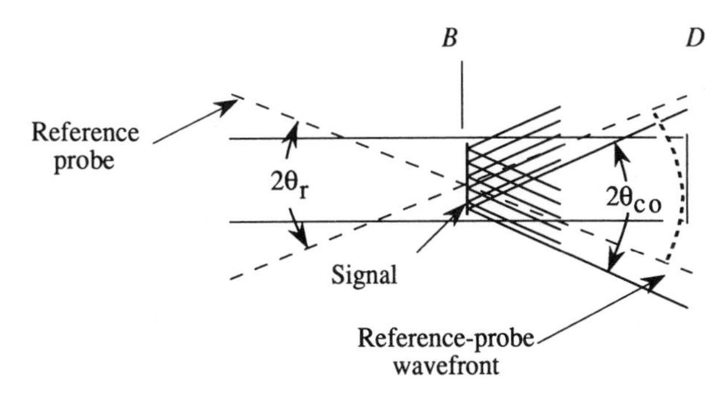

Figure 9.10. Heterodyne action with sinc function signal decomposition.

sufficiently parallel so that photons contribute to the cross-product term. Similarly, we need some photodetector surface at $-M\theta_0 D$ to capture the negative spatial frequencies produced by the signal. A different part of the photodetector therefore collects photons from different spatial frequencies produced by the signal to form the cross-product term. From these considerations, we see that the photodetector must be sufficiently large to capture the overlapping light from both beams, up to the required size of the divergent wavefront representing the probe.

The required size of the probe is clearly determined by the maximum frequency content of the signal. To further explore this relationship, consider a second method for representing the signal, namely, as a set of sinc functions of the form $\text{sinc}(x/d_0)$. In Figure 9.10, we show the reference-beam probe just before the signal at plane B, as a convergent bundle of rays, with $2\theta_r$ as its included angle. The signal, with cutoff frequency α_{co}, is represented by a sequence of sinc functions that propagate as divergent waveforms into the region to the right of plane B, with the marginal rays forming a cone of angle $2\theta_{co}$. The meaning of "nearly parallel beams" is now at its simplest and clearest form; we require that $\theta_r = \theta_{co}$ so that the *reference probe contains a ray that is parallel to every ray that is diffracted by the signal.*

From Figure 9.10 we also conclude that only one sample of the signal plane contributes to the output: the one that coincides with the probe! The size of the photodetector at an arbitrary plane is simply $h = 2\theta_r D$; as $D \to 0$, the value of h tends toward its minimum value of d_0 as set by the diffraction limit. The photodetector size at plane B is therefore also equal to d_0. All the other light, at least from a heterodyne detection viewpoint, is irrelevant. Once again, these results emphasize that light contributes to the cross-product term only if the beams overlap and are collinear.

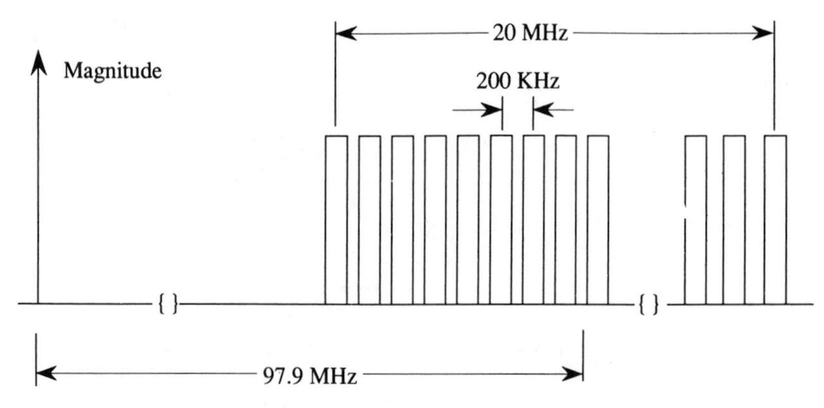

Figure 9.11. Frequency allocation in the FM band.

9.4. THE OPTICAL RADIO

We describe an optical radio to illustrate various techniques for using heterodyne detection in recovering both the magnitude and phase of a time signal. This simple system forms the basis for a discussion of heterodyne spectrum analysis in Chapter 10 because a frequency spectrum analyzer that resolves M frequencies is equivalent to operating M optical radios in parallel. Each radio detects the signal power in an assigned channel.

Consider the FM band of radio frequencies shown in Figure 9.11. In a 20-MHz frequency band, centered at 97.9 MHz, we have 100 possible FM channels, spaced at 200-kHz intervals. We represent the composite signal due to all channels by

$$f(t) = \sum_{n=1}^{N} a_n \cos[2\pi f_n t + \phi_n(t)], \tag{9.28}$$

where a_n is the magnitude of the nth signal and $\phi_n(t)$ is the FM modulation that contains the message signal $m_n(t)$:

$$m_n(t) = \frac{\partial}{\partial t}[\phi_n(t)]. \tag{9.29}$$

If $f(t)$ drives the acousto-optic cell in the interferometer of Figure 9.12, the light distribution at plane P_2 due to the lower branch resembles that shown in Figure 9.11, where the temporal frequencies have been con-

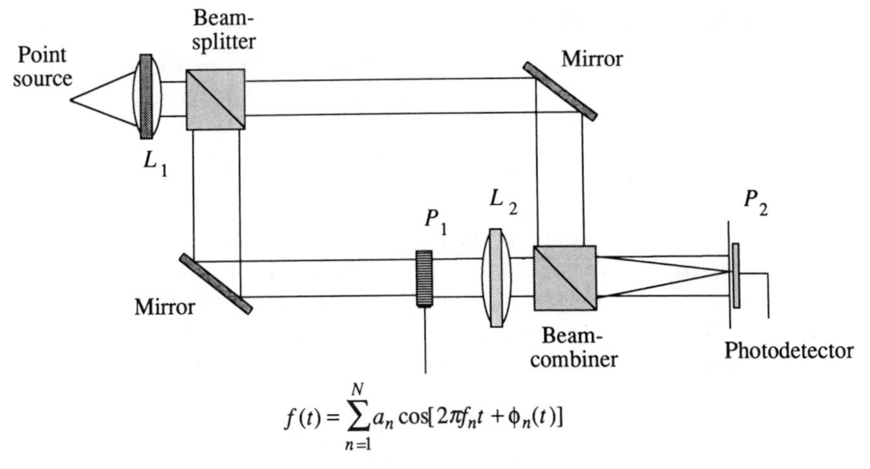

Figure 9.12. An optical radio.

verted to spatial frequencies through the relationship that $\alpha = f/v$. We arrive at this conclusion based on the Fourier-transform properties of coherently illuminated optical systems as described in Chapter 8. In Figure 9.11, we have shown the idealized situation where there are guard bands between the channels and no crosstalk between adjacent channels.

We now consider several potential detection arrangements. Our objective is to produce a signal corresponding to one of the selected channels in the system at the output of a photodetector. Ideally, the signal should occur at an IF frequency so that it can be fed directly to an FM discriminator circuit. The system shown in Figure 9.12 is therefore equivalent to the front end of an FM receiver.

9.4.1. Direct Detection

If we block the reference beam, the lower branch of the interferometer of Figure 9.12 is equivalent to the power spectrum analyzer discussed in Chapter 8. Suppose that a small photodetector element is positioned so that it collects light at the nth channel. The output of the photodetector is then proportional to a_n^2, as suggested by Equation (8.9), because all phase information is lost when light is detected directly. We clearly need heterodyne detection to recover the phase modulation of the input signal; we now consider several possibilities.

9.4.2. Heterodyne Detection

If we unblock the reference beam, interference between the signal and reference beams is restored. At the nth channel the intensity is

$$I(\alpha, t) = |R(\alpha) + S_n(\alpha, t)|^2, \tag{9.30}$$

where $R(\alpha)$ is the response at plane P_2 due to the reference function $r(x)$, and $S_n(\alpha, t)$ is the response due to the signal in the nth channel. When the phase modulation is slowly varying with respect to the cell duration T, we modify Equation (8.9) to find that the positive diffracted order becomes

$$S_n(\alpha, t) = jma_n e^{j[2\pi f_n(t - T/2) + \phi_n(t)]} A(\alpha - \alpha_n), \tag{9.31}$$

so that the resultant intensity produced by the nth channel is represented by

$$I_n(\alpha, t) = \left| R(\alpha) + jma_n e^{j[2\pi f_n(t - T/2) + \phi_n(t)]} A(\alpha - \alpha_n) \right|^2, \tag{9.32}$$

which becomes

$$\boxed{\begin{aligned} I_n(\alpha, t) = {} & |R(\alpha)|^2 + |ma_n A(\alpha - \alpha_n)|^2 + 2ma_n |R(\alpha)| \\ & \times |A(\alpha - \alpha_n)| \cos[2\pi f_n(t - T/2) + \phi_n(t) + \pi/2]. \end{aligned}} \tag{9.33}$$

In Equation (9.33) the reference-beam response $|R(\alpha)|^2$ is a constant in both space and time. The second term of Equation (9.33) is the same as we would obtain for direct detection and shows that the phase information is lost.

The first two terms of Equation (9.33) are at baseband and are easily removed by a bandpass filter whose center frequency is at 97.9 MHz, the midband frequency. The last term of Equation (9.33) has the proper form of an FM modulated signal; it can be fed directly to a discriminator after $I(\alpha, t)$ is detected by the photodetector to produce the output current. The phase factor $\pi/2$ in the argument of the cosine is due to the j factor in Equation (9.31), which reminds us that the diffracted light is in phase quadrature with respect to the undiffracted light. This phase factor merely shifts the phase of the carrier and has no effect on the demodulated signal.

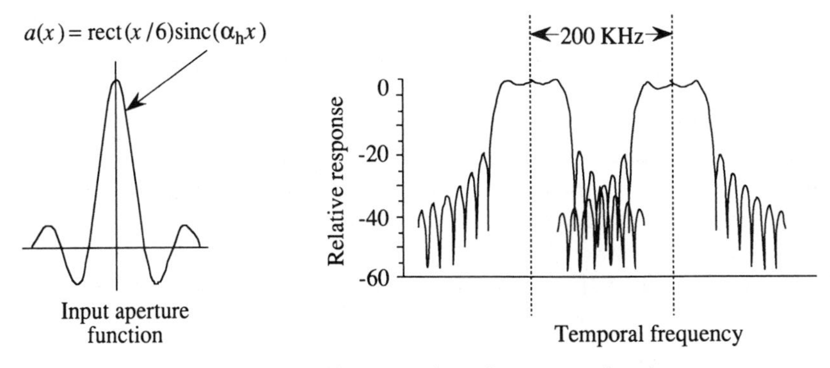

Figure 9.13. Channel response for a sinc aperture function.

To maintain well-formed channels that have steep slopes and low skirt levels, as shown in Figure 9.11, the aperture function $a(x)$ must produce a response in the Fourier plane proportional to $\mathrm{rect}(\alpha/\alpha_h)$, where α_h is the width of the photodetector expressed in terms of a spatial frequency. This means that the aperture weighting function must be $a(x) = \mathrm{sinc}(\alpha_h x)$. At first glance, it seems that this aperture function is difficult to synthesize because we need to generate a mask with negative magnitudes. Recall from Chapter 3, however, that the Fourier transform occurs at an image plane of the source. Therefore, if the primary source in Figure 9.12 is a rect function of the proper dimension, its Fourier transform at plane P_1 must generate the required sinc function and, in turn, the second transform at plane P_2 produces the desired rect function.

The steepness of the channel slopes and the depth of the skirt levels depend on how many sidelobes of the sinc function are passed by the acousto-optic cell. Figure 9.13 shows a candidate aperture function $a(x) = \mathrm{rect}(x/6)\mathrm{sinc}(\alpha_h x)$, which represents a sinc function that has been truncated at the third null on each side of the mainlobe, and its frequency response. The Fourier transform of this truncated function has a reasonably flat bandpass over the channel bandwidth, a region where the response falls rapidly, and a sidelobe level that is at least 20 dB down at the first sidelobe. Crosstalk is about 40 dB down in adjacent channels. Steeper skirts and lower sidelobe levels can be obtained by illuminating the acousto-optic cell with more sidelobes of the sinc function although the rate of gain is not very high.

There are several arrangements of the reference beam that provide the desired signal at the output. Some of these are impractical for various reasons, but it is worthwhile to present them all because we learn some interesting facts about the detection process from each.

Arrangement 1. Given the arrangement of Figure 9.12, we can simply place a small photodetector at the position corresponding to the channel we want to select. The finite size of the photodetector serves to isolate the proper channel and to prevent crosstalk due to neighboring channels. Tuning to a new station is then a matter of moving the photodetector to the new channel position. The disadvantages of this arrangement are the mechanical movement of the photodetector and inefficient use of the reference-beam power because it is spread over the entire FM band. Furthermore, since the output of the photodetector is at the same frequency as the input, we have not brought the output to an IF frequency.

Arrangement 2. We avoid the need to move the photodetector by using an array of photodetector elements at plane P_2. Selecting a channel is then accomplished by switching to the desired photodetector element. The other disadvantages of Arrangement 1 are still present, however.

Arrangement 3. We more efficiently use the reference-beam power and therefore produce a higher signal-to-noise ratio, by modifying the interferometer, as shown in Figure 9.14. Suppose that we move lens L_2 outside the interferometer so that it creates the Fourier transform of the signal and reference beams simultaneously. If we arrange for $r(x) = a(x) = \text{sinc}(\alpha_h x)$, we find that the reference beam is only one channel wide at plane P_2. As a result, the reference beam power is approximately 100 times greater than in the previous two arrangements. Tuning to a new channel is achieved by rotating the beam combiner to an appropriate angle; for a given angle of rotation Ψ, the reflected reference beam is rotated by 2Ψ. The signal from the lower branch is largely unaffected by the rotation; recall from Chapter 2 that when a plane parallel plate is rotated, the rays passing through the plate are simply displaced somewhat. As the angle of rotation is of the order of a few milliradians, the spectrum is not shifted a great deal (see Problem 9.3).

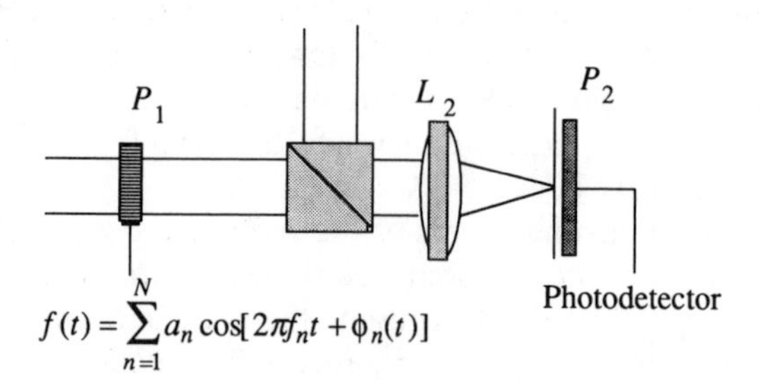

Figure 9.14. Optical radio with the Fourier-transform lens outside the interferometer.

A second advantage of this arrangement is that we can use a large, single photodetector element that covers the entire FM band, as the reference beam is now confined to a single channel. From Section 9.2 we learned that a heterodyne output occurs only if two signals overlap and are collinear. In this case lens L_2 ensures that the beams are sufficiently collinear at the output and the choice of $r(x)$ restricts the region of overlap to the selected channel. Using a single photodetector simplifies the postdetection circuitry at the expense of a somewhat higher shot-noise level because more signal energy than is necessary is falling on the photodetector. As we show in Chapter 10, shot noise is generally dominated by noise introduced by the reference beam so that the additional photodetector size is not a serious drawback. The disadvantages of this arrangement are that the output is still at rf and that we need to mechanically rotate the beam combiner to tune the system.

Arrangement 4. We avoid the mechanical rotation of the beam combiner by further modifying the interferometer as shown in Figure 9.15. Here we use a second acousto-optic cell to provide electronic tuning of the radio. In particular, we drive the second acousto-optic cell with a reference signal $r(t) = \cos(2\pi f_n t)$ to access the nth channel. The channel selection process is fast, limited only by the access time of the reference-beam acousto-optic cell. A new and unfortunate problem has arisen, however; the output, instead of being at rf as in the other arrangements, is now at baseband. This problem becomes apparent when we remember that the reference-beam signal is

$$R(\alpha, t) = jme^{j[2\pi f_n(t-T/2)]}R(\alpha - \alpha_n). \tag{9.34}$$

This reference probe selects just one frequency component from the signal

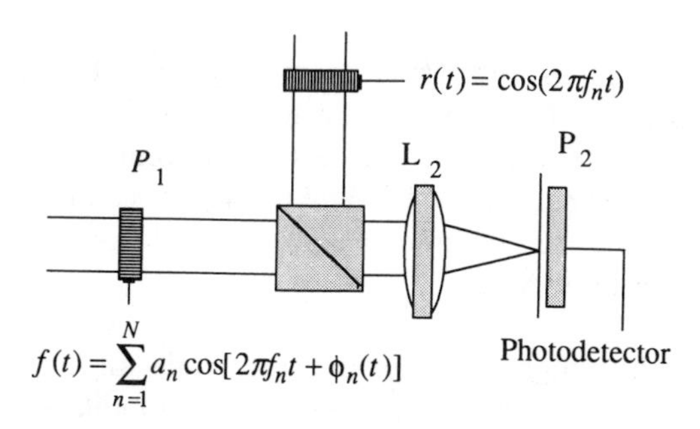

Figure 9.15. Use of an acousto-optic cell to provide the reference beam.

so that the intensity becomes

$$I_n(\alpha, t) = \Big| jme^{j[2\pi f_n(t-T/2)]} R(\alpha - \alpha_n) \\ + jma_n e^{j[2\pi f_n(t-T/2)+\phi_n(t)]} A(\alpha - \alpha_n) \Big|^2. \tag{9.35}$$

When we carry out the expansion, we have

$$I_n(\alpha, t) = |mR(\alpha - \alpha_n)|^2 + |ma_n A(\alpha - \alpha_n)|^2 \\ + 2m^2 a_n R(\alpha - \alpha_n) A^*(\alpha - \alpha_n) \cos[\phi_n(t)], \tag{9.36}$$

which confirms that the cross-product term has been heterodyne shifted to baseband. The cross-product term therefore cannot be separated from the bias terms, rendering this arrangement useless because the output is not at a convenient intermediate frequency.

Arrangement 5. A final modification is to change the reference frequency to f_m so that the reference beam overlaps the mth channel *and* to then rotate the beam combiner to geometrically move the reference probe back to the nth channel. This fixed rotation is performed only when the system is calibrated. In this fashion, we find that the reference and signal beams have slightly different frequencies so that Equation (9.35) becomes

$$I_n(\alpha, t) = \Big| jme^{j[2\pi f_m(t-T/2)]} R(\alpha - \alpha_m) \\ + jma_n e^{j[2\pi f_n(t-T/2)+\phi_n(t)]} A(\alpha - \alpha_n) \Big|^2, \tag{9.37}$$

and the corresponding intensity becomes

$$I_n(\alpha, t) = |mR(\alpha - \alpha_m)|^2 + |ma_n A(\alpha - \alpha_n)|^2 + 2m^2 a_n R(\alpha - \alpha_m) \\ \times A^*(\alpha - \alpha_n) \cos[2\pi f_d t + \phi_n(t) - \pi f_n T], \tag{9.38}$$

where $f_d = |f_m - f_n|$ is the offset frequency. For FM reception, it is convenient to set f_d at 10.7 MHz, the normal IF frequency. The optical system therefore both tunes to the desired channel and simultaneously brings the output to the desired IF. The tuning procedure is simply to add 10.7 MHz to the desired channel frequency f_n to produce the required reference drive frequency f_m.

Arrangement 6. Although Arrangement 5 provides all the desired features of a heterodyne system, we offer a more general modification, as shown in

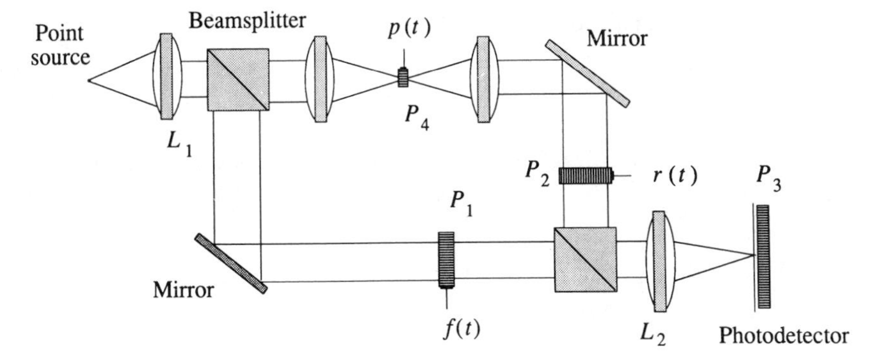

Figure 9.16. A general purpose heterodyne system.

Figure 9.16, in which the lower branch contains an acousto-optic cell driven by a signal $f(t)$. The upper branch contains a similar acousto-optic cell driven by a reference signal $r(t)$. The upper branch may also contain a means for purely time modulating the reference beam at plane P_4 with a signal $p(t)$. The signal $p(t)$ is introduced by means of an acousto-optic point modulator as described in Chapter 7, Section 7.9.1 or by other types of temporal modulators; it is a part of the illumination of the acousto-optic cell in the upper branch that contains $r(t)$.

To illustrate the features of this system, we drive the temporal modulator with the desired offset frequency f_d so that

$$p(t) = \cos(2\pi f_d t). \tag{9.39}$$

Because the offset frequency is provided by the point modulator, the frequency of the reference signal should now be set to that of the channel we want to select. To illustrate the effects of a mistuned radio, we let the reference frequency be f_j:

$$r(t) = \cos(2\pi f_j t). \tag{9.40}$$

The signal at plane P_3 due to the upper branch of the interferometer is

$$R(\alpha, t) = jmR(\alpha - \alpha_j)e^{j2\pi f_j(t - T/2 - x/v)}e^{-j2\pi f_d t}, \tag{9.41}$$

where we have retained the downshifted diffracted order from the point modulator for reasons that will become apparent shortly.

The drive signal to the acousto-optic cell in the lower branch of the interferometer is

$$f(t) = \cos(2\pi f_k t + \phi_k), \tag{9.42}$$

and its Fourier transform at plane P_3 is

$$S(\alpha, t) = jma_k A(\alpha - \alpha_k) e^{j[2\pi f_k(t - T/2) + \phi_k]}. \tag{9.43}$$

Suppose that we use a large-area photodetector at plane P_3 to collect all the light produced by the two branches of the system. The output signal is then

$$\begin{aligned} g(t) &= \int_{-\infty}^{\infty} I(\alpha, t)\, d\alpha = \int_{-\infty}^{\infty} |R(\alpha, t) + S(\alpha, t)|^2\, d\alpha \\ &= g_1(t) + g_2(t) + g_3(t). \end{aligned} \tag{9.44}$$

The cross-product term produces an output

$$\begin{aligned} g_3(t) &= 2\,\mathrm{Re}\left\{\int_{-\infty}^{\infty} S(\alpha, t) R^*(\alpha, t)\, d\alpha\right\} \\ &= 2\,\mathrm{Re}\left\{\int_{-\infty}^{\infty} jma_k A(\alpha - \alpha_k) e^{j[2\pi f_k(t - T/2) + \phi_k(t)]}\right. \\ &\qquad \left.\times (-j) m R^*(\alpha - \alpha_j) e^{-j2\pi f_j(t - T/2)} e^{j2\pi f_d t}\, d\alpha\right\} \\ &= 2\,\mathrm{Re}\left\{m^2 a_k e^{j[2\pi(f_k - f_j)(t - T/2) + 2\pi f_d t + \phi_k]}\right. \\ &\qquad \left.\times \int_{-\infty}^{\infty} A(\alpha - \alpha_k) R^*(\alpha - \alpha_j)\, d\alpha\right\}. \end{aligned} \tag{9.45}$$

Suppose that we set $a(x) = r(x) = \mathrm{rect}(x/L)$ so that we can evaluate the integral on α. The integral is a function of the difference between α_j and α_k as is seen by making a change of variables in which $\gamma = \alpha - \alpha_j$:

$$\begin{aligned} c(\alpha_k - \alpha_j) &= \int_{-\infty}^{\infty} \mathrm{sinc}[\gamma L]\mathrm{sinc}\big[(\gamma + \alpha_k - \alpha_j)L\big]\, d\gamma \\ &= \mathrm{sinc}\big[(\alpha_k - \alpha_j)L\big]. \end{aligned} \tag{9.46}$$

In Equation (9.46), we recognize that the convolution of two sinc functions produces a sinc function of the delay variable. This result shows that the output is maximized when $\alpha_j = \alpha_k$, as we expected; it also shows the rate at which the output drops as the degree of mistuning increases.

To maximize the output, we set $f_j = f_k$ so that Equation (9.45) becomes

$$g_3(t) = 2m^2 a_k \cos[2\pi f_d t + \phi_k]. \tag{9.47}$$

From Equation (9.47) we see that we have recovered the magnitude and the phase of the input signal. If either the magnitude or phase are a function of time, the carrier frequency will be properly modulated by these time-dependent signals. The reason for selecting the downshifted diffraction order from the point modulator is that we retain the proper sign on the phase term. If we had used the upshifted diffracted order, the output would be the conjugate of the desired signal. This option is useful in some signal-processing applications, as we note in later chapters.

The desired signal is available at the output of a bandpass filter centered at f_d, provided that neither of the other two terms from Equation (9.44) have spectral energy in that band. In this example, both the reference and signal are narrowband functions so that the frequency content of both $g_1(t)$ and $g_2(t)$ is concentrated at $f = 0$. We leave it as an exercise for the reader to calculate the frequency content of the two baseband terms when the signal has a finite bandwidth (see Problem 9.2).

Through this sequence of arrangements, we have developed a practical solution for the optical radio. Both Arrangements 5 and 6 have all the desired features: rapid tuning, efficient use of the reference power, and a phase competent output at IF. This study of the optical radio is useful because it leads to an interesting generalization in Section 9.5 from which we can develop other optical computing architectures. This study also leads directly to the development of a heterodyne spectrum analyzer, as discussed in Chapter 10.

9.5. A GENERALIZED HETERODYNE SYSTEM

The optical radio described in Arrangement 5 is the preferred implementation for recovering the magnitude and phase of an arbitrary signal at the output of the system because it is more cost effective than the system given in Arrangement 6. The general-purpose interferometric system shown in Figure 9.16, however, has a higher degree of flexibility and can be configured to perform a wide range of processing operations.

Suppose, for the moment, that the combined beams are Fourier transformed by lens L_2. The Fourier transform of the signal in the lower branch is

$$F_+(\alpha, t) = \int_{-\infty}^{\infty} a(x) f_+\left(t - \frac{T}{2} - \frac{x}{v}\right) e^{j2\pi\alpha x}\, dx. \tag{9.48}$$

Given that the reference beam contains a point modulator $p(t)$ and a Bragg cell driven by $r(t)$, we express the intensity at plane P_3 as

$$I(\alpha, t) = |F_+(\alpha, t) + p(t) R_+(\alpha, t)|^2, \tag{9.49}$$

where $R_+(\alpha, t)$ is defined in a similar fashion to that for $F_+(\alpha, t)$. We expand the intensity to obtain

$$I(\alpha, t) = I_1(\alpha, t) + I_2(\alpha, t) + I_3(\alpha, t), \tag{9.50}$$

where

$$\boxed{\begin{aligned} I_1(\alpha, t) &= |F_+(\alpha, t)|^2, \\ I_2(\alpha, t) &= |p(t) R_+(\alpha, t)|^2, \\ I_3(\alpha, t) &= 2\,\mathrm{Re}\left[F_+(\alpha, t) p^*(t) R_+^*(\alpha, t)\right]. \end{aligned}} \tag{9.51}$$

This general result suggests several processing possibilities. As one example, suppose that the purely time modulation $p(t)$ is given, in analytic form, by

$$p(t) = e^{-j2\pi f_d t}, \tag{9.52}$$

so that the temporal modulation is a simple cw frequency at f_d. Suppose, too, that

$$F_+(\alpha, t) = |F_+(\alpha, t)| e^{j\phi(\alpha, t)} \tag{9.53}$$

and

$$R_+(\alpha, t) = |R_+(\alpha, t)| e^{j\theta(\alpha, t)} \tag{9.54}$$

are completely arbitrary, complex-valued functions, where we explicitly show the magnitude and phase parts of the transforms of the two signals that drive the acousto-optic cells. We express the intensity at plane P_3 as

$$\boxed{\begin{aligned}I(\alpha,t) &= |F_+(\alpha,t)|^2 + |R_+(\alpha,t)|^2 \\ &\quad + 2|F_+(\alpha,t)|\,|R_+(\alpha,t)|\cos[2\pi f_d t + \phi(\alpha,t) - \theta(\alpha,t)].\end{aligned}} \tag{9.55}$$

This is the central result of the analysis of the general-purpose interferometer. The cross-product term of Equation (9.55) reminds us of the general equation, given in many communication texts, that represents phase, magnitude, or frequency-modulated signals:

$$\begin{aligned} v(t) &= A\cos[2\pi f_c t + \phi(t)] \text{ angle modulation} \\ v(t) &= A[1 + m(t)]\cos(2\pi f_c t) \text{ amplitude modulation} \\ v(t) &= Am(t)\cos(2\pi f_c t) \text{ double-sideband modulation} \end{aligned} \tag{9.56}$$

In these representations, f_c is a carrier frequency, $\phi(t)$ is an angle modulation signal, $m(t)$ is a baseband amplitude modulation signal, and A is a constant. By comparing Equation (9.56) with Equation (9.55), we see that the cross-product term from $I(\alpha, t)$ can, with suitable assignment of the reference signals and choice of α, be put into any one of the forms listed for $v(t)$. The main difference between $I(\alpha, t)$ and $v(t)$ is the presence of the two bias terms $|F_+(\alpha, t)|^2$ and $|p(t)R + (\alpha, t)|^2$. This difference disappears when we separate the desired cross-product term from the first two with a bandpass filter. Therefore, a key requirement on the reference functions is that they introduce an offset frequency f_d of suitable value to achieve the separation. In subsequent chapters we consider several applications using this general architecture and signal sources.

PROBLEMS

9.1. The *baseband* signal to an FM optical radio consists of $M = 100$ channels, each one narrowband, so that the input signal can be approximated by

$$s(t) = \sum_{k=1}^{N} b_k(t)\cos(2\pi f_k t),$$

where $b_k(t)$ is a narrow-band modulation. The reference function $r(t) = \cos(2\pi f_j t)$ accesses the jth frequency band of interest and an auxiliary reference function $p(t) = \cos(2\pi f_d t)$ provides the required frequency offset which is the IF frequency (see Figure 9.16). Develop a general relationship for the output $g(t)$ of the system, where

$$g(t) = \int_{-\infty}^{\infty} |R(\alpha, t) + S(\alpha, t)|^2 \, d\alpha$$

and where $R(\alpha, t)$ is the total reference waveform at the output. Calculate the temporal frequency content of the two terms

$$g_1(t) = \int_{-\infty}^{\infty} |R(\alpha, t)|^2 \, d\alpha$$

and

$$g_2(t) = \int_{-\infty}^{\infty} |S(\alpha, t)|^2 \, d\alpha.$$

Calculate the minimum value of f_d to prevent overlap of the spectral terms of the three output signals $g_1(t)$, $g_2(t)$, and $g_3(t)$ if the total *baseband* bandwidth of $s(t)$ is 20 MHz and TW = 1000 for the acousto-optic cell. You are expected to chose a realistic aperture weighting function to support your answer. Sketch the temporal spectrum of all three terms to support your answer.

9.2. Suppose that we use a plane-wave reference probe at the Fourier plane to heterodyne detect the signal produced by the positive diffracted order from an acousto-optic cell. The cell is made of $LiNbO_3$ and operates at a center frequency of $f_c = 1600$ MHz. Suppose that the photodetector size is 25 μ. What is the input frequency range that can be detected if the magnitude of the cross-product term must be within $\frac{1}{2}$ of its maximum value for the following two cases:

(a) when the reference beam is parallel to the optical axis and

(b) when the reference beam is parallel to the chief ray caused by f_c.

Plot the frequency response of the system, when using this photodetector, for a full 1000-MHz bandwidth centered at $f_c = 1600$ MHz. Also, calculate the photodetector size needed to keep the system response to no more than a 3-dB rolloff over the full bandwidth.

9.3. Suppose the beam combiner in the optical radio of Figure 9.12 is rotated 10 mrad. If the combiner is 15 mm on a side and has a refractive index of 1.5, calculate the amount that the spectrum is shifted when it is transmitted *through* the prism, if the distance from the center of the combiner to the Fourier plane is 100 mm?

10

Heterodyne Spectrum Analysis

10.1. INTRODUCTION

The emphasis in this chapter is on achieving more dynamic range in spectrum analyzers by using heterodyne-detection techniques. Other benefits, such as the ability to measure both the magnitude and phase of a signal, also accrue as a result of heterodyne detection. As we learned in Chapter 8, the photodetector current produced by a power spectrum analyzer is proportional to the input rf signal *power*. In heterodyne detection we combine the signal spectrum with a reference beam, called a local oscillator, the magnitude and phase of which are known, so that the photodetector current is proportional to the signal *voltage* instead of to the signal power. The result is a significant increase in the dynamic range.

King et al. (94) described heterodyne detection techniques for recovering both the magnitude and phase information of a light distribution. In their system, shown in Chapter 9, Figure 9.3, the interference of an unmodulated reference beam with a spectrum $F(\alpha, t)$ produces a temporal frequency proportional to the input signal frequency f. Because the wideband input signal is typically centered on a frequency of several hundred megahertz, the interference term occurs at a high temporal frequency that varies as a function of the spatial frequency. As a result, implementing the postdetection filter design for each discrete photodetector element, each with a different center frequency, is not cost effective.

In this chapter we describe a heterodyne spectrum analyzer in which a spatially modulated reference beam is used to reduce the temporal IF frequency to a fixed value over the entire spectrum (99). Discrete element photodetectors with small bandwidths and low-noise equivalent powers are used in the Fourier domain to detect the time-varying signal in each frequency channel. An additional benefit of discrete detectors is that the postdetection processing operations are more flexible and can be performed in parallel to reduce the output data rate. Other advantages of heterodyne spectrum analysis are improved crosstalk rejection, scatter-noise immunity, and uninterrupted evaluation of the spectrum to achieve a 100% probability of intercept.

Self-scanned photodetector arrays cannot be used with this detection technique because they integrate light over a time period that is large compared to $1/W$, where W is the bandwidth of the signal. As a result of the integration, the temporal information contained in the heterodyne output is lost. Although discrete photodetectors are not the most elegant for use in systems whose time bandwidth products are large, arrays with about 100 elements have been implemented; advanced photodetector fabrication techniques may produce integrated devices with attractive operational features (100, 101). Furthermore, there are several applications, such as radar warning receivers, where a reasonably small number of photodetectors are adequate because the channel frequency intervals are fairly large relative to the bandwidth covered.

We begin with a description of the basic theory of the heterodyne spectrum analyzer and the spatial/temporal frequency content of several signal types, including cw and short-pulse signals. We then establish the required characteristics of the reference beam and determine the photodetector geometry and postdetection bandwidth required to achieve a given frequency resolution. We analyze and compare the performance of various reference waveforms based on both their temporal and spatial frequency content. Finally, we calculate the dynamic range obtained by heterodyne detection and compare it with that of a power spectrum analyzer.

10.2. BASIC THEORY

The basic theory of heterodyne spectrum analysis is developed in connection with the interferometric system shown in Figure 10.1. The signal waveform $f(t)$ is applied to the transducer of the acousto-optic cell located in the lower branch of the interferometer at plane P_1. The drive signal may be a carrier frequency modulated by a baseband signal $s(t)$, or a portion of the rf spectrum that is translated to a center frequency f_c. The acousto-optic cell in the signal branch of the interferometer has length L_S, centered at $x = 0$. The reference signal $r(t)$ is applied to the transducer of the acousto-optic cell located in the upper branch of the interferometer at plane P_2; this acousto-optic cell has length L_r centered at $x = 0$. The two acousto-optic cells are illuminated by a collimated source of monochromatic light.

The complex light amplitude $f_+(x, t)$ leaving the signal cell is

$$f_+(x,t) = jma(x)s\left(t - \frac{T_s}{2} - \frac{x}{v}\right)e^{j2\pi f_c(t - T_s/2 - x/v)}, \qquad (10.1)$$

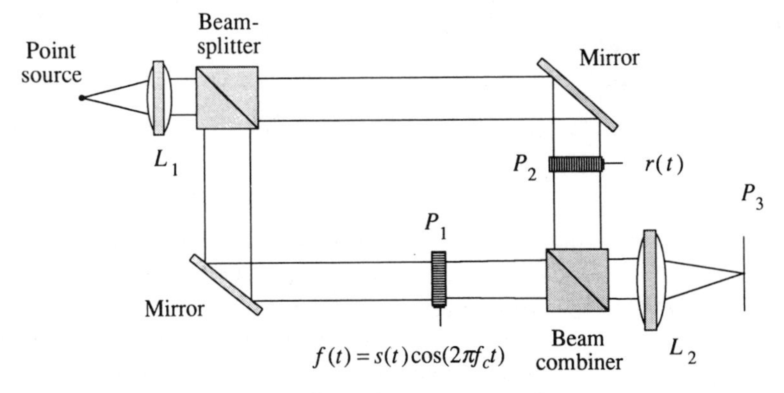

Figure 10.1. Heterodyne spectrum analyzer.

where m is the modulation index, $a(x)$ is the aperture function, T_s is the time duration of the cell, and f_c is the center frequency of the applied rf signal. The spatial Fourier transform of the positive diffracted order is

$$F_+(\alpha, t) = \int_{-\infty}^{\infty} jma(x)s\left(t - \frac{T_s}{2} - \frac{x}{v}\right)e^{j2\pi f_c(t - T_s/2 - x/v)}e^{j2\pi\alpha x}\,dx. \quad (10.2)$$

In a similar fashion, $R_+(\alpha, t)$ is the Fourier transform of the positive diffracted order $r_+(x, t)$. The intensity at plane P_3 is the square of the sum of $F_+(\alpha, t)$ and $R_+(\alpha, t)$:

$$\begin{aligned} I(\alpha, t) &= |F_+(\alpha, t) + R_+(\alpha, t)|^2 \\ &= |F_+(\alpha, t)|^2 + |R_+(\alpha, t)|^2 + 2\,\mathrm{Re}\{F_+(\alpha, t)R_+^*(\alpha, t)\}. \quad (10.3) \end{aligned}$$

After removing the first two bias terms by a bandpass filter centered at f_c, as we discussed in Chapter 9, we detect the cross-product term from Equation (10.3), which contains the desired magnitude, frequency, and phase information.

We relate the spatial Fourier transform given by Equation (10.2) to the temporal transform of $s(t)$ by setting $a(x) = \mathrm{rect}(x/L_s)$ and making a change of variables

$$\begin{aligned} t - T_s/2 - x/v &= u \\ dx &= -v\,du \quad (10.4) \end{aligned}$$

to obtain

$$F_{+}(\alpha, t) = jmve^{j2\pi\alpha v(t - T_s/2)} \int_{t-T_s}^{t} s(u) e^{-j2\pi(f-f_c)u} \, du. \qquad (10.5)$$

We recognize the integral as the Fourier transform $S(f, t)$ of the baseband signal $s(t)$, evaluated over a time window that extends T_s seconds into the past. This spectrum is centered at f_c to reflect the fact that the signal is on a carrier frequency. The spectrum $S(f, t)$ is called the *instantaneous spectrum* of the signal $s(t)$ for two main reasons. First, and most importantly, the limits of integration include the current time variable t. As the signal flows through the cell, the spectrum is recomputed continuously in time. From a sampling viewpoint, however, we preserve information if we confine our attention to the transform computed at time intervals of T_0, where $T_0 = T_s/N$. As usual, $N = 2T_sW$ is the total number of samples for the signal history in the cell. In this sense, the instantaneous spectrum is clearly different from a batch spectrum analyzer, such as an FFT-based system which might compute the spectrum of successive, nonoverlapping blocks of N samples. Second, the time of flight from the signal plane to the Fourier plane is of the order of a few nanoseconds at most. In this sense the spectrum is calculated nearly instantaneously after the most recent N samples are available. This spectrum is also sometimes called the *short-time spectrum* of a signal or the *Gabor transform*.

In a power spectrum analyzer, the intensity of the spatial transform of the signal is sufficient to fully describe the behavior of the system. Heterodyne spectrum analyzers, however, produce temporal signals that can be processed further to extract useful information. It is important, therefore, to examine both the spatial and temporal spectra of the signals to accurately interpret the results and indicate how we can improve the performance of the system.

10.3. SPATIAL AND TEMPORAL FREQUENCIES: THE MIXED TRANSFORM

The temporal frequency content of a signal is an important guideline for designing the postdetection filter that separates the cross-product term from the bias terms in Equation (10.3). Furthermore, after the system has been built, we must verify that its performance meets specifications, using test equipment such as oscilloscopes and electronic spectrum analyzers. The temporal frequency content of the output signal significantly affects our interpretation of the performance of the system. To find both the spatial and temporal frequency content of a signal, we use the *mixed transform*, which is defined as is the simultaneous spatial/temporal Fourier transform of the input space/time signal (102):

$$F_+(\alpha, f) = \iint_{-\infty}^{\infty} f_+(x, t) e^{j2\pi(\alpha x - ft)}\, dx\, dt. \tag{10.6}$$

When displayed as a two-dimensional function, the independent variables represent spatial and temporal frequencies. Another way to view the mixed transform is that it reveals the temporal frequency content at any position α of a probe in the spatial frequency plane. The term *probe* is used here in exactly the same sense as in Chapter 9; it represents a reference beam which, in combination with a photodetector, is used to detect the magnitude, frequency, and phase of light.

We generally first calculate the spatial frequency transform of a signal, as given by Equation (10.2), and then find the mixed transform by completing the temporal part of the transform:

$$F_+(\alpha, f) = \int_{-\infty}^{\infty} F_+(\alpha, t) e^{-j2\pi ft}\, dt. \tag{10.7}$$

The spatial transform part of the mixed transform $F_+(\alpha, t)$ is most useful for visualizing the spectrum detected at the output of the optical system; the mixed transform $F_+(\alpha, f)$ provides additional information about the temporal frequencies and is useful in interpreting the results as displayed on test equipment such as oscilloscopes and spectrum analyzers.

We apply the mixed transform to three signals to demonstrate different relationships between the spatial and temporal frequencies: a cw signal, a short pulse, and a dynamic case of a long pulse that evolves into the system.

10.3.1. The cw Signal

First, suppose that the input waveform is a pure cw signal at frequency f_k with magnitude $c_k = 1$. The spatial part of the mixed transform can be calculated from either Equation (10.2) or Equation (10.5); in this case it is simpler to use Equation (10.2):

$$F_+(\alpha, t) = jme^{j2\pi f_k(t - T_s/2)} \int_{-\infty}^{\infty} a(x) e^{j2\pi(\alpha - \alpha_k)x}\, dx,$$

$$= jme^{j2\pi f_k(t - T_s/2)} L_s \operatorname{sinc}[(\alpha - \alpha_k) L_s], \qquad (10.8)$$

where we have set $a(x)$ equal to $\operatorname{rect}(x/L_s)$ for convenience. This result indicates that the spatial spectrum is a sinc function, centered at α_k, with magnitude proportional to the length of the acousto-optic cell. In addition, the entire spectrum has temporal frequency f_k, as shown by the phasor in Equation (10.8), independent of which spatial frequency α we probe at plane P_3. An exact coupling between spatial and temporal frequencies does not, therefore, exist, as we claimed in Chapter 7. An exact coupling occurs only as $L_s \to \infty$ so that $L_s \operatorname{sinc}[(\alpha - \alpha_k)L_s] \to \delta(\alpha - \alpha_k)$; all the signal energy is then concentrated at one sample position in the spatial frequency plane. The key point is that spatial diffraction is due to integration over a finite range at the space plane. All light associated with diffraction due to the aperture function must therefore have the same temporal frequency as the underlying cw signal.

To complete the mixed transform for the cw input signal, the spatial frequency transform given by Equation (10.8) is used in Equation (10.7) to produce the mixed transform of the cw signal:

$$\boxed{\begin{aligned} F_+(\alpha, f) &= L_s \operatorname{sinc}[(\alpha - \alpha_k) L_s] \int_{-\infty}^{\infty} e^{-j2\pi(f - \alpha_k v)t}\, dt \\ &= L_s \operatorname{sinc}[(\alpha - \alpha_k) L_s] \delta(f - \alpha_k v). \end{aligned}} \qquad (10.9)$$

This result shows that the temporal frequency consists of a delta function centered at $f_k = \alpha_k v$, which implies that the temporal frequency is not dependent on the probe position. The $\operatorname{sinc}[(\alpha - \alpha_k)L_s]$ function shows that the finite spatial aperture causes spectral spreading in the spatial frequency domain, whereas $\delta(f - f_k)$ shows that the temporal frequency content of the signal is pure. The purity of the temporal frequency arises because the detected signal is not truncated in time by the acousto-optic cell or by any other system component. In most experimental instruments, such as in an electronic spectrum analyzer, the temporal integration is constrained to the time range $|t| \leq T_e/2$, so that Equation (10.9) becomes

$$F_+(\alpha, f) = L_s \operatorname{sinc}[(\alpha - \alpha_k) L_s] \int_{-T_e/2}^{T_e/2} e^{-j2\pi(f - f_k)t}\, dt$$

$$= L_s T_e \operatorname{sinc}[(\alpha - \alpha_k) L_s] \operatorname{sinc}[(f - f_k) T_e]. \qquad (10.10)$$

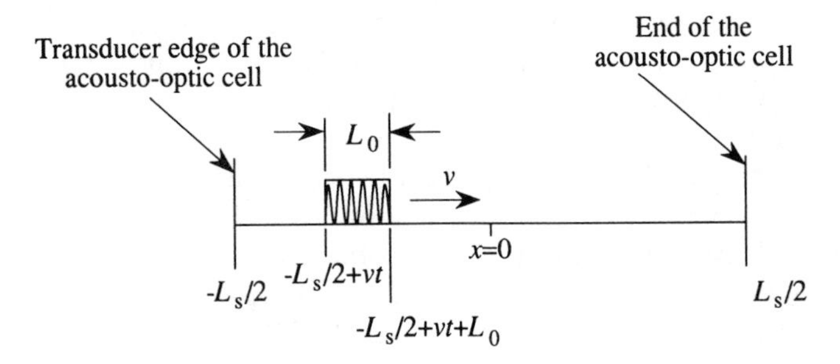

Figure 10.2. Short-pulse input signal.

This result shows that a conventional spectrum analyzer produces "temporal diffraction" if the signal has a finite time duration, similar to the spatial diffraction produced in the optical spectrum analyzer. When $T_e = L/v$, the spatial and temporal frequency spreads are equal.

10.3.2. A Short Pulse

The second signal is a *short pulse* of duration $T_0 \ll T_s$ and carrier frequency f_c that is completely within the acousto-optic cell. This signal, illustrated in Figure 10.2, exists within the cell for the time period $0 < t < (T_s - T_0)$. Using Equation (10.2), we find that the amplitude distribution at plane P_3 at some instant in time is

$$\begin{aligned} F_+(\alpha, t) &= jm \int_{-L_s/2+vt}^{-L_s/2+vt+L_0} e^{j2\pi f_c(t-T_s/2-x/v)} e^{j2\pi\alpha x}\, dx \\ &= jme^{j2\pi\alpha v t} L_0 \operatorname{sinc}[(\alpha - \alpha_c)L_0], \qquad 0 \le t \le (T_s - T_0), \end{aligned} \tag{10.11}$$

where we again ignore unimportant phase terms. The spatial spectrum of a short pulse is a sinc function, centered at α_c, whose magnitude and width are functions of the pulse length L_0.

The mixed transform for a short-pulse is found by substituting the result from Equation (10.11) into Equation (10.7) and noting that the effective integration time is $(T_s - T_0) \approx T_s$ seconds because the signal exists in the cell for only this time period:

$$\boxed{\begin{aligned} F_+(\alpha, f) &= L_0 \operatorname{sinc}[(\alpha - \alpha_c)L_0] \int_0^{T_s} e^{-j2\pi(f-\alpha v)t}\, dt \\ &= L_0 T_s \operatorname{sinc}[(\alpha - \alpha_c)L_0] \operatorname{sinc}[(f - \alpha v)T_s]. \end{aligned}} \tag{10.12}$$

The temporal frequencies for the short pulse are also distributed according to a sinc function because the pulse also has finite time duration. This example shows most clearly that the temporal frequencies for a pulse that exists for a finite *time* duration are spread in the same way as the spatial frequencies are spread when a signal exists over a finite *space* interval. For a probe at any spatial frequency α, the mixed transform reveals that there is a temporal frequency sinc function centered at $f = \alpha v$ and that the temporal frequency spread is given by $\text{sinc}[(f - \alpha v)T_s]$.

10.3.3. The Evolving Pulse

The third signal is a pulse moving into the cell as shown in Figure 10.3; this pulse has duration $T_0 \gg T_s$ and carrier frequency f_c. When t is slightly larger than zero, that part of the signal within the cell behaves as a short pulse, concentrated near the transducer edge of the acousto-optic cell. When $t = T_s/2$ the signal fills half the cell and when $t > T_s$ the signal completely fills the cell so that it behaves as a cw signal. We now discuss the spatial/temporal characterization of the mixed transform as the signal evolves from a short pulse to a cw signal.

If the time of arrival of the leading edge of the pulse is $t = 0$, the leading edge moves to $x = (-L_s/2 + vt)$ for $0 < t \leq T_s$. For a unit amplitude pulse, Equation (10.2) becomes

$$F_+(\alpha, t) = jm \int_{-L_s/2}^{-L_s/2+vt} e^{j2\pi f_c(t - T_s/2 - x/v)} e^{j2\pi\alpha x}\, dx$$

$$= jme^{j2\pi(\alpha - \alpha_c)vt/2}\{vt \,\text{sinc}[(\alpha - \alpha_c)vt]\}, \qquad (10.13)$$

where we ignore unimportant phase terms. When we apply the same

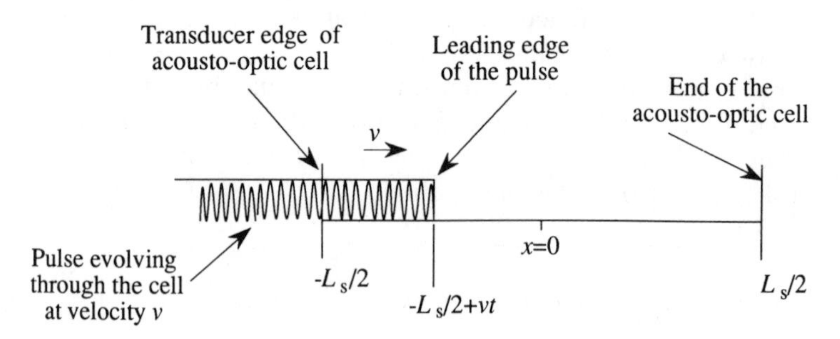

Figure 10.3. Evolving-pulse input signal.

interpretation to Equation (10.13) as we did to the cw signal, we find that the spatial spectrum is a sinc function centered at α_c. However, the magnitude of the sinc function increases linearly as t increases and its width decreases. This behavior reflects the fact that the pulse width, within the cell, increases linearly with time.

The temporal frequency cannot be so easily deduced just from the phasor factor in Equation (10.13), because the linear term vt introduces additional temporal frequencies that are not accounted for by the simple phasor notation. The interesting relationship between the temporal and spatial frequencies can be appreciated only by completing the mixed transform. The mixed transform for the evolving pulse is found by substituting the spatial transform as given by Equation (10.13) into the mixed transform as given by Equation (10.7):

$$F_+(\alpha, f) = \int_0^{T_s} vt \operatorname{sinc}[(\alpha - \alpha_c)vt] e^{j\pi(\alpha+\alpha_c)vt} e^{-j2\pi ft} \, dt, \quad (10.14)$$

where T_s is the time duration of the acousto-optic cell. The integration time is over just that time integral during which the pulse evolves. We express the sinc function as $\sin[\pi(\alpha - \alpha_c)vt]/[\pi(\alpha - \alpha_c)vt]$, cancel the terms in vt, apply the Euler expansion, and perform the integration to produce

$$\boxed{F_+(\alpha, f) = \frac{T_s}{\alpha - \alpha_c}\{\operatorname{sinc}[(f - \alpha v)T_s] - e^{j\phi(\alpha, f)} \operatorname{sinc}[(f - \alpha_c v)T_s]\},} \quad (10.15)$$

where $\phi(\alpha, f)$ is a phase term. For a probe at $\alpha_c + \Delta\alpha$, the result shows two temporal frequencies: one at $f = f_c$ and one at $f_m = f_c + \Delta f$, as shown in Figure 10.4.

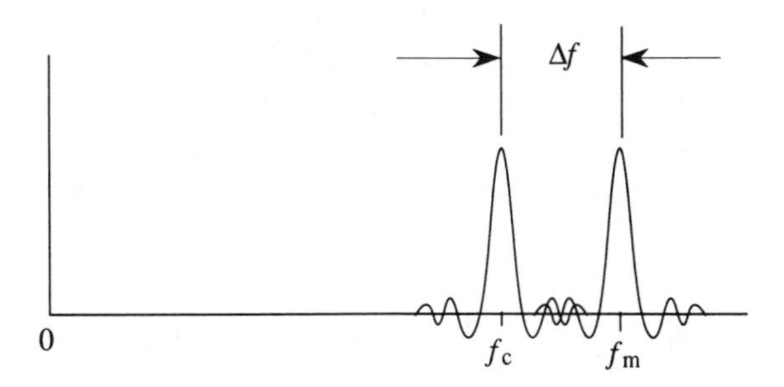

Figure 10.4. Mixed transform of the evolving pulse for a probe at $\Delta\alpha = \Delta f/v$ away from α_c.

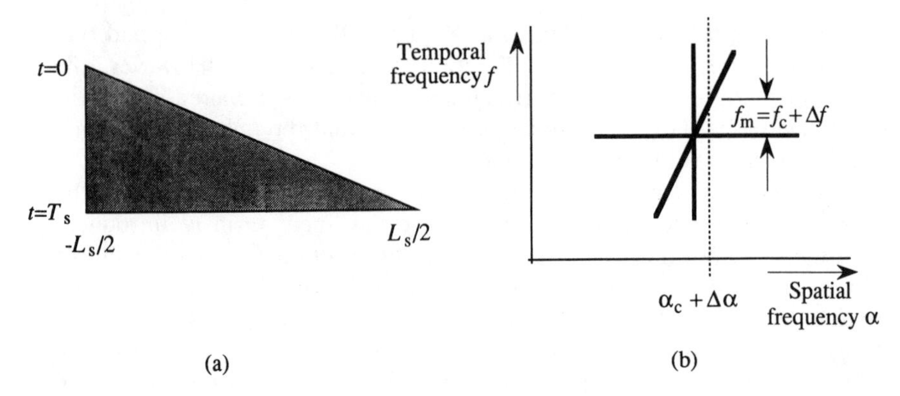

Figure 10.5. Two-dimensional representation of the mixed transform of an evolving pulse: (a) pulse length increases as time increases; (b) the spectrum of the triangle has strong diffraction perpendicular to t.

An alternative way to find the mixed transform is to plot the two-dimensional Fourier transform of the space/time representation of the evolving pulse as shown in Figure 10.5(a). At $t = 0$ the pulse length is zero, and at $t = T_s$ the pulse length is $L_s = vT_s$. The mixed transform of the pulse is found by calculating the two-dimensional transform of the shaded area in Figure 10.5(a) and translating the center of the Fourier transform to the coordinate (α_c, f_c) in the (α, f) domain to account for the spatial and temporal carrier frequencies. The two-dimensional Fourier transform consists of three main ridges of high intensity, as shown in Figure 10.5(b); each ridge is perpendicular to one of the edges of the shaded triangle. If we place a photodetector at $\alpha_c = \Delta\alpha$, as shown by the dotted line in Figure 10.5(b), the distance between the horizontal ridge and the diagonal ridge corresponds to $f_m = f_c + \Delta f$, exactly as predicted by Equation (10.15).

This temporal frequency behavior also becomes apparent if we continue the result from the spatial part of the analysis one more step to reveal the temporal behavior of the signal as it would be displayed on an oscilloscope. For example, if we add a reference beam to $F_+(\alpha, t)$, carry out the square-law detection, and take the real part of Equation (10.13), as required to complete the analysis of the cross-product term, we produce a time signal

$$F_+(\alpha, t) = vt\cos[2\pi(\alpha + \alpha_c)vt/2]\,\mathrm{sinc}[(\alpha - \alpha_c)vt/2], \qquad 0 \le t \le T_s. \tag{10.16}$$

As above, we expand the sinc function to obtain

$$F_+(\alpha, t) = \frac{1}{\alpha - \alpha_c} \sin[2\pi(\alpha - \alpha_c)vt/2] \cos[2\pi(\alpha + \alpha_c)vt/2], \tag{10.17}$$

where we again ignore unimportant constants. Consider the situation where the reference probe is located at $\alpha_c + \Delta\alpha$. We find that

$$F_+(\alpha, t) = \frac{1}{\Delta\alpha} \sin[2\pi(\Delta\alpha/2)vt] \cos[2\pi(\alpha_c + \Delta\alpha/2)vt]; \quad 0 \le t \le T_s, \tag{10.18}$$

which clearly shows that the output has temporal frequency $f_c + \Delta f/2$, modulated by a signal whose frequency is $\Delta f/2$. Its temporal frequency decomposition, by use of trigonometric identities, is exactly that of two frequencies separated by an interval Δf.

To illustrate the time waveform represented by Equation (10.18), we place the photodetector probe outside the mainlobe of the sinc function due to the signal and observe the amplitude modulation on an oscilloscope (103). The acousto-optic cell has a time duration $T_s = 10$ μsec. We use a 20-μsec pulse signal so that we can show (1) the output signal as the pulse evolves into the system for the first 10-μsec interval, (2) its temporal behavior as a cw signal for the next 10 μsec, and (3) its response as it once again becomes an evolving pulse leaving the cell during the last 10-μsec interval. The amplitude modulation for each portion of the output signal is seen in Figure 10.6 for a photodetector position corresponding to $\Delta f = 345$ kHz away from f_c. The frequency of the amplitude modulation at the beginning and end of the output is measured from the oscilloscope as 182 kHz. Because the expected value of the modulation frequency is $\Delta f/2 = 172.5$ kHz, the error of the measured value is 5.2%, which is reasonable for such measurements. The oscilloscope trace also shows a sharp transition to the response expected of a cw signal at 10 μsec, when the pulse fills the cell, and then back to the evolving pulse signal 10 μsec later as the trailing edge of the pulse passes through the cell. This result graphically shows how the temporal frequencies change as the input signal changes its form.

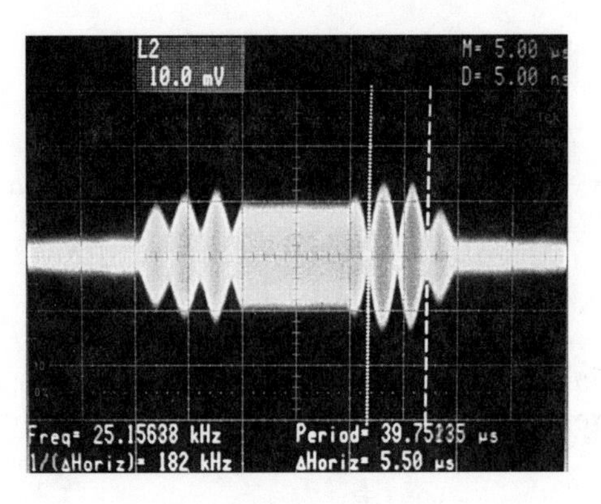

Figure 10.6. Amplitude of output for a 20-μsec pulse (103).

10.4. THE DISTRIBUTED LOCAL OSCILLATOR

At the end of Chapter 9, we discussed methods for implementing an optical radio. If the reference beam is scanned across the spectrum and if the resulting intensity is detected by a single photodetector, the result is a scanning spectrum analyzer. Such a system would be called a superheterodyne receiver if it were implemented electronically; it suffers from the fact that not all frequencies are monitored at all times. To implement a more practical heterodyne spectrum analyzer and to take full advantage of the parallel nature of the optical system, we want to generate a reference beam that produces a constant-output IF frequency for all spatial frequency positions. If we can fix the offset temporal frequency at f_d for all detected spatial frequencies, we can use identical photodetectors and bandpass filters to select the desired cross-product term. Because f_d does not need to be a high frequency to separate the cross-product term from the bias terms, a substantial reduction in the photodetector bandwidth is possible, from several hundred to about 10 MHz, or even less in some applications.

10.4.1. The Ideal Reference Signal

The reference waveform at the Fourier plane plays an important role in the performance of the heterodyne spectrum analyzer. The desired characteristics of $R_+(\alpha, t)$ are that (1) the magnitude should be equal at all photodetector positions so that the magnitudes of the spatial frequencies due to the signal are measured accurately, (2) the spatial and temporal frequencies should be coupled so that, with a relative geometric displacement between the reference waveform and the signal spectrum, equal temporal offset frequencies are produced at all photodetector locations to simplify the postdetection circuitry, as shown in Chapter 9, Section 9.4.2, Method 5, (3) the magnitude should be constant over time to avoid measurement errors, and (4) the duty cycle of the drive signal $r(t)$ should be high so that light is efficiently used.

The reference signal must contain at least N frequency components, where N is of the order of $3M$ and the number of resolvable frequencies M is of the order of $T_s W$. As in Chapter 9, we consider each frequency in the reference signal at plane P_3 of Figure 10.1 as a probe that we use to measure the frequency content of $F_+(\alpha, t)$. The ideal reference signal is generated by summing N equal magnitude frequencies (102):

$$\boxed{r(t) = \sum_{n=N_1}^{N_2} \cos(2\pi n f_0 t - \phi_n),} \qquad (10.19)$$

where f_0 is the fundamental frequency and where $f_1 = N_1 f_0$ and $f_2 = N_2 f_0$ are the lowest and highest frequencies in the signal. By definition, $r(t)$ is composed of exactly $N = N_2 - N_1 + 1$ frequency components, chosen so that there is exactly one probe for each photodetector at the Fourier plane of the spectrum analyzer.

Because $r(t)$ contains N discrete frequencies, each a harmonic of the fundamental frequency f_0, it is a repetitive signal with repetition period $T_r = 1/f_0$. The repetition period is also equal to L_r/v, where L_r is the length of the acousto-optic cell in the reference beam. We can generate a surprising variety of waveforms by specifying the phases appropriately. We begin by finding the waveform that $r(t)$ assumes when the phases are all zero; Equation (10.19) can then be written as

$$r(t) = \sum_{n=0}^{N_2} \cos(2\pi n f_0 t) - \sum_{n=0}^{N_1 - 1} \cos(2\pi n f_0 t), \qquad (10.20)$$

which is the sum of two geometric series. The sum of the terms is given by

$$r(t) = \frac{1}{2}\left\{\frac{1 - e^{j2\pi(N_2+1)f_0 t}}{1 - e^{j2\pi f_0 t}} - \frac{1 - e^{j2\pi N_1 f_0 t}}{1 - e^{j2\pi f_0 t}}\right\} + \text{c.c.}, \qquad (10.21)$$

where we use Euler's formula to expand the cosine terms and c.c.

represents the complex conjugate of all the preceding terms. The terms in Equation (10.21) are combined to give

$$\begin{aligned} r(t) &= \tfrac{1}{2}\frac{e^{j2\pi(N_2+1)f_0t} - e^{j2\pi N_1 f_0 t}}{e^{j2\pi f_0 t} - 1} + \text{c.c.} \\ &= \tfrac{1}{2}\frac{e^{j\pi(N_2+N_1+1)f_0t}\left[e^{j\pi(N_2-N_1+1)f_0t} - e^{-j\pi(N_2-N_1+1)f_0t}\right]}{e^{j\pi f_0 t}\left[e^{j\pi f_0 t} - e^{-j\pi f_0 t}\right]} + \text{c.c.} \\ &= \tfrac{1}{2}e^{j\pi(N_2+N_1+1)f_0t}\frac{\sin[\pi(N_2 - N_1 + 1)f_0 t]}{\sin(\pi f_0 t)} + \text{c.c.} \\ &= \cos[\pi(N_2 + N_1 + 1)f_0 t]\frac{\sin[\pi(N_2 - N_1 + 1)f_0 t]}{\sin(\pi f_0 t)}. \end{aligned} \tag{10.22}$$

We set $N = N_2 - N_1 + 1$ as the number of cosine terms in $r(t)$ and find that Equation (10.22) becomes

$$r(t) = \cos[\pi(N + 2N_1)f_0 t]\frac{\sin(\pi N f_0 t)}{\sin(\pi f_0 t)}. \tag{10.23}$$

This result describes the reference waveform as an impulse train that modulates a temporal carrier frequency of $(N + 2N_1)f_0/2$. As an aside, the impulse train is the temporal equivalent of the spatial Fourier transform of an N-slit grating as derived in classical optical texts (14).

If we set the phases ϕ_n as a linear function of n (e.g., $\phi_n = n\phi_0$), the impulse train is simply advanced or delayed according to the sign and magnitude of ϕ_0. If the phases are quadratic in n,

$$\phi = \frac{d\pi n^2}{N}, \tag{10.24}$$

so that $r(t)$ becomes a repetitive chirp function, the period of which is T_r, the duty cycle of which is d, and the frequency range of which is from f_1 to f_2.

An example of the behavior of $r(t)$, which shows the transition from an impulse train to a chirp train as a function of the parameter d, is given in Figure 10.7. The upper trace shows two periods of $r(t)$ when $d = 0$ so that the phase is zero for all n. The result is an impulse train that modulates a carrier frequency f_c. The presence of the carrier changes the signal

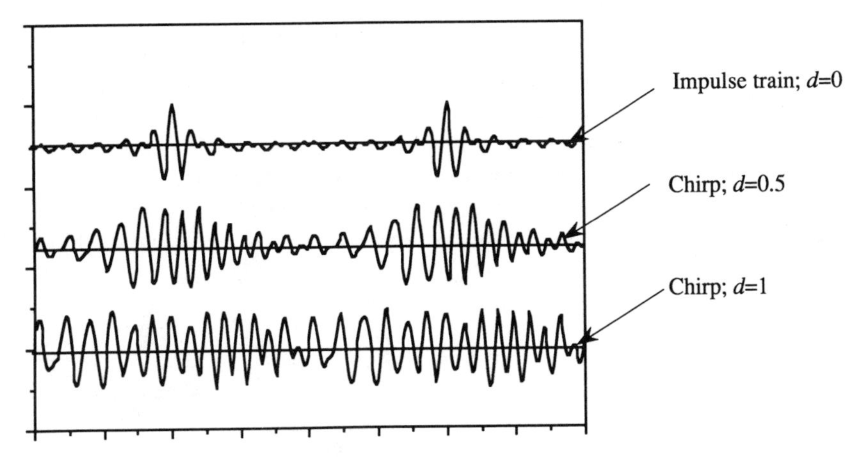

Figure 10.7. Chirp functions with duty cycles of 0, 0.5, and 1.

waveform somewhat from the expected impulse response, because there are only three cycles of the carrier underneath the impulse train envelope. Three carrier cycles per impulse arise from the fact that we require that $N_2 = 2N_1$ to produce an octave bandwidth and that the carrier frequency is $(N_1 + N_2)f_0/2 = (3/2)N_1 f_0$. The traces of Figure 10.7 are therefore accurate representations of the appearance of the signal on a test oscilloscope.

The middle trace is that for a quadratic phase function, according to Equation (10.24), but with $d = 1/2$; we note that the signal has become a repetitive chirp function with a duty cycle of 50%. The lower trace shows the chirp function when $d = 1$; the chirp now has a 100% duty cycle. Thus, we find a smooth progression from an impulse train to a full duty cycle chirp as d is changed from 0 to 1. Each of these signals is a suitable reference signal; the preferred choice is to use a high duty cycle chirp so that the light power is efficiently used. For example, the 100% duty cycle chirp signal produces reference probes at the Fourier plane whose intensities are N times those produced by the impulse train because the impulses intercept only $1/N$ of the input light intensity.

Other useful reference waveforms can be generated by a proper choice of the phases of $r(t)$. A repetitive pseudorandom sequence of length $N = 2^r - 1$, where r is an integer, can be produced if the phases for the various frequencies are suitably chosen (104). If the ϕ_n are random, the resulting signal simulates a bandlimited noise source that nevertheless retains a repetitive feature.

10.4.2. The Mixed Transform of the Reference Signal

The amplitude profile of the illuminating beam, the acoustic attenuation, the size limitations of the cell, and any other weighting factors combine to form a reference-beam aperture weighting function $b(x)$. These features of the interaction can be incorporated with the reference beam in the form $b(x)r(t - x/v)$, where we ignore the time-delay factor $T_r/2$ in this section. The reference signal corresponding to the positive diffracted order is

$$r_+(x, t) = b(x) \sum_{n=N_1}^{N_2} e^{j[2\pi n f_0(t - x/v) - \phi_n]}, \tag{10.25}$$

and the mixed transform of $r(x, t)$ is

$$R_+(\alpha, f) = \iint_{-\infty}^{\infty} r(x, t) e^{j[2\pi(\alpha x - ft)]}\, dx\, dt, \tag{10.26}$$

where α is the spatial frequency and f is the temporal frequency. We calculate the spatial transform first to find that

$$R_+(\alpha, t) = \int_{-\infty}^{\infty} b(x) \sum_{n=N_1}^{N_2} e^{j[2\pi n f_0(t - x/v) - \phi_n]} e^{j2\pi\alpha x}\, dx. \tag{10.27}$$

By separating the time- and space-dependent terms and by performing the integration over space, we find that

$$R_+(\alpha, t) = \sum_{n=N_1}^{N_2} B(\alpha - n f_0/v) e^{j[2\pi n f_0 t - \phi_n]}, \tag{10.28}$$

where $B(\alpha)$ is the Fourier transform of the aperture weighting function $b(x)$. We see that the Fourier transform of the reference signal consists of a sequence of aperture functions $B(\alpha)$ that are centered at each of N photodetector positions. Each probe has an associated pure frequency that is a harmonic of f_0. Because each frequency component of the reference signal behaves as a cw signal, the entire aperture function oscillates at the same temporal frequency.

We complete the mixed transform by calculating the temporal frequency content:

$$R_+(\alpha, f) = \int_{-\infty}^{\infty} R_+(\alpha, t) e^{-j2\pi f t}\, dt. \tag{10.29}$$

We substitute Equation (10.28) into Equation (10.29) to find that

$$R_+(\alpha, f) = \sum_{n=N_1}^{N_2} e^{-j\phi_n} B(\alpha - nf_0/v)\delta(f - nf_0), \qquad (10.30)$$

which can be expressed in an equivalent form as

$$\boxed{R_+(\alpha, f) = B(\alpha - f/v) \sum_{n=N_1}^{N_2} e^{-j\phi_n}\delta(f - nf_0).} \qquad (10.31)$$

We see that $R_+(\alpha, f)$ consists of a two-dimensional function $B(\alpha - f/v)$ that is sampled by a set of phase-weighted delta functions, creating N discrete probes. There is one probe at each photodetector position, shifted by the desired temporal frequency nf_0 and spatial frequency nf_0/v with a shape described by $B(\alpha)$. Because of the nature of the sampled function $B(\alpha - f/v)$, we find that the *magnitude of the mixed transform is independent of* ϕ_n *and, therefore, the specific repetitive reference signal* $r(t)$. This remarkable result states that the performance of the spectrum analyzer is completely independent of the specific reference signal, provided that it can be represented by Equation (10.19).

Figure 10.8 shows the magnitude of $R_+(\alpha, f)$ when $b(x) = \text{rect}(x/L_r)$. For any value n, the spatial frequency response is a sinc function centered at $\alpha = nf_0/v$. The mixed transform shows that the sinc function is also

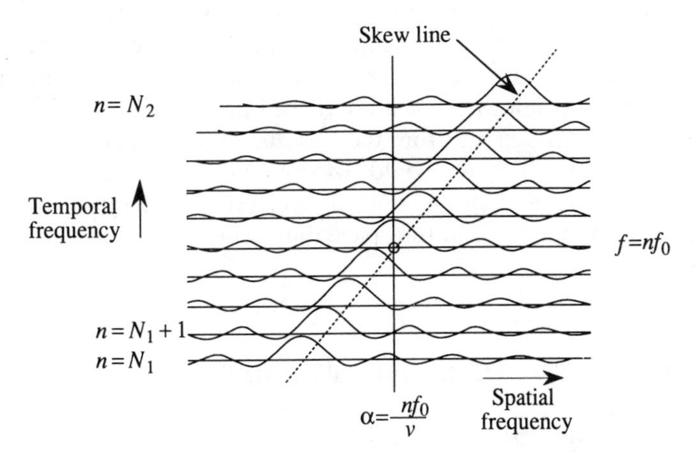

Figure 10.8. Magnitude of mixed transform for an arbitrary signal.

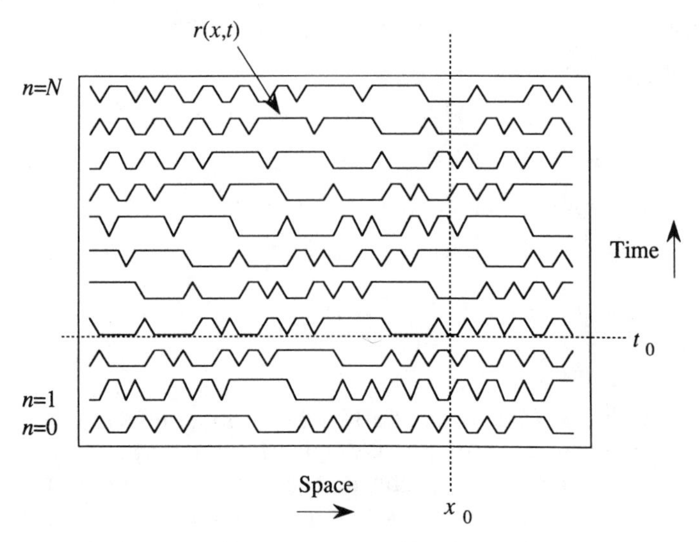

Figure 10.9. Reference signal envelope expressed as a function of space and time.

displayed in the second dimension at $f = nf_0$; as n increases from N_1 to N_2, we find that the sinc functions are centered on a skew line in the α, f plane.

A similar two-dimensional frequency display would result if we were to use the y axis of a conventional Fourier-transforming system to display the time-shifted versions of the reference signal as it passes through the acousto-optic cell. In Figure 10.9, we illustrate such a signal, shifted progressively in space as time increases in the vertical direction. For any spatial position x_0, we find the temporal function by reading the values along a vertical line positioned at x_0. At any time t_0, we find the spatial function resident within the acousto-optic cell by reading the values along a horizontal line through t_0. From the two-dimensional space/time representation of Figure 10.9, we could also obtain the mixed transform of Figure 10.8, provided that the optical aperture of a Fourier-transform system is limited to $\pm L_r/2$ in the space dimension and is unlimited in the time dimension.

10.5. PHOTODETECTOR GEOMETRY AND BANDWIDTH

The mixed transform provides considerable guidance about the design of the photodetector array, the nature of the bandpass filter, and the ex-

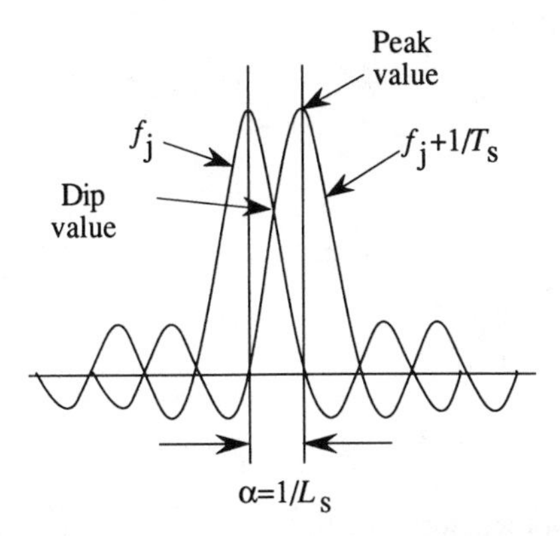

Figure 10.10. Frequency-resolution geometry for a spectrum analyzer.

pected performance levels of the spectrum analyzer. The major task in spectrum analysis is to accurately and reliably detect cw frequencies. We must be able to resolve spatial frequencies at intervals of approximately $1/L_s$, as shown in Figure 10.10. The resolution criterion is generally stated in terms of a required dip between frequencies of about 3 dB. Another consideration is that we wish to measure the signal magnitude to within a specified degree of accuracy. The photodetectors must therefore be spaced so that at least one of them is close to the peak value.

We now discuss how the reference acousto-optic cell length L_r is related to that of the signal cell whose length is L_s. Figure 10.11 shows the light distribution in the Fourier plane caused by a set of equal-magnitude

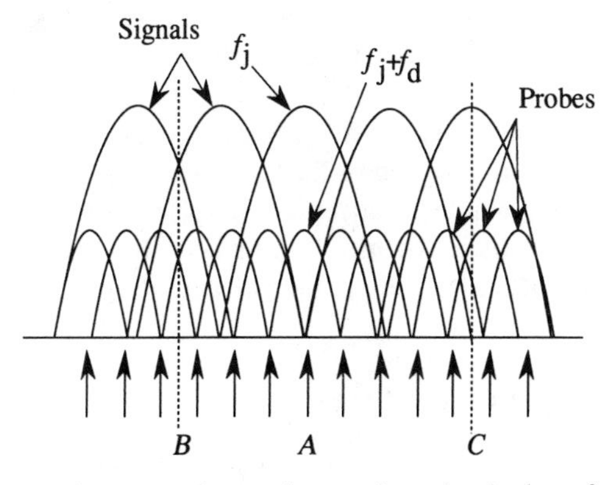

Figure 10.11. Signal spectrum for equal-strength cw signals plus reference probes.

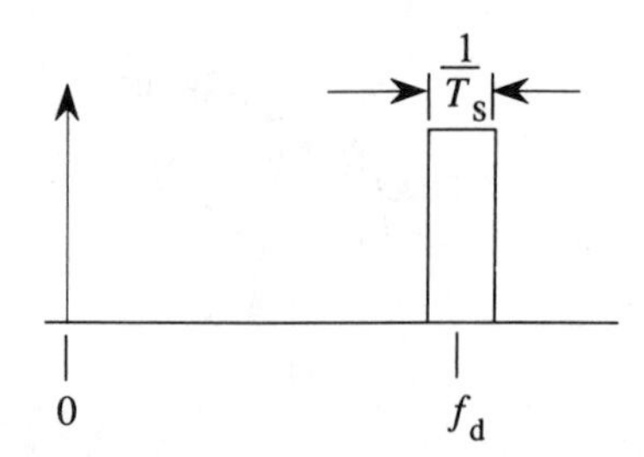

Figure 10.12. Bandpass filter for photodetector output.

cw signals generated from an acousto-optic cell of time duration T_s. For clarity, we show only the central lobes of the Fourier components, which are represented by $\text{sinc}[(f - f_j)T_s]$ in the temporal frequency domain or by $\text{sinc}[(\alpha - \alpha_j)L_s]$ in the spatial frequency domain. The reference beam is represented by a set of sinc-function probes that are spaced at intervals of f_0 in the temporal frequency domain or intervals of $\alpha_0 = f_0/v$ in the spatial frequency domain. The reference probes must be narrower than the cw-signal mainlobes so that the cw signals are sampled accurately. Furthermore, the temporal frequency of the probes is offset by an amount f_d through a geometric displacement at the Fourier plane of the reference beam relative to the signal beam. For example, in the region near point A in Figure 10.11, the cw signal frequency is f_j, whereas that of the central probe is $f_j + f_d$. The frequencies associated with adjacent probes are $f_j + f_d \pm nf_0$, where n is an integer.

We associate one photodetector with each of the reference probes. If we have R photodetectors per resolvable frequency, the interval between them is $1/RT_s$, which, by definition, is equal to f_0. For the moment, we let the photodetectors be point elements to clarify the key principles and identify the centers of the photodetectors by arrows. For this idealized case, a point detector at A detects only the cross product between one probe and the signal because all adjacent probes, as well as adjacent frequencies, have zero crossings at A. The output is therefore at the offset frequency f_d, which is accepted by a bandpass filter as shown in Figure 10.12. The shape of the bandpass filter is a function of the frequency resolution and required measurement accuracy.

10.5.1. The Bandpass Filter Shape

To find the minimum number of photodetectors per frequency to resolve signal frequencies spaced by $1/T_s$ and to accurately measure the signal

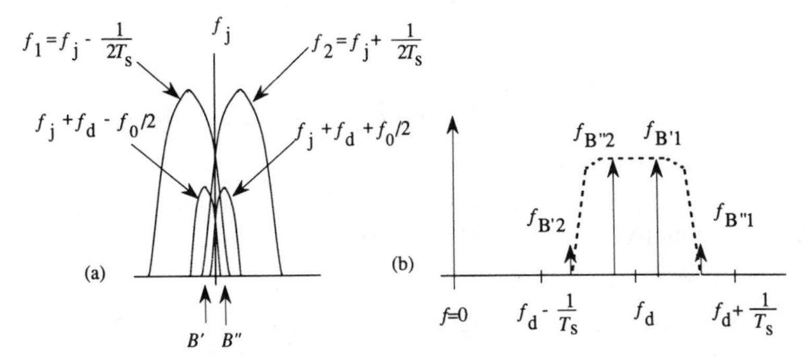

Figure 10.13. Worst-case resolution: (a) position of the probes relative to the signals; (b) the four major frequencies generated by the geometry of part (a).

magnitude is a somewhat more complicated procedure than that used in Chapter 4 in connection with power spectrum analyzers. With heterodyne detection, we must consider how the temporal frequencies generated by the interference between the signal and reference beams affect the output signal. The shape of the postdetection bandpass filter is influenced by the need (1) to control crosstalk, (2) to achieve a given frequency resolution, and (3) to detect the signals with a reasonable degree of accuracy.

In Figure 10.11 we show the case where $R = 2.5$, a value just above that required to satisfy the Nyquist sampling rate criterion. This case conveniently illustrates both the best- and worst-case conditions for detection, depending on the relative positions of the probes with respect to the signals. For example, the worst-case condition for resolving two frequencies occurs when the midpoint between the frequencies coincides with the midpoint between two photodetectors (as near point B in Figure 10.11). Figure 10.13(a) shows this situation with greater detail. Suppose that the temporal frequency associated with the midpoint is f_j. The two frequencies to be resolved are therefore $f_1 = f_j - 1/2T_s$ and $f_2 = f_j + 1/2T_s$. The point photodetectors are identified as being at the positions B' and B''; the frequencies of the associated probes are $f_j + f_d - f_0/2$ and $f_j + f_d + f_0/2$.

The probe/detector at B' produces outputs from both signal frequencies f_1 and f_2. After square-law detection, we find that the associated difference frequencies are

$$f_{B'1} = \left| f_1 - (f_j + f_d - f_0/2) \right| = f_d + \frac{1}{2T_s} - f_0/2 \tag{10.32}$$

and

$$f_{B'2} = \left| f_2 - (f_j + f_d - f_0/2) \right| = f_d - \frac{1}{2T_s} - f_0/2. \qquad (10.33)$$

These two frequencies are sketched in Figure 10.13(b). In a similar fashion, the photodetector at B'' produces outputs at the frequencies

$$f_{B''1} = \left| f_1 - (f_j + f_d + f_0/2) \right| = f_d + \frac{1}{2T_s} + f_0/2 \qquad (10.34)$$

and

$$f_{B''2} = \left| f_2 - (f_j + f_d + f_0/2) \right| = f_d - \frac{1}{2T_s} + f_0/2, \qquad (10.35)$$

which are also shown in Figure 10.13(b). The magnitude of these frequency components are the same as those at B'. Note that both frequencies $f_{B'1}$ and $f_{B''2}$ do not pass through the bandpass filter associated with any one photodetector, but rather that any photodetector must be prepared to accept any frequency within the specified range. When these requirements are met, we can use identical bandpass filters for each photodetector.

One task for the bandpass filter is to control the frequency content of the postdetection signal. The filter, which is identical for each photodetector circuit, must severely attenuate the unwanted frequency components represented by $f_{B''1}$ and $f_{B'2}$, while accepting the desired frequency components represented by $f_{B'1}$ and $f_{B''2}$. Because the maximum frequency spread about f_d produced by a desired signal is $\pm 1/(2T_s)$, the ideal filter shape is a rectangular passband, centered at f_d, with total bandwidth $1/T_s$, as shown in Figure 10.12.

10.5.2. Crosstalk

Crosstalk considerations for a heterodyne spectrum analyzer are much the same as those for a power spectrum analyzer because crosstalk affects the system dynamic range and the accuracy to which the frequency components are measured. Recall that the design procedure in a power spectrum analyzer is to select an aperture function $a(x)$ so that the sidelobes of a strong signal are sufficiently suppressed at the spatial frequency corresponding to a weak signal. The sidelobes in a power spectrum analyzer are

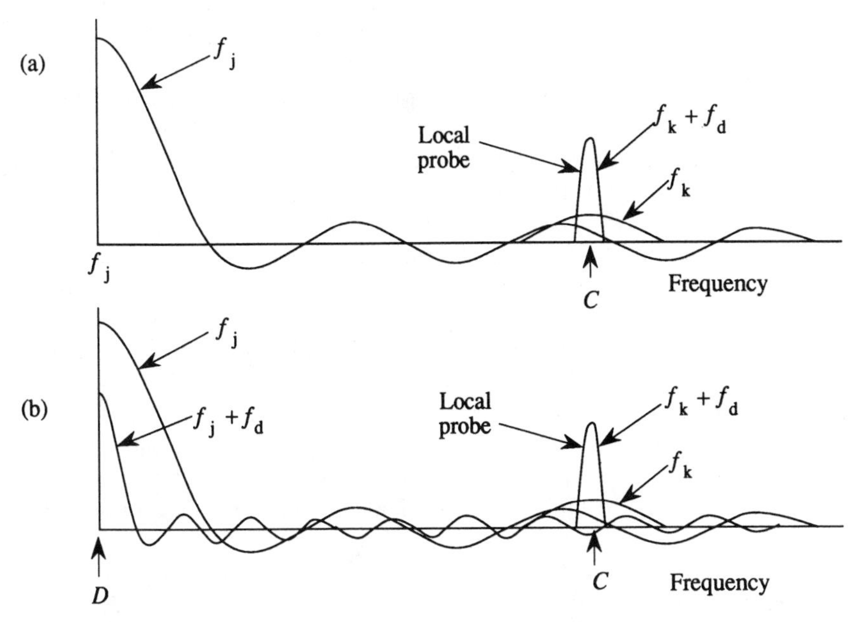

Figure 10.14. Geometry for cross-talk: (a) interference of local probe and strong signal and (b) interference of local probe and reference probe.

proportional to $|A(\alpha)|^2$, and the mainlobe width increases as the sidelobe levels are suppressed. As a result the frequency resolution is reduced.

One difference in the heterodyne spectrum analyzer is that the band-pass filter helps to suppress, in effect, some of the energy due to the sidelobes of a strong signal. The optical intensity that contributes to the crosstalk level in a heterodyne spectrum analyzer is proportional to the product of $A(\alpha - \alpha_j)$, due to the strong-signal frequency at f_j, and of $B(\alpha - \alpha_k)$, due to the reference probe at the weak-signal frequency f_k. Suppose, for example, that $A(\alpha - \alpha_j)$ is a sinc function, as shown in Figure 10.14(a), centered at f_j. At position C, we wish to detect a weak cw signal, whose frequency is f_k, in the presence of the strong signal at f_j. Because the probe associated with the detector at C has frequency $f_k + f_d$, there is no problem in extracting the desired weak-signal information. Even though the sidelobe level from $A(\alpha - \alpha_j)$ is high at C, the bandpass filter will not pass the energy because its frequency $|f_k + f_d - f_j|$ falls outside the passband. There is a special condition, however, when an alias signal slips through the bandpass filter. This condition arises when $f_k - f_j = -2f_d$ (i.e., when the strong signal is at a higher frequency relative to the weak signal). But because the signals are then spread farther apart, the sidelobe levels are lower so that crosstalk is less of a problem.

The bandpass filter does not completely eliminate crosstalk, however. Figure 10.14(b) shows the same situation as in Figure 10.14(a), but it also shows the probe associated with the detector at D. This probe at frequency $f_j + f_d$ along with the signal energy at f_j produces energy at f_d at all spatial frequency positions. The photodetector at C will therefore respond to the output produced by the sidelobes of a strong signal and its associated probe; we still need to use aperture weighting functions to help fight crosstalk as we did in power spectrum analyzers. Fortunately, in the heterodyne spectrum analyzer we have the advantage of being able to independently control $b(x)$ to reduce the sidelobe levels. Because we can control sidelobe levels almost entirely by the aperture weighting function of the reference beam, we do not suffer a loss in system resolution.

The bandpass filter also is effective in reducing the effects of scatter noise, which tends to limit the performance of power spectrum analyzers. The strongest scattered light at the photodetector plane is due to the undiffracted light from the signal and reference acousto-optic cells. These scattered light components do not, however, interfere to produce a signal at the offset frequency f_d. The interference occurs, instead, at baseband where it is removed by the bandpass filter.

10.5.3. Resolution, Accuracy, and Photodetector Size

After the aperture function $b(x)$ has been chosen to control the sidelobe level, we determine the system resolution in much the same way as in Chapter 4. In this case, we begin by convolving the aperture response $A(\alpha)$ from the signal branch with the aperture response $B(\alpha)$ from the reference branch, keeping track of the associated temporal frequencies as a function of the spatial frequency displacement. We then modify the resulting signal by the bandpass characteristics of a filter, accounting for the fact that the band-edge responses of the filter are not perfectly sharp in practice. Based on this information, we determine if the dip between frequencies meets the specification. If not, we iterate the design process until the dip specification is met. We find that $R = 3$ is generally adequate to provide the desired frequency resolution and measurement accuracy.

10.6. TEMPORAL FREQUENCIES OF THE REFERENCE BIAS TERM

So far we have focused on how the characteristics of the cross-product term affect the shape of the bandpass filter and the temporal frequency content of the resulting output signal. The bias term $I_2(\alpha, t)$, due to the reference function $|R_+(\alpha, t)|^2$, can also produce energy within the passband of the filter under certain conditions. We now examine the origin of this energy for the most general case.

From Equation (10.28), we find that the intensity $I_2(\alpha, t)$ due only to the reference beam is

$$
\begin{aligned}
I_2(\alpha, t) &= |R(\alpha, t)|^2 \\
&= \sum_{n=N_1}^{N_2} \sum_{m=N_1}^{N_2} e^{j2\pi(n-m)f_0 t} e^{j(\phi_n - \phi_m)} B(\alpha - nf_0/v) B^*(\alpha - mf_0/v).
\end{aligned}
\tag{10.36}
$$

The mixed transform for the bias term is readily obtained by finding the temporal transform of Equation (10.36):

$$
\begin{aligned}
I_2(\alpha, f) &= \int_{-\infty}^{\infty} \sum_{n=N_1}^{N_2} \sum_{m=N_1}^{N_2} e^{j2\pi(n-m)f_0 t} e^{j(\phi_n - \phi_m)} \\
&\quad \times B(\alpha - nf_0/v) B^*(\alpha - mf_0/v) e^{-j2\pi f t}\, dt \\
&= \sum_{n=N_1}^{N_2} \sum_{m=N_1}^{N_2} e^{j(\phi_n - \phi_m)} B(\alpha - nf_0/v) \\
&\quad \times B^*(\alpha - mf_0/v) \int_{-\infty}^{\infty} e^{-j2\pi[f-(n-m)f_0]t}\, dt \\
&= \sum_{n=N_1}^{N_2} \sum_{m=N_1}^{N_2} e^{j(\phi_n - \phi_m)} B(\alpha - nf_0/v) \\
&\quad \times B^*(\alpha - mf_0/v) \delta[f - (n-m)f_0].
\end{aligned}
\tag{10.37}
$$

The conclusion that we reach from Equation (10.37) is that the bias term, in general, contributes energy at all integer multiples of f_0. Unfortunately, one or more integer multiples of f_0 must fall within the passband of the filter because its width is of the order of $2f_0$ to $3f_0$ if we use two to three probes per signal frequency.

There is a special set of conditions, however, for which $I_2(\alpha, f)$ has energy only at $f = 0$. This set of conditions is

1. That $b(x) = \mathrm{rect}(x/L_r)$.
2. That L_r is equal to an integer multiple of L_s.
3. That point photodetectors are placed at integer multiples of f_0/v so that all photodetectors are at nulls of adjacent sinc functions.

In this special case, we find that the sums in Equations (10.36) and (10.37) have value only for $n = m$; the result is that the power spectrum of $r_+(x, t)$ is *constant in time and at all photodetector positions*. In addition to placing all the point detectors at the nulls of sinc-function probes, we must have an integer number of probes for each resolvable cw signal (for example, $R = 3$). As a result, there is no opportunity for the heterodyne frequency to "walk off" from $f_d \pm nf_0$; this vernier effect can be seen in Figure 10.11, where we set $R = 2.5$.

Because this special set of conditions is impossible to implement in practice, we must find alternative methods to remove energy from the reference bias term. Other practical problems are that we may not be able to generate $N \approx 3M$ reference probes to synthesize $r(t)$ directly if N is large. These and other practical issues have been addressed in the literature (102).

10.7. DYNAMIC RANGE

We turn our attention to a calculation of the dynamic range of the heterodyne spectrum analyzer. The output signal, for a simple cosine of the form $f(x, t) = 0.5[1 + jm_s c_k \cos(2\pi f_k t)]$, is the same as we had in the power spectrum analyzer discussed in Chapter 8, except that we need to account for the beamsplitter and beam-combiner ratios. Suppose that the beamsplitter and beam combiner reflect a fraction γ and ρ of the incident light, respectively, and that the remainder is reflected. Any absorption of light by the beamsplitter or beam combiner is included in the value of A_s, along with other factors that affect the overall system efficiency. The Fourier transform of the signal beam is then

$$F_+(\alpha, t) = \frac{A_s m_s c_k L_s \sqrt{\gamma(1-\rho)}}{4\sqrt{\lambda F}} \operatorname{sinc}\left[(\xi - \xi_k)\frac{L_s}{\lambda F}\right] e^{j2\pi f_k t}, \quad (10.38)$$

where $A_s = \sqrt{P_0 \varepsilon_s / L_s}$ is the effective amplitude of the illumination beam, P_0 is the laser power, ε_s is the overall efficiency of the signal beam, L_s is the length of the acousto-optic cell, and m_s is the modulation index for the signal beam cell necessary to achieve the required spur-free dynamic range. To facilitate the development of analytical solutions, we set $a(x) = \operatorname{rect}(x/L_s)$.

The appropriate reference frequency component for this signal is $r(x, t) = 0.5\{1 + jm_r \cos[2\pi(f_k + f_d)t]\}$, where m_r is the modulation index for the reference beam. We assume that all the frequencies in the reference beam have the same magnitude; for convenience, we include the offset frequency f_d in the reference beam. The Fourier transform of the reference beam is therefore

$$R_+(\alpha, t) = \frac{A_r m_r L_r \sqrt{\rho(1-\gamma)}}{4\sqrt{\lambda F}} \operatorname{sinc}\left[(\xi - \xi_k)\frac{L_r}{\lambda F}\right] e^{j2\pi(f_k + f_d)t}; \quad (10.39)$$

the symbols have meanings similar to those in Equation (10.38). Recall that the intensity at the Fourier plane is $I(\alpha, t) = |F_+(\alpha, t) + R_+(\alpha, t)|^2$ and that the cross-product term produces a current $i_3(t)$:

$$i_3(t) = 2S \int_{-\infty}^{\infty} \operatorname{Re}\{F_+(\xi, t) R_+^*(\xi, t)\} P(\xi)\, d\xi, \quad (10.40)$$

where S is the responsivity of the photodetector. The response $P(\xi)$ accounts for the width and center position of the photodetector and determines the region of integration by the photodetector in the Fourier plane. The integration in Equation (10.40) becomes

$$i_3(t) = \frac{S A_s A_r m_s m_r L_s L_r c_k \sqrt{\gamma \rho (1-\gamma)(1-\rho)}}{8\lambda F} \cos(2\pi f_t)$$
$$\times \int_{-\infty}^{\infty} \operatorname{sinc}\left[(\xi - \xi_k)\frac{L_s}{\lambda F}\right] \operatorname{sinc}\left[(\xi - \xi_k)\frac{L_s}{\lambda F}\right] P(\xi)\, d\xi. \quad (10.41)$$

The photodetector length is equal to $\xi_r/2$, where $\xi_r = \lambda F / L_r$ is the distance to the first zero of the reference probe so that

$$P(\xi) = \operatorname{rect}\left[\frac{(\xi - \xi_k) L_r}{\lambda F}\right]. \quad (10.42)$$

To solve the integral in Equation (10.41), we must find the product of two sinc functions. We express the sinc function in a power series:

$$\operatorname{sinc}(ax) = 1 - \frac{(ax)^2}{3!} + \frac{(ax)^4}{5!} + \cdots \quad (10.43)$$

so that

$$\operatorname{sinc}(ax)\operatorname{sinc}(bx) = 1 - \frac{(ax)^2}{3!} - \frac{(bx)^2}{3!} + \frac{a^2 b^2 x^4}{5!}$$
$$+ \frac{(a^4 + b^4) x^4}{5!} + \cdots \quad (10.44)$$

The integral of the product of two sinc functions is therefore

$$\int \operatorname{sinc}(ax)\operatorname{sinc}(bx)\, dx = x - \frac{(a^2 + b^2)x^3}{3 \cdot 3!} + \cdots, \tag{10.45}$$

where we retain just the leading terms of the expansion. For the photodetector size as given by Equation (10.42), the value of the integral of the two sinc functions given by the leading term of Equation (10.45) is equal to $\lambda F/L_r$. The photocurrent is therefore

$$\boxed{i_3(t) = \tfrac{1}{8} SP_0 \sqrt{\varepsilon_r \varepsilon_s \gamma \rho (1 - \gamma)(1 - \rho)}\, m_s m_r c_k \sqrt{L_s/L_r} \cos(2\pi f_d t).} \tag{10.46}$$

We estimate the value of m_r by noting that the overall efficiency of the reference beam is of the order of 0.5 in intensity. Each frequency will therefore have a diffraction efficiency of $\eta_r = 0.5/N$ so that $m_r = \sqrt{0.5/N}$. The final result for the cross-product term is that

$$i_3(t) = \frac{SP_0 \sqrt{\varepsilon_r \varepsilon_s \gamma \rho (1 - \gamma)(1 - \rho) L_s/L_r}\, m_s c_k}{11\sqrt{N}} \cos(2\pi f_d t). \tag{10.47}$$

We now calculate the photocurrent due to the bias terms; these terms are due to the signal and reference beams acting alone. First, the current $i_2(t)$ due to the reference is obtained by integrating Equation (10.39) over the photodetector area:

$$i_2(t) = S \int_{-\infty}^{\infty} \left| \frac{A_r m_r L_r \sqrt{\rho(1 - \gamma)}}{4\sqrt{\lambda F}} \operatorname{sinc}[(\xi - \xi_k) L_r/\lambda F] \right|^2 P(\xi)\, d\xi, \tag{10.48}$$

where the symbols have the same meanings as before. Following the same procedure as for calculating $i_3(t)$, we find that

$$\boxed{i_2(t) = \frac{SP_0 \varepsilon_r \rho (1 - \gamma)}{32N}.} \tag{10.49}$$

Similarly, the photocurrent due to the signal beam acting alone is obtained by integrating Equation (10.38):

$$i_1(t) = S\int_{-\infty}^{\infty}\left|\frac{A_s m_s c_k L_s\sqrt{\gamma(1-\rho)}}{4\sqrt{\lambda F}}\operatorname{sinc}[(\xi-\xi_k)L_s/\lambda F]\right|^2 P(\xi)\,d\xi, \tag{10.50}$$

which becomes

$$\boxed{i_1(t) = \frac{SP_0\varepsilon_s m_s^2\gamma(1-\rho)L_s/L_r}{16}c_k^2.} \tag{10.51}$$

Recall that $m_s^2 = \eta_f$ is the maximum diffraction efficiency per frequency to avoid exceeding the spur-free dynamic range specification.

The signal-to-noise ratio, following the analysis given in Chapter 4, is

$$\text{SNR} = \frac{\langle i_3^2(t)\rangle}{2eB\left(i_d + \bar{i}_1 + \bar{i}_2\right) + 8\pi kTBf_{\text{co}}c_d}. \tag{10.52}$$

This result is similar to that obtained for the power spectrum analyzer except that the shot noise contains the additional average current term $\bar{i}_2$ produced by the local oscillator. The minimum signal is found by using Equation (10.47) in Equation (10.52) to find that

$$\begin{aligned} c_{k\,\min}^2 &= \frac{2eB\left(i_d + \bar{i}_1 + \bar{i}_2\right) + 8\pi kTBf_{\text{co}}c_d}{\left\langle\left[\frac{1}{8}SP_0\sqrt{\varepsilon_r\varepsilon_s\gamma\rho(1-\gamma)(1-\rho)L_s/L_r}\,m_s m_r\cos(2\pi f_d t)\right]^2\right\rangle} \\ &= \frac{2eB\left(i_d + SP_0\varepsilon_r\rho(1-\gamma)/32N\right) + 8\pi kTBf_{\text{co}}c_d}{S^2P_0^2\varepsilon_r\varepsilon_s\gamma\rho(1-\gamma)(1-\rho)\eta_f L_s/256NL_r}. \end{aligned} \tag{10.53}$$

In passing from the first to the second line of Equation (10.53), we (1) used the fact that $\bar{i}_1$ is negligible relative to the other currents when c_k is small, (2) used the value of $\bar{i}_2$ from Equation (10.49), and (3) distributed the reference-beam energy over the N probes so that the diffraction efficiency m_r^2 is equal to $0.5/N$. The dynamic range is, following the procedure established in Chapter 8, equal to

$$\boxed{\mathrm{DR} = 10\log\left[\frac{S^2P_0^2\varepsilon_r\varepsilon_s\gamma\rho(1-\gamma)(1-\rho)\eta_f L_s/256NL_r}{2eB(i_d + SP_0\varepsilon_r\rho(1-\gamma)/32N) + 8\pi kTBf_{\mathrm{co}}c_d}\right].} \tag{10.54}$$

From Equation (10.54), we see that the system is thermal-noise limited unless the laser power is so large that the shot-noise term prevails. The dynamic range increases as the square of the laser power until the shot-noise limit is reached and increases linearly thereafter. The number of resolvable frequencies plays an important role in the dynamic range calculation, as expected.

From Equation (10.54) we also see that $\rho = \gamma = 0.5$ optimizes the dynamic range when the system is thermal-noise limited. However, when the laser power is increased so that the system becomes shot-noise limited, we find that the dynamic range becomes

$$\mathrm{DR} = 10\log\left[\frac{S^2P_0^2\varepsilon_r\varepsilon_s\gamma(1-\rho)\eta_f L_s/256NL_r}{2eB(SP_0\varepsilon_r/32N)}\right], \tag{10.55}$$

provided that the local-oscillator current dominates the dark current. In this case we find that the factor $\rho(1-\gamma)$ cancels, leaving just the factor $\gamma(1-\rho)$ in the numerator. This result suggests that γ should be made large and ρ should be made small to maximize the dynamic range, a condition that favors directing more of the laser power to the signal branch of the interferometer. Carrying this process to an extreme, however, would return us to the condition where the local-oscillator current no longer dominates the dark current, as can be seen from Equation (10.54). We generally set the beamsplitting and beam-combining ratios to 0.5, which is also the easiest to achieve with commonly available devices, because the maximum improvement in dynamic range is only 3 dB.

As an example of calculating the dynamic range, suppose that we need to detect 100 frequencies separated by 25 kHz. We therefore need 300 probes and discrete photodetector elements in the output array so that $N = 3M = 300$ and $L_s/L_r = \frac{1}{3}$. Consider a system with the following parameters: $P_0 = 100$ mW, $\varepsilon_r = \varepsilon_s = 0.5$, $\gamma = \rho = 0.5$, $S = 0.5$ A/W, $i_d = 10$ nA, $c_d = 1$ pF, $\eta_f = 0.01$, $f_{\mathrm{co}} = 1$ MHz, and $B = 25$ kHz. Substituting these values into Equation (10.52), we confirm that $\bar{i}_1$ can be ignored when calculating the dynamic range. The other values become

$\bar{i}_2 = 5.2\ (10^{-7})$ and $i_3(t) = 1.5(10^{-6})c_k \cos(2\pi f_d t)$ so that the signal-to-noise ratio becomes

$$\text{SNR} = \frac{0.5\left[1.5(10^{-6})c_k\right]^2}{4.2(10^{-21}) + 2.6(10^{-21})}, \tag{10.56}$$

and we see that the shot noise is slightly more than the thermal noise. We solve Equation (10.56) for $c_{k\,\min}^2$ when the signal-to-noise ratio is equal to 1 and find that $c_{k\,\min}^2 = 9.4(10^{-9})$ so that the dynamic range is

$$\text{DR} = 10\log\left[\frac{c_{k\,\max}^2}{c_{k\,\min}^2}\right] = 10\log\left[\frac{1}{9.4(10^{-9})}\right] = 80.2\ \text{dB}. \tag{10.57}$$

10.8. COMPARISON OF THE HETERODYNE AND POWER SPECTRUM ANALYZER PERFORMANCE

If we had used a power spectrum analyzer with the same frequency parameters as given in the last section except for setting $f_{\text{co}} = B$, the system would produce a dynamic range of 55.5 dB. To compare the performance of the heterodyne and power spectrum analyzers under a wide range of conditions, we calculate the gain in the dynamic range, given by Gain $= \text{DR}_{\text{hsa}} - \text{DR}_{\text{psa}}$. Recall that the dynamic range for a power spectrum analyzer, given by Equation (8.35), is

$$\text{DR}_{\text{psa}} = 10\log\left[\frac{0.02P_0\varepsilon S\eta_f}{\sqrt{2eBi_d + 8\pi kTBf_{\text{co}}c_d}}\right]. \tag{10.58}$$

We use Equations (10.54) and (10.58) to find that

$$\text{Gain} = \text{DR}_{\text{hsa}} - \text{DR}_{\text{psa}}$$

$$= 10\log\left[\frac{\dfrac{S^2P_0^2\varepsilon_r\varepsilon_s\gamma\rho(1-\gamma)(1-\rho)\eta_f L_s/256NL_r}{2eB(i_d + SP_0\varepsilon_r\rho(1-\gamma)/32N) + 8\pi kTBf_{\text{co}}c_d}}{0.2P_0\varepsilon S\eta_f/\sqrt{2eBi_d + 8\pi kTBf_{\text{co}}c_d}}\right]. \tag{10.59}$$

We simplify Equation (10.59) by canceling similar factors to find that

$$\boxed{\text{Gain} = 10\log\left[\frac{\{SP_0\varepsilon_r\varepsilon_s\gamma\rho(1-\gamma)(1-\rho)L_s/256NL_r\}\sqrt{2eBi_d + 8\pi kTBf_{\text{co}}c_d}}{0.02\varepsilon\{2eB[i_d + SP_0\varepsilon_r\rho(1-\gamma)/32N] + 8\pi kTBf_{\text{co}}c_d\}}\right].} \tag{10.60}$$

From this general result we can consider several different operating conditions.

10.8.1. Both Systems Thermal-Noise Limited

We first consider the case where both analyzers are thermal-noise limited so that Equation (10.60) becomes

$$\text{Gain} = 10\log\left[\frac{\{SP_0\varepsilon_r\varepsilon_s\gamma\rho(1-\gamma)(1-\rho)L_s/256NL_r\}}{0.02\varepsilon\sqrt{8\pi kTBf_{\text{co}}c_d}}\right]. \tag{10.61}$$

The gain in performance is strongly proportional to the laser power and the number of frequencies that must be resolved, but it is less strongly dependent on the cutoff frequency and postdetection bandwidth.

10.8.2. Both Systems Shot-Noise Limited

Consider the situation when both systems are shot-noise limited so that Equation (10.60) becomes

$$\text{Gain} = 10\log\left[\frac{\{SP_0\varepsilon_r\varepsilon_s\gamma\rho(1-\gamma)(1-\rho)L_s/256NL_r\}\sqrt{2eBi_d}}{0.02\varepsilon\{2eB[i_d + SP_0\varepsilon_r\rho(1-\gamma)/32N]\}}\right]. \tag{10.62}$$

This case can be further subdivided into two cases. In the first and fairly unlikely case, we might find that the current produced by the local oscillator is less than that produced by the dark current. In this case, Equation (10.62) becomes

$$\text{Gain} = 10\log\left[\frac{\{SP_0\varepsilon_r\varepsilon_s\gamma\rho(1-\gamma)(1-\rho)L_s/256NL_r\}}{0.02\varepsilon\sqrt{2eBi_d}}\right], \tag{10.63}$$

so that the gain in performance is most highly dependent on the laser power and the number of frequencies that must be resolved, and the gain is independent of the cutoff frequency f_{co}. The somewhat more likely case occurs when the local-oscillator dominates the dark current in the heterodyne system so that the gain in performance becomes

$$\text{Gain} = 10\log\left[\frac{\{SP_0\varepsilon_r\varepsilon_s\gamma\rho(1-\gamma)(1-\rho)L_s/256NL_r\}\sqrt{2eBi_d}}{0.02\varepsilon\{2eB[SP_0\varepsilon_r\rho(1-\gamma)/32N]\}}\right]$$

$$= 10\log\left[\frac{\{\varepsilon_s\gamma(1-\rho)L_s/256L_r\}\sqrt{2eBi_d}}{0.02\varepsilon(2eB/32)}\right]. \tag{10.64}$$

We now see that the gain in performance is completely independent of the laser power, of the cutoff frequency, and of the number of frequencies to be resolved. The gain in performance in this case is completely determined by the parameters of the system and the postdetection bandwidth.

10.8.3. Power Spectrum Analyzer Thermal-Noise Limited; Heterodyne Spectrum Analyzer Shot-Noise Limited

The most likely scenario is that the power spectrum analyzer will be thermal-noise limited if the readout rate is fairly high, and the heterodyne spectrum analyzer will be shot-noise limited. In this case the gain in performance is

$$\text{Gain} = 10\log\left[\frac{\{SP_0\varepsilon_r\varepsilon_s\gamma\rho(1-\gamma)(1-\rho)L_s/256NL_r\}\sqrt{8\pi kTBf_{co}c_d}}{0.02\varepsilon\{2eB[i_d+SP_0\varepsilon_r\rho(1-\gamma)/32N]\}}\right]. \tag{10.65}$$

Here we see that the general nature of the gain in performance is the same as for Equation (10.61), aside from some different constants.

10.8.4. Power Spectrum Analyzer Using a CCD Array

In all the comparisons made so far, we have assumed that the power spectrum analyzer also uses discrete photodetectors at its output. In many applications, however, the power spectrum analyzer uses a CCD photodetector array, whose characteristics limit the linear dynamic range to about 35 dB, independent of how much laser power is available. Some of the nonlinear CCD arrays now provide dynamic ranges of about 50 dB or so.

The gain comparisons, in these cases, should be based on the dynamic range of a power spectrum analyzer as limited by the CCD array.

In general, we find that the performance of the heterodyne spectrum analyzer is significantly better than that of a comparable power spectrum analyzer. The additional performance comes at the price of needing a discrete photodetector array with more complicated detection circuitry. In Chapter 11, we discuss ways to partially overcome this disadvantage by using a decimated photodetector array.

10.9. HYBRID HETERODYNE SPECTRUM ANALYZERS

Aside from the improved dynamic range that can be obtained from a heterodyne spectrum analyzer, we find that certain postprocessing operations produce further frequency resolution. Suppose, for example, that we want to achieve a 1-kHz frequency resolution over a 100-MHz frequency band. We therefore need to resolve 10^5 frequency components, a performance level beyond that obtainable from conventional one-dimensional acousto-optic cells. In Chapter 4, we showed how two-dimensional falling raster recording formats could easily generate the required frequency resolution, but that system does not operate in real time.

Real-time spectrum analysis with the required resolution can be achieved by using a two-dimensional acousto-optic cell configuration, as discussed in Chapter 15. Here we briefly describe a *hybrid approach* that can perform the spectrum analysis in near-real time. Suppose that we use a heterodyne spectrum analyzer to divide the spectrum of the incoming 100-MHz signal into 100 frequency channels, each 1 MHz wide. We do so by creating 100 reference probes/photodetectors, each with a rectangular shape, to generate the required number of channels. The output of each photodetector is then analog-to-digital converted at a sample rate of 2 MHz; this data is fed to a DFT module that computes the Fourier transform of the narrowband signals. A 1024-point transform with 16-bit magnitude response can be computed in 1 msec, consistent with the required 1-KHz final frequency resolution. Note that the DFT computes the spectrum on a batch processing basis, as opposed to the continuous, sliding window processing available optically, as discussed in Chapter 15.

PROBLEMS

10.1. We design a heterodyne spectrum analyzer which requires a two-tone spur-free dynamic range of 45 dB. Assume that we have 300 reference-beam probes. The other system parameters are

$$\begin{aligned}
P_0 &= 30 \text{ mW}, \\
i_d &= 10 \text{ nA}, \\
c_d &= 4 \text{ pF}, \\
\varepsilon_s &= \varepsilon_r = 0.5, \\
S &= 0.4 \text{ A/W}, \\
T &= 300 \text{ K}, \\
B &= 1 \text{ MHz}, \\
f_{\text{co}} &= 10.7 \text{ MHz}.
\end{aligned}$$

(a) Calculate the local oscillator current. (b) Calculate the value of η_f. (c) Determine whether the system is shot-noise or thermal-noise limited. (d) Calculate the dynamic range for the heterodyne spectrum analyzer.

10.2. We design a heterodyne spectrum analyzer which requires a two-tone spur-free dynamic range of 55 dB and a single-tone dynamic range of 70 dB. We require 150 reference-beam probes. The other system parameters are

$$\begin{aligned}
i_d &= 10 \text{ nA}, \\
c_d &= 2 \text{ pF}, \\
\varepsilon_s &= \varepsilon_r = 0.5, \\
S &= 0.4 \text{ A/W}, \\
T &= 300 \text{ K}, \\
B &= 500 \text{ kHz}, \\
f_{\text{co}} &= 3 \text{ MHz}.
\end{aligned}$$

(a) Calculate the value of η_f. (b) Calculate the required laser power. (c) Calculate the local-oscillator current. (d) Determine whether the system is shot-noise or thermal-noise limited.

10.3. A heterodyne spectrum analyzer, used in a radar threat warning receiver, must resolve 30 frequencies with a resolution of 20 MHz. In this case, two-tone intermodulation products are not important, nor is the crosstalk between channels. You elect to use $LiNbO_3$ as the interaction medium and set the center frequency at 1000 MHz. The Fourier-transform lens has a focal length of 100 mm. On the assumption that the aperture weighting functions are such that

$a(x)$ and $b(x)$ are rectangular functions, calculate (a) the length of the signal and reference-beam cells, (b) the time bandwidth product for each cell, and (c) the complete geometry of the photodetector array (i.e., how many detectors, their size, their spacing, their position at the Fourier plane, and so forth).

10.4. Given the same system parameters as in Problem 10.2, (a) calculate the laser power required to make the shot-noise level twice that of the thermal-noise level, (b) calculate the resulting dynamic range for the heterodyne spectrum analyzer, using the power level found in part (a), (c) calculate the loss in dynamic range had this system been set up as a power spectrum analyzer (i.e., compare DR_{hsa} to DR_{psa}). Is the power spectrum analyzer shot-noise or thermal-noise limited? (d) Calculate the laser power required for the power spectrum analyzer to produce the dynamic range obtained from the heterodyne spectrum analyzer in part (b).

11

Decimated Arrays and Cross-Spectrum Analysis

11.1. INTRODUCTION

As we discussed in Chapter 10, heterodyne spectrum analysis is information preserving because each photodetector is read out on an instantaneous basis. Heterodyne detection, however, places severe demands on the circuitry associated with each photodetector because the rf electronics are not easily implemented with integrated circuits. In this chapter we describe spectrum analyzers that support the sampling rate required to avoid missing signals and yet operate with a significantly reduced number of photodetector elements. The basic idea is to *decimate* an N-element photodetector array by retaining only every Mth element. We then time share the remaining elements by scanning the spectrum across the decimated array. Each photodetector therefore produces, as a time sequence, the spectral content of the received signal over a small frequency range.

In this chapter we also introduce a signal-processing operation called cross-spectrum analysis to determine the angle of arrival of a signal. For example, if we use two antenna elements to receive a signal from a common source, we can use a dual-channel acousto-optic cell to determine the source direction relative to the receiver geometry through a phase-comparison technique. The phase information, obtained by forming the cross spectrum of the signals received, provides a measure of the angle of arrival of cw emitters at each frequency.

11.2. BACKGROUND FOR THE HETERODYNE SPECTRUM ANALYZER

Because the detailed operation of the heterodyne spectrum analyzer was described in Chapter 10, we review only the major points here. The heterodyne spectrum analyzer consists of a conventional spectrum analyzer, modified to include a reference function at the Fourier plane.

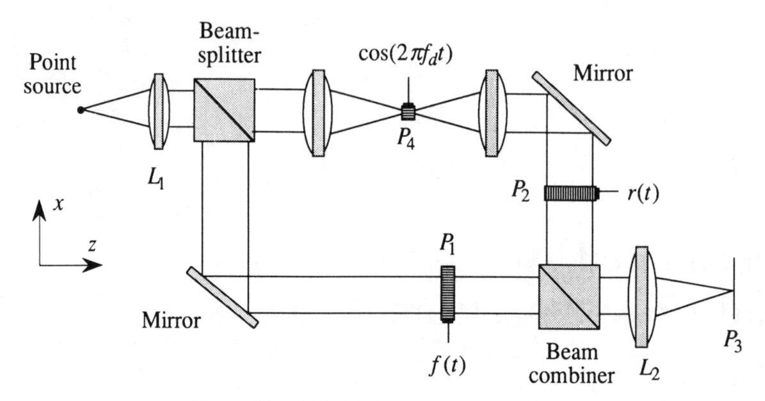

Figure 11.1. Heterodyne spectrum analyzer.

Figure 11.1 shows a Mach-Zehnder interferometer in which the lower branch contains an acousto-optic cell at plane P_1 driven by the signal $f(t)$ that we want to analyze. The Fourier transform of this signal, created at plane P_3, is

$$F_+(\alpha, t) = \int_{-\infty}^{\infty} a(x) f_+\left(t - \frac{T_s}{2} - \frac{x}{v}\right) e^{j2\pi\alpha x}\, dx, \tag{11.1}$$

where $a(x)$ is the aperture weighting function, T_s is the length of the signal acousto-optic cell, α is the spatial frequency variable, and v is the velocity of the acoustic wave. We assume that the rf signal is centered at f_c, the midpoint of the bandpass of the acousto-optic cell. In a similar fashion, the Fourier transform of the reference signal $r(t)$ that drives the acousto-optic cell at plane P_2 in the upper branch is

$$R_+(\alpha, t) = \int_{-\infty}^{\infty} b(x) r_+\left(t - \frac{T_r}{2} - \frac{x}{v}\right) e^{j2\pi\alpha x}\, dx, \tag{11.2}$$

where $b(x)$ is the aperture weighting function and T_r is the length of the reference acousto-optic cell.

The reference waveform $R_+(\alpha, t)$ could be shifted geometrically, relative to the signal spectrum, to provide a fixed temporal offset frequency f_d at each spatial frequency position. An alternative method to produce the offset frequency is to frequency shift the entire reference waveform by f_d, using the acousto-optic point modulator in the upper branch, as shown at plane P_4 in Figure 11.1. This method for providing the offset frequency provides more operating flexibility and is used later in the cross-spectrum analysis application where a geometric shift cannot be tolerated, for reasons that will become apparent later. When we use a point modulator in the reference beam, the intensity at plane P_3 becomes

$$I(\alpha, t) = \left| F_+(\alpha, t) + R_+(\alpha, t) e^{j2\pi f_d t} \right|^2. \tag{11.3}$$

We substitute Equations (11.1) and (11.2) into Equation (11.3) and expand the result to find that the two intensity terms $I_1(\alpha, t) = |F_+(\alpha, t)|^2$ and $I_2(\alpha, t) = |R_+(\alpha, t)|^2$ have frequencies concentrated at baseband. Under the condition that f_d is sufficiently large, the cross-product term does not overlap the baseband terms, and the intensity becomes

$$I_3(\alpha, t) = 2|F_+(\alpha, t)|\,|R_+(\alpha, t)| \cos[2\pi f_d t + \phi(\alpha, t)], \tag{11.4}$$

where $\phi(\alpha, t)$ is the phase difference between $R_+(\alpha, t)$ and $F_+(\alpha, t)$.

The spectral information is measured by a photodetector array consisting of N discrete elements, each followed by a bandpass filter. The photodetector current is given by the integral of $I(\alpha, t)$ over the photodetector area, and the baseband terms are removed by a bandpass filter centered at f_d. The current output of a detector element, centered at a spatial frequency α_k, due to the intensity $I_3(\alpha, t)$ is

$$i_3(t) = A_0 \cos(2\pi f_d t) \int_{-\infty}^{\infty} F_+(\alpha, t) R_+^*(\alpha, t) P(\alpha)\, d\alpha, \tag{11.5}$$

where A_0 is a scaling constant, $P(\alpha) = S \operatorname{rect}[(\alpha - \alpha_k)/\alpha_h]$ is the photodetector response at α_k, S is the responsivity of the photodetector, and α_h is the width of the detector element, expressed as a spatial frequency.

11.3. PHOTODETECTOR GEOMETRY AND DETECTION SCHEME

The spectrum of $f(x, t)$ is centered at a spatial frequency α_c, as shown in Figure 11.2. Each cw signal produces a response $A(\alpha)$ which is the Fourier transform of the aperture weighting function $a(x)$. The number of photodetector elements required in some spectrum-analysis applications may be large. For example, in a channelized receiver, we may measure the energy in each 25-kHz channel for signals that have 40–50-MHz bandwidths, requiring 1600–2000 photodetectors. Each photodetector is followed by extensive of postdetection circuitry, as shown in Figure 11.3. The signal is first amplified, usually by a transimpedance amplifier that converts the photodetector current into a voltage signal, and passed through a

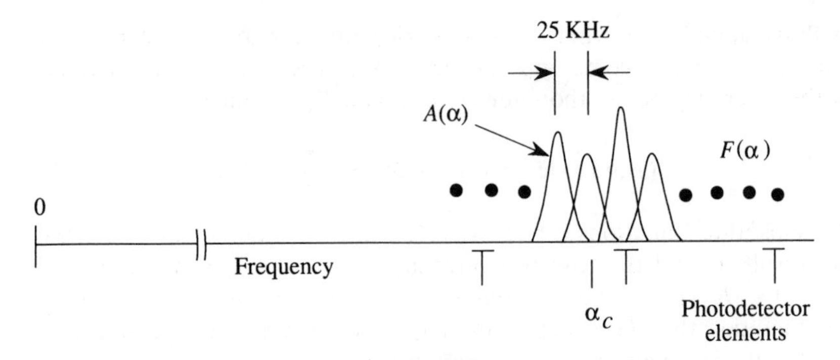

Figure 11.2. Spatial frequency plane and photodetector geometry.

narrowband filter centered at f_d. A nonlinear device compresses the range of magnitudes to facilitate rectification and to display the wide dynamic range of the signal. The outputs from the rectifiers are low-pass filtered, fed to a multiplexer that combines the outputs of the various photodetectors, and sampled before transmission as output data.

A reduction in the number of photodetectors reduces the overall hardware complexity and cost. Suppose that we retain only every M th element of the array so that the total number of elements in the array is $Q = N/\mathsf{M}$. We distinguish between $M = T_s W$ as the *number of resolvable frequencies* and M as the *decimation ratio*. We arrange to scan the signal spectrum, relative to this decimated array, so that each element detects the M frequency components for which it is responsible during a period of time T_m. Thus, each photodetector is sampled M times during each time period T_m to measure its assigned spatial frequencies.

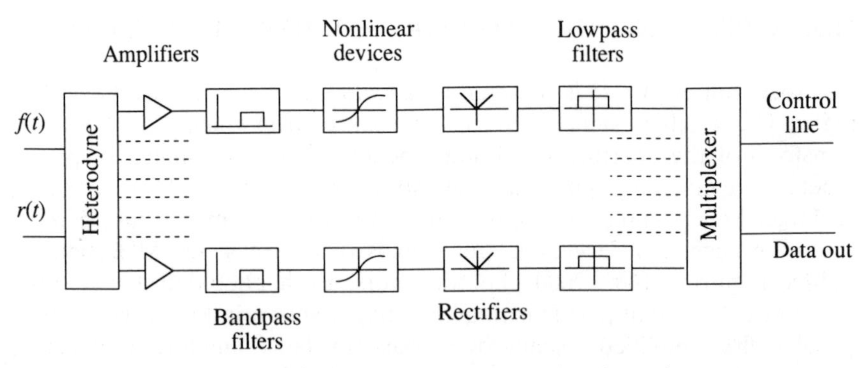

Figure 11.3. Block diagram of post-detection electronics.

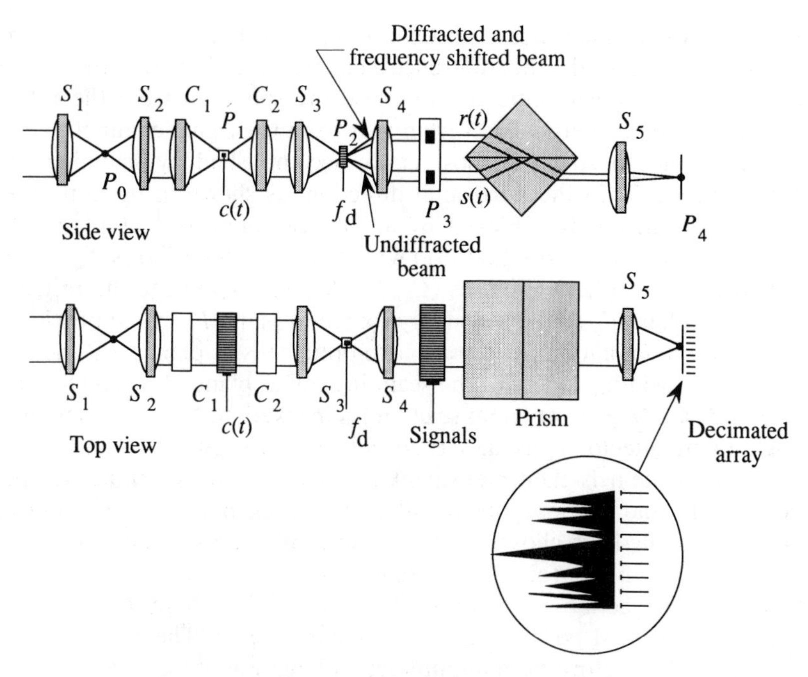

Figure 11.4. Practical implementation of heterodyne spectrum analyzer.

Figure 11.4 shows the top and side views of a spectrum analyzer that incorporates the required scanning action. The two branches of the interferometer are created by using the dual-channel acousto-optic cell at plane P_3 with its transducers stacked in the vertical direction. This cell is illuminated by two parallel beams of light, and the diffracted beams are combined by a prism so that the sum of the spectra occurs at plane P_4. Because the interferometer is compact, the effects of vibrations and thermal variations in the acousto-optic cell are minimized relative to those typically present in a Mach-Zehnder interferometer.

Koontz has described a heterodyne spectrum analyzer in which a beamsplitting prism, similar to that used to combine the beam, generates the desired illumination pattern (105). Because we want to offset one beam by a frequency f_d, an alternative way to generate the two illumination beams is to use an acousto-optic modulator at plane P_2. When the modulator is driven by $\cos(2\pi f_d t)$ at a sufficient input level to produce a 50% diffraction efficiency, the undiffracted and diffracted beams have equal intensities and the diffracted beam is frequency shifted by f_d relative to the undiffracted beam. Another means for achieving the offset is to use a conventional beamsplitter arrangement and to use $f_c + f_d$ as the center frequency in the reference channel. We then require a thin prism after the acousto-optic cell to cause the spectrum from the reference channel to geometrically overlap that from the signal channel.

Scanning the spectrum is most easily accomplished by scanning the source at plane P_2 in the horizontal direction, as shown in the top view. This scanning action is achieved by an acousto-optic cell at plane P_1, which is driven by a chirp signal $c(t)$ whose repetition period is T_m. A set of cylindrical and spherical lenses (C_1, C_2, S_2, and S_3) image the primary source from plane P_0 as the secondary source at plane P_2, while providing the requisite illumination for the acousto-optic cells at planes P_2 and P_3. The time bandwidth product of the scanning cell at plane P_1 is equal to M. The bounds are $\mathsf{M} = 1$, when no scanning is invoked, to $\mathsf{M} = N$, when only a single photodetector is used. We do not, of course, want to set $\mathsf{M} = N$ because the system is then equivalent to an electronic scanning heterodyned system and we lose the parallel processing power of the optical system. Generally M is chosen to be in the range of 8–32, depending on tradeoffs among the performance parameters.

The scanning action of the acousto-optic cell at plane P_1 causes a change in the spatial position of the secondary source. The modulator at plane P_2 must therefore have an interaction length of at least $\mathsf{M}d_0$, where d_0 is the size of the secondary source at plane P_2. The phase responses of the scanning acousto-optic cell and the modulator are not important, because they affect the signals in both branches of the interferometer equally.

11.4. THE REFERENCE AND SCANNING FUNCTIONS

The reference signal is obtained by driving the upper channel of the acousto-optic cell at plane P_3 with a waveform

$$r(t) = \sum_{n=N_1}^{N_2} \cos(2\pi n f_0 t + \phi_n), \tag{11.6}$$

where $N_2 - N_1 + 1 = N$ is equal to the number of probes required and f_0 is the spacing of the frequency components of $r(t)$. The signal represented by Equation (11.6) is a bandpass signal whose center frequency is $f_c = (N_2 + N_1 + 1)f_0/2$. The reference signal $r(t)$ consists of N equal-magnitude, discrete frequencies spaced f_0 apart, as discussed in detail in Chapter 10, Section 10.4. Such a signal is an ideal distributed local oscillator because it causes $R_+^*(\alpha, t)$ in the integrand of Equation (11.5) to have a fixed magnitude at each sample point of the spectrum $F_+(\alpha, t)$; the spectrum is therefore accurately measured.

The scanning action is obtained by driving the acousto-optic cell at plane P_1 with a repetitive chirp signal similar to that used for the reference beam:

$$c(t) = \sum_{m=\mathsf{M}_1}^{\mathsf{M}_2} \cos(2\pi m f_m t + \phi_m), \tag{11.7}$$

where $\mathsf{M} = \mathsf{M}_2 - \mathsf{M}_1 + 1$ is the decimation ratio. This repetitive chirp signal $c(t)$ has period $T_m = 1/f_m$, which is the time allowed between successive samples of the same spatial frequency; we call T_m the *revisit interval*. The scanning action is obtained by modifying Equation (10.24) to

$$\phi_m = \frac{\pi m^2}{\mathsf{M}}, \tag{11.8}$$

so that $c(t)$ is an infinite train of chirp segments with a 100% duty cycle.

The chirp rate depends on the revisit interval T_m and on the decimation ratio M. At the Fourier plane we need to scan across M distinct spatial frequencies, each separated by α_0, in the time interval T_m. We can express the chirp rate in terms of temporal frequencies by noting that $\alpha_0 = f_0 v$. The chirp rate is therefore $\mathrm{CR} = \mathsf{M} f_0/T_m = W/QT_m$. Recall from Chapter 7, Section 7.9.2 that the chirp rate can also be expressed in terms of geometric parameters as

$$\boxed{\mathrm{CR} = \frac{v^2}{\lambda D} = \frac{W}{QT_m},} \tag{11.9}$$

where v is the acoustic velocity and D is the radius of curvature of the wavefront generated by the chirp within the acousto-optic cell.

If the acousto-optic cells at planes P_1 and P_3 in Figure 11.4 have time apertures of the order of T_s, only a fraction T_s/T_m of a given chirp waveform is in the cell at any instant of time. This is the condition for a long-chirp, aperture-limited scanner from Chapter 7, Section 7.9.2. We control the scan velocity by choosing the focal length of S_3, given the value of D, as we showed in Chapter 7, Section 7.9.2. By some straightforward ray tracing, we find that the spot-scanning velocity v_s at plane P_2 is $v_s = -v[F_3/(D - F_3)]$, where F_3 is the focal length of lens S_3 and the negative sign shows that the scan velocity is opposite to the direction of v. In this result, we constrain D to be positive and larger than F_3 to minimize aberrations. The focusing action of the chirp causes the scan plane to fall at plane P_2', located a distance $F_3^2/(D + F_3)$ before the Fourier plane.

The time bandwidth product of the chirp acousto-optic cell at plane P_1 is a factor of T_s/T_m less than the signal-processing cell because of the lower bandwidth required. The bandwidth of the acousto-optic cell at plane P_4 must accommodate the mismatch in the Bragg angle caused by the scanning action. The increased bandwidth is $W_1 = W(1 + \mathsf{M}/N)$ so that M/N represents the fractional increase in W. In most cases the increase in bandwidth is less than 10%.

11.5. SIGNAL-TO-NOISE RATIO AND DYNAMIC RANGE

The sampling rate for each photodetector is found from the requirement that we sample M frequencies during each revisit interval: $R = \mathsf{M}/T_m$. The revisit interval is application dependent and affects the required postdetection bandwidth which, in turn, affects the signal-to-noise ratio and dynamic range. To find the effect of the revisit interval T_m on the dynamic range, recall that we defined the signal-to-noise ratio at the photodetector output as the signal power divided by the sum of the shot-noise and thermal-noise terms. We repeat selected key results from Chapter 10 here for convenience. The signal-to-noise ratio is

$$\text{SNR} = \frac{\langle i_3^2(t) \rangle}{2eB(i_d + \bar{i}_2) + 8\pi kTBf_{\text{co}}c_d}, \tag{11.10}$$

where $i_3(t)$ is the photocurrent produced by the cross-product term, e is the charge on an electron, B is the postdetection bandwidth, i_d is the dark current, $\bar{i}_2$ is the average signal current due to the local oscillator, k is Boltzmann's constant, T is the temperature in degrees Kelvin, f_{co} is the cutoff frequency, and c_d is the capacitance of the photodetector circuit. The minimum detectable signal level occurs when the signal-to-noise ratio is equal to one and establishes the dynamic range as

$$\text{DR} = 10 \log \left[\frac{i_{3\,\text{max}}^2}{i_{3\,\text{min}}^2} \right]. \tag{11.11}$$

The dynamic range is therefore a function of the minimum value of i_3 which is determined by the signal-to-noise ratio.

We now consider how the dynamic range is affected by the decimation ratio. The reference function $R_+(\alpha, t)$ contains N discrete probes, distributed over the analysis bandwidth W. These probes interfere with the signal spectrum $F_+(\alpha, t)$ to produce an output that has a nominal frequency f_d. As we showed in Chapter 10, Section 10.5.1, the postdetection bandwidth in the nonscanning, or *staring*, mode is $B_s = 1/T_s$.

In the scanning mode, the instantaneous carrier frequency f_i within the band $f_d \pm 1/T_s$ is amplitude modulated by the detected spectrum due to the scanning action. We can therefore think of the output signal as being simultaneously amplitude and frequency modulated:

$$i_3(t) = F_+(t)\cos[2\pi f_d t + \phi(t)], \tag{11.12}$$

where $\phi(t)$ is the phase modulation of the carrier as determined by the relative position of the probe to the signal being detected. The amplitude modulation $F_+(t)$ is caused by scanning the spatial spectrum $F_+(\alpha)$ at a scanning velocity v_s. The bandwidth of the amplitude modulation is determined by the need to sample M frequencies in the revisit time interval T_m, based on the Nyquist sampling rate. The amplitude-modulation bandwidth is therefore $B_{\mathrm{am}} = R/2 = \mathsf{M}/2T_m$. A reasonable first-order approximation to the required postdetection bandwidth is the sum of the staring-mode bandwidth B_s and the amplitude-modulation bandwidth B_{am}:

$$B = B_s + B_{am} = \frac{1}{T_s} + \frac{\mathsf{M}}{2T_m} = B_s\left[1 + \frac{\mathsf{M}T_s}{2T_m}\right], \tag{11.13}$$

which relates the bandwidth required in the scanning mode to the bandwidth required in the staring mode. The loss in performance, when using the decimated array, is most easily seen when Equation (11.13) is used in Equation (11.10) so that the signal-to-noise ratio becomes

$$\boxed{\mathrm{SNR} = \frac{\langle i_3^2(t)\rangle}{\{2e(i_d + \bar{i}_2) + 8\pi kTf_{\mathrm{co}}c_d\}B_s[1 + \mathsf{M}T_s/2T_m]}.} \tag{11.14}$$

This result shows that we can have large decimation ratios without a serious loss in performance if the revisit interval T_m is long relative to the signal-processing aperture T_s. In all cases, the loss in performance is less than 3 dB if $B_{\mathrm{am}} \leq B_s$.

As an example, consider a spectrum analyzer whose analysis bandwidth is $W = 50$ MHz and whose effective processing time is $T_s = 40$ μsec. We characterize the performance of the system for a *slow-scan mode*, wherein

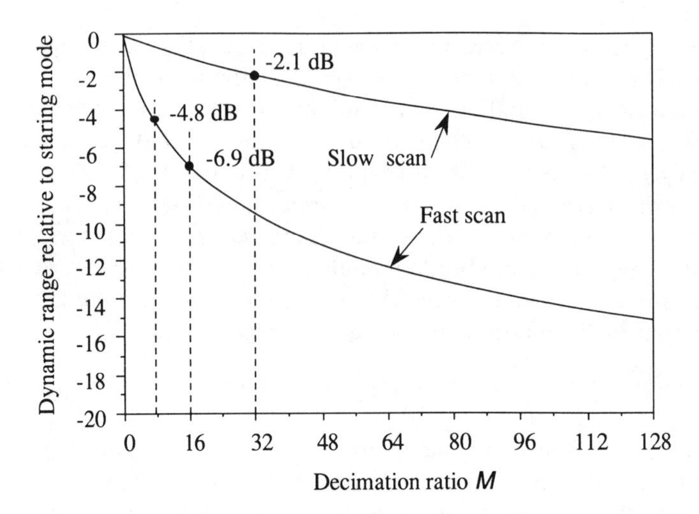

Figure 11.5. Dynamic range for heterodyne spectrum analyzer using a decimated array.

each frequency is revisited in a time interval $T_m = 25T_s$; the slow-scan mode might be used to detect frequency-hopped signals whose dwell times are approximately 1 ms. We also consider a *fast-scan mode*, wherein $T_m \approx 2T_s$; this fast-scan mode might be used, for example, to track rapidly changing signals or to detect short pulses. Figure 11.5 shows the dynamic range versus the decimation ratio M for the fast- and slow-scan modes, referenced to the staring mode for which $B = 1/T_s = 25$ kHz.

The results for the fast-scan mode show that the dynamic range loss is 4.8 dB for M = 8 and 5.9 dB for M = 16. The dynamic range for the slow-scan mode does not deteriorate rapidly as a function of the decimation ratio. For example, when M = 32 in the slow-scan mode, the loss in dynamic range, referenced to the staring mode, is only 2.1 dB; the use of the decimated technique is then an attractive trade.

11.6. IMPROVED REFERENCE WAVEFORM

From the signal-to-noise ratio given by Equation (11.10), we find that an increase in the local-oscillator current $\bar{i}_2$ increases the signal-to-noise ratio when the system is thermal-noise limited because it also increases the signal power $\langle i_2^2(t)\rangle$. The dynamic range therefore increases as well. Suppose that we alter the reference waveform $r(t)$ to more efficiently use the light at the Fourier domain to increase the local-oscillator current $\bar{i}_2$. The geometry of the photodetector elements is the same as that shown in

Figure 11.2, except that the photodetector elements *are now* M *times wider than before so that they are abutting.*

In this case, we cannot scan the signal beam relative to the set of M enlarged photodetector elements because we would not get the required frequency resolution. Instead, we make use of the fact that the crossproduct term given by Equation (11.5) exists only if the two beams overlap. To produce a controlled overlap, we generate a reference waveform that produces exactly $Q = N/\mathrm{M}$ optical probes at plane P_2. Each probe overlaps the spectrum $F_+(\alpha, t)$ for a spectral range that is one part in M of the photodetector width, thus providing the required frequency resolution. Then, instead of scanning the joint spectra across the fixed photodetector array as before, we scan the reference beam across the spectrum $F_+(\alpha, t)$ and the photodetector array, which are both fixed in space. Each reference-beam probe scans the M spatial frequencies associated with its assigned photodetector element before the scan is reset.

The appropriate reference waveform is a chirp function similar to that given by Equation (11.7), where the fundamental frequency of the waveform f_m is now given by the full analysis bandwidth W divided by the number of required probes: $f_m = W/Q$. From this relationship we immediately deduce that, for a 100% duty cycle, the chirp duration is $T_c = T_r/\mathrm{M}$ so that there are exactly M chirp waveforms within the acousto-optic cell, each of time duration T_c. The form of $c(t)$ as given by Equation (11.7) provides the desired waveform even though it is quite different from the one used at plane P_1 of Figure 11.4. Here there are M chirp waveforms *within* the reference acousto-optic cell; this is the condition for a short-chirp, repetition-rate-limited scanner as discussed in Chapter 7, Section 7.9.2.

The reference beam geometry for this case is shown in Figure 11.6. Each of the M chirps in the cell produces a focused spot at plane P_2' that travels with velocity $v_s = -v[F_2/(D - F_2)]$. These focused, scanning spots become secondary sources that completely overlap at the Fourier plane and mutually interfere to produce the required Q stationary probes at plane P_2.

We scan the probe ensemble at the Fourier plane with a controlled velocity by varying the center frequency of the reference signal. The bandwidth of the acousto-optic cell must therefore be increased to $W_1 = W(1 + 1/Q)$; this increase is generally less than 10% or so.

The major advantage of using this modified reference waveform is that it increases the dynamic range when the system is thermal-noise limited. The reference current i_2 is increased by a factor of M, which improves the signal-to-noise ratio and the dynamic range. Figure 11.7 shows that the dynamic range increases, relative to the staring mode, by 6 dB for

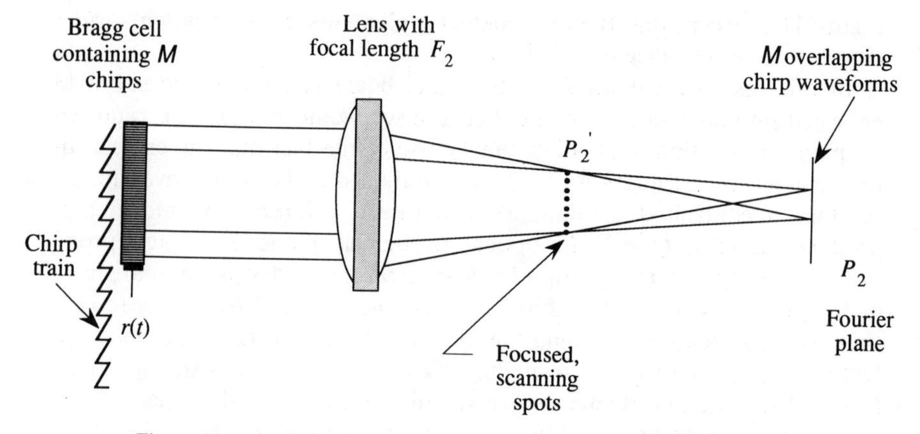

Figure 11.6. Optical subsystem to produce a decimated reference beam.

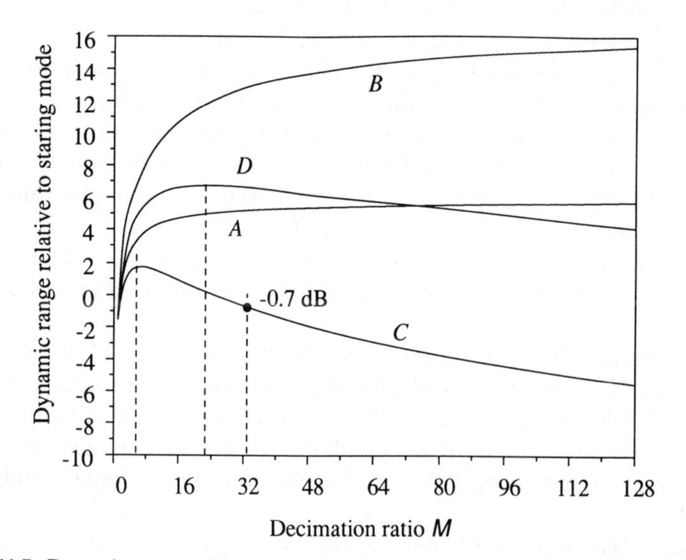

Figure 11.7. Dynamic range with improved reference function: curve A, fast-scan mode; curve B, slow-scan mode; curve C, fast-scan mode with shot noise an order of magnitude less than the thermal noise; curve D, slow-scan mode with shot noise an order of magnitude less than the thermal noise.

the fast-scan mode (curve A) and by 16 dB for the slow-scan mode (curve B). The improvement occurs because the numerator of Equation (11.10) increases more rapidly than the denominator; the increase is more pronounced for the slow-scan mode. For large M, the numerator and denominator of Equation (11.10) are both proportional to M so that the dynamic range approaches an asymptotic value in both cases.

Figure 11.7 also shows the performance of the system when the shot noise is 10 dB less than the thermal noise in the staring mode. As the decimation ratio increases, the dynamic range increases for small values of M, as shown in curve C, for the reasons discussed relative to the thermal-noise-limited case. The shot noise is equal to the thermal noise at $M = 6$ in the fast-scan mode and at $M = 22$ in the slow-scan mode. The dynamic range then begins to decrease as M increases because the signal power is proportional to M, whereas the shot noise, which is now dominant, is proportional to M^2. The results presented in Figure 11.7 show that a significant reduction in the number of photodetectors is obtained without a large penalty in dynamic range. For example, from curve C we see that the loss in dynamic range for $M = 32$ is only 0.7 dB. In all other cases the performance is more favorable.

The system performance using the improved reference beam is even better than that shown in Figure 11.7 because the postdetection bandwidth is purely a function of the amplitude-modulation bandwidth so that the staring-mode bandwidth is not needed. We reach this conclusion because the reference beam frequency is, at some instant in the scanning process, exactly equal to that of the signal beam, independent of the signal frequencies. Therefore, the information about the spectral content of the signal always occurs at the heterodyne difference frequency f_d.

11.7. THE CROSS-SPECTRUM ANALYZER

We now consider the use of decimated arrays in an application such as cross-spectrum analysis in which we want to measure the angle of arrival of signals to assist in a sorting process. Suppose that we want to detect a burst of energy associated with narrowband radio signals of bandwidth B_h, located somewhere in a total rf bandwidth W; the minimum dwell time of the burst signal is T_h. The receiver consists of two antenna elements, as shown in Figure 11.8. The distance between the two elements is D_{ab} and the received signals at the antenna elements are $s_a(t)$ and $s_b(t)$. Suppose that a source is situated at an angle θ with respect to the boresight of the array. The rf energy arrives at antenna element b a time τ later than it

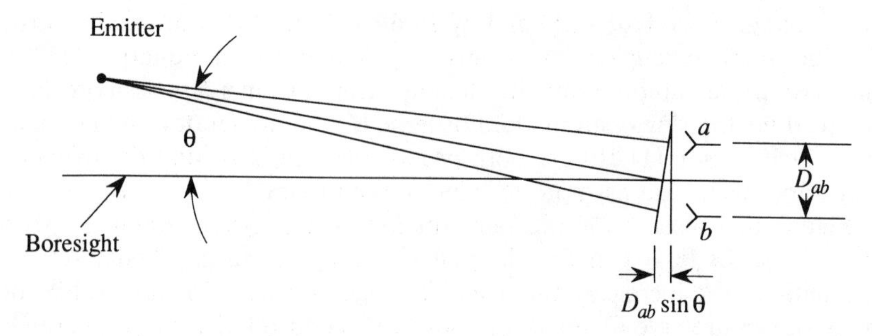

Figure 11.8. Dual-antenna geometry to determine angle of arrival.

arrives at antenna element a. The signals at a and b are thus $s_b(t)$ and $s_a(t - \tau)$, where $\tau = (D_{ab}/c)\sin\theta$ and c is the speed of light.

11.7.1. Cross-Spectrum Analysis with Spatial Heterodyning

One way to measure the angle of arrival is to feed these received signals to the dual-channel acousto-optic cell shown in Figure 11.9. Lens L_2 performs the Fourier transformation in the horizontal direction, centering the Fourier transform at α_c at plane P_2. The side view shows the two channels

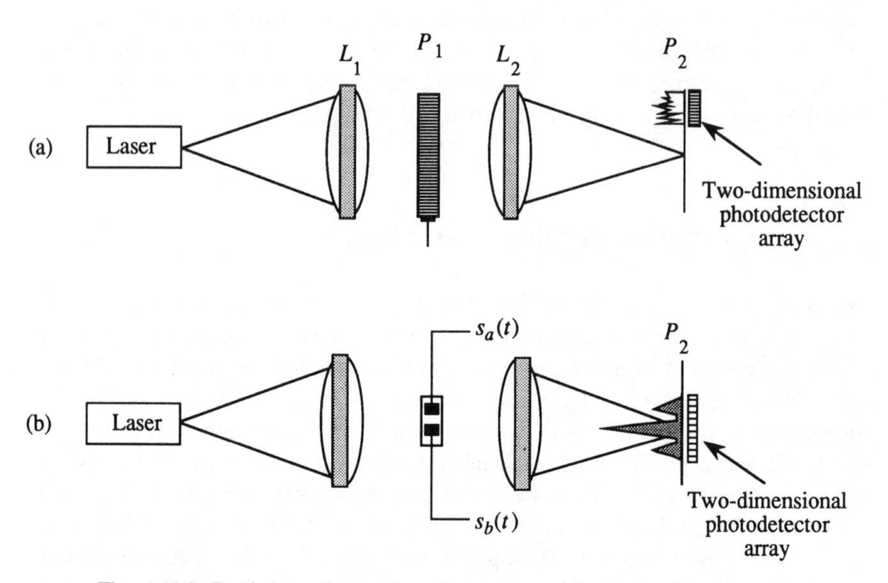

Figure 11.9. Dual-channel acousto-optic processor; (a) top view; (b) side view.

driven by $s_a(t)$ and $s_b(t)$. Lens L_2 also produces the Fourier transform in the vertical direction.

In the horizontal direction we describe the Fourier transform of the first input signal as

$$S_a(\alpha, t) = \int_{-\infty}^{\infty} a(x) s_a\left(t - \frac{T}{2} - \frac{x}{v}\right) e^{j2\pi\alpha x}\, dx. \tag{11.15}$$

We drop the + subscript in favor of using the subscript to identify signals from the two antenna elements; we assume that we retain only the positive diffracted order throughout this section. To obtain the full two-dimensional Fourier transform from the upper channel, we account for the light distribution in the vertical direction by the integral

$$S_a(\beta) = \int_{-\infty}^{\infty} \text{rect}\left[\frac{y - H/2}{h}\right] e^{j2\pi\beta y}\, dy \tag{11.16}$$

$$= \text{sinc}(\beta h) e^{j\pi\beta H}, \tag{11.17}$$

where h is the height of the transducer and H is the distance between the centers of the two channels. The two-dimensional Fourier transform of the upper channel is therefore equal to the product of Equations (11.15) and (11.17):

$$\begin{aligned} S_a(\alpha, \beta, t) &= S_a(\alpha, t) S_a(\beta) \\ &= S_a(\alpha, t)\text{sinc}(\beta h) e^{j\pi\beta H}. \end{aligned} \tag{11.18}$$

For $s_b(t) = s_a(t - \tau)$, we find that the Fourier transform in the horizontal direction for the lower channel is

$$S_b(\alpha, t) = S_a(\alpha, t) e^{j2\pi\alpha v\tau}, \tag{11.19}$$

due to the shift theorem for Fourier transforms. The Fourier transform $S_b(\alpha, t)$ is therefore identical to $S_a(\alpha, t)$, except for a phase term $\phi(\alpha) = 2\pi\alpha v\tau$ that contains the angle-of-arrival information. This result is, in a strict sense, true only if the aperture of the acousto-optic cell is also shifted by τ. But because τ is generally very much less than T, this refinement has little impact on the final results.

By a similar calculation, we find that the Fourier transform in the vertical direction for the lower channel is

$$S_b(\alpha, \beta, t) = S_a(\alpha, t) e^{j2\pi\alpha v\tau}\, \text{sinc}(\beta h) e^{-j\pi\beta H}. \tag{11.20}$$

The total Fourier transform at plane P_2 is the sum of $S_a(\alpha, \beta, t)$ and $S_b(\alpha, \beta, t)$ as given by Equations (11.18) and (11.20). The intensity is the magnitude squared of this sum:

$$I(\alpha, \beta, t) = 2|S_a(\alpha, t)|^2 \operatorname{sinc}^2(\beta h)\{1 + \cos[2\pi H\beta - 2\pi\alpha v\tau]\}. \tag{11.21}$$

This result shows that we have the same output $|S_a(\alpha, t)|^2$ in the α direction as that produced by an instantaneous power spectrum analyzer. In the vertical direction, however, the resultant intensity reveals a sinc(βh) envelope function, due to the height of the transducers, that modulates a *spatial fringe structure* whose frequency is established by the distance H between the transducers. The most important feature of Equation (11.21) is that the *the phase $\phi(\alpha)$ encodes a spatial carrier frequency*.

One way to detect $\phi(\alpha)$ is to use a two-dimensional photodetector array that has N elements in the horizontal direction to sample the spatial frequencies and at least three elements in the vertical direction to sample the phase information. Another detection method is to demodulate the phase by multiplying $I(\alpha, \beta, t)$ with a function $D_c(\beta) = 1 + \cos(h\beta)$ and by integrating the product over all values of β. A second reference function $D_s(\beta) = 1 + \sin(h\beta)$ provides the quadrature component; the details of the quadrature demodulation technique are given in the next section. The disadvantage of these detection schemes is that we need at least two N-element detector arrays to implement the latter scheme and at least one $3 \times N$ element array to implement the former scheme. Furthermore, the available light is not used efficiently; for large N, the arrays may not be sampled often enough to ensure that signals are not missed. We now consider a cross-spectrum analyzer that uses a temporal, instead of a spatial, heterodyne technique to measure the phase. This approach efficiently uses the available light and can do so with fewer photodetector elements.

11.7.2. Cross-Spectrum Analysis with Temporal Heterodyning

The cross-spectrum analyzer is a special case of the heterodyne spectrum analyzer described in Section 11.3. We associate $S_a(\alpha, t)$ and $S_b(\alpha, t)$ with $F_+(\alpha, t)$ and $R_+(\alpha, t)$ so that the heterodyne spectrum analyzer architecture shown in Figure 11.4 is exactly what we need to generate the desired cross-spectral products. The intensity at plane P_4, derived by following the same procedure as in Equation (11.3), is

$$I(\alpha, t)|S_a(\alpha, t) + S_b(\alpha, t)e^{j2\pi f_d t}|^2, \tag{11.22}$$

where $S_a(\alpha, t)$ is given by (11.15). In Equation (11.22) we do not show an explicit functional dependence of the intensity with respect to β because both acousto-optic cells are located on the same vertical axis and no interference occurs in the β direction.

We expand the intensity function to produce

$$I(\alpha, t) = |S_a(\alpha, t)|^2 + |S_b(\alpha, t)|^2 + 2|S_a(\alpha, t)||S_b(\alpha, t)|\cos[2\pi f_d t + \phi(\alpha)]. \tag{11.23}$$

For $s_b(t) = s_a(t - \tau)$, we find that $|S_a(\alpha, t)| = |S_b(\alpha, t)|$ so that

$$\boxed{I(\alpha, t) = 2|S_a(\alpha, t)|^2\{1 + \cos[2\pi f_d t + \phi(\alpha)]\},} \tag{11.24}$$

where $\phi(\alpha, t) = 2\pi\alpha v\tau$ is the phase we seek to measure. The phase term is now associated with a sinusoidal function of time instead of a sinusoidal function of space. Therefore, we need to demodulate $I(\alpha, t)$ in time instead of demodulating $I(\alpha, \beta, t)$ in space as described in Section 11.7.1. As a consequence, we must use discrete-element photodetectors that have sufficient bandwidth to pass the desired signal whose energy is centered at f_d.

As before, we identify three modes of operation: the staring mode, the fast-scan mode and the slow-scan mode. To illustrate the phase-detection technique for all three modes, consider the simple case of energy received from a cw emitter with frequency f_k and magnitude c_k, arriving at angle θ_k with respect to the boresight of the antenna array. The signals from the two antenna elements then become

$$s_a(t) = c_k \cos(2\pi f_k t + \gamma_k), \tag{11.25}$$

and

$$s_b(t) = c_k \cos[2\pi f_k(t - \tau) + \gamma_k]. \tag{11.26}$$

We can set the phase γ_k equal to 0 without loss of generality. From Equations (11.15) and (11.22) we find that the crossproduct term is expressed as

$$g_3(t) = 2c_k^2 \cos(2\pi f_d t + \phi_k) \int_{-\infty}^{\infty} |A(\alpha - \alpha_k + \alpha_c)|^2 P(\alpha)\, d\alpha, \quad (11.27)$$

where $\phi_k = 2\pi f_k t$ is the phase to be measured and $P(\alpha)$ is the photodetector response defined before. The integral is simply a magnitude scaling factor that is combined with several others. For example, we need to account for factors such as the optical efficiency of the system, the acousto-optic cell diffraction efficiency, the beamsplitter and beam-combiner ratios, and the laser power. The output term is then conveniently written as

$$\boxed{g_3(t) = A_0 c_k^2 \cos(2\pi f_d t + \phi_k),} \quad (11.28)$$

where A_0 accounts for all magnitude scaling factors.

The problem of extracting the phase and magnitude information from the temporal carrier f_d is similar to that of demodulating an angle-modulated signal in a communication system. One way to demodulate the signal is through quadrature detection, which recovers the phase angle ϕ_k as shown in Figure 11.10. We first eliminate the baseband signals $g_1(t)$ and $g_2(t)$, produced by the intensity terms $I_1(\alpha, t)$ and $I_2(\alpha, t)$, with a bandpass filter centered at f_d. We then multiply $g_3(t)$ by $2\cos(2\pi f_d t)$ to

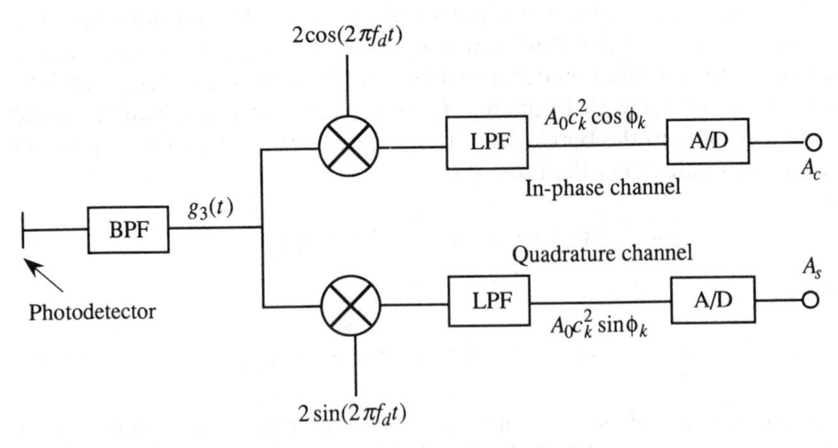

Figure 11.10. Postdetection processing.

obtain the in-phase component of the signal and by $2\sin(2\pi f_d t)$ to obtain the quadrature component of the signal. Because the frequency f_d is derived from the same signal source used to drive the temporal modulator at plane P_2 of Figure 11.4, we guarantee that any instabilities or drifts in the frequency f_d are automatically compensated. The results are then low-pass filtered to provide the signals $A_0 c_k^2 \cos\phi_k$ and $A_0 c_k^2 \sin\phi_k$. After an analog-to-digital conversion, the arctangent operation is applied to obtain the phase ϕ_k; the magnitude c_k is obtained by calculating the square root of the sum of the squares of the output signals A_s and A_c. Finally, the angle of arrival θ is computed from the relationship that $\phi_k = 2\pi f_k \tau$ and $\tau = (D_{ab}/c)\sin\theta$.

There are several ways to implement the circuitry in hardware to extract the phase information, including the use of quadrature detectors, limiter/discriminators, or phase-lock loops. Although these circuits are available in small packages, the challenge of implementing the electronics for 1000–2000 discrete detectors is formidable. The use of a decimated array coupled with a scanning of the spectrum similar to that described before can significantly reduce the hardware complexity.

The significant difference between the cross-spectrum analyzer and the heterodyne spectrum analyzer is that the reference waveform $R_+(\alpha, t)$ in the heterodyne spectrum analyzer serves as a local oscillator that produces a current $\bar{i}_2$ that dominates the dark current i_d. By concentrating all the reference power into exactly M probes, we optimize the system performance. In the cross-spectrum analyzer, however, we do not have an independent reference beam. Instead, the magnitude of the beams from the two branches of the interferometer are identical so that if the received signal magnitude decreases, it does so in both branches of the interferometer. The result is that the performance of the cross-spectrum analyzer more nearly resembles that of a power spectrum analyzer than a heterodyne spectrum analyzer, as suggested by Equation (11.24). There are, however, some significant differences, as Figure 11.11 shows. First, we note that the dynamic range decreases at a slower rate for the fast-scan mode because for any value of M, the *loss* in dynamic range is a factor of 2 less in decibels for the cross-spectrum analyzer versus the heterodyne spectrum analyzer. The reason is that the system loss due to scanning is *allocated between the two input signals* in the cross-spectrum analyzer so that a *system loss* of 6 dB results in a *dynamic range loss* of 3 dB. For the heterodyne spectrumanalyzer, the full system loss is allocated to the received signal because the reference signal has a fixed magnitude. As a result, the cross-spectrum analyzer performance deteriorates less quickly, as a function of the decimation ratio, than does the heterodyne spectrum analyzer performance.

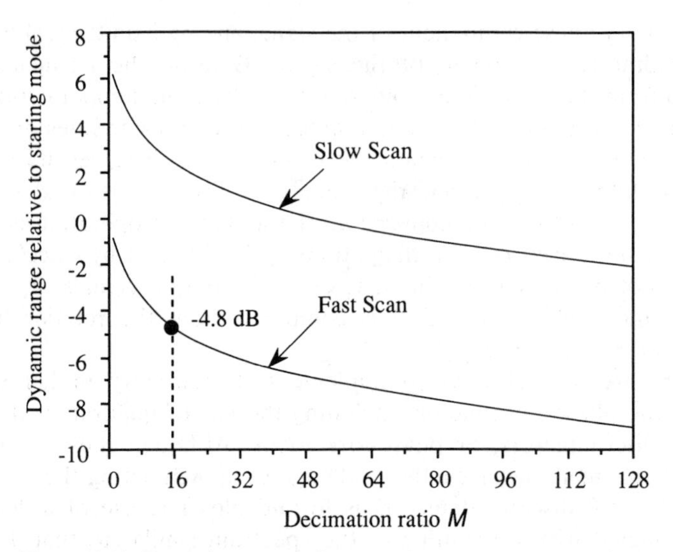

Figure 11.11. Dynamic range performance for cross-spectrum analyzer.

Finally, the slow-scan-mode results are 7 dB better than those for the fast-scan mode because the postdetection bandwidth is simply B_{am}. These results show that rather large savings are made in the photodetector circuitry for more modest penalties in performance. For example, for $M = 32$ we need only 64 photodetector elements instead of 2000. If we are in the fast-scan mode, we can equal the staring-mode performance if the laser power is increased by 6 dB; of course, the dynamic range is reduced by 6 dB if we are already laser-power limited. A 6-dB loss in dynamic range may, however, be a good trade for a 32-fold reduction in the complexity of the readout circuitry.

PROBLEMS

11.1. Suppose that you have two antenna elements that are one meter apart and that there is a cw emitter whose frequency is 200 MHz located 30 mrad off boresight. You build a dual-channel acousto-optic cell with the transducers being 200 μ in height and separated by a center-to-center distance of 2 mm. The system is operated in the *spatial* carrier frequency mode. Calculate the number of fringes between the first zeros of the sinc function in the vertical direction if the Fourier-transform lens has a focal length of 100 mm and

$\lambda = 0.5\ \mu$. How many degrees will the fringe pattern shift for the conditions given relative to the fringe pattern when the emitter is on boresight?

11.2. Calculate the optimum distance between photodetectors for the parameters of Problem 11.1.

11.3. Suppose that the system parameters are as given in Problem 11.1, except that the system is operated in the *temporal* carrier frequency mode with an offset frequency $f_d = 10$ MHz. How many degrees will the fringe pattern shift, for the conditions given, relative to the fringe pattern when the emitter is on boresight? Compare this answer with that from Problem 11.1.

11.4. For the parameters of Problem 11.3, what is the unambiguous angular range for the emitter? That is, how far off boresight can the emitter move before the phase, as measured by the optical system, reaches $\pm\pi$? What steps might you take to improve the unambiguous angular range? List all the relevant variables including the geometry of the array, the emitter frequency, etc., and state how an increase (or decrease) in the variables affects the unambiguous angular range. Provide a consistent set of parameters that provide an unambiguous angular range of just $\pm 15°$.

12

The Heterodyne Transform and Signal Excision

12.1. INTRODUCTION

The generalized concept of the heterodyne transform provides a wealth of insights into how an important class of optical signal-processing systems work. In Chapter 9 we introduced heterodyne-detection concepts that led to a discussion of the heterodyne spectrum analyzer in Chapter 10 and to the use of decimated arrays in Chapter 11. In this chapter we consider generalized heterodyne systems which can be applied to a wide range of signal-processing problems. After we introduce the heterodyne-transform, we discuss applications such as probing three-dimensional light fields and signal excision.

12.2. THE HETERODYNE TRANSFORM

The *heterodyne transform* leads to an unusual method for recovering a signal; we simply integrate the light intensity produced by the sum of the signal and reference functions at the Fourier plane. Consider the interferometer, shown in Figure 12.1, in which lens L_3 in the reference branch creates a point source $r(x, t)$ at some point x_0 at plane P_2. Lens L_2 creates the Fourier transform $R_+(\alpha, t)$ of the reference beam at plane P_3. We express the reference function in terms of a time variable to allow for general filtering operations later. The acousto-optic cell in the signal beam has length L and is driven by $f(t) = s(t)\cos(2\pi f_c t)$. As usual, the signal corresponding to the positive diffracted order is expressed as

$$f_+(x, t) = a(x) s\left(t - \frac{T}{2} - \frac{x}{v}\right) e^{j2\pi f_c(t - T/2 - x/v)}, \qquad (12.1)$$

where the notation of Chapter 7 is used. The jm factor that usually appears in $f_+(x, t)$ is absorbed into $a(x)$. Lens L_2 produces the Fourier

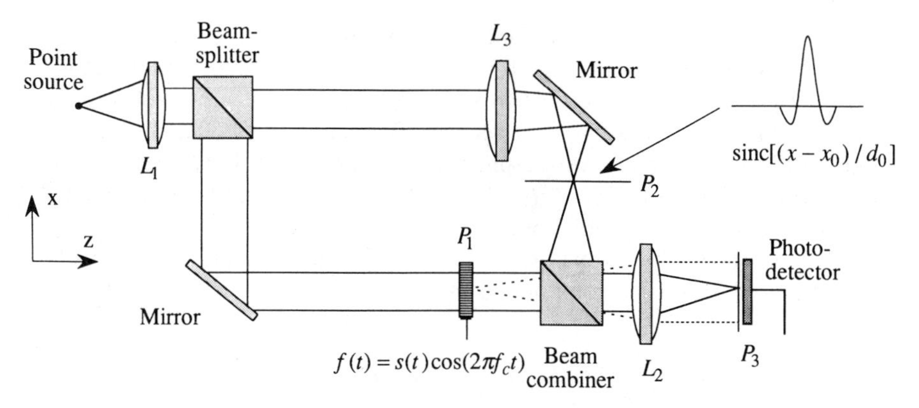

Figure 12.1. Optical system for illustrating the heterodyne transform.

transform $F_+(\alpha, t)$ of $f_+(x, t)$ at plane P_3:

$$F_+(\alpha, t) = e^{j2\pi f_c(t - T/2)} \int_{-\infty}^{\infty} a(x) s\left(t - \frac{T}{2} - \frac{x}{v}\right) e^{j2\pi(\alpha - \alpha_c)x}\, dx. \quad (12.2)$$

The transform $F_+(\alpha, t)$ is centered at α_c at plane P_3 and spans an interval of spatial frequencies $2\alpha_{co}$, where α_{co} corresponds to the cutoff frequency f_{co} of the baseband signal $s(t)$.

The intensity at plane P_3 is

$$I(\alpha, t) = |F_+(\alpha, t) + R_+(\alpha, t)|^2. \quad (12.3)$$

We place a single-element photodetector at plane P_3 and represent it by

$$\begin{aligned} P(\alpha) &= \text{rect}[(\alpha - \alpha_c)/2\alpha_{co}] \\ &= \text{rect}[(\alpha - \alpha_c) d_0], \end{aligned} \quad (12.4)$$

where $\alpha_{co} = 1/(2d_0)$ and d_0 is the Nyquist sampling interval for the baseband signal. The photocurrent, obtained by integrating the intensity over the photodetector surface, is proportional to

$$g(t) = \int_{-\infty}^{\infty} I(\alpha, t) P(\alpha)\, d\alpha. \quad (12.5)$$

We claim that $g(t)$ *is proportional to* $f(t)$. The Fourier-transform relationship implied by Equation (12.5) is not, however, immediately obvious.

First, $I(\alpha, t)$ is an intensity function, not an amplitude function and second, a kernel function of the form $\exp(j2\pi\alpha x)$ is missing from what we expect to be a spatial Fourier-transform relationship. It seems intuitively unlikely, therefore, that we can recover $f(t)$ simply by collecting all the light at the Fourier plane with a photodetector. Nevertheless, we make the assertion that the integral of $I(\alpha, t)$ over the total spatial frequency span produces a signal from which we recover $f(t)$.

To see how the transform is created, we expand $I(\alpha, t)$ from Equation (12.3) into its component parts:

$$g(t) = \int_{-\infty}^{\infty} \{|F_+(\alpha, t)|^2 + |R_+(\alpha, t)|^2 \\ + 2\,\mathrm{Re}[F_+(\alpha, t) R_+^*(\alpha, t)]\} P(\alpha)\, d\alpha. \tag{12.6}$$

The cross-product term of $g(t)$ contains the useful information:

$$g_3(t) = 2\,\mathrm{Re}\left\{\int_{-\infty}^{\infty} F_+(\alpha, t) R_+^*(\alpha, t) P(\alpha)\, d\alpha\right\}. \tag{12.7}$$

The reference branch of the interferometer contains a point source whose position, as seen by an observer at plane P_3, is a distance x_0 from the optical axis at plane P_1. We rotate the beam combiner through an angle θ_c so that the Fourier transform of the reference beam is centered at α_c, the center frequency of the signal spectrum. This rotation, at the plane of the reference signal, is represented as $\exp(-j2\pi\alpha_c x)$ so that

$$R_+(\alpha, t) = \int_{-\infty}^{\infty} \mathrm{sinc}[(x - x_0)/d_0] e^{j2\pi(\alpha - \alpha_c)x}\, dx \\ = \frac{1}{2\alpha_{co}} \mathrm{rect}[(\alpha - \alpha_c)/2\alpha_{co}] e^{j2\pi(\alpha - \alpha_c)x_0}. \tag{12.8}$$

The result given in Equation (12.8) represents a plane wave, bounded by $(\alpha_c - \alpha_{co}) \le \alpha \le (\alpha_c + \alpha_{co})$, directed upward and to the right at plane P_3. We take steps to ensure that the reference-beam power is efficiently used by selecting the focal length of lens L_3 such that, when operating with lenses L_1 and L_2, it constrains the light to the interval $2\alpha_{co}$ at plane P_3. An equivalent way to state this condition is that the point source at plane P_2 has the same value for d_0 as the sample function for the signal at plane P_1.

We substitute $R_+^*(\alpha, t)$ and $F_+(\alpha, t)$ into $g_3(t)$ from Equation (12.7) to find that

$$g_3(t) = \frac{1}{\alpha_{co}} \operatorname{Re}\left\{ e^{j2\pi f_c(t-T/2)} \int_{-\infty}^{\infty} \int_{-\infty}^{\infty} a(x) s\left(t - \frac{T}{2} - \frac{x}{v}\right) e^{-j2\pi\alpha_c x} \right.$$
$$\left. \times \left\{ \operatorname{rect}[(\alpha - \alpha_c) d_0] e^{-j2\pi(\alpha - \alpha_c)x_0} \right\} e^{j2\pi\alpha x} \, dx \, d\alpha \right\}. \quad (12.9)$$

In Equation (12.9) we set $P(\alpha) = \operatorname{rect}[(\alpha - \alpha_c)d_0]$ because the size of the photodetector is essentially governed by the extent of the reference beam at plane P_3 as given by Equation (12.8). The integral over α is performed first and becomes

$$\int_{-\infty}^{\infty} \frac{1}{2\alpha_{co}} \operatorname{rect}[(\alpha - \alpha_c) d_0] e^{j2\pi(\alpha - \alpha_c)(x - x_0)} \, d\alpha = \operatorname{sinc}[(x - x_0)/d_0]. \quad (12.10)$$

We use Equation (12.10) in Equation (12.9) so that the cross-product term becomes

$$g_3(t) = 2 \operatorname{Re}\left\{ e^{j2\pi f_c(t-T/2)} \int_{-\infty}^{\infty} a(x) s\left(t - \frac{T}{2} - \frac{x}{v}\right) \operatorname{sinc}[(x - x_0)/d_0] \, dx \right\}. \quad (12.11)$$

Because the input signal $s(t)$ is strictly bandlimited to $|f| \le f_{co}$, the sinc function is an acceptable form of a delta function, as we showed in Chapter 3, Section 3.5.1. We perform the sifting operation by integrating on x and take the real part of the result to find that

$$\boxed{g_3(t) = a(x_0) s(t - \tau_0) \cos[2\pi f_c(t - T/2)],} \quad (12.12)$$

where $\tau_0 = T/2 + x_0/v$ is a time delay that is dependent on the position of the reference-beam probe. Note that the baseband modulating signal $s(t - \tau_0)$ contained in $g_3(t)$ has exactly the same form as the input signal $s(t)$ as claimed, but it is time delayed by an amount τ_0. The multiplicative constant $a(x_0)$ is simply the signal-beam aperture function evaluated at the equivalent probe position x_0.

The result given by Equation (12.12) is one of the most unexpected and surprising ones in optical signal processing. Even more remarkable is that the axial position of the photodetector is irrelevant, as we show in Section 12.4, provided that it is placed somewhere after the signal and reference beams have been combined. As a partial answer to the mystery of how Equation (12.5) could possibly represent a Fourier transform, note that the missing kernel function is contained in $F_+(\alpha, t)$, as suggested by Equation (12.2).

We note, in passing, that the optical system of Figure 12.1 may have application as a broadband rf delay line. The time delay can vary from $\tau_0 = 0$ when $x_0 = -L/2$ to $\tau_0 = T$ when $x_0 = L/2$. For the acousto-optic cells commonly used, it is possible to obtain delays in the 0–60-μsec range for a 50-MHz bandwidth signal, in the 0–10-μsec range for a 100-MHz bandwidth signal, or in the 0–2-μsec range for a 500-MHz bandwidth signal.

12.3. THE TEMPORAL FREQUENCY RANGE OF THE BASEBAND TERMS

Having demonstrated that the Fourier transform can indeed be obtained from the third term of Equation (12.6), we must prove that the first two terms of $g(t)$ do not contain frequency components that interfere with the recovery of $g_3(t)$. Consider the second term of Equation (12.6):

$$\begin{aligned} g_2(t) &= \int_{-\infty}^{\infty} |R_+(\alpha, t)|^2 P(\alpha)\, d\alpha, \\ &= \int_{-\infty}^{\infty} \left|\text{rect}[(\alpha - \alpha_c)/2\alpha_{co}] e^{j2\pi\alpha x_0}\right|^2 P(\alpha)\, d\alpha. \end{aligned} \tag{12.13}$$

Because the photodetector spans the same frequency range as the reference beam, the integral is simply

$$\boxed{\begin{aligned} g_2(t) &= \int_{-\infty}^{\infty} \text{rect}[(\alpha - \alpha_c)/2\alpha_{co}]\, d\alpha \\ &= \int_{-\alpha_{co}}^{\alpha_{co}} d\alpha = 2\alpha_{co}. \end{aligned}} \tag{12.14}$$

Because the output $g_2(t)$ is not dependent on time, its spectral energy is concentrated at zero frequency, as shown in Figure 12.2. As a result, it will

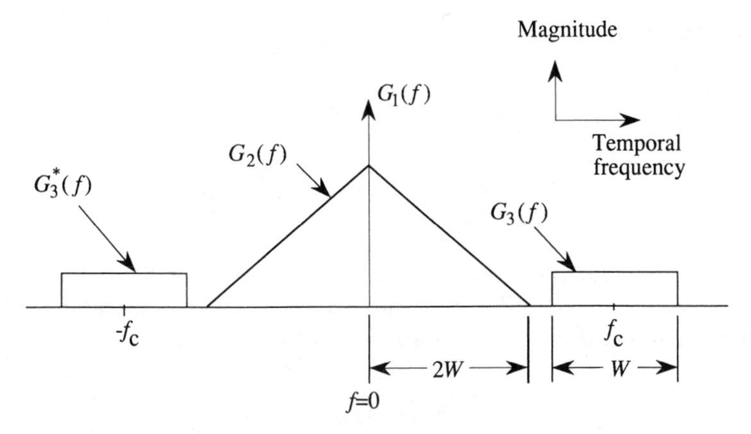

Figure 12.2. Temporal frequency content of interferometer output.

not survive the postdetection bandpass filter and will not corrupt the desired signal $g_3(t)$.

The first term of Equation (12.6), due to the signal beam, is given by

$$g_1(t) = \int_{-\infty}^{\infty} |F_+(\alpha, t)|^2 P(\alpha)\, d\alpha$$
$$= \int_{-\infty}^{\infty} \left| \int_{-\infty}^{\infty} a(x) s\left(t - \frac{T}{2} - \frac{x}{v}\right) e^{j2\pi f_c(t - T/2 - x/v)} e^{j2\pi \alpha x}\, dx \right|^2 P(\alpha)\, d\alpha. \tag{12.15}$$

We expand the squared magnitude of the integral in the integrand by introducing a dummy variable y:

$$g_1(t) = \int_{-\infty}^{\infty} \int_{-\infty}^{\infty} a(x) s\left(t - \frac{T}{2} - \frac{x}{v}\right) e^{j2\pi f_c(t - T/2 - x/v)} e^{j2\pi \alpha x}\, dx$$
$$\times \int_{-\infty}^{\infty} a^*(y) s^*\left(t - \frac{T}{2} - \frac{y}{v}\right) e^{-j2\pi f_c(t - T/2 - y/v)} e^{-j2\pi \alpha y}\, dy\, P(\alpha)\, d\alpha. \tag{12.16}$$

Because the product of $\exp[j2\pi f_c(t - T/2)]$ and its conjugate is unity, the carrier frequency f_c does not survive; the spectrum of $g_1(t)$ must therefore be at baseband. We are still concerned, however, about the total bandwidth of this signal. We rearrange the factors of Equation (12.16) and collect the terms in α:

$$g_1(t) = \int_{-\infty}^{\infty}\int_{-\infty}^{\infty} a(x)a^*(y)s\left(t - \frac{T}{2} - \frac{x}{v}\right) s^*\left(t - \frac{T}{2} - \frac{y}{v}\right)$$
$$\times \left\{\int_{-\infty}^{\infty} e^{j2\pi(\alpha-\alpha_c)(x-y)}P(\alpha)\, d\alpha\right\} dx\, dy. \tag{12.17}$$

The single-element photodetector centered at α_c collects all the light in the positive diffracted order. We use Equation (12.4) in Equation (12.17) so that the integral on α from Equation (12.17) becomes

$$\int_{-\infty}^{\infty} e^{j2\pi(\alpha-\alpha_c)(x-y)} \operatorname{rect}[(\alpha - \alpha_c)/2\alpha_{co}]\, d\alpha. \tag{12.18}$$

We change variables, by letting $\alpha - \alpha_c = z$ so that Equation (12.18) becomes

$$\int_{-\alpha_{co}}^{\alpha_{co}} e^{j2\pi z(x-y)}\, dz = \operatorname{sinc}[(x - y)/d_0]. \tag{12.19}$$

Because Equation (12.19) is a suitable form of a delta function, we use it in Equation (12.17) and apply the sifting theorem to find that

$$\boxed{g_1(t) = \int_{-\infty}^{\infty} |a(x)|^2 \left|s\left(t - \frac{T}{2} - \frac{x}{v}\right)\right|^2 dx.} \tag{12.20}$$

We recognize that $g_1(t)$ results from a convolution operation in which the integration is over the space variable and $|a(x)|^2$ plays the role of the system impulse response. Because the spatial extent of $a(x)$ is much greater than that of the finest spatial detail of $|s(t - T/2 - x/v)|^2$, it behaves as a broad smoothing function in the space domain.

The temporal Fourier transform of $g_1(t)$ is

$$G_1(f) = \int_{-\infty}^{\infty} g_1(t)e^{-j2\pi ft}\, dt$$
$$= \int_{-\infty}^{\infty} |a(x)|^2 \left\{\int_{-\infty}^{\infty} \left|s\left(t - \frac{T}{2} - \frac{x}{v}\right)\right|^2 e^{-j2\pi ft}\, dt\right\} dx. \tag{12.21}$$

The Fourier transform of the magnitude squared of a signal is the convolution of the spectrum of the signal:

$$G_1(f) \propto s(f) * s^*(-f). \tag{12.22}$$

If $S(f)$ has a two-sided spectrum that extends over $|f| \leq W$, the two-sided spectrum of $G_1(f)$ is bounded by $|f| \leq 2W$. Therefore, the carrier frequency must satisfy the inequality that $f_c \geq 3W$, as shown in Figure 12.2.

12.4. PROBING ARBITRARY THREE-DIMENSIONAL FIELDS

The powerful concept of considering the reference beam as a probe, introduced in Chapter 9, is also useful for understanding the heterodyne transform. This concept was developed by Whitman and Korpel for visualizing coherent light fields and probing acoustic surface perturbations (97). In applying this concept to the heterodyne transform, we must answer the important question of where the reference beam overlaps the signal beam. The reference signal has the form of a sinc function whose Fourier transform covers the frequencies spanned by $f(x, t)$. If we observe the input through lens L_3 in Figure 12.3(a), we find that the reference probe overlaps the signal $f(x, t)$ only at $x = x_0$. Because the probe is stationary and because the signal moves at velocity v, Figure 12.3(b) shows that the output of the interferometer is the convolution of $f(x, t)$ and the reference beam. This convolution produces the purely time function $f(t - \tau_0)$.

Suppose, for the moment, that the signal $f(x, t)$ is not a function of time but is a two-dimensional coherent distribution in space. Whitman and Korpel showed that when the reference beam is scanned throughout a

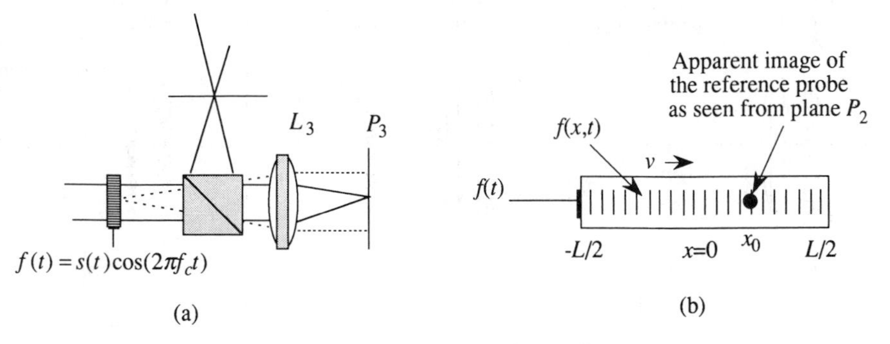

Figure 12.3. Acousto-optic cell and reference-beam geometry.

volume of space, a signal $g_3(t)$ is produced that can be displayed on a TV monitor to visualize the light magnitude throughout the volume. In a remarkable set of experiments, they showed that as the position of the probing beam changes along the optical axis, the display changes from an image of the object, through a succession of Fresnel transforms of the object, and eventually to the Fourier transform of the object.

The position of the photodetector relative to the probe is not important except in applications where filtering operations are formed at the Fourier plane, provided that the photodetector receives all the light in the combined beams. The situation we have just studied is shown in Figure 12.4(a), where we have removed the unnecessary optical components and propagated the reference beam backward towards plane P_1 to show its *apparent* position at x_0 as dotted lines. The photodetector is placed at plane P_3 and centered at α_c so that it collects only the diffracted light.

In Figure 12.4(b), the photodetector has been moved to a plane intermediate between plane P_3 and lens L_3; its size has been increased so that it still collects all the diffracted light, as well as some of the undiffracted light. It is clear that the photodetector is no longer in the Fourier plane of the system; instead, it is in a Fresnel transform plane. Figure 12.4(c) shows the photodetector adjacent to the acousto-optic cell where it collects all of the light, both the diffracted and the undiffracted light. In fact, we do not even use lens L_3, in this case, so that the Fourier transform exists only in terms of an angular spectrum.

We now use the heterodyne transform to show that *all three cases are equivalent and that the cross product $g_3(t)$ of the photodetector output is the same in all cases*. Suppose that the photodetector is placed at *any* plane P_4 positioned a distance D to the right of plane P_3 in Figure 12.1. The

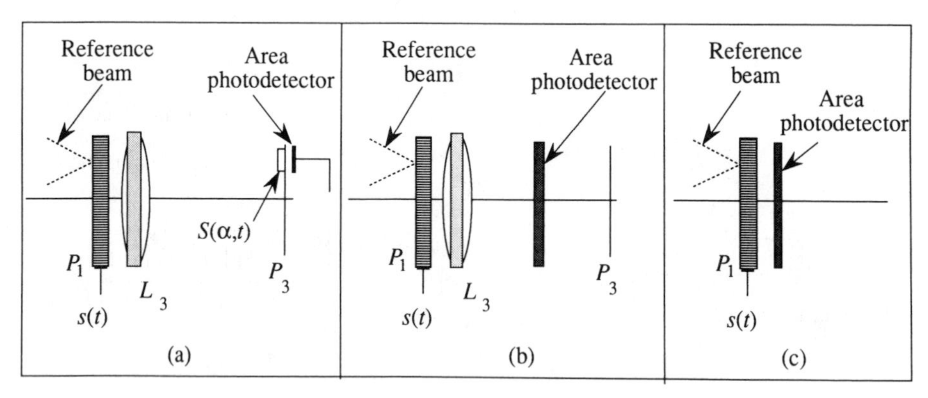

Figure 12.4. Equivalent heterodyne-detection systems.

new light distribution $F(\gamma, t)$ is obtained from a Fresnel transform of $F_+(\alpha, t)$:

$$F_+(\gamma, t) = \int_{-\infty}^{\infty} F_+(\alpha, t) e^{-j(\pi/\lambda D)(\gamma-\alpha)^2} \, d\alpha, \tag{12.23}$$

where γ is the coordinate at plane P_4 and we ignore scaling factors. We expand the exponential to find that

$$F_+(\gamma, t) = \int_{-\infty}^{\infty} F_+(\alpha, t) e^{-j(\pi\gamma^2/\lambda D)} e^{-j(\pi\alpha^2/\lambda D)} e^{j(2\pi\gamma\alpha/\lambda D)} \, d\alpha. \tag{12.24}$$

In a similar fashion, the reference beam becomes

$$R_+(\gamma, t) = \int_{-\infty}^{\infty} R_+(\alpha, t) e^{-j(\pi\gamma^2/\lambda D)} e^{-j(\pi\alpha^2/\lambda D)} e^{j(2\pi\gamma\alpha/\lambda D)} \, d\alpha. \tag{12.25}$$

A photodetector at plane P_4 detects

$$h(t) = \int_{-\infty}^{\infty} \left| F_+(\gamma, t) + R_+(\gamma, t) \right|^2 P(\gamma) \, d\gamma, \tag{12.26}$$

where $P(\gamma)$ defines the photodetector boundaries and ensures that all the light is collected. We substitute $F_+(\gamma, t)$ and $R_+(\gamma, t)$ into Equation (12.26), using dummy variables as appropriate.

Consider first the cross-product term $h_3(t)$:

$$h_3(t) = 2\,\mathrm{Re}\left\{ \int_{-\infty}^{\infty} \int_{-\infty}^{\infty} F_+(\alpha, t) e^{-j(\pi\gamma^2/\lambda D)} e^{-j(\pi\alpha^2/\lambda D)} e^{j(2\pi\gamma\alpha/\lambda D)} \, d\alpha \right.$$

$$\left. \times \int_{-\infty}^{\infty} R_+^*(\beta, t) e^{j(\pi\gamma^2/\lambda D)} e^{j(\pi\beta^2/\lambda D)} e^{-j(2\pi\gamma\beta/\lambda D)} \, d\beta \, P(\gamma) \, d\gamma \right\}. \tag{12.27}$$

We integrate the terms in γ first to produce

$$\int_{-\infty}^{\infty} P(\gamma) e^{j(2\pi(\alpha-\beta)\gamma/\lambda D)} \, d\gamma = \mathrm{sinc}\big[(\alpha - \beta)\, d_0\big], \tag{12.28}$$

where the sinc function has the properties of a delta function because the photodetector collects all the light produced by the signal; the size of the photodetector is changed to ensure this condition. We use Equation (12.28) in Equation (12.27) and perform the sifting operation on

β to find that

$$h_3(t) = 2\,\mathrm{Re}\left\{\int_{-\infty}^{\infty} F_+(\alpha,t)R_+^*(\alpha,t)P(\alpha)\,d\alpha\right\}, \tag{12.29}$$

which is identical to $g_3(t)$ as given by Equation (12.7). This analysis proves, therefore, that the position of the photodetector is not important. Similar arguments apply, of course, for the photodetector positions shown in Figure 12.4(b) and Figure 12.4(c).

There is, however, a subtle difference between $h_3(t)$ and $g_3(t)$ as given by Equations (12.29) and (12.7). The difference is a time-delay factor $t_D = D/c$, where c is the velocity of light and D is the distance between planes P_3 and P_4. Therefore it is more appropriate to say that

$$h_3(t) = g_3(t - t_D). \tag{12.30}$$

By a similar analysis, we can show that

$$h_1(t) = \int_{-\infty}^{\infty} |R_+(\gamma, t - t_D)|^2 P(\gamma)\,d\gamma = g_1(t - t_D), \tag{12.31}$$

and that

$$h_2(t) = \int_{-\infty}^{\infty} |F_+(\gamma, t - t_D)|^2 P(\gamma)\,d\gamma = g_2(t - t_D). \tag{12.32}$$

We have shown, then, that the photodetector output, to within a small time delay, is the same for every photodetector position. In most applications that we consider, the value of t_D is not important; it is dropped from our notation for the same reasons that we have dropped unimportant phase terms. Generally the maximum value of t_D is very much less than $1/W$ so that we can consider the envelope of the propagating light wave to have the same temporal variation everywhere in the system. For example, if we are processing a signal with a 50-MHz bandwidth, we find that one cycle of the modulation waveform has a time period of 20 nsec which represents an optical path length of about 6 m. If the optical system is of the order of 100 mm in total length, we say that the same temporal information exists everywhere in the optical system simultaneously. At the other extreme, if the signal bandwidth is of the order of 1 GHz, the period is reduced to 1 nsec (300 mm) so that we may need to account for the time delay t_D in applications such as those involving feedback.

The results given in this section are a generalization of the *conservation of cross-power* theorem which states that

$$\boxed{\int_{-\infty}^{\infty} f(x)g^*(x)\,dx = \int_{-\infty}^{\infty} F(\alpha)G^*(\alpha)\,d\alpha.} \tag{12.33}$$

This theorem is also a generalization of Parseval's theorem. As we showed in Chapter 5, Problem 5.2, we find that the generalization includes Fresnel transforms so that

$$\boxed{\int_{-\infty}^{\infty} f(x)g^*(x)\,dx = \int_{-\infty}^{\infty} F(\alpha)G^*(\alpha)\,d\alpha = \int_{-\infty}^{\infty} F(\xi)G^*(\xi)\,d\xi,} \tag{12.34}$$

where $F(\xi)$ is the Fresnel transform of $f(x)$.

12.5. SIGNAL EXCISION

In communication systems it is fairly common that a wideband signal is corrupted by one or more strong narrowband interference signals. These narrowband jammers may be intentional, created by an adversary who wishes to render a receiver inoperative; or they may be unintentional, such as TV or radio stations. We often want to remove these narrowband interference signals before further processing the desired wideband signal. In some cases, the narrowband interference signals are fixed in frequency. In other cases, they may randomly hop from one frequency to another.

The received signal, including one or more narrowband jammers, is

$$f(t) = s(t) + \sum A_j \cos(2\pi f_j t + \phi_j), \tag{12.35}$$

where $s(t)$ is the wide bandwidth signal we wish to retain and where the A_j, f_j, and ϕ_j are the magnitudes, frequencies and phases of the narrowband jammers. Figure 12.5(a) shows the optical system we described before, except that here we use a spatial light modulator to provide an arbitrary filter function $H(\alpha)$ in the Fourier plane P_3, just before the photodetector. The input signal $f(t)$, given by Equation (12.35) and illustrated in Figure 12.5(b), is dominated by two strong narrowband interference signals so that the wideband signal $s(t)$ is not obvious in $f(t)$.

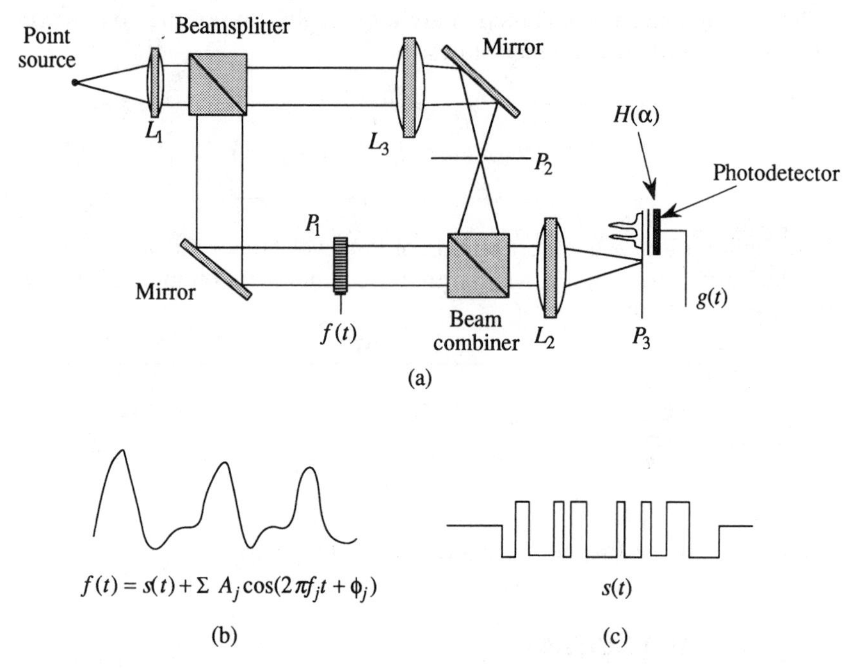

Figure 12.5. Signal excision or notch filtering: (a) optical system, (b) signal dominated by jammers, and (c) signal after notch filtering.

We want to excise the jammers to increase the output signal to noise ratio, as shown in Figure 12.5(c), in which the desired signal $s(t)$ is obvious.

Figure 12.6(a) shows the spectral content of the received signal, with the energy corresponding to the strong narrowband interference signals concentrated at two spatial frequencies. We excise the offending narrowband interferers with a filter whose response is shown in Figure 12.6(b), a process sometimes called *notch filtering*. Because the jammer signal energy is concentrated in small spatial regions in the Fourier plane, it is the most effective plane at which to notch the jammers. This application clearly shows why placing the photodetector at, or just beyond, the Fourier-transform plane is advantageous (106).

A typical filter function $H(\alpha)$ for a single narrowband interference signal at f_j has the idealized shape given in Figure 12.7. We define the desired filter function as

$$H(\alpha) = \text{rect}\big[(\alpha - \alpha_c)/2\alpha_{\text{co}}\big] - \text{rect}\big[(\alpha - \alpha_j)/\alpha_n\big], \quad (12.36)$$

where α_{co} is the cutoff spatial frequency and α_n is the full spatial

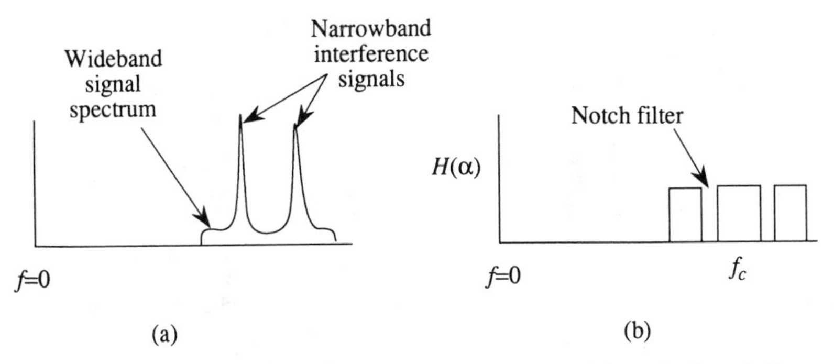

Figure 12.6. Signal excision: (a) narrowband interferers and (b) required notch filter.

frequency span of the notch. The first term of $H(\alpha)$ defines the system bandwidth, and the second term defines a notch that is centered at α_j and has a width α_n. The high time bandwidth product of the system allows for many notches to be set in a parallel processing environment.

The ideal spatial light modulator is one that could form many notches with an infinite on/off ratio. Some candidate devices are liquid crystal light valves, magneto-optic devices, or membrane light modulators, as discussed in Chapter 4. In practice, these devices are limited to intensity contrast ratios of the order of 10^3, although some techniques have been developed to increase the contrast ratio (107). For the moment, we assume that the contrast ratio is infinite.

In general, we use three spatial light modulator elements to notch an isolated frequency. The notching strategy is also affected by the aperture weighting function $a(x)$ whose response in the Fourier plane is $A(\alpha)$. Suppose that we have a notch filter, as shown in Figure 12.8(a), and a

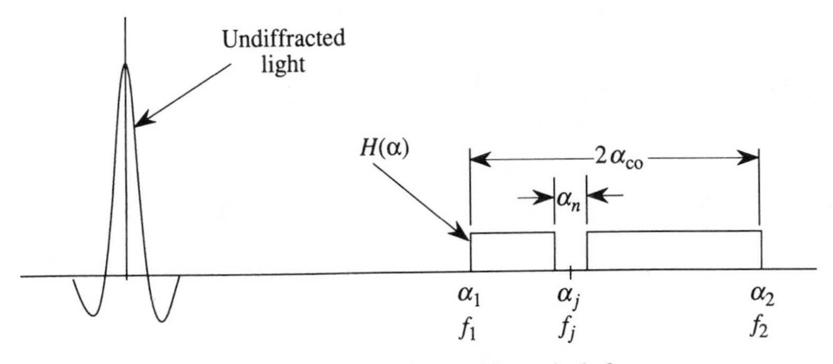

Figure 12.7. Filter function for notching a single frequency.

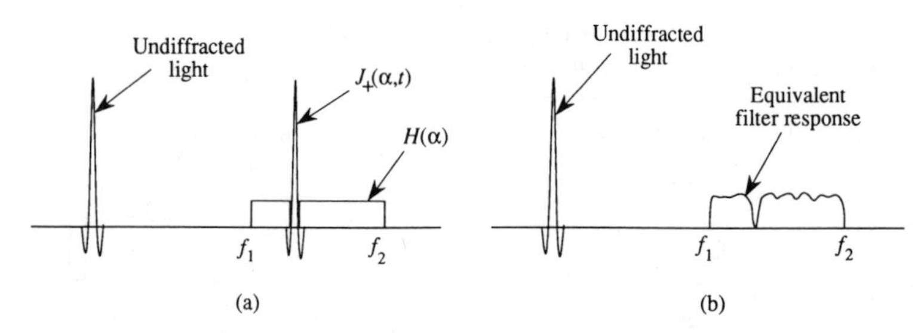

Figure 12.8. Filter response: (a) ideal response and (b) effective filter response.

jammer response given by $J_+(\alpha, t)$. Suppose that we sweep the jammer frequency over the entire passband of the system and plot the magnitude of the output signal $g_3(t)$ as a function of frequency. The cross-product term is

$$g_3(t) = 2\,\mathrm{Re}\left\{\int_{-\infty}^{\infty} J_+(\alpha, t) R_+^*(\alpha, t) |H(\alpha)|^2 P(\alpha)\, d\alpha\right\}, \quad (12.37)$$

where $P(\alpha)$ is the photodetector response. Because the filter transmittance $H(\alpha)$ is applied to both spectral distributions, its effect enters Equation (12.37) as $|H(\alpha)|^2$. The phase of $H(\alpha)$ is therefore not important.

The reference signal is given by

$$R_+(\alpha, t) = e^{j2\pi\alpha x_0}, \quad (12.38)$$

and the jammer response is given by

$$J_+(\alpha, t) = e^{j2\pi f_j(t-T/2)} A(\alpha - \alpha_j). \quad (12.39)$$

We substitute Equations (12.38) and (12.39) into Equation (12.37), take the temporal term outside the integral, and find that

$$g_3(t) = 2\,\mathrm{Re}\left\{e^{j2\pi f_j(t-T/2)} \int_{-\infty}^{\infty} A(\alpha - \alpha_j) e^{-j2\pi\alpha x_0} |H(\alpha)|^2 P(\alpha)\, d\alpha\right\}. \quad (12.40)$$

The output signal has the same frequency content as the input and its amplitude is the convolution of the notch filter function and the Fourier

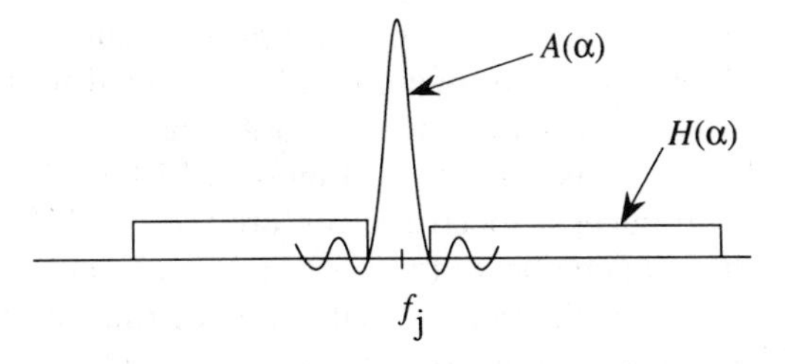

Figure 12.9. Origin of lack of notch depth.

transform of the weighting function with its associated exponential multiplier that indicates the position of the probe.

As we carry out the convolution, it is clear that the sharp transitions associated with the notch filter become smoothed and that the notch depth is not as great as expected; this result is shown in Figure 12.8(b). The details of the origin of the lack of expected notch depth are illustrated in Figure 12.9. When the swept frequency is exactly at f_j, we remove all the energy in the central lobe of the cw signal for the case where $a(x) = \text{rect}(x/L)$. The energy in the sidelobes, however, still passes through those portions of $H(\alpha)$ with unity response so that much of the sidelobe energy is not removed from the system. The effective notch depth R is given by the ratio

$$R = \frac{\left| \int_{-\infty}^{\infty} A(\alpha - \alpha_j) H(\alpha) \, d\alpha \right|^2}{\left| \int_{-\infty}^{\infty} A(\alpha - \alpha_j) \, d\alpha \right|^2}. \tag{12.41}$$

One method for getting a deeper notch is to modify $a(x)$ so that $A(\alpha)$ does not have as much energy in its sidelobes. This result is achieved by making $a(x)$ a Gaussian function, as we showed in Chapter 4. The penalty is that the frequency resolution is reduced and the notch widths are increased somewhat to suppress the now-widened central lobe of the jammer. A second method for increasing the notch depth is to activate more of the elements of the spatial light modulator in the vicinity of f_j. A third method is to shape the notch so that it has the same shape as the jammer response, as discussed in Chapter 15 in connection with adaptive filtering; the result is that the mainlobe and all the sidelobe energy tends to be removed.

Figure 12.10(a) shows a short pulse of 100-nsec duration. This pulse modulates a 350 MHz carrier frequency and was added to a narrowband interferer at 307 MHz to produce the composite signal in which the pulse is completely obscured, as shown in Figure 12.10(b). In the Fourier domain the signal produces a spectrum, centered at 350 MHz, resembling a sinc function whose mainlobe width is 20 MHz and whose sidelobes are 10 MHz wide. Figure 12.10(c) shows that the spectrum of the composite signal. The sinc function is due to the short pulse and the narrowband interferer is about 30 dB above the noise floor. After the spatial light modulator elements corresponding to the frequency of the interferers are switched, the output of the system is demodulated to produce the output pulse shown in Figure 12.10(d). The output pulse has a shape similar to that of the reference pulse, except that the rise-time overshoot, apparent in the reference pulse, is partially removed when we excise the narrowband interferer. This example qualitatively illustrates that the optical processor is capable of removing narrowband interferers without seriously distorting the pulse shape.

There are various methods for measuring signal distortion. We use the root-mean-square-error criterion as a measure of distortion:

$$\mathrm{RMSE} = \sqrt{\frac{1}{T}\int_0^T [b(t) - d(t)]^2 \, dt} \tag{12.42}$$

where the undistorted pulse at the output of the system in the absence of excision is indicated by $b(t)$ and the distorted signal is indicated by $d(t)$.

Because most of the pulse signal energy is in the mainlobe of its sinc function, we are concerned with the resulting pulse shape when the narrowband interferer is inside the mainlobe of the sinc function. Suppose that we increase the frequency of the narrowband interferer so that it passes through the mainlobe from left to right, distorting the spectrum of the pulse as the interferer is excised. Figure 12.11 shows the root-mean-square error as a function of the frequency of the interferer. The root-mean-square error is small when the interferer frequency is located near the low-frequency edge of the mainlobe because the mainlobe has a low magnitude in this region. As we increase the frequency of the narrowband interferer so that it passes through the mainlobe, the root-mean-square error increases until the frequency reaches the center frequency of the pulse. The root-mean-square error is largest when the interferer is centered in the mainlobe of the signal because the spatial light modulator elements then remove a considerable amount of signal energy in addition

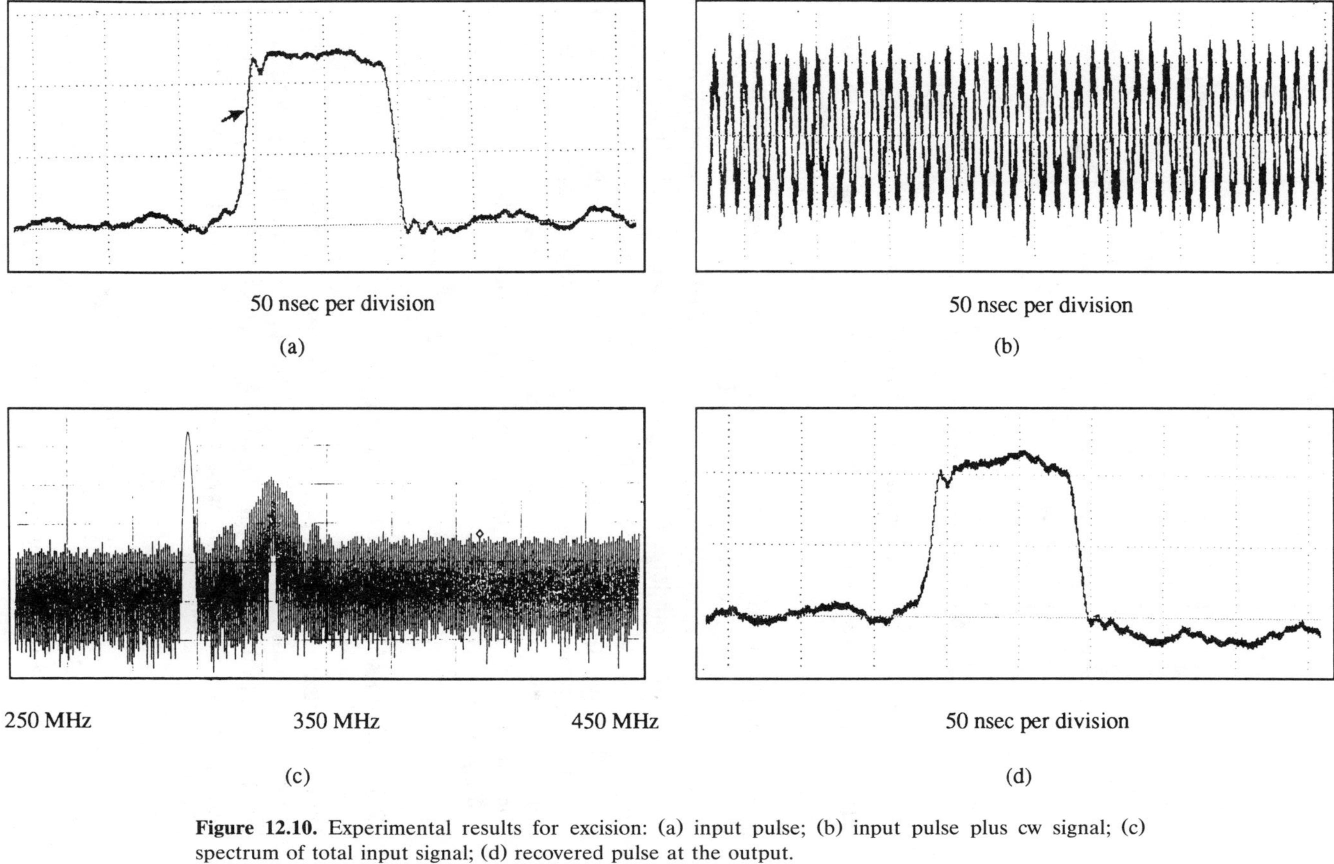

Figure 12.10. Experimental results for excision: (a) input pulse; (b) input pulse plus cw signal; (c) spectrum of total input signal; (d) recovered pulse at the output.

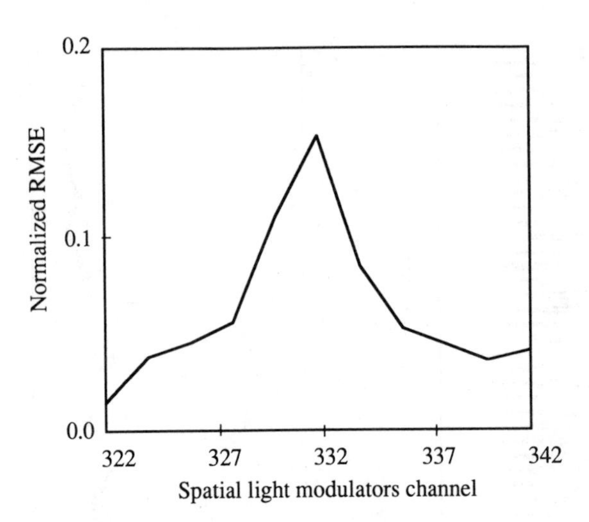

Figure 12.11. Normalized root-mean-square error versus spatial light modulator position.

to excising the interferer. The error then decreases until the first null of the sinc function is reached at 342 MHz.

An alternative to using a spatial light modulator as the notch filter is to use a special photodetector array in which detector elements can be switched to one of two summing buses. If a particular frequency element is to be shut off, it is switched from bus A to bus B; the signal information therefore appears on bus A and the jammer information appears on bus B (108). Each element in the photodetector array, as well as the entire signal bus, must have the same bandwidth as the wideband signal to ensure proper signal recovery.

12.6. ARBITRARY FILTER FUNCTION

When the filter function is placed in the Fourier domain as in Figure 12.5, for example, both the reference and signal beams pass through the filter. In this situation, filters with phase responses cannot be realized in the Fourier plane. Suppose that we attempt to implement the complex-valued filter

$$H(\alpha) = |H(\alpha)|e^{j\phi(\alpha)}, \tag{12.43}$$

where $|H(\alpha)|$ is the magnitude response and $\phi(\alpha)$ is the phase response of the filter. We find the system output as

$$g(t) = \int_{-\infty}^{\infty} \left| F_+(\alpha, t) H(\alpha) e^{j\phi(\alpha)} + R_+(\alpha, t) H(\alpha) e^{j\phi(\alpha)} \right|^2 d\alpha. \quad (12.44)$$

When the integrand is expanded into its component terms the output becomes

$$g(t) = \int_{-\infty}^{\infty} \Big\{ \left| F_+(\alpha, t) \right|^2 + \left| R_+(\alpha, t) \right|^2 + 2\,\mathrm{Re}\big[F_+(\alpha, t) R_+^*(\alpha, t) \big] \Big\} \left| H(\alpha) \right|^2 d\alpha, \quad (12.45)$$

and the phase response disappears. One of the important principles of heterodyne detection, therefore, is that once two beams are combined, the *phase responses of any optical elements that follow are irrelevant* because the phase affects both the reference and signal waveforms in the same way.

Consider the question of how to generate an arbitrary filtering operation, using heterodyne detection techniques in which we can control the phase. Figure 12.12 shows that one way to implement the filter is to place it at plane P_0, which is imaged into the Fourier plane, so that its information is carried only by the reference beam. In this case the filter function $|H(-\alpha)| \exp[j\phi(-\alpha)]$ is placed at plane P_0 so that the reference beam becomes

$$R_+(\alpha, t) = \left| H(\alpha) \right| e^{j\phi(\alpha)}. \quad (12.46)$$

The negative sign associated with α in the filter function compensates for

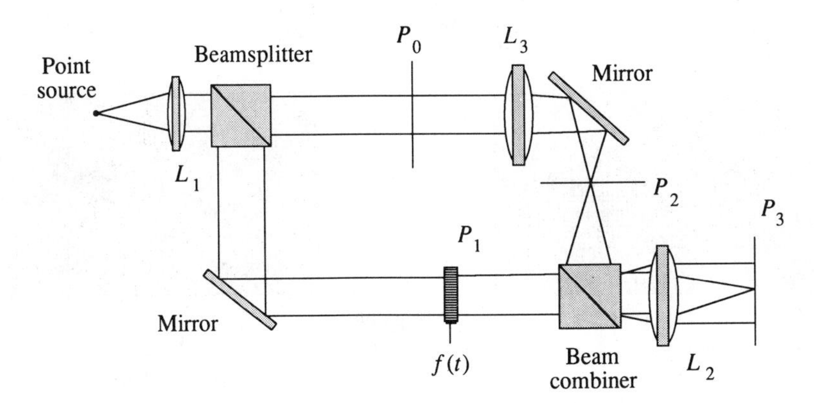

Figure 12.12. System for generating arbitrary filter functions.

the coordinate inversion due to the imaging operation between planes P_2 and P_3. The detected intensity of the cross-product term between the reference beam and the positive diffracted order from the signal beam becomes

$$g_3(t) = 2\,\mathrm{Re}\left[\int_{-\infty}^{\infty} F_+(\alpha, t)|H(\alpha)|e^{-j\phi(\alpha)}\,d\alpha\right], \tag{12.47}$$

and we see that the spectrum $F_+(\alpha, t)$ is modulated by the desired filter function.

The arbitrary phase function $\exp[j\phi(\alpha)]$ remains, as always, difficult to implement. One way to sidestep the phase-synthesis operation is to implement the impulse response $h(x)$ at plane P_2 of the interferometer shown in Figure 12.1. Although the magnitude of the impulse response could be implemented by a transmittance function, it is sometimes difficult to retain the desired degree of linearity. Because the signals in this system are basically one dimensional, we have a useful degree of freedom to use area-modulation techniques for generating the impulse response. In the *area-modulation technique*, we take advantage of the fact that when a two-dimensional function $t(x, y)$ whose transmittance is either zero or one is integrated in the vertical direction, we obtain an analog signal $h(x)$ that has no amplitude distortion. To illustrate this procedure, suppose that we create the two-dimensional function $t(x, y)$ shown in Figure 12.13(a). We place this function at plane P_2 of the interferometer shown in Figure 12.1,

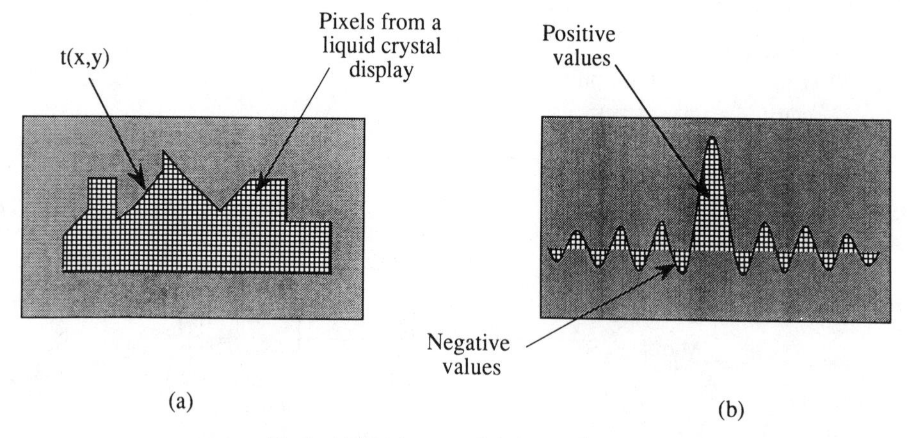

Figure 12.13. Area-modulation technique.

having first removed lens L_3 so that $t(x, y)$ is illuminated by a collimated beam of light (109).

Lens L_2 simultaneously creates the Fourier transform of the light distribution at planes P_1 and P_2; the Fourier transform of $t(x, y)$ is

$$T(\alpha, \beta) = \iint_{-\infty}^{\infty} t(x, y) e^{j2\pi(\alpha x + \beta y)}\, dx\, dy. \tag{12.48}$$

We evaluate $T(\alpha, \beta)$ at $\beta = 0$ and rearrange the integral in the following form:

$$T(\alpha, 0) = \int_{-\infty}^{\infty} \left\{ \int_{-\infty}^{\infty} t(x, y)\, dy \right\} e^{j2\pi\alpha x}\, dx. \tag{12.49}$$

The integral on y is the desired impulse response $h(x)$:

$$h(x) = \int_{-\infty}^{\infty} t(x, y)\, dy, \tag{12.50}$$

whose Fourier transform becomes

$$\boxed{R_+(\alpha) = T(\alpha, 0) = H(\alpha) = \int_{-\infty}^{\infty} h(x) e^{j2\pi\alpha x}\, dx.} \tag{12.51}$$

Thus, we see that the light distribution in the Fourier plane along the line $\beta = 0$ is, in fact, the same as would be obtained if $h(x)$ were a one-dimensional transmittance function at plane P_2. When this function is used, the cross-product term is the same as that given by Equation (12.47).

Negative values for the impulse response can be created by placing a phase-shifting optical element in contact with a spatial light modulator such as a liquid-crystal display. A sketch of how this technique implements a sinc function is shown in Figure 12.13(b). The upper part of the spatial light modulator, representing the positive part of the sinc function, is not modified; the lower parts of the spatial light modulator, representing the negative part of the sinc function, are covered by an optical element that retards the light by exactly one-half a wavelength. The integration in the y direction is then performed in the same way as described above. Because the impulse response is generally real valued, this technique is sufficient to construct an arbitrary filtering function. Another technique for introducing negative values in the filter function is to phase-shift a carrier frequency that supports the desired impulse response function (110).

PROBLEMS

12.1. Suppose that we start with a signal $s(t)$ whose spectrum was initially flat over the band from 50 to 100 MHz. After passing through several subsystems, the signal spectrum is no longer flat; the effective transfer function of the combined systems is

$$H(f) = \frac{e^{j2\phi(f)}}{1 + (f/f_0)^2},$$

where $f_0 = 100$ MHz and $\phi(f) = 4\pi(f/f_0)^2$ is the phase response of the combined system. We want to use a general interferometric system to equalize the spectrum of the detected signal. Find a suitable filter function that will perform this function if we use an acousto-optic cell for which $TW = 1000$ and $v = 4.2$ km/s. Assume that $\lambda = 0.5\ \mu$ and the focal length of the final lens is $F = 100$ mm. Calculate the physical scale of the filter (in millimeters) in terms of the parameters given and show what auxiliary lenses you may need. For example, if you elect to implement the required phase response with a lens, state the required focal length and aperture. Show how the filter overlaps with the spectrum at the proper scale. Remember that the filter must be passive.

13

Space-Integrating Correlators

13.1. INTRODUCTION

In Chapter 5, we described matched filtering techniques for the detection of two-dimensional signals. The basic technique is to correlate a received signal with a known signal which is stored either as a matched filter or as a reference function. These system architectures are of the *space-integrating type* because light that contributes to the correlation peak is integrated over a well-defined spatial region. In this chapter we examine space-integrating correlators that use acousto-optic cells for processing wideband time signals in real time. Because correlation gain, in general, is equal to the space or time bandwidth product of the signal and because the time bandwidth product for most acousto-optic cells is of the order of 1000–2000, the correlation gain is of the order of 30–33 dB, an acceptable figure for processing signals used in some radar systems and communication systems. In Chapter 14, we describe time-integrating architectures that provide more correlation gain.

An early reference to the use of acousto-optic cells for correlation was made by Rosenthal (86). Because Rosenthal described correlation only briefly, his publication did not receive as widespread attention as did the work of Slobodin (95) and Arm et al. (111), whose early work related mostly to processing radar signals. Later papers by Felstead (112, 113), by Atzeni and Pantani (114, 115) and by King et al. (94) illustrate the wide range of configurations and applications that we cover in this chapter.

We first describe a correlator that uses an acousto-optic cell to input the received signal and a fixed reference mask to store a replica of the signal to be detected. We show how the correlation function is detected in *intensity* by direct photodetection and then discuss methods to detect the correlation function in *amplitude* by using heterodyne-detection techniques. We next compare detection schemes in both the frequency domain and the time domain. We also describe systems wherein the fixed reference mask is replaced by an acousto-optic cell so that programmable filtering becomes possible.

Recall that correlation is used to detect a known signal $s(t)$ in noise $n(t)$. The received signal is represented by $f(t) = s(t) + n(t)$, and when the noise spectral density is uniform, the appropriate processing operation is to convolve $f(t)$ with a stored signal $s^*(-t)$. This convolution results in a *cross-correlation function*

$$\boxed{c(\tau) = \int_{-\infty}^{\infty} f(t)s^*(t+\tau)\,dt,} \tag{13.1}$$

where τ is a displacement variable between the received and stored signals. Because correlation is not communicative, the correlation function is normally subscripted to show how it relates to the functions being correlated. In this way, we might use $c_{fs}(\tau)$ to indicate cross correlation and $c_{ff}(\tau)$ and $c_{ss}(\tau)$ to represent the *autocorrelation functions* for the received and stored signals. We do not follow this subscripting practice here; the order of the functions is generally obvious from the associated integral and we consistently use the order given by Equation (13.1).

13.2. REFERENCE-FUNCTION CORRELATORS

In Chapter 5, we compared the merits of reference-function correlators with those of frequency-plane correlators. Because we want to process entire images in parallel, without the need to scan a two-dimensional signal over the entire object plane, a frequency-plane correlator is required. One-dimensional time signals that travel through acousto-optic cells, however, produce a natural scanning of the received signal so that reference-function correlators are easily implemented.

The configuration shown in Figure 13.1 is used extensively throughout this chapter and all references are to it unless otherwise noted. We use the

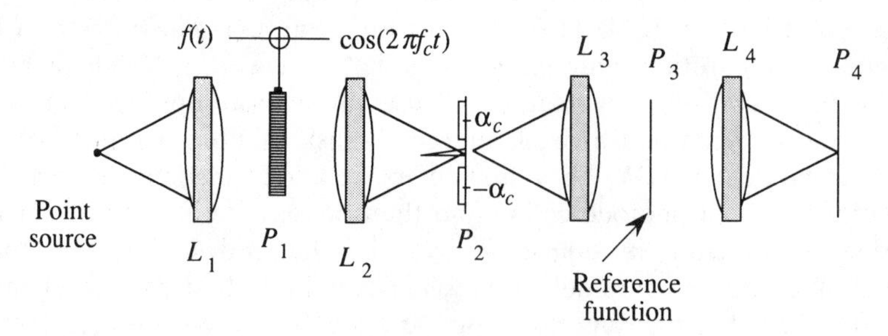

Figure 13.1. Basic space-integrating correlator.

Raman-Nath mode of illumination to support our analyses so that both the upshifted and the downshifted diffracted orders are available to easily obtain the conjugate form of a signal as needed to satisfy Equation (13.1). The illumination is normal to the acousto-optic cell and the diffracted light shows symmetry at plane P_2. We arrange the transducer opposite to its normal position to indicate that the acoustic wave is propagating in the negative x direction; we adopt this arrangement to simplify the notation.

Lens L_2 produces the Fourier transform at plane P_2 of the drive signal $f(t)$:

$$F(\alpha, t) = \int_{-\infty}^{\infty} a(x)\left\{1 + jmf\left(t - \frac{T}{2} + \frac{x}{v}\right)\right.$$
$$\left.\times \cos\left[2\pi f_c\left(t - \frac{T}{2} + \frac{x}{v}\right)\right]\right\} e^{j2\pi\alpha x}\, dx, \quad (13.2)$$

where α is the spatial frequency, $a(x)$ is the aperture weighting function, m is the modulation index, T is the time duration of the acousto-optic cell, and v is the velocity of the acoustic wave. The first term of $F(\alpha, t)$, produced by the undiffracted light, is simply $F_0(\alpha, t) = A(\alpha)$, the Fourier transform of the aperture function. The other two terms of $F(\alpha, t)$ are obtained by using the Euler formula to expand $\cos[2\pi f_c(t - T/2 + x/v)]$. The Fourier transform of the positive diffracted order is

$$F_+(\alpha, t) = e^{-j2\pi f_c(t - T/2)} \int_{-\infty}^{\infty} a(x) f\left(t - \frac{T}{2} + \frac{x}{v}\right) e^{j2\pi(\alpha - \alpha_c)x}\, dx, \quad (13.3)$$

where we have dropped scaling constants. In a similar fashion we find that the third term of $F(\alpha, t)$ represents the negative diffracted order:

$$F_-(\alpha, t) = e^{j2\pi f_c(t - T/2)} \int_{-\infty}^{\infty} a(x) f\left(t - \frac{T}{2} + \frac{x}{v}\right) e^{j2\pi(\alpha + \alpha_c)x}\, dx. \quad (13.4)$$

The Fourier transform, as developed here, is similar to that developed in connection with spectrum analysis in Chapter 8. There we kept $A(\alpha)$ in the equations explicitly because it plays such an important role in controlling sidelobe levels and frequency resolution. In correlation applications, the sidelobe levels associated with $A(\alpha)$ are of no particular concern; we prefer, therefore, to let $a(x) = \mathrm{rect}(x/L)$, both because it is an ideal window function that weights all contributions to the correlation function equally and because it renders the mathematics more tractable. We return to consider the physical importance of $a(x)$ on the correlation process in Section 13.2.3.

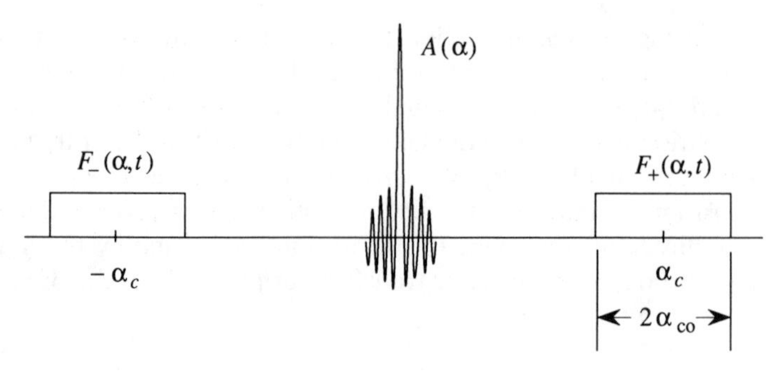

Figure 13.2. Spectral content, for Raman-Nath mode, at first Fourier plane.

The total result for $F(\alpha, t)$ is that

$$F(\alpha, t) = F_0(\alpha, t) + F_+(\alpha, t) + F_-(\alpha, t), \tag{13.5}$$

as sketched in Figure 13.2. We see that the Fourier transform $F_+(\alpha, t)$ of the windowed input signal $f(t)$ is centered at α_c, and $F_-(\alpha, t)$ is centered at $-\alpha_c$.

Suppose that we use a spatial filter $H(\alpha)$ at plane P_2 to eliminate all but the positive diffracted order centered at α_c. We define the filter as

$$\begin{aligned} H(\alpha) &= 1, \qquad (\alpha_c - \alpha_{co}) \le \alpha \le (\alpha_c + \alpha_{co}) \\ &= 0, \qquad \text{elsewhere,} \end{aligned} \tag{13.6}$$

where $2\alpha_{co} = W/v$ is the system bandwidth, expressed as a spatial frequency. Only $F_+(\alpha, t)$ propagates beyond plane P_2 because the other two terms have been removed by the filter $H(\alpha)$.

Lens L_3 produces the Fourier transform of $F_+(\alpha, t)$ at plane P_3. As a convenient alternative to calculating the second Fourier transform explicitly, we recognize that two successive transforms reproduce the input function with reversed coordinates. We therefore represent the system transfer function between planes P_1 and P_3 as $\text{sinc}[(x + u)/d_0]$:

$$f_+(u, t) = \int_{-\infty}^{\infty} f_+(x, t) \text{sinc}\left[\frac{x + u}{d_0}\right] dx, \tag{13.7}$$

where u is the spatial coordinate at plane P_3. For bandlimited signals, this sinc function is equivalent to $\delta(x + u)$; we use the sifting theorem to find that the light amplitude at plane P_3 is

$$f_+(u,t) = f\left(t - \frac{T}{2} - \frac{u}{v}\right)e^{-j2\pi f_c(t-T/2-u/v)}. \tag{13.8}$$

The signs attached to the space variable u in Equation (13.8) show that the signal at plane P_3 is traveling in the positive u direction and that the wave is propagating downward and to the right.

13.2.1. Real-Valued Impulse Responses

We begin by considering those applications in which the desired filtering operation has a real-valued impulse response $h(t)$. To implement this impulse response, we place a reference function $h(u/v)$ at plane P_3. We use a notation in which we divide the space coordinate by the acoustic velocity to emphasize the temporal nature of the output of the system. We want to produce the correlation function formed by $f_+(u,t)$, as given by Equation (13.8), and the reference function $h(u/v)$. The required integration is performed optically by producing the Fourier transform of the product by means of lens L_4 and evaluating the transform at plane P_4 over a narrow spatial frequency interval with a point photodetector, similar to the technique discussed in connection with area modulation in Chapter 12, Section 12.6. With the time signal bounded by the duration of the acousto-optic cell, we find

$$G(\gamma,t) = e^{-j2\pi f_c(t-T/2)}\int_{-\infty}^{\infty} f\left(t - \frac{T}{2} - \frac{u}{v}\right)h\left(\frac{u}{v}\right)e^{j2\pi(\gamma+\gamma_c)u}\,du, \tag{13.9}$$

as the Fourier transform of the product which, when evaluated at $-\gamma_c$, becomes

$$G(-\gamma_c,t) = e^{-j2\pi f_c(t-T/2)}\int_{-\infty}^{\infty} f\left(t - \frac{T}{2} - \frac{u}{v}\right)h\left(\frac{u}{v}\right)du. \tag{13.10}$$

The integral in Equation (13.10) is clearly of the convolutional type. To produce a correlation function, we first change variables in which $t - T/2 - u/v = z$:

$$G(-\gamma_c,t) = e^{-j2\pi f_c(t-T/2)}\int_{-\infty}^{\infty} f(z)h(-z + t - T/2)\,dz. \tag{13.11}$$

Suppose that we now set $h(u) = s^*(-u + L/2)$, which indicates that the impulse response is equal to the conjugate of the signal we want to detect with a reversed spatial coordinate and the equivalent of a time delay of $T/2$. We then find that Equation (13.11) becomes

$$G(-\gamma_c, t) = e^{-j2\pi f_c(t-T/2)} \int_{-\infty}^{\infty} f(z) s^*(z + t - T)\, dz. \tag{13.12}$$

The integration in Equation (13.12) is similar to $c(\tau)$ as given in Equation (13.1). Some differences are worth noting, however. The usual mathematical representation of correlation is to use τ as the time delay between two signals $f(t)$ and $s(t)$. In the optical implementation, the time signal $f(t)$ is moving past a fixed spatial signal $h(u/v)$ with velocity v. The variable associated with the output signal is naturally $(t - T)$, the current time variable. The total time-delay factor T, along with the limits of integration imposed by the acousto-optic cell, show that the earliest time that the peak correlation can occur is T seconds later than that instant in time at which the signal just entered the acousto-optic cell.

13.2.2. Complex-Valued Impulse Responses

In the development of Section 13.2.1, we assumed that we can set $h(u)$ equal to a complex-valued function. In optical systems we must, of course, implement complex-valued functions using the spatial carrier frequency technique discussed in Chapter 5, Section 5.9, in which we record a function on a spatial light modulator so that the transmittance of the reference function at plane P_3 is

$$t(u) = a_1 + 2a_2 |h(u)| \cos[2\pi\gamma_1 u + \phi(u)], \tag{13.13}$$

where $|h(u)|$ is the magnitude and $\phi(u)$ is the phase of a general impulse response $h(u)$. This function is placed at plane P_3 of the optical system shown in Figure 13.1. The constant a_1 is chosen to ensure that the filter transmittance $t(u)$ is non-negative. The carrier frequency γ_1 ensures that the terms are separated at the output plane. We rewrite Equation (13.13) as

$$\begin{aligned} t(u) &= a_1 + a_2 |h(u)| e^{j[2\pi\gamma_1 + \phi(u)]} + a_2 |h(u)| e^{-j[2\pi\gamma_1 + \phi(u)]} \\ &= a_1 + a_2 h(u) e^{j2\pi\gamma_1 u} + a_2 h^*(u) e^{-j2\pi\gamma_1 u}. \end{aligned} \tag{13.14}$$

We follow the same procedures as in the previous section in which we integrated the product $f_+(u,t)t(u)$ over all space. The Fourier transform of the product at plane P_3 is

$$G(\gamma,t) = \int_{-\infty}^{\infty} f\left(t - \frac{T}{2} - \frac{x}{v}\right) t(u) e^{-j2\pi f_c(t - T/2 - u/v)} e^{j2\pi\gamma u}\, du, \quad (13.15)$$

where γ is the spatial frequency variable at plane P_4. When we substitute Equation (13.14) into Equation (13.15), we find that the output consists of three terms. The output due to the bias a_1 of $t(u)$ is

$$G_0(\gamma,t) = a_1 e^{-j2\pi f_c(t-T/2)} \int_{-\infty}^{\infty} f\left(t - \frac{T}{2} - \frac{u}{v}\right) e^{j2\pi(\gamma+\gamma_c)u}\, du, \quad (13.16)$$

where $\gamma_c = f_c/v$. We see that $G_0(\gamma,t)$, which is located at plane P_4, is simply the image of $F_+(\alpha - \alpha_c, t)$, which is located at plane P_2. This result shows that the reference function does not affect the image of $F_+(\alpha - \alpha_c, t)$, except for a scaling factor a_1, which we ignore. The term $G_0(\gamma,t)$ has a total spatial frequency extent $2\alpha_{co}$ in the spatial frequency domain, centered at $-\gamma_c$ as shown in Figure 13.3.

The output due to the second term of Equation (13.14) is

$$G_1(\gamma,t) = e^{-j2\pi f_c(t-T/2)} \int_{-\infty}^{\infty} f\left(t - \frac{T}{2} - \frac{u}{v}\right) h\left(\frac{u}{v}\right) e^{j2\pi(\gamma+\gamma_c+\gamma_1)u}\, du, \quad (13.17)$$

which is the Fourier transform of the product of $f(u,t)$ and $h(u)$, as they slide by one another as a function of time. This term is centered at $\gamma = -\gamma_1 - \gamma_c$ and has spatial frequency extent $4\alpha_{co}$, as shown in Figure 13.3. The output at plane P_4 due to the final term of the mask

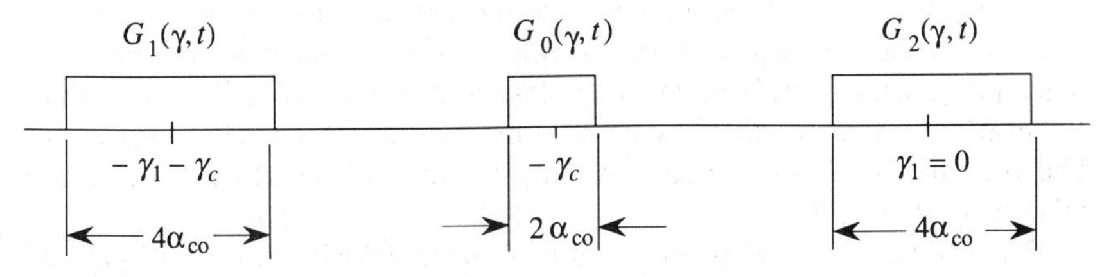

Figure 13.3. Spectral content in second Fourier plane.

represented by Equation (13.14) is

$$G_2(\gamma, t) = e^{-j2\pi f_c(t-T/2)} \int_{-\infty}^{\infty} f\left(t - \frac{T}{2} - \frac{u}{v}\right) h^*\left(\frac{u}{v}\right) e^{j2\pi(\gamma+\gamma_c-\gamma_1)u}\, du. \tag{13.18}$$

Suppose that we set $\gamma_1 = \gamma_c$ in Equation (13.18) so that the spatial carrier frequency for the reference function is equivalent to the temporal carrier frequency of $f(t)$. We then find that

$$G_2(\gamma, t) = e^{-j2\pi f_c(t-T/2)} \int_{-\infty}^{\infty} f\left(t - \frac{T}{2} - \frac{u}{v}\right) h^*\left(\frac{u}{v}\right) e^{j2\pi\gamma u}\, du. \tag{13.19}$$

This function is centered on the optical axis. When we evaluate the integral at $\gamma = 0$ by using a small photodetector at that point, we find that

$$G_2(0, t) = e^{-j2\pi f_c(t-T/2)} \int_{-\infty}^{\infty} f\left(t - \frac{T}{2} - \frac{u}{v}\right) h^*\left(\frac{u}{v}\right) du. \tag{13.20}$$

This result has essentially the same form as Equation (13.10), with the exception that we already have the conjugate of $h(u)$. We reverse the space coordinate and shift the signal so that $h(u) = s(-u + L/2)$ to obtain the desired correlation function:

$$\boxed{G(0, t) = e^{-j2\pi f_c(t-T/2)} \int_{-\infty}^{\infty} f(z) s^*(z + t - T)\, dz.} \tag{13.21}$$

13.2.3. A Wavefront View of Matched Filtering

The physical nature of correlation is nicely connected to the physics of the optical system. Suppose that $s(t)$ is a *pseudorandom sequence* that contains $N = 2^n - 1$ samples, where n is an integer. We say that a *frame* of the pseudo-random sequence consists of N elements, generally called *chips*, of duration T_0. In spread spectrum communication systems these N chips are used to encode just one bit of the message signal $m(t)$. A space-integrating correlator must be designed so that NT_0 is no less than T, because only then will all N chips contribute to the correlation process. The bit rate R of the message signal and the cell length are therefore related by $R = 1/T$.

The received signal at plane P_3, moving with velocity v, is multiplied by the reference signal component $s^*(-u)$ as shown in Figure 13.4. For a

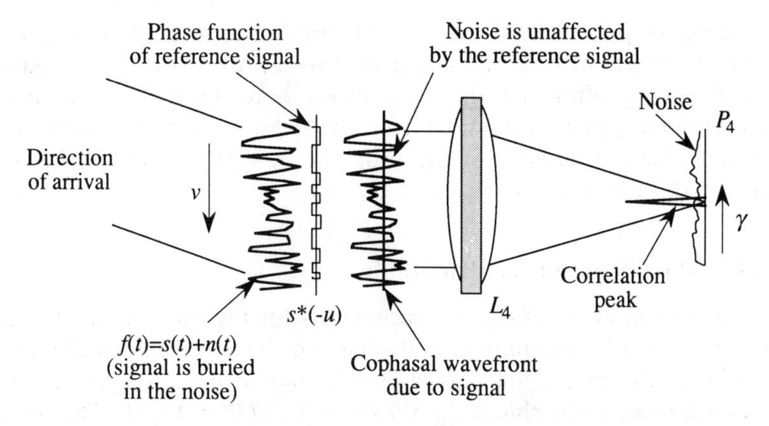

Figure 13.4. Physical description of correlation or matched filtering operation.

pseudorandom sequence, the phase is a simple binary function that changes the optical path difference by $\lambda/2$. The action of the reference signal on the image of the received signal is to conjugate the pseudorandom-sequence part of the received signal. The signal portion of $f(t)$ is therefore rendered into a cophasal wavefront so that L_4 produces the light amplitude $\text{sinc}(\gamma L)$ at plane P_4 with its peak intensity at $\gamma = 0$. In a direct-detection system, the photodetector output is therefore proportional to the integral of $\text{sinc}^2(\gamma L)$ over some region $\Delta\gamma$, centered at $\gamma = 0$. The statistics of the noise portion of $f(t)$ are unaffected by the reference function $s^*(-u)$. Recall from Equation (5.25) that the mean-square noise level at plane P_4 is the same as if the phase function $s^*(-u)$ were absent.

We now easily see the effect of an aperture weighting function $a(x)$. The correlation operation, for an arbitrary aperture weight function, produces a plane wave that has magnitude $a(x)$. At plane P_4, the spatial light distribution is therefore $A(\gamma)$ instead of $\text{sinc}(\gamma L)$. Because the sinc function has the maximum possible intensity at $\gamma = 0$, relative to all possible weighting functions, the effect of using a weighting function for which $a(x) \neq \text{rect}(x/L)$ is the loss of some correlation gain. From another viewpoint, any nonuniform illumination function prevents all the samples of the functions in the integrand of the correlation function from contributing equally to the correlation peak.

The final step is to determine the optimum size of the photodetector that maximizes the signal-to-noise ratio. If the noise is uniformly distributed at the correlation plane, the noise increases linearly as the photodetector size increases. As we showed in Chapter 10, Section 10.7, the signal is proportional to the integral of a sinc^2 function which increases linearly with the photodetector size, provided that its argument is less than $\pm \frac{1}{2}$; after that, the signal grows more slowly than the noise so that the signal-to-noise ratio no longer increases. The physical width of the photodetector is therefore of the order of $h = \lambda F_4/L$, where F_4 is the focal length of lens L_4, so that $h = d_0$.

13.2.4. The Photodetector Bandwidth

The correlation peak, of course, occurs only for the short interval in time when the position, magnitude, and phase of the received signal matches that of the reference signal. For a pseudorandom sequence, the duration T_0 of each rectangular chip is $T_0 = T/N = T/2TW = 1/2W$. The correlation function $c(t)$ is a triangular function whose total width is $2T_0$. The bandwidth of the correlation function is, therefore, of the order of W. A small, high-speed photodetector is therefore required to support the detection of the correlation peak.

13.2.5. Correlation in the Presence of Doppler Frequency Shifts

A powerful feature of an optical correlator is that we can use one of its spatial coordinates to detect Doppler shifts at the same time that the system is correlating signals. The system therefore simultaneously processes the entire time delay and Doppler search space. Suppose that the carrier frequency of the received signal $f(t)$ arrives with an unknown Doppler frequency shift f_d due to a relative motion between the transmitter and receiver. The Doppler causes the carrier frequency f_c to change to $f_c + f_d$; we assume that the Doppler is not so severe as to change the time scale of the baseband signal $s(t)$. The only change in our analysis, then, is to replace f_c by $f_c + f_d$ in the expression for $f_+(u, t)$. This change propagates through to Equation (13.19), which shows that the correlation peak is now located at the spatial frequency $\gamma = -f_d/v$.

In terms of the physical picture presented in Figure 13.2, the Doppler frequency changes the position of the spatial frequency content of the positive diffracted order at plane P_2 of Figure 13.1. This change in position changes the angle of the illumination incident on the reference function at plane P_3 and, in turn, the angle of the cophasal wavefront leaving plane P_3. It is important to note that the Doppler shift has absolutely no effect on the relative position of the signals at plane P_3, and therefore has no effect on the accuracy of the time-of-arrival measurement. In this fashion, the output plane P_4 of the correlator displays the correlation function in *time* and the Doppler frequency of the received signal in *space*. The correlation for a Doppler-shifted signal is therefore simply shifted to a new spatial frequency coordinate at plane P_4.

That part of the optical system from plane P_3 to P_4 in Figure 13.1 is a spectrum analyzer, while the first section of the system is a spatial parallel multiplier. As a spectrum analyzer, the second stage of the system both performs the integration of the product of the two spatial functions present at plane P_3 and detects Doppler shifts of the received signal. We need an array of photodetectors at plane P_4 to detect the correlation function for any Doppler frequency, one for each resolvable Doppler bin. These detectors lie along the γ axis which is parallel to the x axis at the input. The outputs of these discrete photodetectors are thresholded and monitored to determine when the signal arrives.

The display of Doppler information is shown in Figure 13.5 in the horizontal or spatial frequency direction, and the correlation function $c(\tau)$ associated with each Doppler frequency is displayed in time. The expanded scale part of Figure 13.5 shows that the correlation function for a pseudorandom sequence and $a(x) = \text{rect}(x/L)$ has a sinc^2 envelope in the Doppler direction and a triangular envelope in the time-delay direction.

The maximum signal-to-noise ratio is obtained when the channel bandwidth in the Doppler direction is as small as possible relative to the theoretical limit of $1/T$. The photodetector spacing ratio and duty cycle are therefore essentially the same as those discussed in connection with the heterodyne spectrum analyzer in Chapter 10, Section 10.5. By using an array of N_d photodetectors, we have implemented a set of N_d correlators by using a previously unused degree of freedom, namely, one of the space variables, to parametrically process signals in parallel. We note, in passing, that the bandwidth of the acousto-optic cell may need to be increased

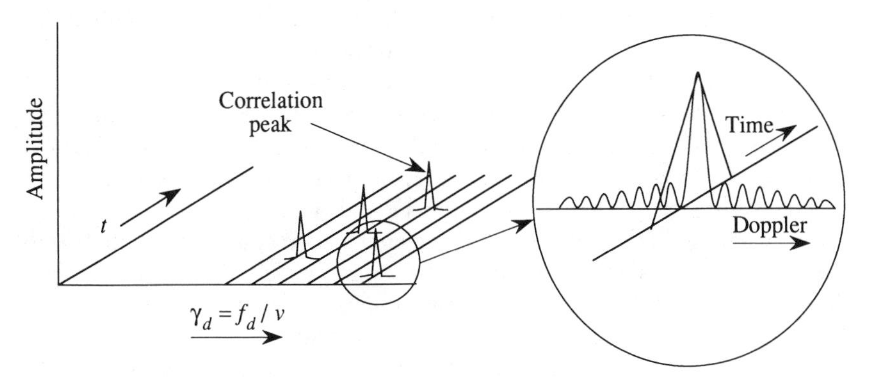

Figure 13.5. Ambiguity function representation.

somewhat to handle the Doppler shifts; but the fractional increase is small because the Doppler shift is generally small relative to f_c.

13.2.6. Programmable Matched Filter

In some applications, we may need to switch quickly from one reference function to another. We can implement *programmable* matched-filtering operations if we replace the fixed reference mask by an acousto-optic cell. The key change in the analysis is that the reference signal is traveling through the optical system at a velocity v in a direction *opposite* to that of the received signal. The typical correlation function, obtained by techniques similar to those used earlier in this chapter, is now $c(2t - T)$ for the situation where $f(t)$ and $s^*(-t)$ are counterpropagating. The factor of 2 arises because the signals are traveling at a relative velocity of $2v$ or at $v_1 + v_2$ if the acoustic velocities are different. Because the correlation peak has only half the width it had when the reference function is fixed in space, the photodetector bandwidth must be increased by a factor of 2. All other features of the system have been explained as before, except for the minor point that the diffracted light from the second acousto-optic cell is in phase quadrature with the diffracted light from the first cell. This difference is observable only with heterodyne detection as a phase change of the carrier frequency at the output (see Problem 13.6).

13.3. MULTICHANNEL OPERATION

When the received signal $f(t)$ contains one of a set of possible signals $s_i(t)$, $i = 1, 2, \ldots, N$, we need to perform multiple correlations. In one-dimensional electronic systems this correlation must be done serially or by an N-fold duplication of the hardware. An optical processor provides the needed degrees of freedom if we use the vertical direction to correlate $f(t)$ with a *library* of reference functions in parallel. Suppose that the reference mask at plane P_3 is divided into channels in the vertical direction as shown in Figure 13.6. Each channel contains one of N possible signals or parametric variations of a signal.

For example, $f(t)$ might be a chirp signal whose chirp rate is unknown. The sequence of signals is then a set of chirp functions, each with a different chirp rate. If we increase the number of channels without limit, the variation in the reference signal from one channel to the next becomes small; we can then replace the reference function, with its N discrete channels, by a conical lens whose focal length is a smooth continuous

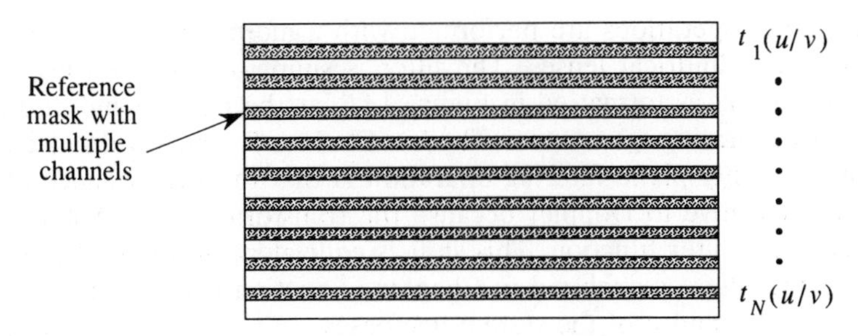

Figure 13.6. Multichannel reference-function format.

function of the vertical dimension. This is the key technique for processing radar data, as we showed in Chapter 5, Section 5.6.

We need to modify the optical system of Figure 13.1 somewhat to implement the multichannel correlation operation. Lens L_3 must now include a cylindrical lens to expand $f_+(u, t)$ in the vertical direction to cover the reference-signal format shown in Figure 13.6. Lens L_4 also includes a cylindrical lens to image plane P_3 onto plane P_4 in the vertical direction. The photodetector array now must have N_c detectors in the vertical direction, one for each channel, and N_d detectors in the horizontal direction, one for each resolvable Doppler frequency. If Doppler is not a problem, $N_d = 1$ so that the array is a linear one.

An alternative processing technique is to incorporate a reference function, either single channel or multichannel, into a complex-valued filter function whose transmittance we indicate by $T(\alpha, \beta)$. Such a filter function is placed at plane P_2 of the system shown in Figure 13.7. The advantage of the spatial filter relative to the reference function is that

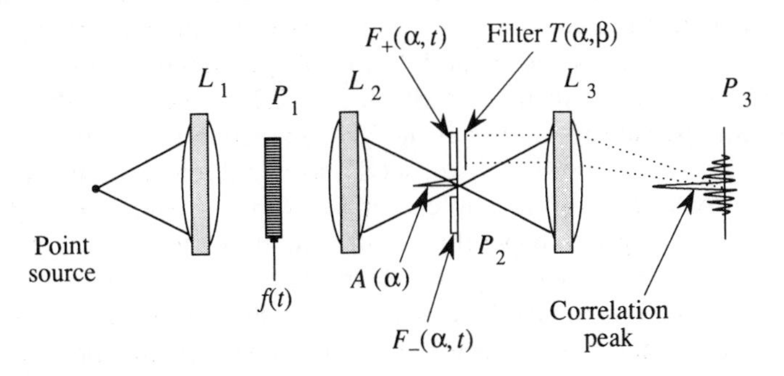

Figure 13.7. Matched-filter mode of operation.

multichannel operations are performed with a more compact system that contains no cylindrical lenses. The filter is simply constructed using the reference mask, as formatted in Figure 13.6, as the input signal in one of the filter-generating systems described in Chapter 5. The key disadvantage of this frequency-plane filtering operation is that the performance of the system is sensitive to Doppler because the transform $F_+(\alpha, t)$ shifts with respect to the filter function. This shift is equivalent to a lateral displacement of the filter as analyzed in Chapter 6, Section 6.8.1, with the result that the performance of the system decreases.

In this matched-filtering configuration we have traded a useful degree of freedom, the horizontal space coordinate that could display Doppler information, for the ability to locate the signal in the input plane. Such a trade is highly sought after in a pattern-recognition application because the signal location is generally unknown. Here, however, the signal location will eventually become known; all we need to do is wait until that moment when the scanning action at plane P_1 produces the correlation peak. Trading the ability to locate the function by using a spatial dimension, when we are already using the time dimension for this purpose, is a poor use of the capability of the system.

A useful architecture emerges, however, if we interchange the position of the spatial filter and the acousto-optic cell. If we place the filter $T(\alpha, \beta)$ at plane P_1, one part of its transform is $s_i^*(-u)$, which is then convolved with $f_+(u, t)$ at plane P_2. In this case, correlation is achieved even when Doppler is present because f_d will produce a simple spatial translation of $c(t)$ at plane P_3, as described in Section 13.2.4.

13.4. HETERODYNE / HOMODYNE DETECTION

We have seen that discrete-element photodetectors having bandwidths of W are needed for direct detection of the correlation intensity function $c^2(t)$. To extend the dynamic range and to provide for some new processing operations, such as the demodulation of a communication signal, we consider heterodyne detection. In Section 13.2.2, we saw that the correlation function $c(t)$ always occurs at the $\gamma = 0$ position in the output spatial frequency plane if $f_d = 0$. All we need to do, therefore, is to provide a reference probe at $\gamma = 0$ to implement heterodyne detection. A particularly interesting way to achieve this result is to allow the undiffracted light, which was blocked by $H(\alpha)$ at plane P_2 in Figure 13.1, to pass through the system. This light is imaged at plane P_4, where it is represented by $\text{sinc}(\gamma L)$, given that $a(x) = \text{rect}(x/L)$ is the aperture function. We sum this function and $G_2(\gamma, t)$ from Equation (13.19) to produce the intensity function

$$I(\gamma, t) = |\text{sinc}(\gamma L) + G_2(\gamma, t)|^2, \tag{13.22}$$

which we integrate in the vicinity of $\gamma = 0$. After the expansion is made, we consider the cross-product term

$$I_3(\gamma, t) = 2\,\text{Re}\{\text{sinc}(\gamma L) G_2(\gamma, t)\}. \tag{13.23}$$

Because the sinc(γL) probe function overlaps $G_2(\gamma, t)$ only in the vicinity of $\gamma = 0$, the integration as required by Equation (13.20) is optimally performed. The size of the photodetector is not particularly important because sinc(γL) ideally constrains the detection region.

The sinc function, when multiplied by $G_2(\gamma, t)$ as in Equation (13.23), produces $G_2(0, t)$. It is an ideal integrator function because its inverse transform is a rect function over the input space variable so that all samples from the signal are equally weighted. In contrast, the effect of a small photodetector whose response is a rect function in the spatial frequency domain is equivalent to a sinc weighting function in the input plane. In this case, the signal at the ends of the acousto-optic cell do not contribute as much to the correlation function as do those at the center of the cell.

The integration on γ produces an output

$$\begin{aligned} g_3(t) &= 2\,\text{Re}\{c(t - T)e^{j[2\pi f_c(t - T/2) + \pi/2]}\}, \\ &= 2c(t - T)\cos[2\pi f_c(t - T/2) + \pi/2], \end{aligned} \tag{13.24}$$

where we have used the result for $G_2(0, t)$ from Equation (13.21). In the heterodyne-detection process, the correlation function $c(t)$ is the envelope of a $\cos(2\pi f_c t)$ carrier function and the photodetector output is now linearly related to the *amplitude* $c(t)$ of the correlation function instead of to the *intensity* $c^2(t)$. The $\pi/2$ factor in Equation (13.24) arises because the diffracted light from the acousto-optic cell is in phase quadrature with the undiffracted light. We could, of course, drop the $\pi/2$ and replace the cosine function by a sine function. The desired output, as given by the cross-product term of Equation (13.24), is separated by a bandpass filter from the baseband bias terms given by $g_1(t)$ and $g_2(t)$ from the expansion of Equation (13.22).

The example given here is more appropriately called a *homodyne technique* because the input acousto-optic cell provides both the reference beam as well as the diffracted light. The optical system is then a nearly *common path interferometer* and is fairly insensitive to vibrations. When we compare the architecture given in Figure 13.1 to that of a Mach-Zehnder interferometer as shown in Figure 8.1, we find that the input acousto-optic cell plays the role of the beamsplitter, and the spatial carrier frequency contained in the reference function plays the role of the beam combiner.

Another way to obtain the homodyne action is to remove $H(\alpha)$ from plane P_2 altogether. The negative diffracted order from plane P_2 is now allowed to propagate to plane P_3, where it interacts with the total reference function $t(u)$ given in Equation (13.14). When all three terms from plane P_2 are allowed to interact with all three terms of $t(u)$, nine terms appear at the output: three terms are centered at $\gamma = 0$, two terms are centered at both γ_c and $-\gamma_c$, one term is centered at $2\gamma_c$, and one is centered at $-2\gamma_c$. These terms are most easily accounted for by showing just their principal rays.

We begin by showing the principal rays for the positive diffracted order from $f(x, t)$ in Figure 13.8(a). The spatial frequency signal $F_+(\alpha, t)$ forms $f_+(u, t)$ at plane P_3, where it is multiplied by $t(u)$. Of the three terms generated, the desired correlation function, given by the convolution of $f_+(u, t)$ with $s^*(-u)$, is centered at $\gamma = 0$. The image of the spectrum of $f_+(u, t)$ is centered at $-\gamma_c$ and the convolution of $f_+(u, t)$ with $s(u)$ is centered at $-2\gamma_c$.

The output due to the undiffracted light from the acousto-optic cell is shown in Figure 13.8(b). The output is, in essence, the impulse response of the reference function similar to that described in Chapter 5. It consists of the spectrum $S(\gamma)$ centered at γ_c, an image of the delta function located at $\gamma = 0$, and the conjugate spectrum $S^*(\gamma)$ centered at $-\gamma_c$.

The output due to the negative diffracted order of the input is shown in Figure 13.8(c). The convolution of $f^*(u, t)$ with $s(u)$ is centered at $\gamma = 0$, the spectrum $F_-(\gamma, t)$ is centered at γ_c, and the convolution of $f^*(u, t)$ and $s^*(-u)$ is centered at $2\gamma_c$.

Figure 13.8 forms the basis for several new and interesting concepts, because it is the most general case possible, except for when we replace the fixed reference signal by a programmable acousto-optic cell as discussed in Section 13.2.4. We now return to the basic homodyne configuration and let all the light through the system and perform the detection in the vicinity of $\gamma = 0$. The intensity in this region is

$$I(\gamma, t) = |\text{sinc}(\gamma L) + G_2(\gamma, t) + G_2^*(\gamma, t)|^2. \tag{13.25}$$

The sinc(γL) probe function overlaps $G_2(\gamma, t)$ and $G_2^*(\gamma, t)$ only in the vicinity of $\gamma = 0$ to perform the spatial integration over the input plane.

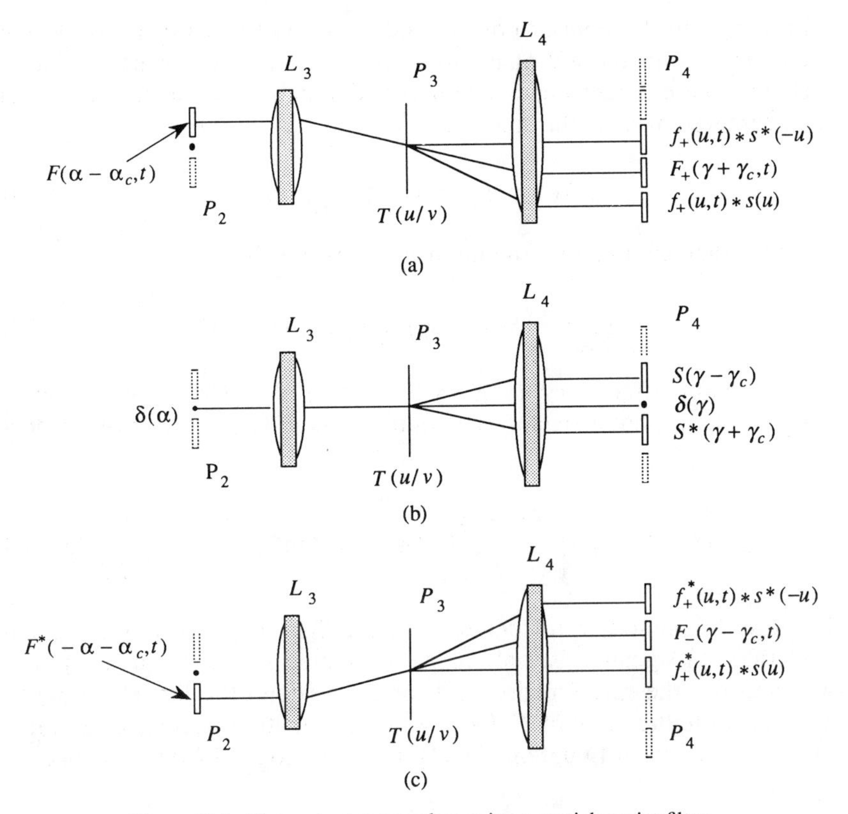

Figure 13.8. Nine output terms when using a spatial carrier filter.

We expand the intensity function to get nine terms. It can be shown that three of these terms are at baseband, two passband terms are centered at $\pm f_c$, and one passband term is centered at $\pm 2f_c$ (see Problem 13.5). The desired term at f_c is proportional to $c(t - T)\cos[2\pi f_c(t - T/2) + \pi/2]$ and is separated from the others by bandpass filtering. The same output results if the filter passes only the negative diffracted order and the undiffracted light.

13.5. HOMODYNE DETECTION IN THE FOURIER DOMAIN

We can also achieve homodyne detection by placing the photodetector at positions in the Fourier plane P_4 other than at $\gamma = 0$. Recall from Chapter 12, Section 12.2 that we used a large-area photodetector to collect all the light in the Fourier domain and discovered that we recovered the input signal with the aid of heterodyne detection. We can use a similar technique here. From Figures 13.8(a) and 13.8(b), we find that the total light centered at γ_c is the sum

$$S(\gamma - \gamma_c) + F_-(\gamma - \gamma_c, t). \tag{13.26}$$

Suppose that we integrate the intensity

$$I(\gamma, t) = |S(\gamma - \gamma_c) + F_-(\gamma - \gamma_c, t)|^2 \tag{13.27}$$

over a spatial frequency range $\gamma_c - \alpha_{co} < \gamma < \gamma_c + \alpha_{co}$ with a large, single-element photodetector. The output signal due to the cross-product term is

$$g_3(t) = 2\,\mathrm{Re}\left\{\int_{-\infty}^{\infty} \mathrm{rect}\left[\frac{\gamma - \gamma_c}{2\alpha_{co}}\right] F_-(\gamma - \gamma_c, t) S^*(\gamma - \gamma_c)\, d\gamma\right\}. \tag{13.28}$$

Because the photodetector detects only the light in the vicinity of γ_c and because both signals exist only in the stipulated spatial frequency range, we can drop the rect function and formally extend the limits to infinity. We substitute the value of $F_-(\gamma - \gamma_c)$ from Equation (13.4) and the value of $S^*(\gamma - \gamma_c)$ from Equation (13.14) into Equation (13.28) to produce

$$g_3(t) = 2\,\mathrm{Re}\left\{\int_{-\infty}^{\infty} e^{j2\pi f_c(t-T/2)} \int_{-\infty}^{\infty} a(x) f\left(t - \frac{T}{2} + \frac{u}{v}\right) e^{j2\pi(-\gamma+\gamma_c)u}\, du \right.$$
$$\left. \times \int_{-\infty}^{\infty} h^*\left(\frac{y}{v}\right) e^{j2\pi(-\gamma+\gamma_c)y}\, dy\, d\gamma\right\}, \tag{13.29}$$

where y is a dummy variable. We first integrate on γ to produce a sinc function that behaves as a delta function $\delta(u + y)$. After performing the sifting operation with the delta function, we find that

$$\boxed{g_3(t) = e^{-j2\pi f_c(t-T/2)} \int_{-\infty}^{\infty} f\left(t - \frac{T}{2} - \frac{u}{v}\right) h^*\left(\frac{u}{v}\right) du,} \tag{13.30}$$

which is identical to the result given in Equation (13.20), except for the negative sign on the exponential which has no effect on the detection process. Through a similar line of analysis, the same results are obtained by centering an area photodetector at $-\gamma_c$ at plane P_4.

We have therefore found two alternative methods for implementing homodyne detection: the first method is to use a point probe at $\gamma = 0$; the second method is to use an area detector centered at either $+\gamma_c$ or $-\gamma_c$. Because the point probe overlaps the signals only at $\gamma = 0$, we can also use an area detector to detect this term. Hence, we can place one *grand photodetector* at plane P_4 to collect all the light and still have an output of the form $c(t - T)\cos[2\pi f_c t + \phi_0]$. Because the terms at $2\gamma_c$ and $-2\gamma_c$ do

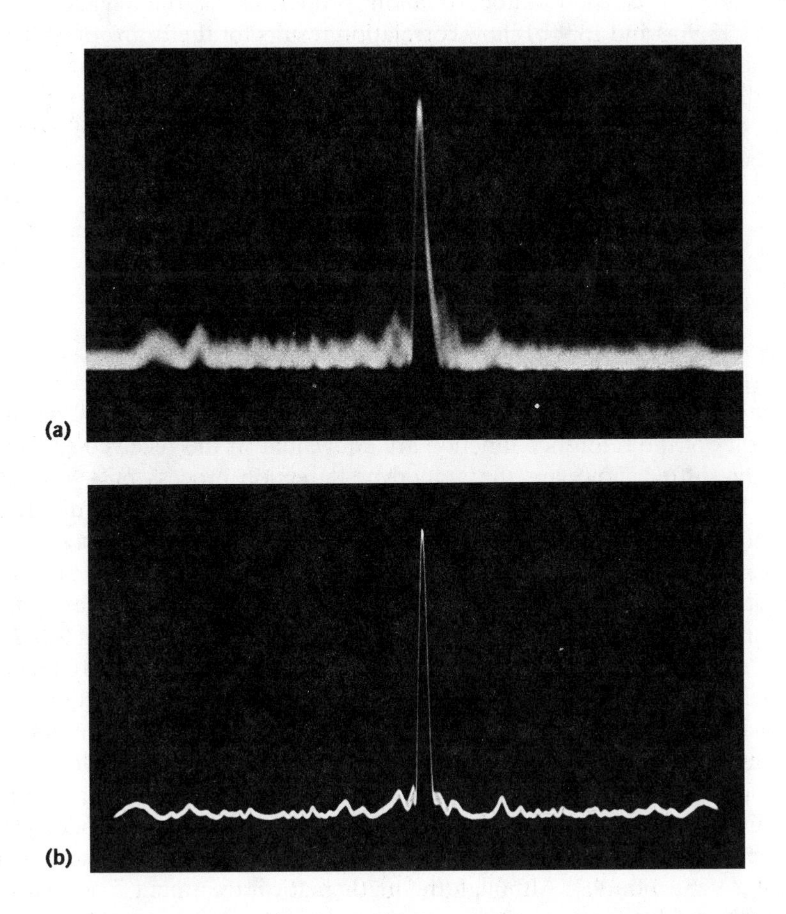

Figure 13.9. Experimental results for a 36 Frank code: (a) by the optical matched filter; (b) by the electronic matched filter (courtesy E. B. Felstead) (112) (copyright © IEEE, 1967).

not overlap with any reference beams, they are not within the passband of the filter.

An interesting extension of this idea is that the photodetector can be placed *anywhere* after plane P_3 in a fashion similar to our discussion in Chapter 12, Section 12.3 in connection with Figure 12.4. The photodetection process selects only those wavefront segments from the signal and reference beams that overlap and propagate in the same direction. Once the photodetector is placed beyond the beam combiner, it does not matter where the light is collected.

Felstead provides some nice experimental results by using polyphase codes that have good autocorrelation properties as the signal (113). Figures 13.9(a) and 13.9(b) show correlation results for the Frank polyphase code after envelope detection. Figure 13.9(a) shows the result obtained optically; Figure 13.9(b) shows the result obtained electrically. We see that the results are in good agreement except for a slightly reduced signal-to-noise ratio in the optical case.

13.6. HETERODYNE DETECTION

The coherent detection techniques used in Sections 13.4 and 13.5 involve the principle of homodyne detection. The underlying assumption is that the temporal carrier frequency of the signal and the spatial carrier frequency of the reference function are equivalent. If the received signal is Doppler shifted, however, the homodyne-detection process fails to operate. The physical insight as to the cause of this failure is different depending on the detection method. If the detection is in the vicinity of $\gamma = 0$, the Doppler frequency shift causes the correlation peak to move while the probe remains fixed. As the relative displacement increases, the performance decreases due to the lack of overlap. For $a(x) = \text{rect}(x/L)$, the performance degrades as $\text{sinc}(\gamma_d L)$, where γ_d is the spatial frequency equivalent of the Doppler shift f_d (see Problem 13.4). If the detection is in the vicinity of $\pm\gamma_c$, the reason for the loss of performance is different. In this case, the received signal spectrum is displaced relative to the reference spectrum and there is a linear phase induced across the spectrum. The linear phase induces a spatial frequency over the photodetector surface that causes a reduction in the spatial integral, as we showed in Chapter 9, Section 9.3. Although the mathematical treatment is somewhat different, the loss in performance is the same, as a function of f_c, as for detection at $\gamma = 0$.

If we wish to heterodyne-detect signals with Doppler shifts, we need to provide a completely separate reference beam derived from the same laser source. We can use one of the schemes discussed in Chapters 9–12; the general principles of operation are the same and will not be repeated here.

13.7. CARRIER FREQUENCY REQUIREMENTS

In Chapter 5, Section 5.13 we calculated the spatial carrier frequency requirements for matched filters and in Chapter 10, Section 10.5 we calculated the temporal carrier frequency requirement for separating bias terms from the desired output of a heterodyne spectrum analyzer. In each case the requirements were based on different criteria. On the surface it may seem that the requirements on the carrier frequency for heterodyne correlation are similar to those for heterodyne spectrum analysis. The key difference is that the correlation function may have a very short duration and we need several cycles of the carrier to properly support the function. We can see from the temporal Fourier transform of Equation (13.25) that the bias terms are centered at $f = 0$ and have bandwidths of the order of $2W$. The desired cross-product signal also has bandwidth of the order of W. Although we do not require complete *spatial frequency separation* of all terms if we detect at $\gamma = 0$, as shown in Figure 13.3, we do require complete *temporal frequency separation* of the terms in Equation (13.25). From this line of reasoning, we might conclude that a carrier frequency $f_c > 3W/2$ is sufficient to separate the terms. This is true. In general, however, the carrier must be much higher than $3W/2$ to support the correlation function. In Figure 13.10(a) we show the envelope of the correlation function for a pseudorandom sequence whose chips are of duration T_0. The correlation function therefore has a triangular shape with its mainlobe width being of the order of $2T_0$. Figures 13.10(b), 13.10(c), and 13.10(d) show the effects of using successively lower carrier frequencies; the case of Figure 13.10(d) represents the case where f_c is of the order of $3W/2$ so that only a few cycles of the carrier support the envelope. The envelope is then difficult to measure accurately both with respect to its magnitude and time of arrival. From an electronic demodulation point of view, the issue is the steepness of the skirts of the filter function; a high carrier frequency helps to obtain greater detection accuracy. The penalty, of course, is that the bandwidth of the photodetector must increase proportionally; this increase adversely impacts the signal-to-noise ratio.

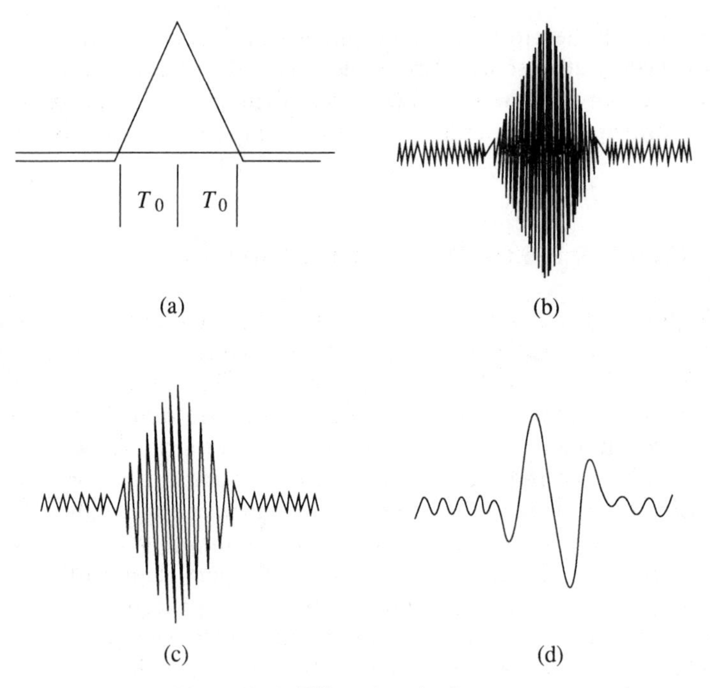

Figure 13.10. Effect of carrier frequency.

13.8. ILLUMINATION REQUIREMENTS

So far we have assumed that a point light source provides a plane wave of monochromatic light at plane P_1. The implication is that the source has a high degree of both spatial and temporal coherence. We now show that both these requirements can be relaxed, subject to some constraints imposed by the need to satisfy the Bragg illumination condition.

Suppose that the source contains more than one wavelength. The position of a spatial frequency component in the Fourier-transform plane has a linear dependence on wavelength so that for a given spatial frequency α, the position of the diffracted light is $\xi = \alpha\lambda F$. Thus the effect of a polychromatic source is to spread a Fourier spatial frequency component over a distance proportional to $\lambda + \Delta\lambda$, where $\Delta\lambda$ is the range of wavelengths. At plane P_3, however, the inverse Fourier transform causes all spectral components to form an image without any evidence of spatial spreading, provided that the optics are free of chromatic aberrations. Once the multiplication by the reference signal occurs, other effects due to

$\Delta\lambda$ are not important because the light from the two beams have been interferometrically combined; they overlap and are collinear. Therefore, although we used Fourier analysis to obtain the key results, the fundamental correlation operation is based on what happens in the space domain, not on what happens in the frequency domain.

The degree of temporal coherence is unimportant when the Raman-Nath mode of operation is used; but the wavelength spread may cause a departure from the Bragg-angle illumination condition [$\sin\theta_B = (\lambda/2\Lambda)$] when the Bragg mode is used. Injection laser diodes are now available with spectral widths of the order of $\Delta\lambda/\lambda = 0.1\%$; these fractional widths have a small impact on the range for $\sin\theta_B$ which, in turn, has a negligible effect on the diffraction efficiency of the acousto-optic cell.

As the source increases in size, the Fourier transform spreads, but by the same arguments as given above, this spreading has no effect on $c(t)$ other than a small change in its magnitude. A finite source does, however, produce a spectrum of plane waves that also affects the Bragg condition. Typical injection laser diodes have sizes ranging from 1 to 20 μ; when used with a 50-mm focal-length collimating lens, the resulting angular spread about the Bragg angle is less than 10^{-3} rad, with a negligible effect on the diffraction efficiency.

13.9. INTEGRATE AND DUMP

So far we have described a true matched-filtering mode of operation. Such an operation is rarely performed electronically because the number of computations required per unit time is excessive when processing a wide-band signal. For example, if a signal has a time duration T and a bandwidth W, the correlation operation requires $2TW$ complex multiplications and summations during each sample time period $1/(2W)$ for a computational rate of $4TW^2$. For a system having a bandwidth of 100 MHz and a 10-μsec time duration, the computational rate is $400(10^9)$ complex operations per second. When the bandwidth increases to 1000 MHz while the time bandwidth product remains the same, the computational rate is increased by a factor of 100, to $40(10^{12})$ operations per second. Even very high-speed electronic systems cannot sustain such high computational rates. Furthermore, these calculations do not include the time needed to perform the Doppler search, which is calculated in parallel in the optical system.

For signal acquisition and tracking, electronic systems use a correlation technique called *integrate and dump*. This technique involves integrating

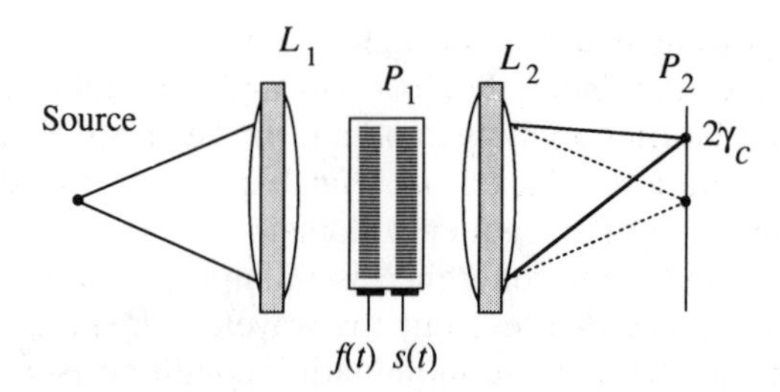

Figure 13.11. Integrate-and-dump architecture.

the product $f(t)s^*(t)$ under the assumption that $\tau = 0$. If so, the correlation peak exceeds threshold after some integration time; if not, one signal is slipped in time, generally by $T_0/2$ sec, and the correlation and thresholding process is repeated. On the assumption that the signal duration is T sec, it clearly takes T sec to integrate and threshold the product for each half chip displacement. If the signals have a large time difference uncertainty, the acquisition process can take a long time. The search time is even longer when Doppler is present; when the Doppler is rapidly changing, the search procedure may not converge.

Although it is not a particularly computationally intensive operation, we now show, for completeness, how to perform the integrate-and-dump operation optically. Consider the system shown in Figure 13.11 in which the signals $f(t)$ and $s(t)$ travel in the same direction but with some time delay τ between them. The simplest way to perform the integrate-and-dump operation is to use direct detection. Suppose that the received signal $f(t)$ is at a carrier f_c and that it upshifts the light frequency. We modulate a second carrier f_c with $s(t)$ and arrange the light interaction to also upshift the light frequency. The combined double upshift produces the correlation peak at $2\gamma_c$ at the Fourier plane P_2. A small photodetector then senses $c^2(t)$. We are not particularly interested in coherent detection here because $c^2(t)$ has only very low-frequency content of the order of $1/T$.

13.10. SOME MORE CONFIGURATIONS

So far we have described members of a family of space-integrating correlators wherein the signals $f(u, t)$ and $s^*(-u)$ convolve by moving one signal past the other and integrating the product. The structure of the system from Figure 13.1 shows that the space signals $f(u, t)$ and $t(u)$ are multiplicative when we detect light in the region of $\gamma = 0$. We close this

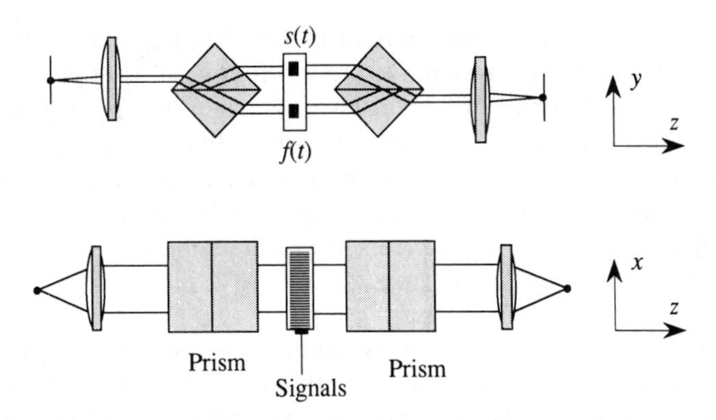

Figure 13.12. A dual-port correlator.

chapter by describing an architecture, similar to that shown in Figure 11.4 in connection with heterodyne spectrum analysis, in which the multiplication is performed is a different way. Figure 13.12 shows two views of a dual-channel interferometer in which the signals $f(t)$ and $s(t)$ do not multiply in space. Instead, two transducers are stacked in the vertical direction, and prisms are used to split and recombine the light beams. Lens L_2 produces the Fourier transforms of the two signals and the light intensity is

$$\begin{aligned} I(\alpha, t) &= |S(\alpha - \alpha_c, t) + F(\alpha - \alpha_c, t)|^2 \\ &= |S(\alpha - \alpha_c, t)|^2 + |F(\alpha - \alpha_c, t)|^2 \\ &\quad + 2\,\mathrm{Re}\{F(\alpha - \alpha_c, t) S^*(\alpha - \alpha_c, t)\} \end{aligned} \tag{13.31}$$

in which the cross-product term, when integrated over all α, produces an output proportional to $c(2t - T)\cos[4\pi f_c(t - T)]$ by virtue of the analysis given in Section 13.5 where we integrated light in the vicinity of $\pm\gamma_c$.

There are four general ways to perform correlation:

1. We can multiply $f(x)$ and $s^*(x + u)$ in the space domain and detect at $\gamma = 0$ in the Fourier domain with a point detector to perform the required integration over the spatial coordinate x.
2. We can multiply $F(\alpha, t)$ and $S^*(\alpha, t)$ in the Fourier domain, where one of these functions is in the form of a spatial filter, and detect at $x = 0$ in the space domain with a point detector.

3. We can add $F(\alpha, t)$ and $S(\alpha, t)$ in the Fourier plane, square-law detect using a large-area detector, and retain only the cross-product term to implement the heterodyne transform. The conjugate of $S(\alpha, t)$ is provided by the square-law operation.
4. We can add $f(x)$ and $s(x+u)$ in the space plane, square-law detect with a large-area detector, and retain only the cross-product term to implement the heterodyne transform. The conjugate of $s(x+u)$ is provided by the square-law operation.

With methods 1 and 2, we can use direct detection, homodyne detection, or heterodyne detection. With methods 3 and 4, we must use the heterodyne-transform technique because we must separate the desired cross-product term from the bias terms. The general rule, then, is that we use point detectors (or point probes) when we produce correlation through multiplication of functions, but we use area detectors when we produce correlation through a nonlinear operation on the sum of functions.

PROBLEMS

13.1. You receive a signal whose form is given by

$$s(t) = \sum_{n=1}^{N} \cos(2\pi n f_0 t + \phi_n),$$

where the ϕ_n are arbitrarily chosen phases.

(a) Calculate the mean value and the average power (the variance) of the signal $s(t)$ and indicate how they depend on the phase.

(b) Calculate the autocorrelation function for $s(t)$ and indicate how it depends on the phase:

$$c(\tau) = \lim_{T\to\infty} \frac{1}{2T} \int_{-T}^{T} s(t) s^*(t+\tau)\, dt.$$

(c) Compare the ratio of the square of the peak value of the autocorrelation function with the variance from (a). This is the correlation gain.

13.2. You have a space-integrating correlator that uses a fixed reference mask $s(u)$ set up to receive a signal of the form $f(t)\cos(2\pi f_c t)$.

Suppose that the acousto-optic cell has $T = 20$ μsec, $W = 40$ MHz, and $v = 630$ m/sec. Suppose that $f(t) = \text{rect}[(t - 200)/20]$ where the times are in microseconds and that the Doppler frequency is 50 kHz. Calculate and sketch the proper form for $s(u)$. When and where will the maximum correlation peak occur if all lens focal length are 500 mm and $\lambda = 0.5$ μ? Sketch and label the output in the α and t coordinate axes. If the total Doppler range is ± 100 kHz, what detector size would you use and how many do you need? Assume that the signal is uniformly illuminated and that the noise spectrum is flat.

13.3. You have designed and built a space-integrating correlator in which you drive an acousto-optic cell with an input signal $f(t)$ and use a carrier frequency reference function $s(u/v)$, all based on using a solid-state laser with $\lambda = 830$ nm. A much more powerful laser has become available, but its wavelength is $\lambda = 750$ nm. You want to retrofit the system with this laser, if possible. What changes would you need to make to the system if it operates (1) in the direct detection mode and (2) in the homodyne mode? Provide sketches and analyses as appropriate to support your answer.

13.4. Derive the relationship for the rate at which $|c(t - T/2)|^2$ drops as a function of the Doppler frequency γ_d in a space-integrating homodyne-detection system. If we allow a 1.5-dB loss in correlation-peak intensity, what is the maximum Doppler that can be permitted if the acousto-optic cell has a time duration of 10 μsec?

13.5. Expand the intensity function

$$I(\gamma, t) = \left|\text{sinc}(\gamma L) + G_2(\gamma, t) + G_2^*(\gamma, t)\right|^2,$$

calculate and sketch the temporal frequency response of the terms produced by a photodetector.

13.6. Prove that the output of a programmable matched-filtering system (one which uses acousto-optic cells for both the received signal and the reference function) is of the form $c(2t - a)\cos(4\pi f_c t + \phi)$, where a is some constant. Determine the value of a. Where is the detector placed in the system to get this result? (See Section 13.2.6.)

14

Time-Integrating Systems

14.1. INTRODUCTION

With the exception of the radiometer, all the systems described so far in this book are of the *space-integrating type*. The name derives from the fact that operations such as spectrum analysis and correlation have their ranges of integration limited by the spatial extent of the input signal. There are, in some applications, additional integration processes implied by bandpass filtering of the electronic signal after detection. However, this postdetection processing integration does not affect the intrinsic frequency resolution or correlation gain produced by the space-integrating nature of the optical system.

In the 1970's, *time-integrating* optical signal-processing systems were developed. The early work was due to Montgomery (116), who described a time-integrating system that uses chirp signals to implement operations such as spectrum analysis and correlation. Later, Sprague and Koliopoulos introduced a time-integrating correlator that did not require the use of auxiliary chirp signals (117). From these two major building blocks, new and useful techniques evolved for performing operations such as spectrum analysis, correlation, ambiguity and Wigner-Ville function generation, and range/Doppler processing.

The ability to integrate in time, as well as in space, provides for improved performance. For example, a time aperture of the order of 33 msec is required to provide a 30-Hz frequency resolution in a spectrum analyzer. A one-dimensional acousto-optic cell to support such a requirement must have a physical aperture of about 20 m, which is clearly unacceptable. In other applications, we may need correlation gains of the order of 60–80 dB, far beyond the 30-dB or so available from one-dimensional correlators. Sometimes the increase in performance can be obtained by using a two-dimensional architecture, as discussed in Chapter 15; in other cases, the gain can be achieved only through time integration, as discussed in this chapter.

We begin by describing basic methods for improving the temporal resolution of a spectrum analyzer. We then describe two basic types of time-integrating correlators. We conclude this chapter by describing several systems that combine these two major operations.

14.2. SPECTRUM ANALYSIS

We can use a time-integrating technique to improve the frequency resolution of a space-integrating spectrum analyzer. Figure 14.1 shows a photodetector array at plane P_5 that integrates light over a long time period relative to the period of any of the temporal signals in the system. We want to display the spectrum at plane P_5 of a baseband signal $s(t)$. Suppose that we establish a reference signal at plane P_5 that behaves as a *distributed local oscillator* with a different temporal frequency at each spatial position at plane P_5. Sum and difference frequencies are generated when the local oscillator is multiplied by a temporal frequency component f_j contained in the signal $s(t)$; the difference frequency is zero only at that spatial position where the local-oscillator frequency is f_j. The response of the photodetector therefore increases linearly over the integration time period to produce the proper magnitude coefficient $S(f_j)$ of the spectrum. The product of the frequency component f_j and all other local oscillators at the output plane produces difference frequencies whose magnitudes integrate to zero.

Various methods are available for creating the desired distributed local oscillator that provides both the spatial carrier frequency and the necessary temporal probes. Because a chirp waveform has an instantaneous frequency that ranges from f_1 to f_2, it is an obvious source of the linear distributed frequencies required of the local oscillator. But if such a chirp

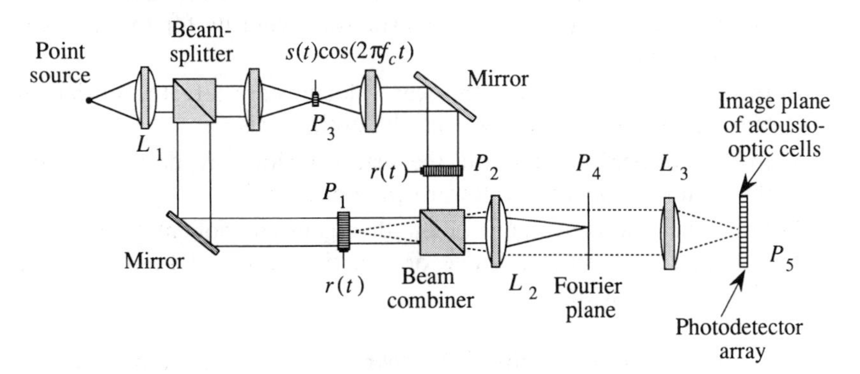

Figure 14.1. Interferometric time-integrating architecture.

is used to drive the acousto-optic cell in the lower branch of the Mach-Zehnder interferometer shown in Figure 14.1, its image at plane P_5 moves in the opposite direction with velocity v when the magnification is unity. The distributed local oscillator therefore does not have a fixed temporal frequency at each spatial frequency location because the chirp waveform moves. The local oscillator can be stabilized in space by driving the acousto-optic cell in the upper branch of the interferometer with a similar chirp signal, but in the opposite direction at plane P_5 relative to the first chirp. As the two chirps move through the acousto-optic cells, they mix at plane P_5 to provide the distributed local oscillator, as we explain in the next section.

14.2.1. Requirements on the Reference Signals

For the moment, we remove the signal modulator at plane P_3 so that we concentrate on how the two chirp signals produce the required distributed local oscillator. From the discussion of acousto-optic scanners given in Chapter 7, Section 7.9.2, we recall that a chirp signal at plane P_1 produces a focused spot just before or just after plane P_4, depending on whether we use an upchirp or a downchirp signal and on whether we use the positive or negative diffracted order. Also, the temporal frequencies are upshifted or downshifted depending on which diffracted order is used. These relationships are summarized in Figure 14.2; the upchirp and downchirp conditions are shown in parts (a) and (b) when the acoustic wave propagates in the positive x direction. The corresponding conditions when the acoustic wave propagates in the negative x direction are shown in parts (c) and (d). From these four sketches we conclude the following:

1. The scanning velocity is in the same direction as the acoustic wave velocity for the convergent component of the chirp and is in the opposite direction from the acoustic wave velocity for the divergent component of the chirp.
2. In the upchirp condition, the scan velocities are directed so that the scanning spots move away from the optical axis.
3. In the downchirp condition, the scan velocities are directed so that the scanning spots move toward the optical axis.
4. For either the convergent or the divergent component of the chirp, we can select either positive or negative temporal frequencies for either scan velocity.

To help visualize the origin of the distributed local oscillator and to set the stage for determining the proper configuration for the second

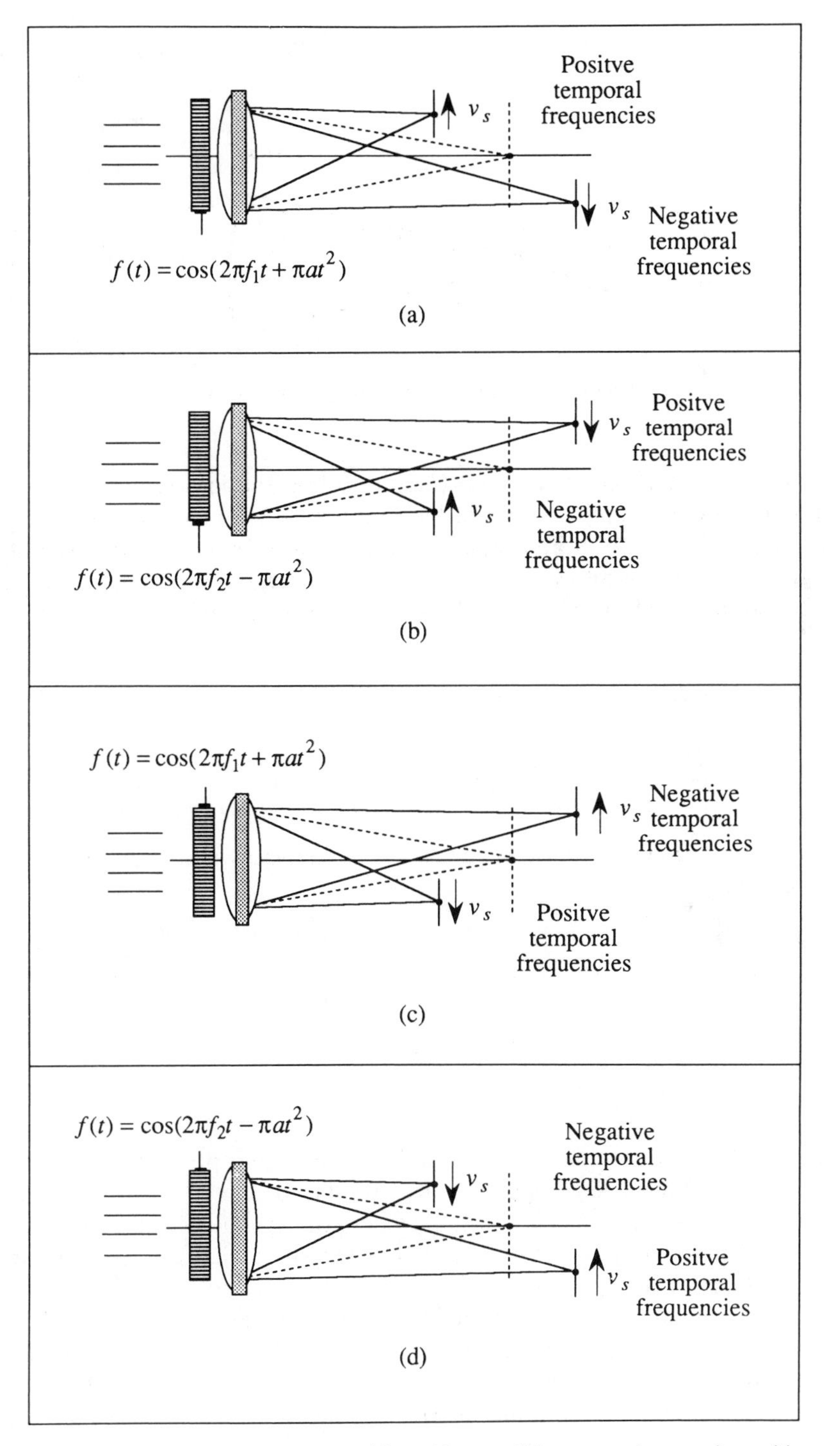

Figure 14.2. Possible chirp conditions: (a) upchirp condition, acoustic wave in positive x direction; (b) downchirp condition, acoustic wave in positive x direction; (c) upchirp condition, acoustic wave in negative x direction; and (d) downchirp condition, acoustic wave in negative x direction.

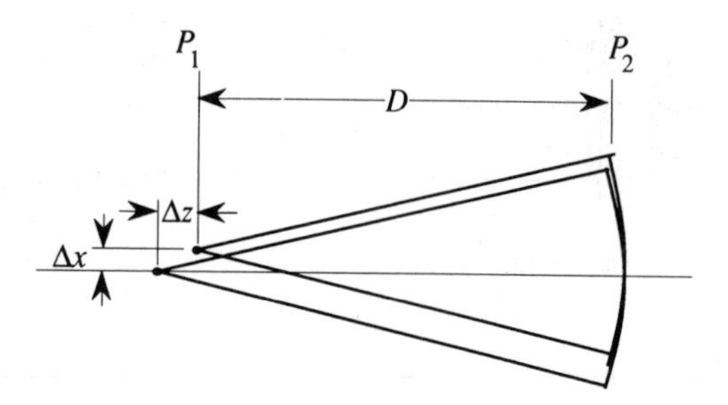

Figure 14.3. Stationary linear frequency pattern.

acousto-optic cell, we recall from Chapter 3 that the interference pattern produced by two samples, spaced by an interval Δx, produces the linear fringe pattern shown in Figure 3.6. To achieve a *linearly increasing* spatial frequency, we must introduce a separation between the two samples in the z direction as well, as shown in Figure 14.3. The intensity at plane P_2 is

$$\begin{aligned} I(\xi) &= \left| e^{-j(\pi/\lambda D)\xi^2} + e^{-j[\pi/\lambda(D+\Delta z)](\xi-\Delta x)^2} \right|^2 \\ &= 2\left[1 + \cos\left\{\frac{\pi}{\lambda D}\xi^2 - \frac{\pi}{\lambda(D+\Delta z)}(\xi - \Delta x)^2\right\}\right]. \end{aligned} \tag{14.1}$$

The spatial frequency of $I(\xi)$ is given by the derivative of the argument of the cosine:

$$\begin{aligned} \text{spatial frequency} &= \frac{1}{2\pi}\frac{\partial}{\partial \xi}\left[\frac{\pi}{\lambda D}\xi^2 - \frac{\pi}{\lambda(D+\Delta z)}(\xi - \Delta x)^2\right] \\ &= \frac{\Delta x D + \Delta z\, \xi}{\lambda D(D+\Delta z)}. \end{aligned} \tag{14.2}$$

The spatial frequency therefore has a fixed component proportional to Δx and a linearly variable component proportional to Δz.

This exercise helps us choose the proper configuration for the acousto-optic cell from the possible configurations shown in Figure 14.2. Suppose that we operate the cell at plane P_1 in the upchirp mode, with the acoustic wave traveling in the positive x direction. The two scanning spots will then be as shown by the solid lines in Figure 14.4. The scan velocity for the divergent part of the wave is in the negative x direction, so that the scanning spot moves away from the optical axis, consistent with the sketch

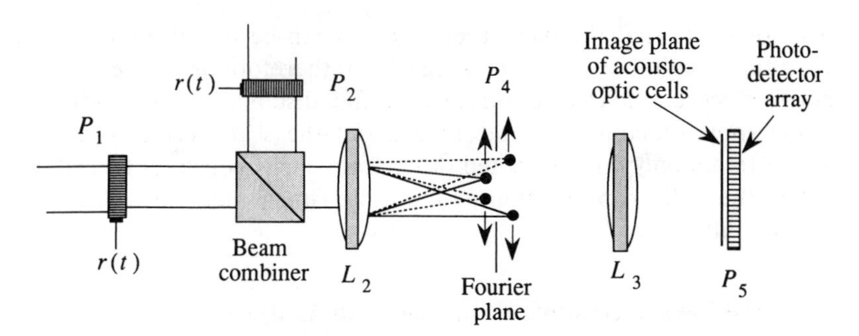

Figure 14.4. The required acousto-optic cell configuration.

of Figure 14.2(a); the scan velocity for the convergent part of the wave is in the positive x direction so that this scanning spot also moves away from the optical axis.

An inspection of the possibilities shows that the appropriate condition for the second cell is given in part (c). The upchirp signal propagates in the negative x direction and produces a scanning spot that has the same velocity as that produced by the first cell as well as the axial displacement required to produce the linear spatial frequency for the local oscillator. The fixed displacement in the horizontal direction between this spot and the one produced by the first cell provides a fixed spatial carrier frequency at plane P_5.

The appropriate cell configurations are shown in Figure 14.4, in which dotted lines show rays from the cell in the upper branch. For the Raman-Nath mode of operation, we find two pairs of spots scanning in the same direction, with the same velocity, but displaced slightly in the z-direction. It is this displacement in z, as well as the intrinsic displacement in the spatial frequency position, that produces a stable distributed local oscillator at plane P_5. When we operate in the Bragg mode, we arrange the illumination so that we select the negative (or positive) diffracted order from each of the cells. This choice results in only two scanning spots that are on the same side of the Fourier plane and are moving in opposite directions. Although this setup appears to contradict the criteria established above, recall from Chapter 13, Section 13.10, that one way to perform mutiplication optically is to add two functions in a space plane and to square-law detect the result. The squaring operation produces the conjugate of one of the two functions so that the behavior of the negative diffracted order assumes that of a positive diffracted order. Thus, the scanning spots are, in effect, on the same side of the Fourier plane and are moving in the same direction with equal velocities.

It is fortunate that the basic requirements can be met using the same upchirp function in both cells. This signal can therefore be derived from a common electrical source, ensuring that the distance between the two scanning spots remains the same throughout the scan period. Also, because we retain only one diffracted order from each acousto-optic cell, we see that the cells can be operated in the Bragg mode for increased diffraction efficiency.

14.2.2. The Basic Operation of the Spectrum Analyzer

Having shown how the distributed local oscillator is provided by a pair of counterpropagating chirp signals, we examine the behavior of the time-integrating spectrum analyzer in detail. Suppose that the acousto-optic cell at plane P_3 in the upper branch of the interferometer is driven by the signal $s(t)\cos(2\pi f_c t)$ and that we retain only the positive diffracted order. The chirp signal that drives the acousto-optic cell at plane P_2 is

$$r(t) = \cos\left[2\pi f_a t + \pi a t^2\right], \qquad 0 \le t \le kT, \tag{14.3}$$

where f_a is the starting frequency of the chirp waveform, a is the chirp rate, and kT is the period of the chirp. The total waveform when we retain only the negative diffracted order from the acousto-optic cell containing the chirp siganl in the upper branch of the interferometer at plane P_5 is

$$r_1(u,t) = s(t)e^{j2\pi f_c t}e^{-j[2\pi f_a(t-T/2+u/v)+\pi a(t-T/2+u/v)^2]}, \qquad T \le t \le kT, \tag{14.4}$$

where u is the spatial coordinate at plane P_5 and we ignore unessential magnitude and phase factors. The laser source is blanked for the time period $0 \le t \le T$ while the cell is being filled with the chirp signal.

The acousto-optic cell in the lower branch is driven with a similar chirp waveform so that the waveform due to the lower branch of the interferometer at plane P_5 is

$$r_2(u,t) = e^{-j[2\pi f_b(t-T/2-u/v)+\pi a(t-T/2-u/v)^2]}, \qquad T \le t \le kT, \tag{14.5}$$

where f_b is the starting frequency of the chirp and we retain only the negative diffracted order, as required from the discussion in

Section 14.2.1. The intensity at plane P_5 is

$$I(u,t) = |r_1(u,t) + r_2(u,t)|^2. \tag{14.6}$$

We integrate $I(u,t)$ over the time interval from T to kT:

$$g(u) = \int_T^{kT} I(u,t)\,dt. \tag{14.7}$$

In the time-integrating architecture, the value of k determines the integration time. Because the performance of space-integrating systems are proportional to T, we anticipate that the performance of time-integrating architectures will be proportional to kT, with $k \gg 1$. We can therefore more easily make performance comparisons with space-integrating architectures where the integration is typically over a time interval T.

We substitute Equations (14.4) and (14.5) into Equation (14.6) and then into Equation (14.7) to find the three terms at the output. The first term is

$$\begin{aligned} g_1(u) &= \int_T^{kT} |r_1(u,t)|^2\,dt \\ &= \int_T^{kT} |s(t)|^2\,dt = (k-1)T\langle |s(t)|^2 \rangle, \end{aligned} \tag{14.8}$$

which we recognize as being proportional to the time average of the magnitude squared of the input signal $s(t)$. If $s(t)$ is a reasonably well-behaved signal, its average power is a constant in time and so too is $g_1(u)$. The second term is

$$\begin{aligned} g_2(u) &= \int_T^{kT} |r_2(u,t)|^2\,dt \\ &= \int_T^{kT} dt = (k-1)T, \end{aligned} \tag{14.9}$$

which is also a constant. Having shown that the first two terms are constant bias terms, we concentrate on the cross-product term:

$$\begin{aligned} g_3(u) = 2\,\mathrm{Re}\Bigg\{ \int_T^{kT} & s(t) e^{j2\pi f_c t} e^{-j[2\pi f_a(t-T/2-u/v) + \pi a(t-T/2-u/v)^2]} \\ & \times e^{+j[2\pi f_b(t-T/2+u/v) + \pi a(t-T/2+u/v)^2]}\,dt \Bigg\}, \qquad T \le t \le kT. \end{aligned} \tag{14.10}$$

We expand the exponentials and collect terms to find that

$$g_3(u) = 2\,\mathrm{Re}\left\{e^{j[2\pi(f_a+f_b-aT)u/v-\pi(f_b-f_a)T]}\right.$$
$$\left.\times\int_T^{kT} s(t)e^{j[2\pi(f_c+f_b-f_a)t]}e^{+j4\pi atu/v}\,dt\right\}, \qquad T \le t \le kT. \tag{14.11}$$

Because we expect the output to include a spatial carrier frequency, we express (14.11) as

$$g_3(u) = 2\,\mathrm{Re}\left\{e^{+j[2\pi(\alpha_a+\alpha_b-\alpha_0)u-\phi_0]}\right.$$
$$\left.\int_T^{kT} s(t)e^{j[2\pi(2au/v+f_c+f_b-f_a)t]}\,dt\right\}, \qquad T \le t \le kT, \tag{14.12}$$

where α_a and α_b are spatial frequencies corresponding to f_a and f_b and $\phi_0 = \pi(f_b - f_a)T$. The spatial frequency span is related to the duration of the acousto-optic cell by

$$\alpha_0 = \frac{aT}{v}. \tag{14.13}$$

Suppose that we make an association for the temporal frequency variable f in which

$$f = \frac{2au}{v} + f_c + f_b - f_a, \tag{14.14}$$

so that Equation (14.12) becomes

$$g_3(u) = 2\,\mathrm{Re}\left\{e^{j[2\pi(\alpha_a+\alpha_b-\alpha_0)u-\phi_0]}\int_T^{kT} s(t)e^{j2\pi ft}\,dt\right\}, \qquad T \le t \le kT. \tag{14.15}$$

The integral is in the form of a Fourier transform over a time window of duration $(k - 1)T$. The sign of the kernel in Equation (14.15) simply indicates the direction of the positive frequencies at plane P_5. The transform part of Equation (14.15) can, by definition, be written as $S(f)$. If

$S(f)$ is complex valued, we find that

$$\boxed{g_3(u) = 2|S(f)|\cos[2\pi(\alpha_a + \alpha_b - \alpha_0)u - \phi_0 + \theta(f)], \qquad T \leq t \leq kT,} \tag{14.16}$$

where $\theta(f)$ is the phase of the spectrum $S(f)$. The complex-valued information is therefore encoded on a spatial carrier frequency, just as with spatial carrier frequency filters. The association in Equation (14.14) shows that the spatial variable u at plane P_5 represents the temporal frequency variable f.

14.2.3. The Key Features of the Time-Integrating Spectrum Analyzer

Some of the important features of this time-integrating architecture follow.

1. The frequency resolution is inversely proportional to the integration time $(k-1)T$ and can, in principle, be made as small as we wish by increasing k.
2. The range of temporal frequencies Δf is determined by the *range of the spatial variable* Δu at plane P_5. From Equation (14.14) we find that the frequency range is

$$\Delta f = \frac{2a\,\Delta u}{v} = \frac{2aL}{v} = 2aT, \tag{14.17}$$

so that by changing the chirp rate a we can change the frequency range that is covered. Note that aT is exactly equal to the range of temporal frequencies of the chirp signal that is present in the acousto-optic cell at any instant. The chirp rate is equal to the bandwidth of the acousto-optic cell divided by the chirp period: $a = W/kT$, so that

$$\boxed{\Delta f = \frac{2W}{k}.} \tag{14.18}$$

This result shows that *as we increase the value of k to achieve better frequency resolution, we find that the available frequency range is reduced by the same factor*.

3. The time bandwidth product at the output plane is $kT\,\Delta f$. Because the signals in the two acousto-optic cells are counterpropagating, the time bandwidth product at plane P_5 is $2TW$. We rearrange the factors in Equation (14.18) and multiply both sides by T to find that

$$kT\,\Delta f = 2TW. \tag{14.19}$$

Furthermore, the frequency resolution is

$$\boxed{f_0 = \frac{\Delta f}{2TW} = \frac{2W/k}{2TW} = \frac{1}{kT},} \tag{14.20}$$

as we asserted in feature (1).

4. We can shift the position of the spectrum at plane P_5 to increase the useful frequency range because half of the information in a symmetric spectrum is redundant. From Equations (14.14) and (14.17) we see that the position of the zero frequency at the output plane is at

$$\boxed{u = \frac{L}{\Delta f}(f - f_c - f_b + f_a),} \tag{14.21}$$

and that positive temporal frequencies correspond to positive spatial positions, provided that $(f_c + f_b - f_a) \geq 0$. If we want to maximize the frequency range, we want $f = 0$ to occur at $u = -L/2$. To force this event, we must satisfy the relationship that

$$\boxed{f_c + f_b - f_a = \Delta f.} \tag{14.22}$$

Under this condition, the local oscillator ranges from 0 to $2\,\Delta f$; the frequency range is doubled relative to the previous case where we displayed the two-sided spectrum of $s(t)$. Furthermore, because the zero frequency is now located at $-L/2$, only the positive frequencies of $s(t)$ are displayed. If we use a baseband electro-optic modulator to introduce the signal $s(t)$, the value of f_c is zero. In this case, we can set $f_b - f_a = \Delta f$, or we can set $f_b = f_a$ and delay one of the chirps slightly relative to the other to provide the frequency offset.

5. The value of the spatial carrier frequency is found from the exponential coefficient from Equation (14.15):

$$\boxed{\begin{aligned} \alpha_a + \alpha_b - \frac{aT}{v} &= \alpha_a + \alpha_b - \frac{T}{v}\frac{W}{kT} \\ &= \alpha_a + \alpha_b - \frac{\alpha_{co}}{k}, \end{aligned}} \tag{14.23}$$

where α_{co} is the cutoff spatial frequency corresponding to the bandwidth W of the acousto-optic cell. Although the spatial carrier frequency is a function of the integration factor k, it tends to be more strongly influenced by the values of α_a and α_b. The carrier frequency will automatically be high enough to support the spectral information in $s(t)$ because it is high enough to support the chirp signals in the acousto-optic cells.

6. The photodetector array must have enough elements to resolve the spatial carrier frequency as given by Equation (14.23). The Nyquist sampling rate of two detector elements per cycle of the spatial carrier frequency is sufficient. The detection process typically consists of reading out the video line from the photodetector array so that the spatial signal is converted to a temporal signal. A bandpass filter, with its center frequency matched to the temporal equivalent of the spatial carrier frequency, is used to remove the bias terms from the cross-product term. The rest of the processing operations were described in Chapter 10 in connection with heterodyne spectrum analyzers.

To illustrate the operation of a time-integrating spectrum analyzer, suppose that we have an acousto-optic cell with a 100-MHz bandwidth and a time duration of 10 μsec; the time bandwidth product is therefore 1000. Suppose that we want to design a spectrum analyzer that achieves a 1-kHz frequency resolution. From Equation (14.20) we find that the integration time factor is $k = 1 \text{ ms}/10\ \mu\text{sec}$, so that $k = 100$. Equation (14.18) shows that the frequency range is $\Delta f = 2$ MHz and Equation (14.17) shows that the chirp rate must be $a = W/kT = (10^{11})$ Hz/sec. The time bandwidth product at plane P_4 is 2 MHz/1 kHz = 2000, just double the intrinsic system time bandwidth product of the acousto-optic cells.

14.3. TIME-INTEGRATING CORRELATION

In spread spectrum communication systems, the bandwidth is much larger than that needed to transmit just the data and the average transmitter

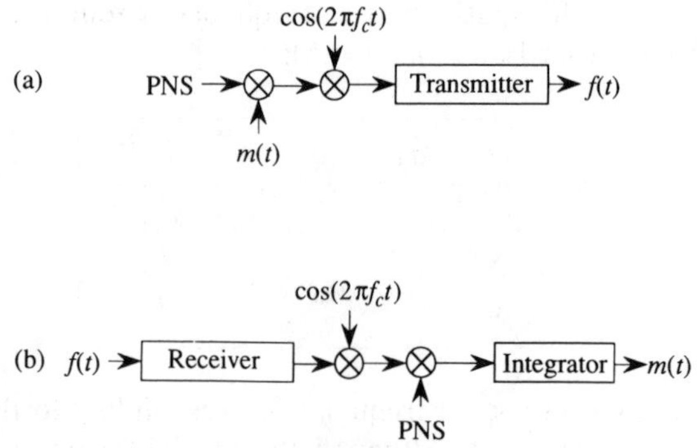

(a) PNS → ⊗ → ⊗ → Transmitter → $f(t)$; $m(t)$; $\cos(2\pi f_c t)$

(b) $f(t)$ → Receiver → ⊗ → ⊗ → Integrator → $m(t)$; $\cos(2\pi f_c t)$; PNS

Figure 14.5. Transmitter/receiver for a communication system.

power is therefore significantly reduced. Correlation is then used at the receiver to integrate the signal energy and to produce the signal-to-noise ratio needed for reliable detection. As shown in Figure 14.5(a), the message signal $m(t)$ is multiplied by any code, such as a pseudorandom sequence, that has good correlation properties. Because we may use a code that has 10^3–10^6 elements for each message symbol, the transmission bandwidth required is increased by a similar factor. The product of the code and the message data is then multiplied by the carrier frequency $\cos(2\pi f_c t)$ and transmitted to the receiver.

The receiver shown in Figure 14.5(b) first multiplies the received signal by $\cos(2\pi f_c t)$ and then by the pseudorandom sequence. The product is integrated for a period of time corresponding to the sequence length to complete the correlation operation. When the receiver is synchronized to the transmitter, the output of the system is the desired message signal $m(t)$. The advantage of such a coding scheme is that the signal $m(t)$ is protected by the coding scheme, and the probability of interception is therefore lower than if no coding were used. The integration also provides correlation gain so that lower average transmitter power is needed; the coding also provides immunity to certain types of noise sources, including some jammers.

A key problem with such communication systems is that the synchronization process may require a long time when the coding sequences are long and when no common clock sources are available to both the transmitter and receiver. Furthermore, the received signal may be Doppler shifted so that the carrier frequency is no longer f_c at the receiver. The allowable Doppler shift is quite small if the integration time is long; a shift of just one cycle in the carrier frequency over the integration time is

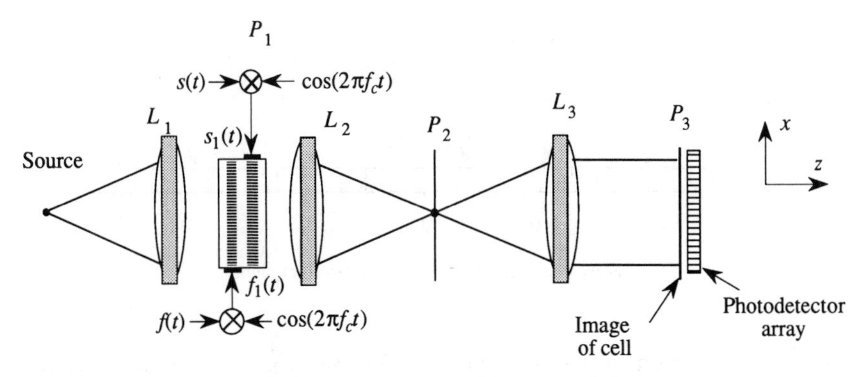

Figure 14.6. Basic time-integrating architecture due to Montgomery.

sufficient to reduce significantly the correlation gain, if not to destroy it altogether.

A search is needed over all possible time delays and all possible Doppler frequencies to achieve synchronization. In Chapter 13, we discussed space-integrating correlators that provide the simultaneous search over all possible time delays and Doppler frequencies. In space-integrating architectures, however, the integration time is limited by the signal resident in the acousto-optic cell; we may not have enough correlation gain to detect a weak signal buried in noise. We therefore turn our attention to time-integrating optical correlators.

14.3.1. Time-Integrating Correlator due to Montgomery

The earliest detailed description of a time-integrating correlator is available in the patent literature, where Montgomery describes a system that can be used for either correlation or spectrum analysis (116). The basic structure for a time-integrating correlator, as given by Montgomery, is shown in Figure 14.6. Suppose that the acousto-optic cells are operated in the Raman-Nath mode. The two signals have the form $f(t)\cos(2\pi f_c t)$ and $s(t)\cos(2\pi f_c t)$.

Figures 13.8(a)–13.8(c) in Chapter 13 showed how the spectra of the signals are generated and how they propagate to the Fourier plane. A similar process applies here. The result in the Fourier plane P_2 is shown in Figure 14.7, where we use F_+ as a shorthand notation for

$$F_+(\alpha, t) = \int_{-\infty}^{\infty} a(x) f\left(t - \frac{T}{2} - \frac{x}{v}\right) e^{j2\pi f_c(t-T/2-x/v)} e^{j2\pi\alpha x}\, dx, \quad (14.24)$$

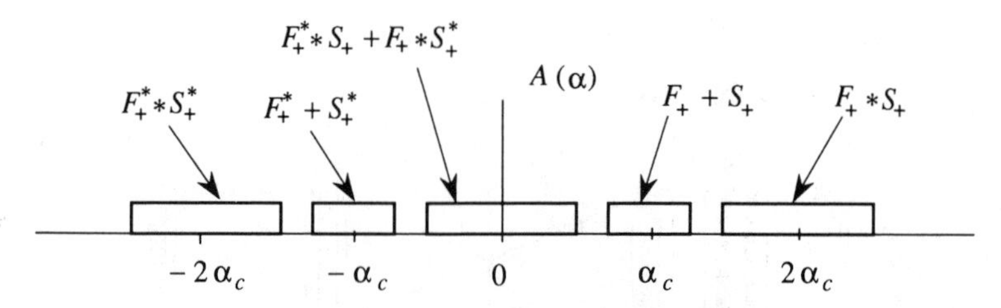

Figure 14.7. Spectral display of product signal.

and S_+ is defined in a similar fashion. We see that the spatial spectrum consists of nine terms. Three terms are centered on the optical axis; the term $A(\alpha)$ is the Fourier transform of the aperture weighting function $a(x)$ and the other two terms are the convolution of the spectra $F_+(\alpha, t)$ and $S_+^*(\alpha, t)$, plus the complex conjugate of the convolution.

Suppose that we place a spatial filter $H(\alpha)$ at plane P_2 that passes only the information in the bands centered at $\pm\alpha_c$. Lens L_3 performs a second Fourier-transform operation. The overall operation of the system is equivalent to imaging plane P_1 onto plane P_3, after removing the bias and cross-product terms centered at $\alpha = 0$ and at $\pm 2\alpha_c$. The result is equivalent to convolving the input signal with $\delta(x + u)$, where u is the coordinate at plane P_3. The amplitude at plane P_3, produced by the two positive diffracted orders, is therefore

$$a(u)\left[f\left(t - \frac{T}{2} + \frac{u}{v}\right)e^{j2\pi f_c(t - T/2 + u/v)} + s\left(t - \frac{T}{2} - \frac{u}{v}\right)e^{j2\pi f_c(t - T/2 - u/v)}\right]. \tag{14.25}$$

If we set $a(u) = \mathrm{rect}(u/L)$ for convenience, the light intensity at the output plane is

$$\begin{aligned} I(u,t) &= \left| f\left(t - \frac{T}{2} + \frac{u}{v}\right)e^{j2\pi f_c(t - T/2 + u/v)} + s\left(t - \frac{T}{2} - \frac{u}{v}\right)e^{j2\pi f_c(t - T/2 - u/v)}\right|^2, \\ &= I_1(u,t) + I_2(u,t) + I_3(u,t). \end{aligned} \tag{14.26}$$

The key difference between this system and the correlators we studied before is that the integration by the photodetector array is now over a time variable. Suppose that we use a CCD detector whose basic integration time is kT, where k is a constant and where $T = L/v$ is the time aperture of the cell. We therefore integrate Equation (14.26) over the time interval kT:

$$g(u) = \int_0^{kT} I(u,t)\,dt. \tag{14.27}$$

We use Equation (14.26) in Equation (14.27) and find that the first term has the form

$$g_1(u) = \int_0^{kT} \left| f\left(t - \frac{T}{2} + \frac{u}{v}\right)\right|^2 dt. \tag{14.28}$$

If kT is long relative to the lowest frequency in $f(t)$, $g_1(u)$ will be a constant for all $|u| \le L/2$. The second term of the output, $g_2(u)$, is also a constant by virtue of a similar analysis.

The third term of $g(u)$ is the desired cross-product term:

$$\begin{aligned} g_3(u) &= 2\,\mathrm{Re}\left\{ e^{j4\pi\alpha_c u} \int_0^{kT} f\left(t - \frac{T}{2} + \frac{u}{v}\right) s^*\left(t - \frac{T}{2} - \frac{u}{v}\right) dt \right\} \\ &= 2\cos[4\pi\alpha_c u] \int_0^{kT} f\left(t - \frac{T}{2} + \frac{u}{v}\right) s^*\left(t - \frac{T}{2} - \frac{u}{v}\right) dt. \end{aligned} \tag{14.29}$$

This integral has the form of $c(\tau)$, the correlation function for $f(t)$ and $s(t)$. To see this relationship more clearly, we make a change of variables by letting $t - T/2 + u/v = z$:

$$\boxed{g_3(u) = 2\cos(4\pi\alpha_c u) \int_a^b f(z) s^*(z - 2u/v)\,dz,} \tag{14.30}$$

where $a = -(T/2 - u/v)$ and $b = kT + a$. We recognize the integral as the correlation function and the displacement variable as $\tau = -2u/v$; the negative sign is due to the imaging operation in propagating from plane P_1 to P_3. Thus, the spatial coordinate in the output plane carries the correlation-function information. In Chapter 13, Section 13.2.4, we noted a time compression by a factor of 2 when two signals counterpropagate; here we note a *space compression* by a factor of 2. For $\tau = -2u/v$, the range of available time delays is $|\tau| \le T$ for $|u| \le L/2$.

If $k \gg 1$, the lower limit in Equation (14.30) is much less than the upper limit for any space coordinate u. We can then restore the limits to

0 and kT. The final form of the cross-product term is therefore

$$\boxed{g_3(\tau) = 2|c(\tau)|\cos(2\pi\alpha_c\nu\tau),} \tag{14.31}$$

which shows that the correlation function is an envelope that modulates a *spatial carrier frequency*. The cross-product term can be separated from the two bias terms by reading out the photodetector array every kT seconds. The spatial carrier term in Equation (14.31) is thereby converted to a *temporal carrier frequency* so that $g_3(\tau)$ can be separated from $g_1(\tau)$ and $g_2(\tau)$ by a bandpass filter. The envelope $c(\tau)$ can then be recovered by the use of conventional demodulation techniques.

Note that each point on the photodetector array represents the product of $f(t)$ and $s(t)$ with a relative time delay that is a linear function of distance. The integrating action of the photodetector array over a time interval kT produces the correlation function. Because k is much greater than unity, the correlation time can be much longer relative to the integration time T of a space-integrating correlator. Furthermore, since the system produces $N = 4TW$ correlation values in parallel, we can dismiss $2T$ sec of signal if no correlation occurs. Because the optical system is equivalent to 2000–4000 integrate-and-dump electronic correlators operating in parallel, the processing speed is increased by the same factor.

The amount of integration time at the array is subject to constraints similar to those discussed in connection with the radiometer in Chapter 8, Section 8.10. We read out the detector to avoid saturation and accumulate the results in a digital postprocessing system.

The fact that $c(\tau)$ is formed by integrating the intensity formed by the sum of $f(t)$ and $s(t)$ in a space plane has two important ramifications. First, imaging the signal from the acousto-optic cell onto the photodetector array does not require temporal coherence; we make no direct use of the Fourier transform of the signal. Second, spatial coherence does not affect the imaging properties of the system.

14.3.2. Time-Integrating Correlator due to Sprague and Koliopoulos

An entirely different form of a time-integrating architecture is due to Sprague and Koliopoulos (117). The basic system shown in Figure 14.8 is a building block for many other time-integrating architectures and is the first system discussed in this book in which the source is modulated. The source modulator at plane P_0 can be an acousto-optic device, or it can be an electrically modulated source such as an injection laser diode or a

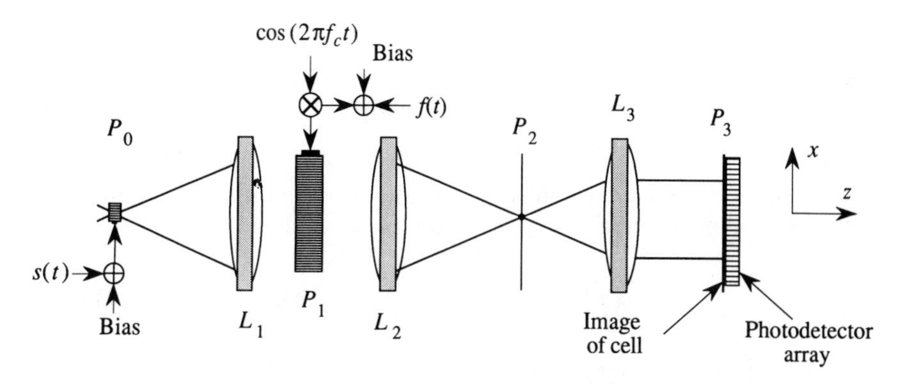

Figure 14.8. Time-integrating correlator due to Sprague and Koliopoulos.

light-emitting diode. In each case, we attempt to achieve a high power output with reasonably stable operation.

There are broad classes of applications where the reference signal $s(t)$ is binary. In these instances the linearity of the source modulation is not important, provided that the two amplitude values remain the same. Furthermore, severe distortion of $s(t)$, such as hard clipping, has a relatively small effect on correlation gain (118, 119). We shall therefore ignore the linearity requirements on the point-modulating source.

The temporally modulated light is collimated by lens L_1 and illuminates an acousto-optic cell that is also operated linearly in intensity. Recall from Chapter 7 that the intensity of the diffracted light is

$$\eta = \frac{I_d}{I_0} = \sin^2 \sqrt{BP_0}, \tag{14.32}$$

where I_d is the intensity of the diffracted light, I_0 is the intensity of the incident light, B is a constant that represents the figure of merit and other parameters of the interaction material, and P_0 is the rf drive power. We bias the acousto-optic cell at the 50% response point to achieve the best linearity, as we showed in Chapter 7. Whereas the linearity of the source modulation is not important, it is important to process the received signal $f(t)$ linearly, because the signal and noise can combine to produce unwanted cross-product terms when the process is nonlinear.

Suppose that we represent the drive signal to the acousto-optic cell as the sum of a bias term and a baseband signal $f(t)$ that modulates a carrier frequency f_c. In Chapter 7, we showed that when the bias term is chosen correctly we can represent $I_2(x,t)$ as

$$I_2(x,t) = 1 + a_2 f\left(t - \frac{T}{2} + \frac{x}{v}\right)\cos\left[2\pi f_c\left(t - \frac{T}{2} + \frac{x}{v}\right)\right], \quad (14.33)$$

where a_2 is a scaling constant that ensures that the intensity is non-negative. We follow a similar analysis for the point modulator to show that

$$I_1(t) = 1 + a_1 s(t)\cos(2\pi f_c t), \quad (14.34)$$

where a_1 is also a scaling constant.

The light intensity at plane P_3 is the product of the point modulator intensity $I_1(t)$ and the image of the intensity distribution due to the acousto-optic cell at plane P_1. This intensity distribution, when referenced to plane P_3, is $I_2(u, t)$. The intensity distribution at plane P_3 is integrated over time to produce an output signal

$$g(u) = \int_0^{kT} I_1(t) I_2(u,t)\, dt, \quad (14.35)$$

where $I_2(u,t)$ is the image of the acousto-optic cell intensity. From Equations (14.33) and (14.34) we find that the product of I_1 and I_2 contains terms whose frequencies are centered at 0, f_c, and $2f_c$. Because the integration is zero for all terms except those at baseband, we find that the output can be partitioned into two terms:

$$g_1(u) = \int_0^{kT} \left\{1 + a_1 s(t)\cos(2\pi f_c t) + a_2 f\left(t - \frac{T}{2} - \frac{u}{v}\right)\cos\left[2\pi f_c\left(t - \frac{T}{2} - \frac{u}{v}\right)\right]\right\} dt \quad (14.36)$$

and

$$g_2(u) = \cos(2\pi\alpha_c u)\int_0^{kT} s(t) f\left(t - \frac{T}{2} - \frac{u}{v}\right) dt, \quad (14.37)$$

where we ignore scaling constants. The first term integrates to a constant so that $g_1(u) = kT$, which is proportional to the integration period. The second term is the desired correlation function. Through a change of variables, we find that

$$\boxed{g_2(u) = \cos(2\pi\alpha_c u)\int_0^{kT} f(z)\,s\left(z + \frac{u}{v} + \frac{T}{2}\right) dz.} \tag{14.38}$$

We make the association that $\tau = u/v + T/2$ so that

$$g_2(\tau) = c(\tau)\cos[2\pi f_c(\tau - T/2)]. \tag{14.39}$$

The coordinate of the output plane, therefore, represents time delays. The position of the correlation function, when there is zero relative time delay between $s(t)$ and $f(t)$, occurs when $\tau = 0 = u/v + T/2$ or at $u = -vT/2 = -L/2$. Thus, the $\tau = 0$ position corresponds to the extreme lower part of the output plane. This position is the image of the transducer of the acousto-optic cell at plane P_1 and is where we expect the correlation peak to be located if there is no time delay between the signals.

We use a photodetector array to transfer the information to an electronic postprocessing system. The range of time delays in this implementation is T, which is just one-half that obtainable when the signals counterpropagate as described in Section 14.3.1. In a typical application this technique might be used to correlate a coded binary signal, such as a pseudorandom sequence, for a time period kT consistent with the amount of correlation gain required to raise the signal-to-noise ratio to the level at which reliable detection can be made. In this case the source modulation is binary, thus simplifying the choice of light sources. Binary signals allow us to perform both pre- and postprocessing to overcome certain noise sources.

An example of postprocessing is carried out by removing the effects of noise components such as (1) fixed pattern noise due to variations in photodetector responsivity from element to element, (2) clock noise associated with the binary signal, (3) scatter noise due to stray light, and (4) video line readout noise from the array. A *complementary cancellation scheme* to remove these effects was developed by Montgomery and implemented by Rotz (120). Suppose that we correlate $f(t)$ and $s(t)$ for some time interval $T' \ll kT$ and then correlate $f(t)$ with the complement of the binary signal $s(t)$ for an equal time period. When we use $s(t)$, we have

$$c_1(\tau) = c_0(\tau) + c(\tau). \tag{14.40}$$

When we use the complement of $s(t)$, the correlation function changes sign so that

$$c_2(\tau) = c_0(\tau) - c(\tau), \tag{14.41}$$

where $c_0(\tau)$ is the contribution to the output from all the fixed noise sources. We subtract $c_2(\tau)$ from $c_1(\tau)$ after each signal has been read out from the array and digitized. The result, as we see, is simply $2c(\tau)$. A nice advantage of this approach is that it also removes the strong bias level that is present, even if all other noise sources are absent. The accuracy of the technique depends on $s(t)$ having the same statistical properties for each of the integration time intervals, a likely assumption for codes such as pseudorandom sequences.

An example of this complementary cancellation scheme is shown in Figure 14.9. The signal bandwidth is 100 MHz at baseband, the acousto-optic cell has a time duration of 1 μsec and the photodetector array has 1024 elements. Wideband Gaussian noise is added to a pseudorandom sequence to produce a -20-dB signal-to-noise ratio at the input. Clock noise subsequently reduces the input signal-to-noise ratio to about -35 dB.

Figure 14.9(a) shows the result when the pseudorandom sequence chip rate is 10.6 Mc/sec. The correlation function has the familiar triangular waveform characteristic of pseudorandom sequence codes; the base of the correlation function is two chip durations wide. The entire horizontal span corresponds to 1 μsec of delay time with $\tau = 0$ at the extreme left; we see that the relative delay between the two signals is about 210 nsec. Also noteworthy is the uniform baseline, which demonstrates how well the noise-cancellation scheme works.

Figure 14.9(c) shows the result when $s(t)$ is *high-pass filtered* before correlation. Note that the chip rate is 15.3 Mc/sec and that $c(\tau)$ has a strong positive peak showing a 310-nsec relative delay between the two inputs. There are also negative peaks on either side of the main peak, a response typical of signals that have been differentiated. These results show how an optical system that is linear in intensity can still produce signals with negative values and illustrate the usefulness of electronic postprocessing to remove some of the artifacts of optical processing.

The correlation gain in the examples given is about 55 dB for integration times of the order of a few milliseconds. Note that the time bandwidth product of the system is of the order of only about 100, so that the best we could do using these components in a space-integrating architecture would be a correlation gain of about 20 dB. The additional correlation gain, less some implementation losses, is due to integration over a longer period of time.

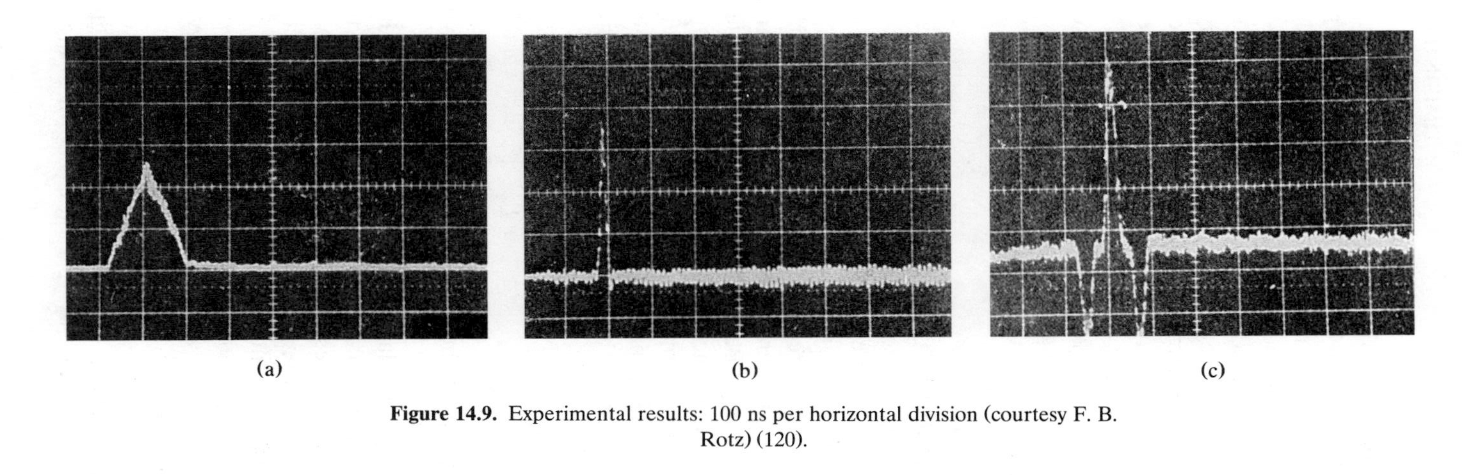

Figure 14.9. Experimental results: 100 ns per horizontal division (courtesy F. B. Rotz) (120).

Because we can image the 1-μsec time duration represented by the acousto-optic cell onto a 2048-element photodetector array, we can easily obtain about 800 resolvable time delays so that the correlation peak can be located to a precision of about 1.25 nsec. Suppose we were to use such a system for acquiring a coded signal in a communication system. We might start correlating with a large uncertainty in the time delay between signals. If no correlation peak occurred in the first kT sec, we could advance the reference signal by 800 chips and integrate for another kT sec. When a correlation is found, the reference signal clock can be set to within ± 0.6 nsec which, for a 100-Mc/sec signal, is well within one chip duration so that conventional tracking and servo loops can take over. (Problem 14.4 shows how Doppler affects this system.)

14.4. ELECTRONIC REFERENCE CORRELATOR

Electronic reference architectures provide the features of systems whose response is linear in light amplitude, but the systems still use incoherent illumination because the required reference signal is inserted electronically. The reference and signal functions are, therefore, mixed in space but separated in frequency. Because there are no path-length differences between the two signals, incoherent light can be used.

Consider the same system architecture as shown in Figure 14.8, except that we now use an electronic offset frequency as a part of the drive signal (121). We want to achieve a cross-product term at plane P_3 of the form

$$s(t)\cos(2\pi f_0 t) f\left(t - \frac{T}{2} - \frac{x}{v}\right) \cos\left[2\pi f_0\left(t - \frac{T}{2} - \frac{x}{v}\right)\right], \quad (14.42)$$

so that when we perform the time integration, the surviving term is of the form $c(\tau)\cos(2\pi f_0 \tau)$.

Because the output at plane P_3 is the product of two intensity functions $I_1(t)$ and $I_2(x, t)$, we must cause each of these signals to have a $\cos(2\pi f_0 t)$ carrier when they are expressed as intensity functions. A clue as to how to produce such a signal is given in Chapter 5, in connection with creating spatial carrier frequency filters. A sketch of the input plane of the filter generator is shown in Figure 14.10(a). Suppose that we could create an amplitude function given by $\sqrt{I_2(x, t)}$ whose frequency content had the same form as that shown in Figure 14.10(a). Because the acousto-optic cell cannot support a zero-frequency function, we substitute a reference frequency f_c for that purpose. We now want to shift the signal spectrum to a new center which we denote by $f_c + f_0$, as shown in Figure 14.10(b).

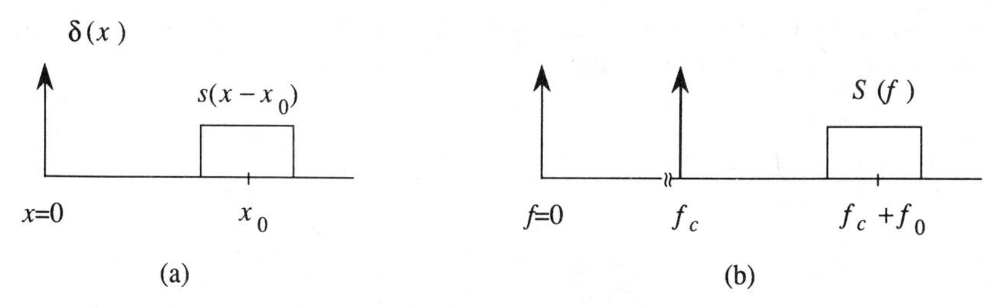

Figure 14.10. Equivalence of electronic reference and spatial filter construction.

The analogy, then, is that the reference tone at f_c in Figure 14.10 (b) plays the same role as the delta function at $x = 0$ in Figure 14.10 (a); both functions are considered as the reference signals. The frequency offset f_0 is therefore analogous to the offset distance given by x_0 in Figure 14.10(a). The intensity of the Fourier transform of $\delta(x) + s(x - x_0)$ provides the desired function as given by Equation (5.60). The intensity of the signal that generates the spectrum shown in Figure 14.10(b) should therefore produce the desired results.

The modulation scheme for the acousto-optic cell is modified as shown in Figure 14.11(a). The received signal is translated to $f_c + f_0$ by mixing it with an oscillator signal at $f_c + f_0$. The reference signal at f_c is then added to the modulated signal; the total signal $f_1(t)$ is used to modulate the acousto-optic cell. The amplitude leaving the acousto-optic cell can be expressed as

$$
\begin{aligned}
f_1(x,t) &= Be^{j[2\pi f_c(t - T/2 - x/v)]} + f\left(t - \frac{T}{2} - \frac{x}{v}\right)e^{j[2\pi(f_c + f_0)(t - T/2 - x/v)]} \\
&= \left\{B + f\left(t - \frac{T}{2} - \frac{x}{v}\right)e^{j[2\pi f_0(t - T/2 - x/v)]}\right\}e^{j[2\pi f_c(t - T/2 - x/v)]},
\end{aligned}
\tag{14.43}
$$

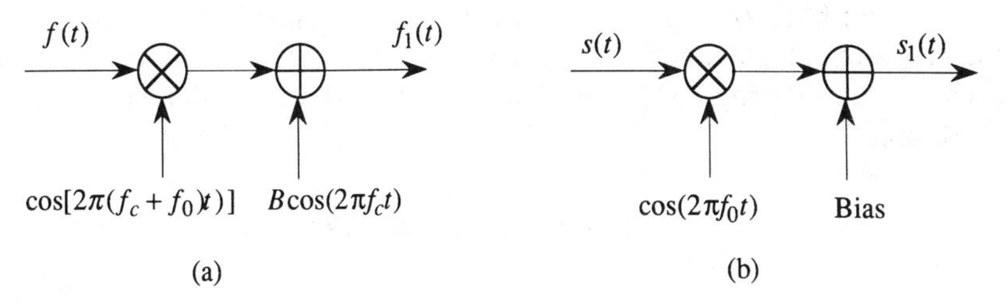

Figure 14.11. Electronic reference modulation.

where, as usual, we retain only the positive frequency terms from the acousto-optic cell. We find that

$$I_2(x,t) = |f_1(x,t)|^2 = B^2 + \left| f\left(t - \frac{T}{2} - \frac{x}{v}\right) \right|^2 + 2Bf\left(t - \frac{T}{2} - \frac{x}{v}\right)$$
$$\times \cos\left[2\pi f_0\left(t - \frac{T}{2} - \frac{x}{v}\right)\right]. \tag{14.44}$$

The last term of Equation (14.44) has the desired form as specified by Equation (14.42). Because f_c does not appear in the intensity function, it can be any frequency within the acousto-optic cell bandwidth. The frequency f_c does not appear in the intensity function because its value merely contributes a linear phase shift to the amplitude function as given by Equation (14.43). A similar phenomenon results from our analogy to spatial filtering. For example, if we were to move the entire function shown in Figure 14.10(a) by some distance x_1, the Fourier transform would contain an additional factor $\exp(j2\pi x_1 \alpha)$. When square-law detected, however, this factor disappears. The recorded filter function is therefore dependent only on the relative distance x_0 between the reference function and the signal. So too is the intensity function $I_2(x,t)$ dependent only on the frequency difference f_0.

If we use an acousto-optic cell as a point modulator to handle the reference signal $s(t)$, we would use exactly the same modulation scheme as shown in Figure 14.11(a) to produce the proper intensity function $I_1(t)$. For variety, then, we consider the use of an injection laser diode or light-emitting diode source whose intensity we wish to modulate directly. As before, we need to use a bias term to keep the modulation positive as shown in Figure 14.11(b). The stored signal $s(t)$ is translated to f_0 and added to a bias term C. The intensity of the source is linearly proportional to the applied drive voltage. Thus, we have that

$$I_1(t) = C + s(t)\cos(2\pi f_0 t). \tag{14.45}$$

A spatial filter $H(\alpha)$ at plane P_2 removes the undiffracted light so that the intensity at plane P_3 is $I_1(t)I_2(u,t)$, where $I_2(u,t) = I_2(-x,t)$ because of the imaging condition. The output of the photodetector array is therefore

$$g(u) = \int_0^{kT} I_1(t) I_2(u,t)\, dt. \tag{14.46}$$

When we substitute for $I_1(t)$ and $I_2(u,t)$, we generate several terms which are grouped according to whether they are the desired term, are dependent on the bias terms, or are signal dependent.

We let $\tau = -T/2 + u/v$ and express the relevant terms as

$$g_1(\tau) = \int_0^{kT} CB^2\,dt = kTCB^2,$$

$$g_2(\tau) = \int_0^{kT} C|f(t+\tau)|^2\,dt,$$

$$g_3(\tau) = B\cos(2\pi f_0\tau)\int_0^{kT} s(t)f(t+\tau)\,dt$$

$$= Bc(\tau)\cos(2\pi f_0\tau). \tag{14.47}$$

Several other terms appear in the process of multiplying $I_1(t)$ and $I_2(u,t)$; these terms all contain factors of $\cos(2\pi f_0 t)$ or $\cos(4\pi f_0 t)$ which integrate to zero for the integration time kT which is long relative to $1/f_0$. As with the spatial filtering method, the offset frequency must be chosen so that $f_0 > 3W$, where W is the baseband signal bandwidth. This requirement, seen from Equation (14.44), allows us to separate the desired signal from the bias terms. The cross product once again contains the correlation function on a carrier frequency as desired.

We compare the features of the incoherent and the electronic reference approaches. The incoherent approach achieves linearity by operating at a low signal-to-bias ratio, and it requires postdetection processing schemes to recover the desired correlation function. The full bandwidth of the acousto-optic cell is used and the number of photodetector elements is at a minimum for a given time bandwidth product. This approach is insensitive to Doppler. The electronic reference approach does not require a bias term; linearity is achieved by using a low diffraction efficiency. The acousto-optic cell bandwidth is not as efficiently used; if we place f_c at one band edge, $f_c + f_0 + W$ is at the other band edge. Because $f_0 \geq 3W$, the useful part of the acousto-optic cell bandwidth is reduced by a factor of 4. The range of time delays is the same in both cases. The electronic reference approach is also sensitive to Doppler.

14.5. COMPARISON OF FEATURES

Time-integrating correlators offer interesting features relative to space-integrating correlators. Recall that space-integrating correlators have wide bandwidths, good dynamic ranges, single-element detectors, unlimited time-delay ranges, but limited correlation gains. Furthermore, these systems are of the convolving type so that we need to use time-reversed reference signals; this is typically not a problem unless $s(t)$ is an analog signal. They easily handle those complex-valued signals that may arise in the optical domain.

In contrast, time-integrating correlators have narrower bandwidths at the output (most notable when we need to handle complex-valued signals), require multielement photodetector arrays, may need more complex post-detection processing to combat the bias terms, do not require time reversal of the reference signal, have processing gain that is limited only by the time bandwidth product of $s(t)$, but have limited time-delay ranges.

14.6. INTEGRATED OPTICAL SYSTEMS

In the second half of this book we have concentrated on the use of acousto-optic technology to process wide-bandwidth signals in real time. We can also use acousto-optic phenomena in thin-film waveguide structures that incorporate the laser, processing elements, and photodetectors on a single substrate. These devices, referred to as *optical integrated circuits*, are classified as *monolithic* optical integrated circuits, in which all components are integrated on a common substrate, or *hybrid* optical integrated circuits, in which the components are fabricated from different materials and assembled into the final product by external coupling techniques.

Considerable effort has been expended toward the development of integrated optical signal-processing devices to implement computationally intensive tasks such as correlation, Fourier transformation, radar signal processing, and vector matrix multiplication. As an illustration of these efforts, consider the integrated optical rf spectrum analyzer sketched in Figure 14.12 (122). Space- and time-integrating correlators can be made by a fairly obvious extension of the basic building blocks used in this spectrum analyzer (123–125). The spectrum analyzer consists of a laser source that diverges within a planar waveguide until it reaches a waveguide lens. This lens is incorporated into the waveguide and renders the diverging light into a parallel beam of light. The next element in the system is an acousto-optic component consisting of an rf input signal, an amplifier, and an interdigital transducer array. The interaction of the light and the acoustic waves causes the light to be diffracted at an angle proportional to the frequency of the input rf signal, just as with bulk acousto-optic cells. The second lens focuses the diffracted light on a photodetector array so that the linear position of a focused beam is proportional to the rf

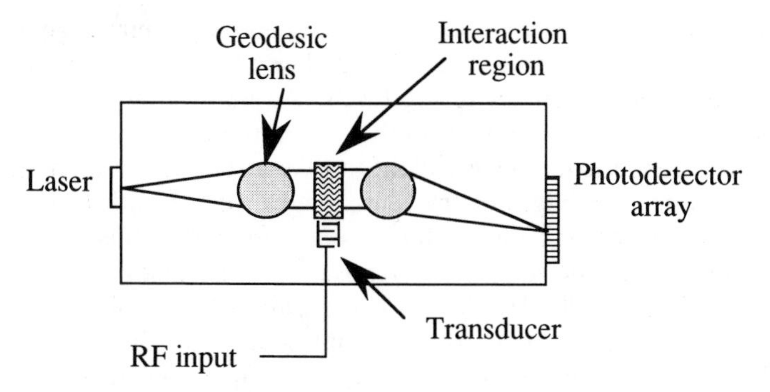

Figure 14.12. A guided-wave rf spectrum analyzer.

frequency. In this fashion, the spectrum of the rf signal is produced at the plane of the photodetector array. All the elements of this device are not, unfortunately, integrated on the same substrate as desired. The rf spectrum analyzer is therefore a hybrid instead of a monolithic integrated optical circuit. We now briefly review the key elements of this system.

Double heterostructure injection lasers with cw outputs are ideal sources for integrated optical applications. These lasers are, themselves, examples of integrated optics because they are made from a sandwich structure of thin films. Unfortunately, the III–V materials most useful for making good lasers are not the most suitable for fabricating the other elements required in an optical signal-processing system because they lack the strong electro-optic effects needed for acousto-optic interaction. The lasers must therefore be coupled to the substrate, the most common method being a butt coupled joint.

Integrated optical circuits often require several lenses on the waveguide substrate to control the light within the waveguide. To achieve the high level of performance needed for spectrum analysis, these lenses must be nearly diffraction limited, a level of performance difficult to obtain over large field angles even with multielement lenses. There are three basic types of lenses used in integrated optical circuits: the modified Luneburg lens, the Fresnel lens, and the geodesic lens. In a *modified Luneburg lens*, the index of refraction varies according to the radius of the lens. The index is lower near the edges than at the center of the lens, so that the effective path length through the edge of the lens is shorter relative to the paths near the center of the lens. A large index of refraction change is needed to obtain a lens of short focal length, and the tolerances are so extreme that this method for forming lenses is not attractive. A *Fresnel lens* is formed by depositing or embedding a phase shifting grating in the waveguide.

Such lenses have restricted fields of view due to their large off-axis aberrations, which results in a small number of resolvable frequencies, poor efficiency, and high scattering levels. A *geodesic lens* consists of a depression or protrusion formed on the substrate before the waveguide is formed. The geodesic lens causes light to take a shorter path length when traveling through the edges of the lens, relative to the path taken by the light that passes through the center. A disadvantage of this technique is that the lens cannot be formed by deposition or etching operations. The major problem with all these lenses is that, although the first lens needs merely to collimate the light from a point source, the second one must focus light to as many points as there are frequencies to be resolved. Aberrations must therefore be well controlled at large off-axis angles. Modern-day acousto-optic spectrum analyzers can resolve 1000–2000 frequencies; the system illustrated here can resolve only 75–80 frequencies, so that the difference in performance is significant.

The most common form of interaction between light and acousto-optic devices is the diffraction of one guided wave into another guided wave by interaction with a surface acoustic wave device. The interaction between guided optical waves and surface acoustic wave devices is considerably more complex than is acousto-optic interaction in discrete devices because of the nonuniform profiles of the guided optical mode in its waveguide and the additional diffraction effects caused by the interdigital pattern of the fingers of the transducer. One way to achieve a wide bandwidth is to use several separate transducers, each operating at a different center frequency. The fingers of each of the transducers are varied in length and spacing to provide efficient diffraction and bandwidth. Bandwidths of several hundred megahertz are possible at a drive power requirement of about 1 mW of rf power per megahertz of bandwidth. Broadband operation can also be achieved by using *phased array transducers* consisting of a set of transducers operating at the same center frequency but with different finger spacings. The logical extension of this idea is a chirp frequency transducer in which the finger spacings change smoothly from a low to high value; they provide good acoustic coupling over a broad band of frequencies.

The ideal material for fabricating photodetector arrays is silicon; the high speed provided by GaAs is usually not needed in these applications because the optical system is performing many computations in parallel and the data rate *per detector* is low. In some correlation applications, however, a single high-speed detector is needed; GaAs may then be the material of choice. Butt coupling these arrays to the lithium niobate substrate is a formidable task and one that is not easily automated.

While nonintegrated laser sources are adequate for laboratory research and for some integrated optics applications, much of the benefit of integrated optical circuit technology cannot be obtained without monolithically integrated light sources and photodetectors. Furthermore, the performance parameters of the overall system are likely to be compromised in one or more of the areas if a single substrate material is used. Efforts to build truly integrated optical circuits have produced disappointing results, particularly when compared to competing technologies. For example, the rf spectrum analyzer used to illustrate how a typical integrated optical circuit would perform, even though it was not fully integrated, had a 400-MHz bandwidth on a 600-MHz center frequency. A key performance parameter of spectrum analyzers is the number of frequencies that can be resolved over the bandwidth. This device has a 5.3-MHz half-power frequency resolution, but the operational resolution is reduced because of the need to sample the spectrum so that a measurable dip between frequencies is made. The operational resolution is therefore more like 4–8 MHz, leading to 50–100 resolvable frequencies. Dynamic range was measured in the 25–30-dB range. This system used a photodetector array with 140 elements and was packaged in a volume of $1 \times 2 \times 5$ in.

It is interesting to compare the performance figures for the integrated optical circuit spectrum analyzer and one fabricated using discrete (bulk) elements. A typical acousto-optic spectrum analyzer having a 400-MHz bandwidth has a frequency resolution of 400 kHz leading to 1000 resolvable frequencies, a factor of 20 larger than the integrated optical circuit described here. Furthermore, acousto-optic cells having much higher bandwidths are available, while retaining the same number of resolvable elements. An acousto-optic spectrum analyzer for use as a threat warning receiver has been packaged in a volume of $1 \times 1 \times 5$ in, a smaller volume than the integrated optical spectrum analyzer.

There are several reasons why the performance/cost ratio of discrete acousto-optic systems is currently superior to that of integrated optical circuits. First, each element in a discrete system can be optimized for performance. Frequency resolution and dynamic range are the most important parameters of a spectrum analyzer. Frequency resolution is obtained by optimizing the acousto-optic cell in terms of the material choice and configuration. For example, although lithium niobate is frequently used as the interaction material in an acousto-optic cell, other materials such as tellurium dioxide and gallium phosphide provide significantly better performance in certain frequency ranges. The performance of photodetector arrays is driven by the video camera industry; these arrays can be quite easily adapted to discrete optical systems without the concern for how they need to be modified to be butt coupled to an integrated optical circuit.

Second, a commonly overlooked element in signal-processing systems is free space. Air is the optimum free-space medium because its index of refraction is low and it contains no scattering particles that reduce the performance of systems such as spectrum analyzers. Because the index of refraction of lithium niobate is already high, forming lenses in an integrated optical circuit with an even higher effective refractive index is difficult; any imperfections in the planar waveguide causes light to scatter in an uncontrolled fashion (92, 126). The high index of refraction has another negative effect: it actually causes the integrated optical circuit to be larger than its discrete component counterpart.

Third, discrete systems can take full advantage of the two-dimensional propagation of light through free space. This feature is useful not only when processing imagery and multichannel electronic signals, it is extremely important when processing one-dimensional time signals to display more than one parameter. For example, a range Doppler radar system must calculate and display the position and velocity of targets within its field of view. In a planar technology such as integrated optical circuits, there is no way to display more than two parameters, such as one in time and one in space, because the thickness of the waveguide provides only one resolvable element in the orthogonal spatial direction. In a discrete component system, displaying parameters in two dimensions and detecting them with a two-dimensional photodetector array is easy, as we discuss in the next chapter.

PROBLEMS

14.1. We have a Bragg cell for which $W = 400$ MHz, $T = 2$ μsec, and $v = 5$ km/sec. We want to design a time-integrating spectrum analyzer having 50-Hz resolution.

(a) Calculate kT and k.

(b) Calculate the new frequency range Δf.

(c) Calculate the chirp rate.

(d) Calculate the period of the chirp relative to the acousto-optic cell time duration.

(e) What frequency range of the chirp is in the cell at any time?

(f) How many frequencies are displayed at the output?

14.2. Suppose that the spectrum analyzer from Problem 14.1 is adjusted so that $f_a = f_b = 200$ MHz and that $f_c = 50$ MHz. Calculate the value of the spatial carrier frequency at the output plane. How many photodetector elements are needed in a linear array to readout the information? What is the required separation of the elements?

14.3. We have a Bragg cell for which $W = 50$ MHz, $T = 40$ μsec, and $v = 5$ km/sec. We want to achieve a correlation gain of 10^6 in a time-integrating correlator of the Sprague and Koliopoulus type. Calculate the value of k. Suppose that the maximum time uncertainty between the two signals is 1 sec. How many threshold decisions are needed to find the signal, and what length of signal history can be dismissed after each decision?

14.4. Does the performance of the time-integrating correlator depend on Doppler? If not, what range of Dopplers can be detected? If so, how small must the Doppler frequency be? Justify your answer by analysis. Also, what interpretation do you give to the integral part of the result of your analysis?

14.5. In a time-integrating spectrum analyzer, we have the following parameters:

$$\text{frequency range} = 100 \text{ kHz},$$

$$\text{integration time} = 10 \text{ msec},$$

$$\text{chirp rate} = 10^{11} \text{ Hz/sec}.$$

What is the temporal resolution of the system? What is the time duration T of the cell? What is the bandwidth W of the cell?

14.6. You have designed and built a time-integrating spectrum analyzer in which you drive an acousto-optic cell with an input signal $f(t)$ and use a solid-state laser with $\lambda = 830$ nm as the source. A much more powerful laser has become available, but its wavelength is $\lambda = 750$ nm. You want to retrofit the system with this laser, if possible. How does the change in wavelength affect the system performance? Provide sketches and analyses as appropriate to support your answer.

15

Two-Dimensional Processing

15.1. INTRODUCTION

In Chapters 7–14 we concentrated on the use of one-dimensional acousto-optic cells used for processing wideband time signals, and we showed how multichannel versions of these devices are used to perform certain two-dimensional operations. The emphasis in this chapter is on the use of two or more acousto-optic cells, generally in orthogonal configurations, to more fully use the second dimension of the optical system. More powerful signal-processing operations are therefore possible. We describe both coherently and incoherently illuminated systems used in applications such as correlation, spectrum analysis, and range/Doppler processing (127–131).

In previous chapters we showed that the correlation between two signals is a powerful and useful operation. For example, we showed how a one-dimensional acousto-optic cell processor, as shown in Figure 14.8, is used to form the correlation function

$$c(\tau) = \int_0^{kT} f(t)s(t+\tau)\,dt, \tag{15.1}$$

where T is the time duration of the acousto-optic cell and kT is the integration interval. The correlation function reveals the degree of similarity between two signals and, in a general way, can be extended to include spectral analysis of signals. For example, if $s(t) = \exp[-j2\pi ft]$, we find that

$$\boxed{\begin{aligned} c(f,\tau) &= \int_0^{kT} f(t)e^{-j2\pi f(t+\tau)}\,dt \\ &= e^{-j2\pi f\tau}\int_0^{kT} f(t)e^{-j2\pi ft}\,dt \\ &= e^{-j2\pi f\tau}F(f). \end{aligned}} \tag{15.2}$$

When we detect the squared magnitude $|c(f,\tau)|^2$ of the correlation function at an arbitrary value of τ, the result is the same function $|F(f)|^2$ as obtained from a conventional power spectrum analyzer. One interpretation of spectrum analysis is that it shows the degree of similarity between a function such as $f(t)$ and an exponential function of a frequency variable. In effect, this correlation process determines how much energy $f(t)$ has at that frequency component. The spectral components in the Fourier plane can then be considered as "correlation peaks" which give the frequency and magnitude of the spectral components. A broad class of processing operations are therefore unified when represented by the correlation function.

A natural extension of correlation between two signals (a *two-product* operation) is to produce the *triple-product* operation given by

$$c(\tau_1,\tau_2) = \int_0^{kT} s_1(t)s_2(t+\tau_1)s_3(t+\tau_2)\,dt. \tag{15.3}$$

The integrand is the product of three signals, with two independent variables, to provide an even broader range of processing operations. We now show how we use the second dimension of the optical system to implement the triple-product operation.

15.2. TRIPLE-PRODUCT PROCESSING CONCEPT

One way to implement the generalized triple-product operation given by Equation (15.3) is to use two time-integrating correlators in an orthogonal configuration as shown in Figure 15.1. We considerably simplify the optical details of the system in this figure so that we can concentrate on the concepts. The optical modules are similar to those studied in previous chapters. In the horizontal direction we use a time-integrating correlator of the Sprague and Koliopolous type to form the product $s_{1_x}(t)s_2(t+\tau_1)$ at the output plane, where $\tau_1 = x/v - T/2$. We spread the light in the vertical direction to cover all elements of the two-dimensional output detector array.

A similar arrangement in the vertical direction yields the product $s_{1_y}(t)s_3(t+\tau_2)$ at the detector array, where $\tau_2 = y/v - T/2$. This distribution is spread in the horizontal direction to cover all elements in the photodetector array. When the two beams are joined and square-law

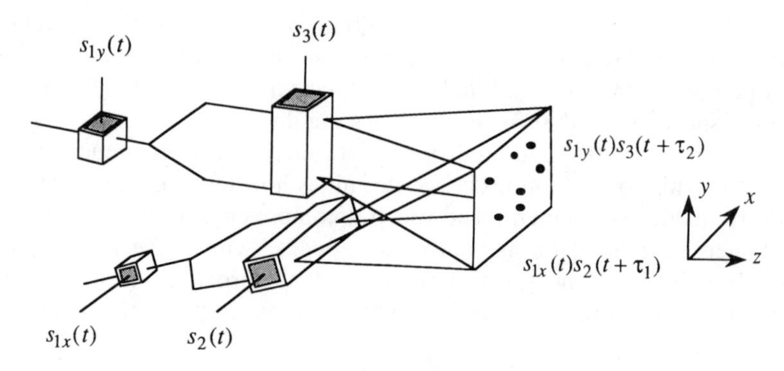

Figure 15.1. Basic form of the triple-product processor.

detected, the intensity becomes

$$\begin{aligned} I(\tau_1, \tau_2) &= \left| s_{1x}(t) s_2(t+\tau_1) + s_{1y}(t) s_3(t+\tau_2) \right|^2 \\ &= \left| s_{1x}(t) s_2(t+\tau_1) \right|^2 + \left| s_{1y}(t) s_3(t+\tau_2) \right|^2 \\ &\quad + 2\,\mathrm{Re}\left\{ s_{1x}(t) s_{1y}(t) s_2(t+\tau_1) s_3(t+\tau_2) \right\}. \end{aligned} \quad (15.4)$$

As usual, the time integral of the first two terms of Equation (15.4) contribute to bias terms at baseband that are of no interest. The time integral of the cross-product term becomes

$$g_3(\tau_1, \tau_2) = 2\,\mathrm{Re}\left\{ e^{j2\pi\alpha_c x} \int_0^{kT} s_{1x}(t) s_{1y}(t) s_2(t+\tau_1) s_3(t+\tau_2)\, dt \right\}, \quad (15.5)$$

where α_c is a spatial carrier frequency induced by the angle formed by the optical axes of the two branches of the system. The integrand of the cross-product term given by Equation (15.5) is a product of four functions; but because the product $s_{1x}(t)s_{1y}(t)$ is a function of only one variable, the combination is usually written as $s_1(t)$.

The cross-product term is of the form

$$g_3(\tau_1, \tau_2) = 2c(\tau_1, \tau_2)\cos(2\pi\alpha_c x). \quad (15.6)$$

The real-valued function $c(\tau_1, \tau_2)$ is a generalized two-dimensional correlation function, supported by a one-dimensional spatial carrier frequency. We use bandpass filtering techniques described in earlier chapters to separate the cross-product term $g_3(\tau_1, \tau_2)$ from the baseband terms $g_1(\tau_1, \tau_2)$ and $g_2(\tau_1, \tau_2)$

We recognize the architecture of Figure 15.1 as one in which we achieve the multiplication of two signals by means of addition and square-law detection. These systems must, therefore, be coherently illuminated so that the amplitudes add before the detection process. An alternative architecture allows for either coherent or incoherent operation by the placement of two acousto-optic cells and a light-emitting diode in series so that multiplication takes place in cascaded stages; such a system is shown in Figure 15.2. In this configuration, the point source is modulated by $s_1(t)$. The second signal drives the acousto-optic cell in the horizontal direction to produce the product $s_1(t)s_2(t + \tau_1)$. We place an optical system between the horizontal and vertical acousto-optic cells to spread light in the vertical direction and to focus light in the horizontal direction. The signal $s_1(t)s_2(t + \tau_1)$ is then multiplied by $s_3(t + \tau_2)$ as provided by the vertical acousto-optic cell. A second lens system mutually images the two acousto-optic cells at the detector array plane. The undiffracted light and unwanted components are eliminated at intermediate planes and the output therefore has the same form as given by Equation (15.3).

A triple-product processing architecture that does not require source modulation is shown in Figure 15.3. Signal $s_2(t)$ drives an acousto-optic cell that is imaged onto a second acousto-optic cell that is driven by $s_3(t)$. The two signals are counterpropagating so that the product, referenced to the plane of the second acousto-optic cell, is $s_2(t + x/v)s_3(t - x/v)$; we have omitted the $T/2$ delay factor to simplify the mathematics. The remainder of the system is the same as that shown in Figure 15.2. The output of the system is

$$c(x', y') = \int_0^{kT} s_1(t + y'/v)s_2(t + x'/v)s_3(t - x'/v)\,dt, \quad (15.7)$$

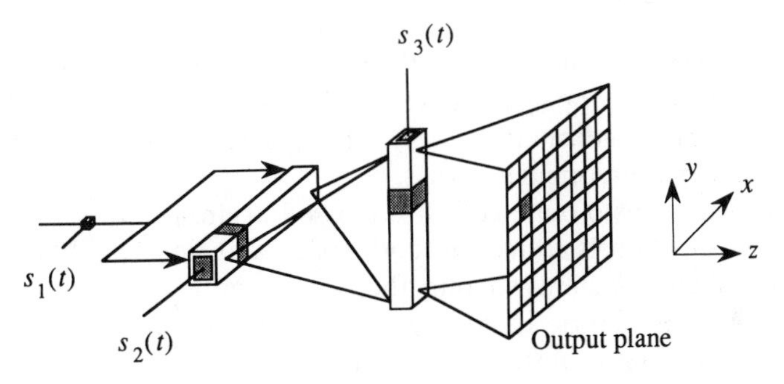

Figure 15.2. Alternative configuration of triple-product processor.

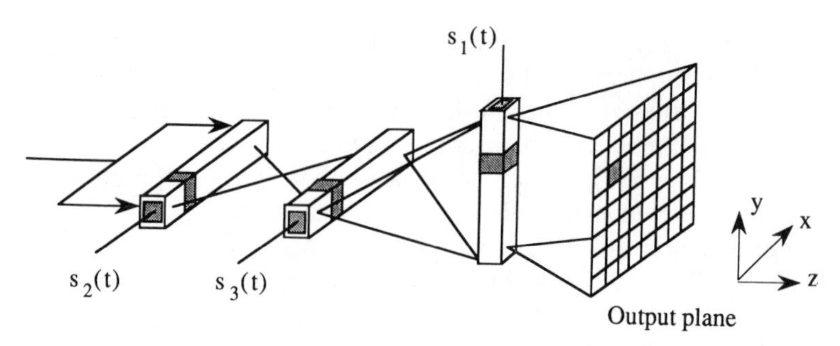

Figure 15.3. Modified triple-product processor.

where x' and y' are coordinates at the output plane. Equation (15.7) is in the proper form of the triple-product operation if we make a change of variables in which $t + y'/v = z$. We then have that

$$c(x', y') = \int_{y'/v}^{kT+y'/v} s_1(z) s_2\left(z + \frac{x' - y'}{v}\right) s_3\left(z + \frac{-x' - y'}{v}\right) dz. \quad (15.8)$$

This result is modified to resemble Equation (15.3) when we make a change in variables in which we rotate the coordinates of the output plane. This procedure of rotating the output plane coordinates to provide the desired result will now be discussed in more detail.

15.3. CROSSED ACOUSTO-OPTIC CELL GEOMETRY

In Chapter 13, Section 13.9 we showed that only a single value of the correlation function $c(\tau)$ is displayed when the acoustic wavefronts propagate in the same direction; we refer to this configuration as the integrate-and-dump mode of operation. The variable τ occurs in "slow" time because the rate at which various time delays are generated is low. If the two signals counterpropagate, however, the correlation function occurs in "fast" time and is indicated by $c(2t)$. The next step is to consider how we can obtain some intermediate cases.

When two acousto-optic cells are crossed at an arbitrary angle θ, the correlation function is formed in both space and time. Said and Cooper (132) described a system wherein the acousto-optic cells are crossed so that $\theta = 45°$ as shown in Figure 15.4. The special cases of co- and counterpropagating acoustic signals, with $\theta = 0$ and $\theta = 180°$, were discussed in earlier chapters. The other obvious angle to use is $\theta = 90°$, as shown in Figures 15.2 and 15.3. We now consider the benefits of placing the acousto-optic cells at other angles.

We represent the duration between the independent time samples along the acousto-optic cells by T_0. To obtain the one-dimensional function $c(\tau)$, we integrate the product of the two signals $f(t)$ and $s(t)$ along lines parallel to the bisector of the arbitrary rotation angle θ between the cells (133). All signal contributions along a line parallel to the bisector represent a fixed time delay between the two signals. The intersection of the time samples are parallelograms; the appropriate samples, corresponding to no relative time delay between the two signals, are shown as shaded areas. We can represent a time-delay axis by extending the line of integration to some convenient point outside the overlapping regions and by drawing an axis perpendicular to the line of integration. We note that this axis forms an angle ϕ with respect to the original y axes.

A crossed-cell geometry is used to display the correlation function in *space* instead of *time* or, in some special cases, in both space *and* time. We connect the spatial coordinates with a time delay coordinate by referring to the spatial displacement axis as the time delay or τ axis. The intervals between the τ values can be related to T_0 through the following procedure. When $s(t)$ is delayed by one time sample T_0 relative to $f(t)$, the shaded area representing the correlation-peak line is shifted upward;

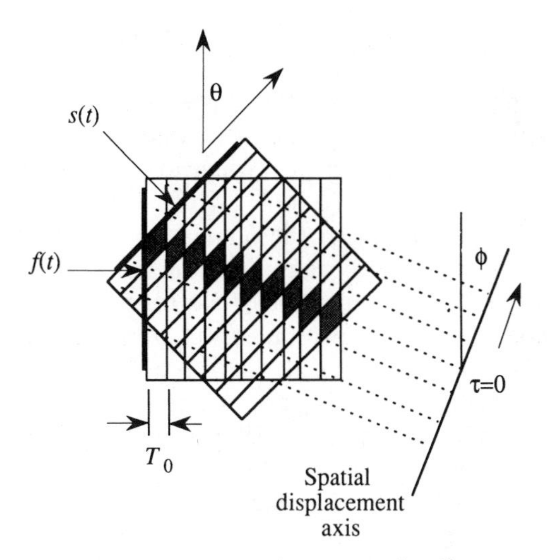

Figure 15.4. Crossed acousto-optic cells.

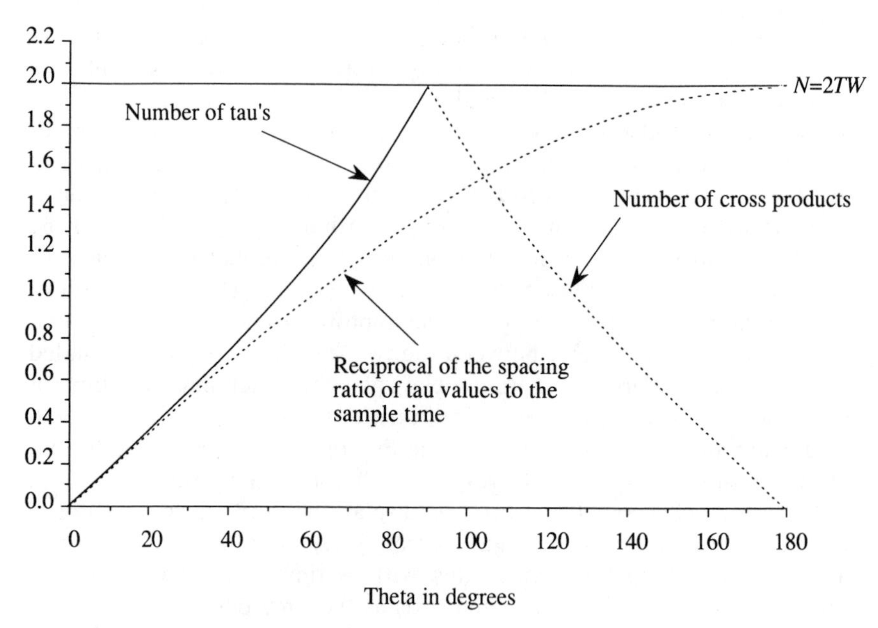

Figure 15.5. Key performance parameters as a function of θ.

the intersection with the τ axis now occurs for the general case at

$$\tau_d = \frac{T_0}{2\sin\phi}, \qquad 0 \le \phi \le 90^\circ. \tag{15.9}$$

The effective time samples therefore have durations that range from infinity for $\phi = 0^\circ$ when we implement the integrate-and-dump mode of correlation, to $\tau = T_0/2$ for $\phi = 90^\circ$ when we implement the matched-filtering mode.

From Figure 15.4 we note that the number of samples that contribute to $c(\tau)$ is dependent on τ because the region of integration is the octagonal boundary defining the region of overlap for the two acousto-optic cells. This boundary, in effect, introduces weighting functions that change from a rectangular function when $\phi = 0^\circ$ to a triangular function when $\phi = 45^\circ$. Furthermore, as ϕ changes, so too does the number of τ samples available change. The relationship of these parameters is shown in Figure 15.5. We see that the number of available time delays starts at zero when $\theta = 0^\circ$ and reaches N, which is just twice the time bandwidth product of the acousto-optic cells when $\theta = 90^\circ$. The number of cross products available at $\tau = 0$ is constant at N for $0^\circ \le \theta \le 90^\circ$; the number then begins to drop, which is the primary reason that we do not consider

$\theta > 90°$ as a viable system architecture. The region for $\theta > 90°$, which represents the counterpropagating mode, also implies that we need to use a time-reversed signal to restore the peak value of the correlation function.

The connection between this more general analysis and the version of $c(\tau_1, \tau_2)$ as given in Equation (15.6) is that we need to rotate the coordinate axes by 45° (because $\theta = 90°$ and therefore $\phi = 45°$). We then have

$$c(\tau_1, \tau_2) = \int_0^{kT} s_1(t) s_2(t + \tau_1) s_3(t + \tau_2)\, dt, \tag{15.10}$$

which agrees with Equation (15.3).

15.4. THE BISPECTRUM

The cross bispectrum is defined as the Fourier transform of the triple correlation function $c(\tau_1, \tau_2)$. By applying the two-dimensional Fourier transform to Equation (15.10), we find that

$$\begin{aligned}
C(\alpha, \beta) &= \iint_{-\infty}^{\infty} c(\tau_1 \tau_2) e^{j2\pi(\alpha\tau_1 + \beta\tau_2)}\, d\tau_1\, d\tau_2 \\
&= \iint_{-\infty}^{\infty} \left[\int_0^{kT} s_1(t) s_2(t + \tau_1) s_3(t + \tau_2)\, dt \right] e^{j2\pi(\alpha\tau_1 + \beta\tau_2)}\, d\tau_1\, d\tau_2 \\
&= \int_{-\infty}^{\infty} \int_0^{kT} s_1(t) s_2(t + \tau_1) e^{j2\pi\alpha\tau_1} \\
&\quad \times \left[\int_{-\infty}^{\infty} s_3(t + \tau_2) e^{j2\pi\beta(t+\tau_2)}\, d\tau_2 \right] e^{-j2\pi\beta t}\, d\tau_1\, dt \\
&= \int_{-\infty}^{\infty} \int_0^{kT} s_1(t) s_2(t + \tau_1) e^{j2\pi\alpha\tau_1} S_3(\beta) e^{-j2\pi\beta t}\, d\tau_1\, dt \\
&= \int_0^{kT} s_1(t) e^{-j2\pi\beta t} S_3(\beta) \\
&\quad \times \left[\int_{-\infty}^{\infty} s_2(t + \tau_1) e^{j2\pi\alpha(t+\tau_1)}\, d\tau_1 \right] e^{-j2\pi\beta t}\, dt \\
&= S_2(\alpha) S_3(\beta) \int_0^{kT} s_1(t) e^{-j2\pi(\alpha+\beta)t}\, dt \\
&= S_1(-\alpha - \beta) S_2(\alpha) S_3(\beta).
\end{aligned} \tag{15.11}$$

The cross bispectrum of the triple product therefore consists of the product of the Fourier transform of $s_3(t)$ with respect to the frequency variable β, the Fourier transform of $s_2(t)$ with respect to the frequency variable α, and the Fourier transform of $s_1(t)$ with respect to the frequency variable $-\alpha - \beta$. The third transform is therefore displayed along a diagonal axis in the two-dimensional Fourier-transform plane.

The auto-bispectrum for the auto triple correlation, for which we have $s_1(t) = s_2(t) = s_3(t)$, has the symmetry properties that

$$\boxed{C(\alpha, \beta) = C(\beta, \alpha) = C(-\alpha - \beta, \alpha)}. \tag{15.12}$$

If the signal is real valued, the auto-bispectrum is Hermitian:

$$\boxed{C(\alpha, \beta) = C^*(-\alpha, -\beta),} \tag{15.13}$$

as is typically true of the spectra of ordinary two-dimensional signals. The auto-bispectrum has several interesting properties. As an example, Figure 15.6(a) shows the auto-bispectrum of a signal $s(t) = \cos(2\pi f_j t)$. Each of the three spectra that comprise Equation (15.11) consist of a pair of line delta-functions. Because all three spectra do not overlap at any point in the frequency domain, the auto-bispectrum is zero. This curious result shows that a pure sinusoidal signal cannot be detected from the auto-bispectrum. If a bias is added to the sinusoidal signal so that $s(t) = 1 + \cos(2\pi f_j t)$, the auto-bispectrum contains three additional line delta functions, as shown in Figure 15.6(b). We find that there are now six locations where one of the line delta functions from each spectrum overlaps, so that the product is not zero. The auto-bispectrum therefore is

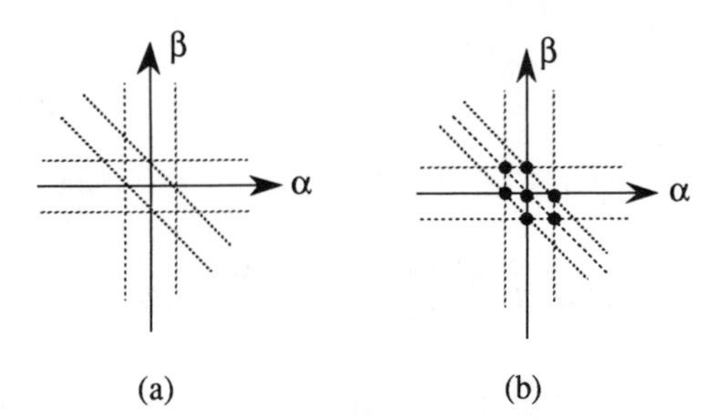

Figure 15.6. Auto-bispectrum of (a) an unbiased cosine waveform and (b) a biased cosine waveform.

a sensitive method for sorting unbiased and biased sinusoids. The bispectrum is sensitive to other signal features as well, and the reader is referred to the literature for a fuller discussion of the properties of the bispectrum (131).

15.5. SPECTRUM ANALYSIS

The concepts associated with the bispectrum and with increasing the dimensionality of time-integrating correlators suggest that we can also perform spectrum analysis using the triple-product processor. The major increase in performance is more frequency resolution for a given analysis bandwidth. Recall from Chapter 4, Section 4.7 that a wideband signal can be recorded on photographic film and used as the input to a two-dimensional spectrum analyzer to produce fine frequency resolution. We now show how the triple-product processor with some suitable modifications can perform the same operations in real time.

15.5.1. Real-Time Raster-Format Spectrum Analysis

Figure 15.7 shows the same optical system as in Figure 15.1, but the drive signals are modified to produce a two-dimensional spectrum analyzer. In

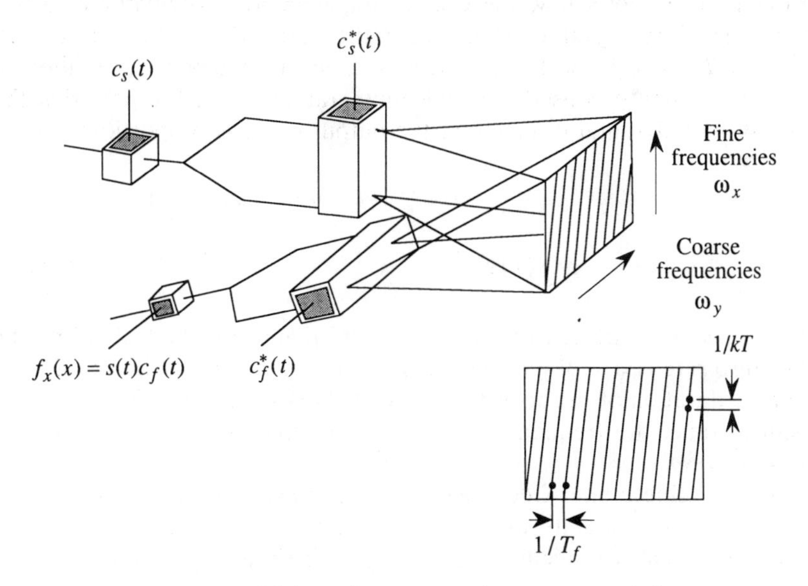

Figure 15.7. Triple-product processor for spectrum analysis.

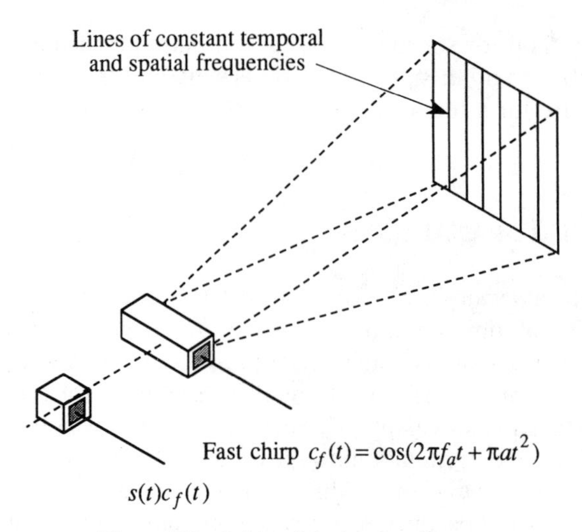

Figure 15.8. Origin of the local oscillator.

the horizontal direction, we implement spectrum analysis by using a chirp signal as a premultiplier of $f(t)$ and a conjugate chirp signal to drive the acousto-optic cell. This configuration basically implements a one-dimensional Fourier transform as explained in Chapter 14, Section 14.2.

Figure 15.8 shows how the coarse frequency local oscillator is generated. The chirp signal $c_f(t)$ is called the *fast chirp signal* because its duration T_f is reciprocally related to the normal temporal resolution of the system when it is used in a space-integrating mode. We note that the light in the horizontal direction at the output plane is proportional to

$$\begin{aligned} r_x(x,t) &= s(t)e^{-j2\pi a t^2}e^{j[2\pi f_a(t-x/v)+\pi a(t-x/v)^2]} \\ &= s(t)e^{-j2\pi a t x/v}e^{j2\pi f_a(t-x/v)}e^{j\pi a(x/v)^2}, \end{aligned} \tag{15.14}$$

where f_a is the starting frequency of the chirp and a is the fast chirp rate. The chirp rate is $a = W/T_f$, where W is the bandwidth of the cell; if T is the time duration of the acousto-optic cell, $T_f = T$. When we integrate this result in time, the frequency resolution corresponds to $1/T_f$, the *coarse frequency resolution*.

We can achieve finer frequency resolution if we use a pair of slow chirp signals $c_s(t)$ to modulate the acousto-optic cells in the vertical direction, using an optical arrangement similar to that shown in Figure 15.8, but rotated 90°. The result at the output plane from this part of the system is

$$r_y(y,t) = e^{j\pi bt^2} e^{-j[2\pi_b(t-y/v)+\pi b(t-y/v)^2]}, \tag{15.15}$$

where f_b is the starting frequency for the chirp and the slow chirp rate is b, with $b \ll a$. This slow chirp determines the fine frequency resolution $f_0 = 1/T_s$, where T_s is the duration of the slow chirp. We combine this signal with that from Equation (15.14) to produce a signal

$$\begin{aligned} g(x,y,t) &= r_x(x,t) + r_y(x,t) \\ &= s(t) e^{j2\pi atx/v} e^{-j\pi a(x/v)^2} e^{j2\pi f_a(t-x/v)} \\ &\quad + e^{-j2\pi bty/v} e^{j\pi b(y/v)^2} e^{-j2\pi f_b(t-y/v)}. \end{aligned} \tag{15.16}$$

When we integrate the intensity $I(x, y, t) = |g(x, y, t)|^2$ over time and take the real part of the output, we produce a cross-product term of the form

$$\boxed{g_3(f_x, f_y) = 2\cos[2\pi(\alpha_a x + \beta_b y)]\,\mathrm{Re}\left\{\int_0^{kT} s(t) e^{-j2\pi(f_x+f_y)t}\, dt\right\},} \tag{15.17}$$

where α_a is the spatial frequency in the x direction due to f_a and β_b is the spatial frequency in the y direction due to f_b. This form of the output most clearly shows that the system performs a spectrum analysis for a time period $T_s = kT$ and that the spectrum modulates a spatial carrier so that the spectrum can be separated from the bias terms. We see immediately that the frequency resolution is of the order of $1/kT$; the chirp rates are adjusted to change the frequency resolution, as we showed in Chapter 14, Section 14.2.

15.5.2. Frequency Resolution

The methodology for finding the resolution in the coarse and fine frequency directions is the same as that given in Chapter 4, Sections 4.7.1 and 4.7.2; the reader may wish to review those sections before proceeding. We provide a corresponding analysis here so that the roles of the slow- and fast-chirp rates become clear. For convenience, we set the bandwidths of the cell in both directions equal to W so that the number of resolvable frequencies in each direction are equal: $M_x = M_y$. The coarse frequency resolution δf_x in the horizontal direction is

$$\boxed{\delta f_x = \frac{1}{T_f} = \frac{a}{W},} \qquad (15.18)$$

and the fine frequency resolution δf_y in the vertical direction is

$$\boxed{\delta f_y = \frac{1}{T_s} = \frac{b}{W},} \qquad (15.19)$$

where T_f is the duration of the fast chirp and T_s is the duration of the slow chirp. By definition, $f_0 = \delta f_y$ is the overall system resolution so that the slow-chirp duration is $T_s = kT$. Suppose that we want to analyze a signal with bandwidth B to a frequency resolution of $f_0 = 1/kT$. If we assume that $M = M_x M_y$ is the total number of resolvable frequencies, we can readily establish that $\delta f_y = B/M$ is the fine frequency resolution and that $\delta f_y = 1/T_s = 1/kT$. Therefore, we have that $T_s = kT = M/B$.

The fine frequency *range* is equal to $\Delta f_y = M_y B/M$ and this frequency range must equal the coarse frequency *resolution* δf_x to avoid any holes or redundancy in the spectrum. Therefore $\delta f_x = 1/T_f = BM_y/M$, so that $T_f = M/BM_y$. The ratio of the slow-chirp duration to the fast-chirp duration is therefore $T_s/T_f = M_y$. This technique is a powerful spectrum-analysis tool because it produces arbitrarily small frequency resolution, with approximately 10^5 resolvable frequencies in real time.

15.5.3. Experimental Results

An example of the system performance is shown in Figure 15.9. The upper-left panel shows the output light distribution for an input signal consisting of frequencies f_1 and f_2 that are 435 Hz apart. We see two spots of light, supported by a spatial carrier frequency; the carrier has components in both directions, as noted from Equation (15.17); the coarse resolution is of the order of 500 Hz. The upper right-hand panel shows the result when $f_2 = f_1 + 900$ Hz; the first frequency component f_1 remains fixed relative to its local oscillator, whereas frequency f_2 mixes with a local oscillator in the next coarse frequency column. When we increment the frequency by 30 Hz so that $f_2 = f_1 + 930$ Hz, the bright spot moves to the

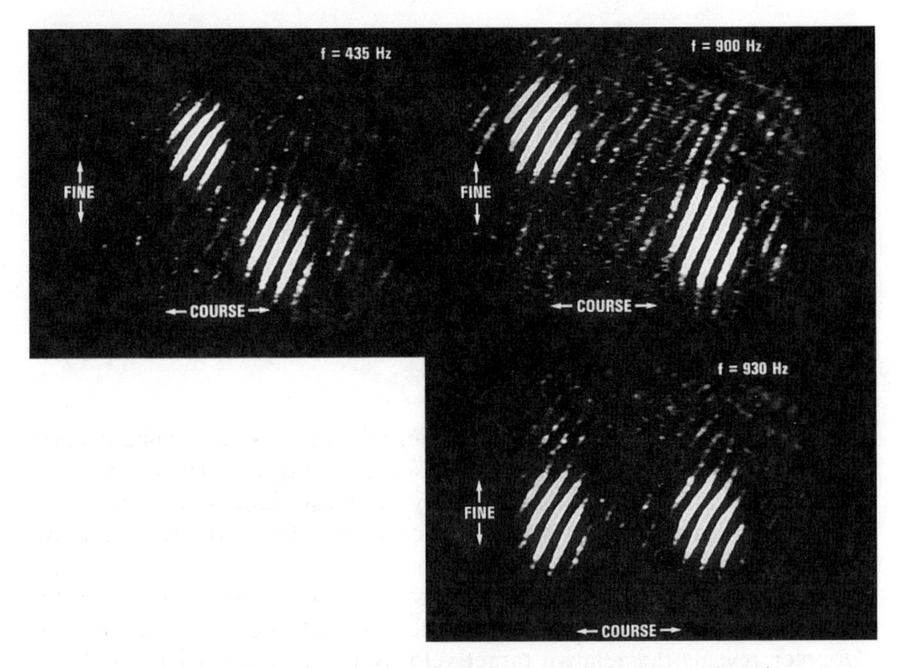

Figure 15.9. Two-dimensional spectrum analysis (courtesy T. M. Turpin).

next vertical oscillator position, showing that the fine frequency resolution of the system is better than 30 Hz.

15.6. AMBIGUITY FUNCTION GENERATION

One application of correlation is to synchronize a pseudorandom sequence used for direct spreading modulation in a communication system. For each application we need to consider the condition for which a space-integrating correlator is appropriate and for which a time-integrating correlator is more appropriate. For example, most satellite communication systems require large correlation gains to overcome jamming and other noise sources. In these systems the synchronization is done only infrequently so that rapid acquisition is not a severe problem. Furthermore, the sequence may be very long before it repeats so that we cannot record a frame of the pseudorandom sequence on a fixed mask; nor can the sequence be contained within the acousto-optic cell. In these applications we need time integrating architectures such as those discussed in Chapter 14 to obtain the necessary correlation gain. In a rapidly changing environment such as an air-to-air or air-to-ground communication system, rapid acquisition is important. Fixed reference signals and space-integrating architectures are practical in such applications because the codes tend to be shorter.

Synchronization is difficult to achieve in a dynamic environment because the received signal also has an unknown Doppler frequency. Doppler effects arise when the transmitter and receiver have relative motion. If the radial velocity between the two points is V_r, the Doppler frequency is

$$f_d = \frac{2V_r}{\Lambda_m}, \tag{15.20}$$

where Λ_m is the wavelength of the carrier frequency of the communication link. The received signal therefore has uncertainty with respect to both time of arrival and Doppler frequency. The two-dimensional ambiguity function determines how well a particular waveform can resolve these uncertainties.

The ambiguity function also has extensive application to waveforms used in radar systems, where the time delay reveals the target location and Doppler reveals the relative target velocity (134). In its symmetric form, the *cross-ambiguity function* for signals $f(t)$ and $s(t)$ is defined as

$$\boxed{A(\tau, \phi) = \int_{-\infty}^{\infty} f\left(t + \frac{\tau}{2}\right) s^*\left(t - \frac{\tau}{2}\right) e^{-j2\pi\phi t}\, dt,} \tag{15.21}$$

where τ is the time delay and ϕ is the Doppler shift between the two signals. As an integral in the frequency domain, we have the equivalent form

$$\boxed{A(\tau, \phi) = \int_{-\infty}^{\infty} F\left(f + \frac{\phi}{2}\right) S^*\left(f - \frac{\phi}{2}\right) e^{-j2\pi f\tau}\, df.} \tag{15.22}$$

From either Equation (15.21) or Equation (15.22), we see that $A(\tau, \phi)$ reduces to the ordinary cross-correlation function $c(\tau)$ when the Doppler shift is zero.

It can be shown that the integral of $|A(\tau, \phi)|^2$ over the entire range of time delays and Doppler frequency shifts is equal to 1, independent of the signal waveform if the ambiguity function is normalized by $|A(0, 0)|^2$ (see Problem 15.1). The waveform is usually chosen to favor either time

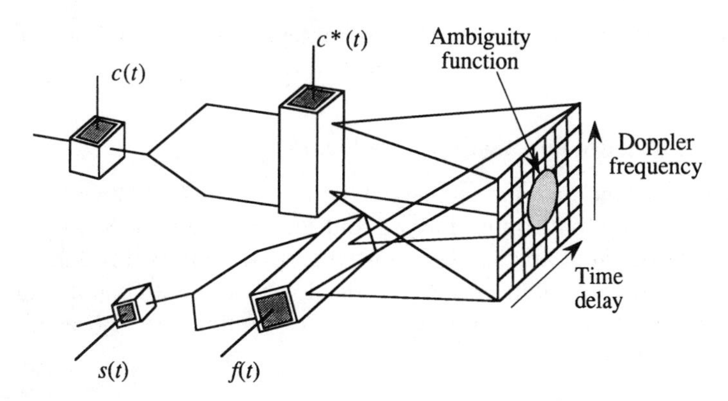

Figure 15.10. Triple product processor configured for ambiguity function generation.

resolution or frequency resolution. An increase in one, however, is accompanied by a decrease in the other so that the volume under the ambiguity function remains equal to 1. We therefore achieve fine resolution in time at the expense of frequency resolution and vice versa; we cannot achieve both simultaneously.

We discussed time-integrating correlation in Section 15.2 and we discussed spectrum analysis in Section 15.5. By combining the basic elements of both systems, we can generate the ambiguity function generating system as shown in Figure 15.10. In the horizontal branch of the system we use a correlator of the Sprague and Koliopoulos type to produce the product $s(t)f^*(t + x/v)\exp[-j2\pi f_a(t + x/v)]$ at the output plane, spreading the light in the vertical direction to cover the photodetector array. In the vertical direction we generate a local oscillator of the form

$$e^{j\pi bt^2} e^{-j2\pi bty/v} e^{j\pi b(y/v)^2} e^{j2\pi f_b(t-y/v)} \tag{15.23}$$

by the use of chirp waveforms. These two functions are added at the output plane and square-law detected. The cross-product term is

$$g_3(\tau, \phi) = 2\,\mathrm{Re}\left\{e^{j2\pi(f_a\tau + f_b\phi)} \int_0^{kT} f(t)s^*(t+\tau)e^{-j2\pi\phi t}\,dt\right\}, \tag{15.24}$$

where we have made the associations that $\tau = x/v$ and $\phi = by/v$, and have ignored the quadratic phase factors in τ and f. These quadratic phase factors can be removed by the use of lenses if necessary. The integral is equivalent to the ambiguity function $A(\tau, f)$ as given by Equation (15.21) except for a shift in the time origin. As usual, the spatial carrier frequency provides the means to separate $A(\tau, f)$ from the bias terms.

15.6.1. Ambiguity Function for a cw Signal

Suppose that the transmitted signal is $s(t) = \cos(2\pi f_j t)$ and that the received signal is both time delayed and Doppler shifted. We ignore the receiver noise and any scaling constants that appear in this analysis because we want to concentrate on the form of the ambiguity function itself. We begin by using the integral part of Equation (15.24) as given by Equation (15.21):

$$\begin{aligned} A(\tau, \phi) &= \int_{-\infty}^{\infty} f\left(t + \frac{\tau}{2}\right) s^*\left(t - \frac{\tau}{2}\right) e^{-j2\pi\phi t}\, dt \\ &= \int_{-\infty}^{\infty} \cos\left[2\pi f_j\left(t + \frac{\tau}{2}\right)\right] \cos\left[2\pi f_j\left(t - \frac{\tau}{2}\right)\right] e^{-j2\pi\phi t}\, dt. \end{aligned} \tag{15.25}$$

We use the sum and difference formula (AII.25) to find that

$$A(\tau, \phi) = \int_{-\infty}^{\infty} \{\cos(4\pi f_j t) + \cos(2\pi f_j \tau)\} e^{-j2\pi\phi t}\, dt. \tag{15.26}$$

The integration on time is now easily performed to produce

$$\begin{aligned} A(\tau, \phi) &= \tfrac{1}{2}\delta(\phi - 2f_j) + \tfrac{1}{2}\delta(\phi + 2f_j) \\ &\quad + \tfrac{1}{2}e^{j2\pi f_j \tau}\delta(\phi) + \tfrac{1}{2}e^{-j2\pi f_j \tau}\delta(\phi). \end{aligned} \tag{15.27}$$

The output of the optical system is found by using (15.27) in (15.24):

$$\begin{aligned} g_3(\tau, \phi) &= 2\,\mathrm{Re}\Big\{e^{j2\pi f_a \tau}\big[\tfrac{1}{2}\delta(\phi - 2f_j) + \tfrac{1}{2}\delta(\phi + 2f_j) \\ &\qquad + \tfrac{1}{2}e^{j2\pi f_j \tau}\delta(\phi) + \tfrac{1}{2}e^{-j2\pi f_j \tau}\delta(\phi)\big]\Big\} \\ &= \cos[2\pi f_a \tau]\delta(\phi - 2f_j) + \cos[2\pi f_a \tau]\delta(\phi + 2f_j) \\ &\quad + \cos[2\pi(f_a + f_j)\tau]\delta(\phi) + \cos[2\pi(f_a - f_j)\tau]\delta(\phi). \end{aligned} \tag{15.28}$$

This output light distribution is sketched in Figure 15.11 and the envelope

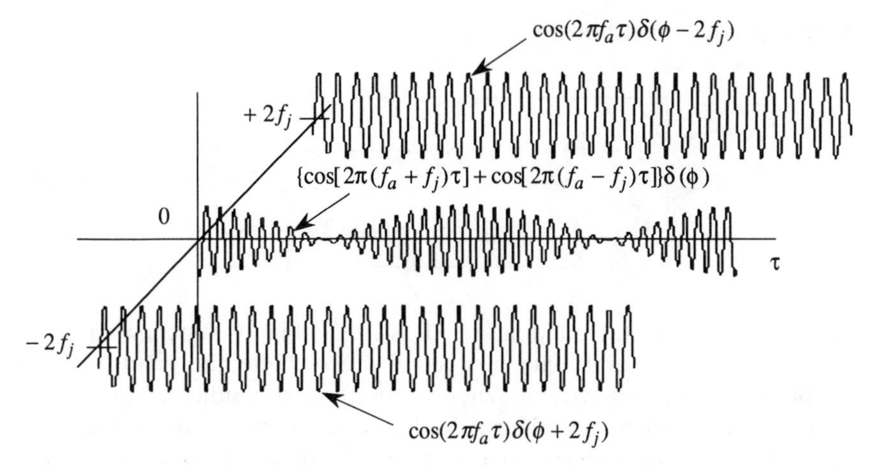

Figure 15.11. Ambiguity function for a cw signal.

of the carrier frequency shows that a cw signal provides extremely good doppler resolution but no time-delay resolution. One way to obtain time-delay resolution is to use a short-pulse signal.

15.6.2. Ambiguity Function for a Short-Pulse Signal

Consider a short-pulse burst signal whose center frequency is f_j. The ambiguity function for the short pulse becomes

$$\begin{aligned} A(\tau, \phi) &= \int_{-\infty}^{\infty} \operatorname{rect}\left[\frac{t + \tau/2}{T_0}\right] \cos\left[2\pi f_j\left(t + \frac{\tau}{2}\right)\right] \operatorname{rect}\left[\frac{t - \tau/2}{T_0}\right] \\ &\quad \times \cos\left[2\pi f_j\left(t - \frac{\tau}{2}\right)\right] e^{-j2\pi\phi t}\, dt \\ &= \int_{-\infty}^{\infty} \operatorname{rect}\left[\frac{t + \tau/2}{T_0}\right] \operatorname{rect}\left[\frac{t - \tau/2}{T_0}\right] \\ &\quad \times \left\{\cos(4\pi f_j t) + \cos(2\pi f_j \tau)\right\} e^{-j2\pi\phi t}\, dt, \end{aligned} \tag{15.29}$$

where T_0 is the pulse duration. We can use Equation (15.28) if we multiply it by an envelope function $E(\tau, \phi)$ in the τ direction and convolve it with $E(\tau, \phi)$ in the ϕ direction. We therefore concentrate on solving for the envelope function $E(\tau, \phi)$, in which the overlap of the two rect functions produces a new rect function:

$$\boxed{\begin{aligned} E(\tau, \phi) &= \int_{-\infty}^{\infty} \operatorname{rect}\left[\frac{t}{T_0 - |\tau|}\right] e^{-j2\pi\phi t}\, dt \\ &= \{T_0 - |\tau|\} \operatorname{sinc}[\phi(T_0 - |\tau|)]. \end{aligned}} \tag{15.30}$$

In this development, Equation (15.30) is the ambiguity function envelope of the spatial carrier frequency needed to complete the picture of the output of the optical system. When we perform the indicated operations between the envelope and the carrier frequency f_j, the resulting output of the optical system is shown in Figure 15.12. For clarity, we have set the Doppler equal to zero and show only that portion of the output that is centered at $\phi = 0$. The complete picture of the ambiguity function is to impose the envelope as given by Equation (15.30) on each of the terms sketched in Figure 15.11.

From Equation (15.30) we see that, in the τ direction, the envelope $E(\tau, \phi)$ is a triangular function of total width $2T_0$. The short pulse significantly improves time-delay resolution, but at the expense of less Doppler resolution. In the limit as T_0 tends to zero, the time-delay resolution becomes very good and the Doppler resolution becomes very poor. At the other extreme, in the limit as T_0 tends to infinity, the results degenerate into that of the cw signal discussed in the last section. Thus, the price we pay for improved performance in one of the parameters is a reduced performance in the other parameter, no matter what the waveform.

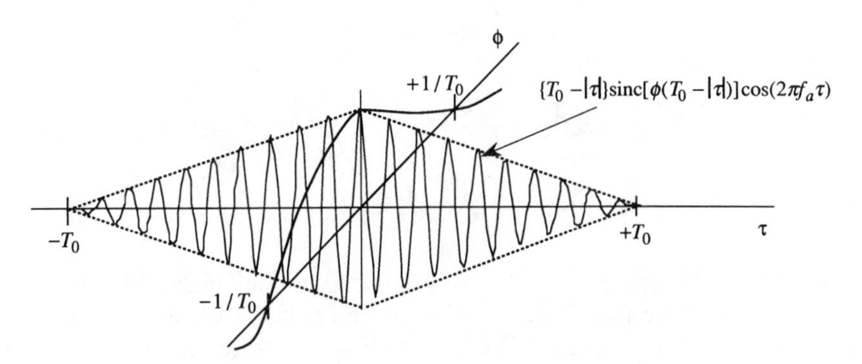

Figure 15.12. Ambiguity function for a short pulse.

15.6.3. Ambiguity Function for an Infinite Time Duration Chirp Signal

The final waveform we consider is that of a chirp waveform. Suppose that $s(t) = \cos(2\pi f_j t + \pi a t^2)$ and that we do not initially bound this waveform in time. The frequency content of $s(t)$ therefore ranges from $-\infty$ to $+\infty$. The ambiguity function becomes

$$A(\tau,\phi) = \int_{-\infty}^{\infty} \cos\left[2\pi f_j\left(t+\frac{\tau}{2}\right) + \pi a\left(t+\frac{\tau}{2}\right)^2\right] \times \cos\left[2\pi f_j\left(t-\frac{\tau}{2}\right) + \pi a\left(t-\frac{\tau}{2}\right)^2\right] e^{-j2\pi\phi t}\,dt$$
$$= \int_{-\infty}^{\infty} \left\{\cos\left[4\pi f_j t + 2\pi a\left(t^2+\frac{\tau^2}{4}\right)\right] + \cos(2\pi a t\tau)\right\} e^{-j2\pi\phi t}\,dt. \tag{15.31}$$

We confine our attention to the baseband part of Equation (15.31) and find that

$$\boxed{\begin{aligned} A(\tau,\phi) &= \int_{-\infty}^{\infty} \cos(2\pi a t\tau) e^{-j2\pi\phi t}\,dt \\ &= \delta(\phi - a\tau) + \delta(\phi + a\tau). \end{aligned}} \tag{15.32}$$

We substitute Equation (15.32) into Equation (15.24) to produce the final output of the optical system. This time, however, we display the ambiguity function as a projection onto the τ-ϕ plane, as shown in Figure 15.13. As with the other ambiguity surfaces, we have a basic uncertainty in the time-delay and Doppler parameters. We expect that an infinite-chirp waveform should provide extremely high time-delay information because

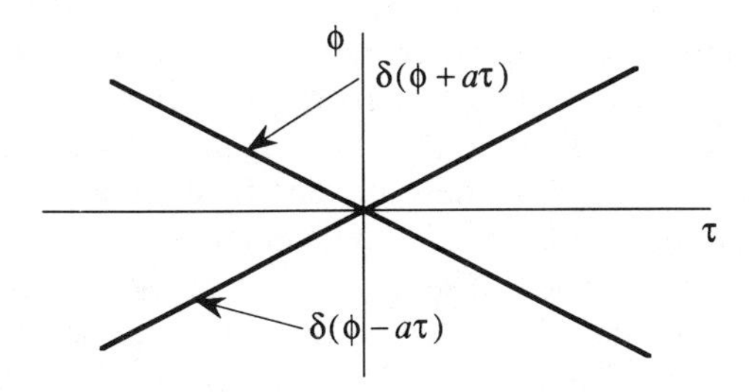

Figure 15.13. Ambiguity function of an infinite-chirp waveform.

the chirp can be compressed into a very short time interval. Suppose that a target at a certain range has a radial velocity component with respect to the transmitter/receiver so that the chirp is frequency shifted. This combination of time delay and Doppler cannot be distinguished from another target with a somewhat different time delay and Doppler. We leave it to the reader to derive the ambiguity function for a chirp of finite time duration (see Problem 15.2).

15.7. WIGNER-VILLE DISTRIBUTIONS

The Wigner-Ville distribution is defined as

$$W(t,f) = \int_{-\infty}^{\infty} f\left(t+\frac{\tau}{2}\right) s^*\left(t-\frac{\tau}{2}\right) e^{-j2\pi f\tau}\, d\tau, \tag{15.33}$$

and it can also be shown that it is the two-dimensional Fourier transform of the ambiguity function:

$$W(t,f) = \iint_{-\infty}^{\infty} A(\tau,\phi) e^{j2\pi(\phi t - f\tau)}\, d\tau\, d\phi. \tag{15.34}$$

This distribution is a function of *time* and *temporal frequencies* instead of *delay time* and *difference frequencies* (Doppler). In Figure 15.14 we show the Wigner-Ville distribution for chirp signals with various chirp rates (see Problem 15.3). In the upper-left panel, the chirp rate is zero. The signal is, therefore, a cw tone that has a fixed frequency (vertical axis) for all possible time delays. The spatial carrier frequency is clearly evident in the horizontal direction which represents real time, and a sinc function in the vertical direction determines the frequency resolution.

In the upper right-hand panel, the chirp signal has a low chirp rate. We see that the same carrier frequency and sinc function are present; the key difference is that the Doppler and time delay are now coupled in the argument of the sinc function. The peak value of the sinc function occurs along the $f - at = 0$ line. The third panel shows the Wigner-Ville distribution for a signal with approximately twice the chirp rate of the second signal.

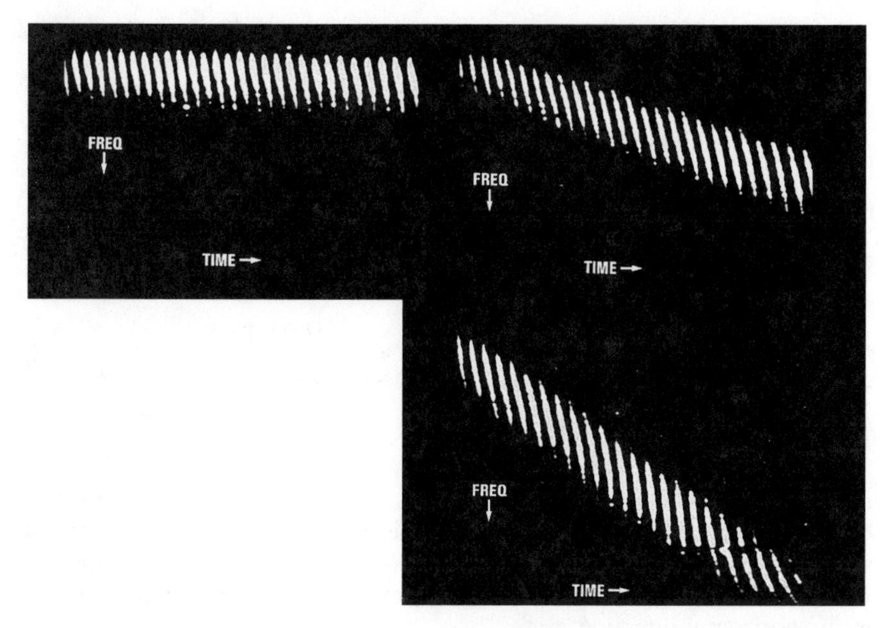

Figure 15.14. Wigner-Ville function for a chirp waveform (courtesy T. M. Turpin).

15.8. RANGE AND DOPPLER SIGNAL PROCESSING

Processing pulse Doppler radar signals is another example of a computationally intensive task. The basic problem is to determine the range and velocity for each target in the field. We therefore need to implement a correlator to determine the target range and a time-integrating spectrum analyzer to determine the Doppler frequency from which we derive the target velocity. To simplify the analysis we initially consider a transmitted short-pulse waveform whose duration T_0 is equal to one range resolution element R_0 so that $T_0 = R_0/c$, where c is the speed of light. The received signal $f(t)$ consists of successive scans of the total target range $R = N_r R_0$ where N_r is the number of range elements processed.

We begin by attacking the Doppler processing part of the problem. The optical processor of Figure 15.15 contains two acousto-optic cells, driven by the chirp waveforms $\cos(2\pi f_a t + \pi a t^2)$ and $\cos(2\pi f_b t + \pi a t^2)$ so that the signals counterpropagate. These two cells are imaged at the output plane of the system; in the horizontal direction they provide the distributed local oscillator, as described in Chapter 14, Section 14.2, needed for time-integrating spectrum analysis. The signal $f(t)\cos(2\pi f_c t)$ drives the acousto-optic cell oriented in the vertical direction, and this cell is

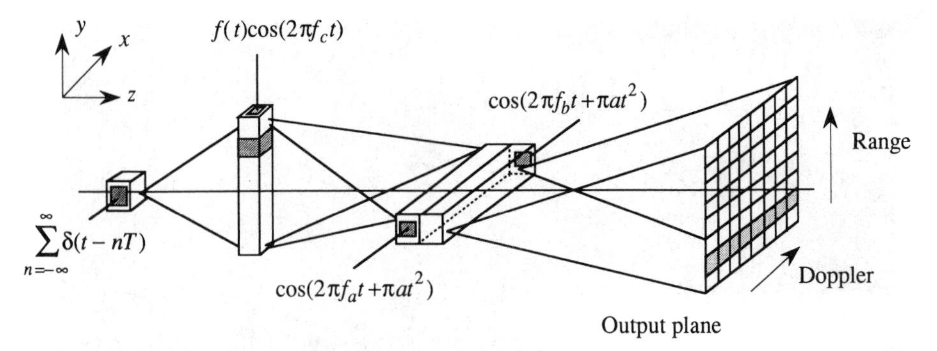

Figure 15.15. Basic elements of a range/Doppler processor.

imaged through the two acousto-optic cells onto the output plane. Because the acousto-optic cell converts time to distance, the vertical axis of the system output represents the range of the targets. The mixing of the signal $f(t)$ with the distributed local oscillator provides the Doppler information in the horizontal direction.

Because the signal is moving through the acousto-optic cell with velocity v, it appears that we cannot integrate the returns for a sufficiently long time over a given range channel to achieve fine Doppler resolution. We overcome this difficulty by *range gating* the signal so that the same range cells always appear at the same range positions at the output plane. Range gating is accomplished by pulsing the light source with an impulse train, whose repetition period is $T = R/c$, as provided by the first acousto-optic cell in Figure 15.15. An alternative method for obtaining the impulse train is to cavity dump the laser source directly, thus conserving average laser power. Either of these techniques temporarily "freeze" the signal in the acousto-optic cell so that, at the times the impulses occur, the same range elements are always imaged to the same positions at the output plane.

The pulse-repetition interval T_r is set by the highest Doppler frequency f_d and the Nyquist sampling rate. The highest expected Doppler frequency is

$$f_d = \frac{2V}{\Lambda}, \tag{15.35}$$

where V is the maximum radial velocity to be detected, and Λ is the radar wavelength. The pulse-repetition interval that satisfies the Nyquist condition is therefore

$$T_r = \frac{1}{2f_d}, \tag{15.36}$$

and the total number of range channels is

$$N_r = \frac{T_r}{T_0}. \tag{15.37}$$

The number of Doppler filters required is a function of how long the target is in the radar beam. Suppose that we rotate the antenna, as in an air defense system or an airport traffic monitoring system, at an angular rate Ω expressed in rad/sec and use an antenna whose beamwidth is ϕ rad. The target is therefore illuminated for a time interval

$$T_i = \frac{\phi}{\Omega}, \tag{15.38}$$

and the number of Doppler resolution elements is

$$N_d = \frac{T_i}{T_r} = 2f_d T_i. \tag{15.39}$$

The distributed local-oscillator section of the time-integrating spectrum analyzer must provide N_d probes to produce the required Doppler resolution. Recall from Chapter 14 that the useful output of the spectrum analyzer is given by Equation (14.15):

$$\boxed{g_3(u) = 2\,\mathrm{Re}\left\{e^{j[2\pi(\alpha_a+\alpha_b-\alpha_0)u-\phi_0]}\int_T^{kT} s(t)e^{j2\pi ft}\,dt\right\}.} \tag{15.40}$$

As before, we relate the temporal frequency variable f to the spatial variable u in the horizontal direction of the output by

$$f = \frac{2au}{v} + f_c + f_b - f_a, \tag{15.41}$$

so that the Doppler information is displayed in this direction. Because the

Doppler may be either positive or negative, we want $f = 0$ to occur at $u = 0$ so that we must arrange the various carrier frequencies to ensure that $f_c + f_b - f_a = 0$.

The transmitted radar waveform is generally not a short pulse because it requires high peak power to achieve good performance. More typically, the transmitted waveform is a long coded signal with good correlation properties. The received signal is then correlated with a replica of the transmitted signal to achieve the required range resolution. This process can be done optically, using the space-integrating correlator described in Chapter 13 to produce the input signal $f(t)$ shown in Figure 15.15. Such a correlator could be integrated into the front end of the system, shown here, with the periodic correlation peaks providing the self-synchronizing impulse train.

Another application of acousto-optics is to real-time synthetic radar processing (135). The optical processing system described in Chapter 5, Section 5.6 for processing synthetic radar data used a set of lenses to provide the required matched filter to focus the chirp waveforms in both the range and aximuth directions and thereby generate a radar map. In this acousto-optic processing system, the chirp waveforms generated in the range direction are correlated with a simple space-integrating architecture, while the Fresnel zone patterns generated in the azimuth direction are processed and collected on a two-dimensional CCD array. The array is read out sequentially in parallel to generate the radar map. Higher performance systems that include programmable architectures have been described (136).

15.9. OPTICAL TRANSVERSAL PROCESSOR FOR NOTCH FILTERING

Adaptive filtering, using transversal tapped delay lines with feedback, has been applied to problems such as redundancy removal in data, reduction of intersymbol interference through equalization, noise cancellation, self-tuning, and speech processing. Because the data rate or the signal bandwidth is limited by the processing speeds of digital circuits, we investigate the use of optical processing to extend the bandwidths of signals that can be processed. In Chapter 12, Section 12.5, we discussed a notch filtering application to excise unwanted narrowband interference signal. In this section we describe how this process can be implemented adaptively with direct feedback control.

15.9.1. Sampled Time Analysis

A general form of an adaptive linear predictor is shown in Figure 15.16. Let $s(t)$ be a sampled signal that drives a delay line. The delayed samples are multiplied by N weights c_j to form an estimate $\hat{s}(t)$ of the received signal:

$$\hat{s}(t) = \sum_{j=1}^{N} c_j s(t - jT_0), \tag{15.42}$$

where T_0 is the time delay of each element of the tapped delay line. The optimum weights are determined by our criterion for how well the estimated signal $\hat{s}(t)$ represents $s(t)$. The least mean-square error is a useful criterion; the energy in the residual signal $z(t) = s(t) - \hat{s}(t)$ is then

$$E_z = \sum_K \left| s(t) - \sum_{j=1}^{N} c_j s(t - jT_0) \right|^2, \tag{15.43}$$

where K is the number of samples accumulated to form the result. We minimize E_z with respect to a given tap weight c_j by forming the derivative

$$\begin{aligned} \frac{\partial E_z}{\partial c_j} &= -2 \sum_K \left[s(t) - \sum_{j=1}^{N} c_j s(t - jT_0) \right] s(t - jT_0) \\ &= -2 \sum_K z(t) s(t - jT_0). \end{aligned} \tag{15.44}$$

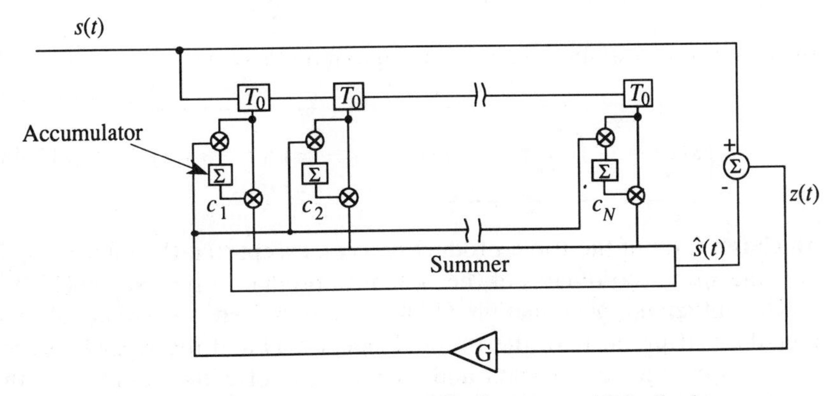

Figure 15.16. Discrete transversal filtering system with feedback.

From Equation (15.44) we see that the derivative is calculated by multiplying the residual signal by the delayed values of the received signal, and we then use this result to update the tap weights; this process is shown in Figure 15.16.

15.9.2. Continuous-Time Analysis

In the optical processor, the delay line is the acousto-optic cell which is tapped photonically so that the taps are continuous, instead of discrete. The optimum weights for the continuous-time situation are represented by

$$c(\tau) = G\int_{t-T_1}^{t} z(u)s(u-\tau)\,du, \tag{15.45}$$

where G is the gain in the feedback loop, T_1 is the integration time of the accumulator, and t represents the continuous-time equivalent of the discrete delay. We let $T_0 = T/N$ be the minimum discernible time delay in an acousto-optic cell, where $N = 2TW$. The estimate of the received signal is

$$\hat{s}(t) = \int_{0}^{T} c(\tau)s(t-\tau)\,d\tau, \tag{15.46}$$

so that, by substituting Equation (15.46) into Equation (15.45), we have the estimate

$$\hat{s}(t) = G\int_{t-T_1}^{t}\int_{0}^{T} z(u)s(u-\tau)s(t-\tau)\,du\,d\tau. \tag{15.47}$$

By a change of variables we rewrite Equation (15.47) as

$$\boxed{\hat{s}(t) = G\int_{0}^{T}\int_{T}^{0} z(t+q)s(t+q-\tau)s(t-\tau)\,dq\,d\tau,} \tag{15.48}$$

which is a form of the triple-product process, except that the integration is over the space coordinates of the system to produce a time signal (137).

The integrand of Equation (15.48) requires that we create all the time-delayed products of the residual signal $z(t)$ and the received signal $s(t)$ to form the tap weights and to then convolve the result with the received signal. Figure 15.17 shows one way to implement the triple product using acousto-optic cells. The functions $z(t)$ and $s(t)$ drive two

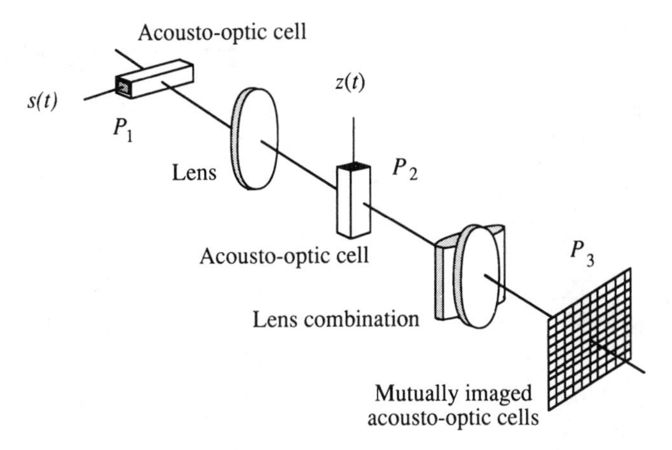

Figure 15.17. Acousto-optic cell system for producing tap weights.

acousto-optic cells in the orthogonal directions shown. These two acousto-optic cells are mutually imaged onto plane P_3 with the diffracted light from the first acousto-optic cell passing through the second acousto-optic cell.

At plane P_3 we have the light distribution necessary to provide the tap weights, as is more clearly seen from Figure 15.18(a), in which the tap weight plane has been rotated by 45 degrees for convenience. The numbers in each delay cell represent the successive time indices for the two functions. If we integrate the light in the vertical direction as shown by the dotted lines, we obtain the function $c(\tau)$. This function is then convolved with the received signal $s(t)$ which drives a third acousto-optic cell, as we shall show shortly.

Although Figure 15.17(a) shows only a small number of taps, the number of tap weights available is equal to twice the time bandwidth product of the acousto-optic cell. From the figure we see that the central tap weight is the sum of N products, whereas the end tap weights contain only one product. The end tap weights are therefore noisy and reduce the performance of the system. The number of taps required depends on the application. For example, to reduce intersymbol interference, the number of taps required is determined by the extent of the channel impulse response which may be of the order of only 50–100 symbol periods. For adaptive notch filtering, the number of taps required is determined by the number of frequencies that need to be removed simultaneously; at least two taps per frequency are required. If the time bandwidth product of the acousto-optic cells is of the order of 1000–2000, the use of only 50–100

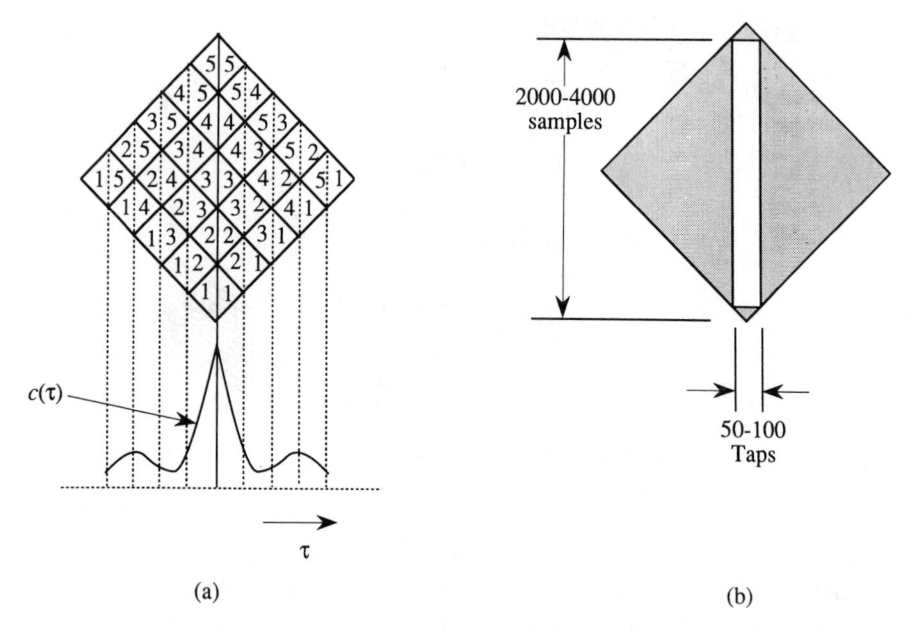

Figure 15.18. Space-plane representation of tap weights.

taps means that the triangular weighting of $c(\tau)$ is less significant; the region of integration can be truncated to the rectangular window as shown in Figure 15.17(b). Each of the 50–100 taps will then contain approximately 2000–4000 samples which represent the finite length of the accumulators in each of the correlation cancellation loops.

15.9.3. A Frequency Plane Implementation

From Figure 15.17 and from Equation (15.45), we see that the required tap weights $c(\tau)$ are obtained by integrating the product of $s(t)$ and $z(t)$ over a range of delays. Recall from Chapter 12, Section 12.6 that one way to integrate light is to create the Fourier transform of an area-modulated function and evaluate the transform along the $\beta = 0$ axis. In preparation for applying this technique to produce $c(\tau)$, we note that the Fourier transform of the triple product given by Equation (15.48) with respect to the temporal frequency is given by the double convolution theorem in Equations (5.7) and (5.8). As a result, we have the frequency plane representation of the signal estimate

$$\hat{s}(t) = \int_{-\infty}^{\infty} Z(f,t)|S(f,t)|^2 e^{j2\pi ft}\, df, \tag{15.49}$$

where $Z(f,t)$ and $S(f,t)$ are the Fourier transforms of the most recent T sec of the residual and received signals.

When we append the optical system shown in Figure 15.19 to that shown in Figure 15.17, the resultant system simultaneously integrates the light in the vertical direction and creates the Fourier transform of the tap weights in the horizontal direction. The light distribution along the horizontal axis is proportional to $A(f,t) = Z(f,t)S(f,t)$. Suppose that the combined optics in Figure 15.17 and in Figure 15.19 are placed in one branch of an interferometer (137). The interferometer is arranged so that $S(f,t)$ is added to $A(f,t)$. This sum is then square-law detected to provide the integrand of Equation (15.49), and the output of a single-element photodetector provides the estimate $s(t)$ by virtue of the heterodyne transform discussed in Chapter 12.

An example of the performance of an adaptive system is shown in Figure 15.20. The lower trace shows the spectrum of the received signal which consists of a 50-MHz wideband signal and a narrowband jammer at 90 MHz. The upper trace shows the notch that has formed adaptively and shows a notch depth of about 32 dB (138). The adaptation time is dependent on the specific application, although it appears that most

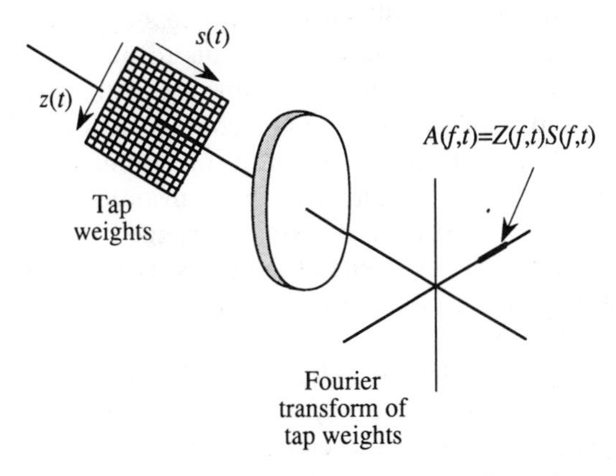

Figure 15.19. Optical system to produce Fourier transform of tap weights.

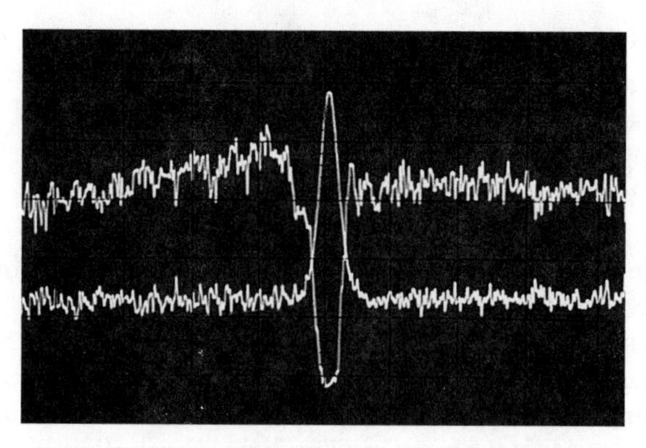

Figure 15.20. Experimental result of notch filtering (138).

systems cited in the literature adapt within a time interval corresponding to 500–2000 samples. The adaptation time is determined, in part, by the required values of T and T_1; these values also influence noise in the tap weights and, therefore, the accuracy at convergence. The gain factor G also affects both the rate of adaptation and the accuracy. Finally, the dynamic range of the photodetector determines the minimum increment by which the tap weights can be set. Generally, at least 10^3 distinct levels provide adequate performance.

15.10. PHASED ARRAY PROCESSING

We close this chapter on two-dimensional processing by describing some applications of multichannel acousto-optic cells to process data produced by phased arrays. These processing operations are of the space-integrating type. Consider the phased array antenna system sketched in Figure 15.21. Suppose that the array elements monitor electromagnetic emitters at unknown frequencies f_j and angles ϕ_j. The conventional way to monitor the field is to sweep a narrow beam over the range of azimuth angles by weighting the phases of the elements of the array. Because of this scanning process, the probability of intercepting emitters that are pulsing in short bursts may be small.

A useful technique is to let each element of the array drive a channel of a multichannel acousto-optic cell and to Fourier transform the resulting signal. In this way we recreate the emitter field at light wavelengths but with a λ/Λ scale factor, where λ and Λ are the light and microwave frequencies. The system shown in Figure 15.22 behaves as a spectrum

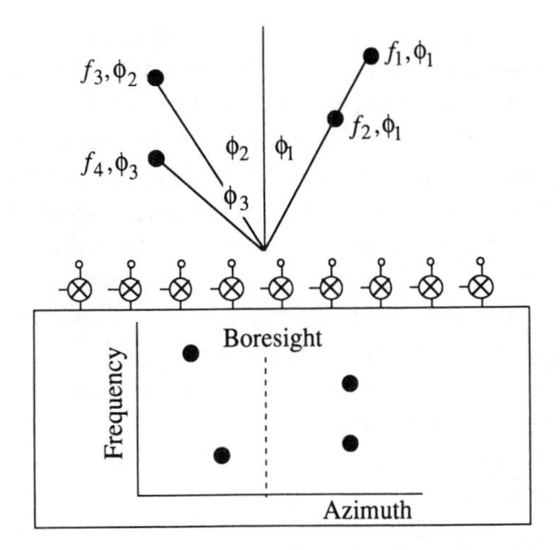

Figure 15.21. Phased array configuration for sorting emitters.

analyzer in the horizontal direction. A new feature is that we also Fourier transform *across* the channels to decompose the information at a particular f_j into its azimuth components at angles ϕ_j. In effect, we perform the beamforming operation *after detection*, with the aid of the Fourier-transform operation, instead of using the beam scanning as a predetection operation.

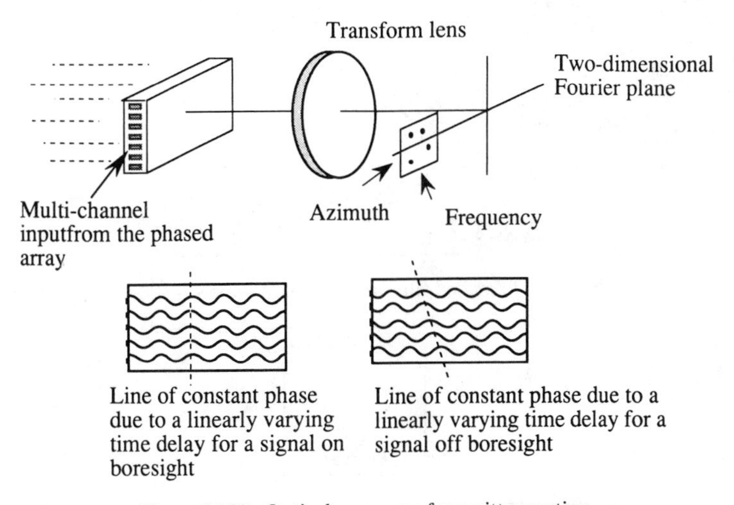

Figure 15.22. Optical processor for emitter sorting.

We do not suffer a signal-to-noise ratio loss in this process because the local signal-to-noise ratio at the two-dimensional output plane is the same as it is with scanning. It is the enormous parallel processing power of the optical system that makes the postdetection beamforming attractive, and the probability of intercept is now 100%. This emitter location technique is an extension of the cross-spectrum analysis discussed in Chapter 11. There we used just two antenna elements to find the angle of arrival of energy from an emitter. The advantage of using many antenna elements is that the information can be displayed as a two-dimensional spatial function over a larger unambiguous angular interval.

Another application is processing monostatic or bistatic radar data. Figure 15.23 shows a monostatic radar system in which the central element of the array is a high-power transmitter that emits a wideband waveform. All other elements in the array operate only in the receive mode. The basic idea is that the target returns are compressed to provide good range resolution and are sorted in azimuth to fix the target location. The broadband signal may be a pseudorandom sequence, a coded signal, or a chirp waveform.

We need to generate the appropriate matched filter for the broadband signal, place it in the Fourier plane, and use a line of high-speed photodetectors as shown in Figure 15.24. What we have done, in effect, is to create

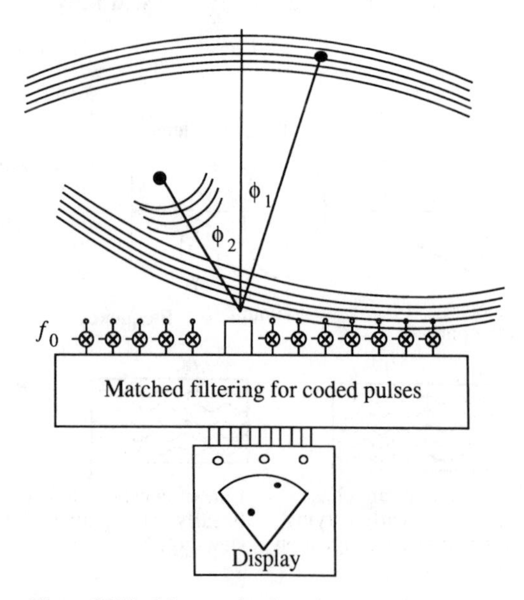

Figure 15.23. Monostatic phased array radar system.

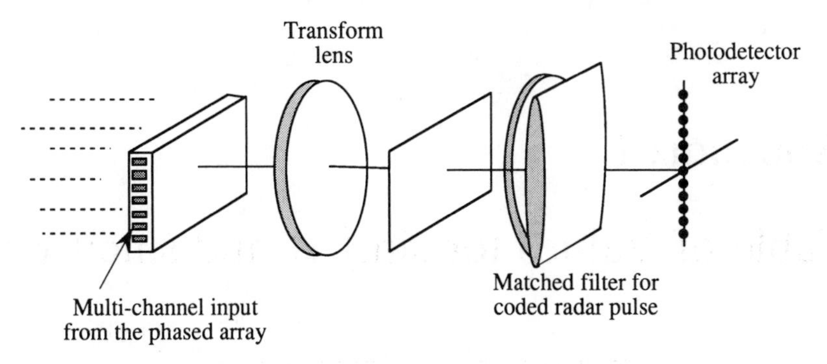

Figure 15.24. Optical processor for radar signals.

a multichannel space-integrating correlator. The times of arrival of the correlation peaks at a detector at some azimuth angle ϕ_j give the ranges to all targets along that line.

PROBLEMS

15.1. Calculate *and sketch* the squared magnitude for the auto ambiguity function

$$|A(\tau, \phi)|^2 = \left| \int_{-\infty}^{\infty} s\left(t + \frac{\tau}{2}\right) s^*\left(t - \frac{\tau}{2}\right) e^{-j2\pi\phi t}\, dt \right|^2$$

for a Gaussian waveform $s(t) = \exp(-t^2/T^2)$, where T controls the waveform duration. For a normalization in which $|A(0,0)|^2 = 1$, calculate the volume V under this function:

$$V = \int\int |A(\tau, \phi)|^2\, d\tau\, d\phi.$$

For $T = 1$ μsec, what is the best Doppler and range resolution that can be expected? This result will clearly show the tradeoff between range and Doppler resolution.

15.2. Derive the ambiguity function for a chirp waveform of a finite duration found by multiplying the infinite-duration chirp by a rect(t/T) function.

15.3. Derive the Wigner-Ville transform for the infinite-chirp signal, using the two-dimensional Fourier-transform relationship.

Appendix I

Table of Values for sinc(x) and sinc^2(x)

x	sinc(x)	sinc^2(x)
0.00	1.00	1.00
0.04	1.00	0.99
0.08	0.99	0.98
0.12	0.98	0.95
0.16	0.96	0.92
0.20	0.94	0.88
0.24	0.91	0.82
0.28	0.88	0.77
0.32	0.84	0.71
0.36	0.80	0.64
0.40	0.76	0.57
0.44	0.71	0.50
0.48	0.66	0.44
0.52	0.61	0.37
0.56	0.56	0.31
0.60	0.50	0.25
0.64	0.45	0.20
0.68	0.40	0.16
0.72	0.34	0.12
0.76	0.29	0.08
0.80	0.23	0.05
0.84	0.18	0.03
0.88	0.13	0.02
0.92	0.09	0.01
0.96	0.04	0.00
1.00	0.00	0.00
1.04	−0.04	0.00
1.08	−0.07	0.01
1.12	−0.10	0.01
1.16	−0.13	0.02
1.20	−0.16	0.02
1.24	−0.18	0.03
1.28	−0.19	0.04

Appendix II

Temporal Fourier transform / inverse:

$$F(f) = \int_{-\infty}^{\infty} f(t) e^{-j2\pi ft}\, dt$$

$$f(t) = \int_{-\infty}^{\infty} F(f) e^{j2\pi ft}\, df \qquad \text{(AII.1)}$$

Spatial Fourier transform / inverse:

$$F(\alpha) = \int_{-\infty}^{\infty} f(x) e^{j2\pi\alpha x}\, dx$$

$$f(-x) = \int_{-\infty}^{\infty} F(\alpha) e^{j2\pi\alpha x}\, d\alpha \qquad \text{(AII.2)}$$

Shift theorem:

$$\int_{-\infty}^{\infty} f(x - x_0) e^{j2\pi\alpha x}\, dx = F(\alpha) e^{j2\pi\alpha x_0} \qquad \text{(AII.3)}$$

Convolution:

$$g(x) = \int_{-\infty}^{\infty} f(u) h(x - u)\, du \qquad \text{(AII.4)}$$

Convolution theorem:

$$G(\alpha) = \int_{-\infty}^{\infty} \left\{ \int_{-\infty}^{\infty} f(u) h(x - u)\, du \right\} e^{j2\pi\alpha x}\, dx$$

$$= F(\alpha) H(\alpha) \qquad \text{(AII.5)}$$

Correlation:

$$c_{fs}(u) = \int_{-\infty}^{\infty} f(x)s^*(x+u)\,dx \qquad \text{(AII.6)}$$

Fresnel transform:

$$F(\xi) = \int_{-\infty}^{\infty} f(x)e^{-j(\pi/\lambda D)(\xi-x)^2}\,dx, \qquad \text{(AII.7)}$$

where D is the distance between the planes containing $f(x)$ and $F(x)$.

Generalization of Parseval's Theorem (Cross-Product Theorem):

$$\int_{-\infty}^{\infty} f(x)g^*(x)\,dx = \int_{-\infty}^{\infty} F(\alpha)G^*(\alpha)\,d\alpha = \int_{-\infty}^{\infty} F(\xi)G^*(\xi)\,d\xi, \qquad \text{(AII.8)}$$

where $f(x)$ is a spatial function, $F(\alpha)$ is its Fourier transform, and $F(\xi)$ is its Fresnel transform.

Sifting property of the delta function:

$$\int_{-\infty}^{\infty} f(u)\delta(x-u)\,du = f(x). \qquad \text{(AII.9)}$$

The sifting property also applies to a sinc function of the form $\mathrm{sinc}(2\alpha_{\mathrm{co}}x)$:

$$\int_{-\infty}^{\infty} f(u)\,\mathrm{sinc}[2\alpha_m(x-u)]\,du = f(x), \qquad \text{(AII.10)}$$

provided that $f(x)$ is bandlimited to α_{co} and that $\alpha_m \geq \alpha_{\mathrm{co}}$.

Other useful relationships:

$$\sum_{n=-\infty}^{\infty} \delta(x-nd_0) = \frac{1}{d_0}\sum_{n=-\infty}^{\infty} e^{j2\pi n\alpha_0 x}, \qquad \alpha_0 = \frac{1}{d_0}, \qquad \text{(AII.11)}$$

$$\int_{-\infty}^{\infty} \mathrm{sinc}(x)\,dx = \int_{-\infty}^{\infty} \mathrm{sinc}^2(x)\,dx = 1, \qquad \text{(AII.12)}$$

$$\mathrm{sinc}(x) = 1 - \frac{x^2}{3!} + \frac{x^4}{5!} - \frac{x^6}{7!} + \cdots \qquad \text{(AII.13)}$$

$$\int e^{-(ax^2+2bx+c)}\,dx$$

$$= \tfrac{1}{2}\sqrt{\frac{\pi}{a}}\, e^{(b^2-ac)/a}\, \operatorname{erf}\left[\sqrt{a}\,x + \frac{b}{\sqrt{a}}\right] + \text{const}; \qquad a \neq 0, \quad \text{(AII.14)}$$

$$\operatorname{erf}(z) = \frac{2}{\sqrt{\pi}} \int_0^{\infty} e^{-t^2}\,dt = \frac{2}{\sqrt{\pi}} \sum_{n=0}^{\infty} \frac{(-1)^n z^{2n+1}}{n!(2n+1)}, \quad \text{(AII.15)}$$

$$\operatorname{erf}(z) \to 1 \quad \text{as} \quad z \to \infty, \quad \text{(AII.16)}$$

$$J_\nu(z) = \left(\tfrac{1}{2}z\right)^\nu \sum_{n=0}^{\infty} \frac{\left(-\tfrac{1}{4}z^2\right)^n}{n!\,\Gamma(\nu+n+1)}, \quad \text{(AII.17)}$$

$$\sin(x) = x - \frac{x^3}{3!} + \frac{x^5}{5!} - \frac{x^7}{7!} + \cdots, \quad \text{(AII.18)}$$

$$\cos(x) = 1 - \frac{x^2}{2!} + \frac{x^4}{4!} - \frac{x^6}{6!} + \cdots, \quad \text{(AII.19)}$$

$$\sin(x+y) = \sin x \cos y + \cos x \sin y, \quad \text{(AII.20)}$$

$$\sin(x-y) = \sin x \cos y - \cos x \sin y, \quad \text{(AII.21)}$$

$$\cos(x+y) = \cos x \cos y - \sin x \sin y, \quad \text{(AII.22)}$$

$$\cos(x-y) = \cos x \cos y + \sin x \sin y, \quad \text{(AII.23)}$$

$$\sin x \sin y = \tfrac{1}{2}[\cos(x-y) - \cos(x+y)], \quad \text{(AII.24)}$$

$$\cos x \cos y = \tfrac{1}{2}[\cos(x-y) + \cos(x+y)], \quad \text{(AII.25)}$$

$$\sin x \cos y = \tfrac{1}{2}[\sin(x-y) + \sin(x+y)]. \quad \text{(AII.26)}$$

References

1. N. Wiener, "Optics and the Theory of Stochastic Processes," *J. Opt. Soc. Am.*, Vol. 43, p. 225 (1953).
2. P. Elias, "Optics and Communication Theory," *J. Opt. Soc. Am.*, Vol. 43, p. 229 (1953).
3. P. Fellget, "Concerning Photographic Grain, Signal-to-Noise Ratio, and Information," *J. Opt. Soc. Am.*, Vol. 43, p. 271 (1953).
4. E. H. Linfoot, "Information Theory and Optical Imagery," *J. Opt. Soc. Am.*, Vol. 43, p. 808 (1955).
5. G. Toraldo di Francia, "The Capacity of Optical Channels in the Presence of Noise," Opt. Acta, Vol. 2, p. 5 (1955).
6. E. L. O'Neill, "Spatial Filtering in Optics," *IRE Trans. Inf. Theory*, Vol. IT-2, p. 56 (1956).
7. A. Marachal and P. Croce, "Un Filtre de Frequencies Spatiales Pour l'Amelioration du Contraste des Images Optiques," *C.R. Acad. Sci.*, Vol. 127, p. 607 (1953).
8. L. J. Cutrona, E. N. Leith, C. J. Palermo, and L. J. Porcello, "Optical Data Processing and Filtering Systems," *IRE Trans. Inf. Theory*, Vol. IT-6, p. 386 (1960).
9. A. VanderLugt, "Signal Detection by Complex Spatial Filtering," *IRE Trans. Inf. Theory*, Vol. IT-10, p. 139 (1964).
10. M. Born and E. Wolf, *Principles of Optics*, Pergamon Press, New York, 1964.
11. F. A. Jenkins and H. E. White, *Fundamental of Optics*, McGraw Hill, New York, 1957.
12. R. S. Longhurst, *Geometrical and Physical Optics*, Longmans, Green, & Co., London, 1967.
13. H. H. Hopkins, *Wavefront Theory of Aberrations*, University Microfilms, Ann Arbor, MI, 1952.
14. B. D. Guenther, *Modern Optics*, Wiley, New York, 1990.
15. A. Papoulis, *Systems and Transforms with Applications in Optics*, McGraw Hill, New York, 1968, p. 322.
16. A. Sommerfeld, *Optics*, Academic, New York, 1964.

17. R. N. Bracewell, *The Fourier Transform and its Applications*, McGraw-Hill, New York, 1965.

18. J. D. Gaskill, *Linear Systems, Fourier Transforms, and Optics*, Wiley, New York, 1978.

19. D. Gabor, "Microscopy by Reconstructed Wave-Fronts," *Proc. R. Soc. London, Ser., A*, Vol. 197, p. 454 (1951); "Microscopy by Reconstructed Wave-Fronts—II," *Proc. Phys. London*, Vol. 64, p. 449 (1951).

20. E. N. Leith and J. Upatnieks, "Reconstructed Wavefronts and Communication Theory," *J. Opt. Soc. Am.*, Vol. 52, p. 1123 (1962).

21. J. W. Goodman, *Introduction to Fourier Optics*, McGraw-Hill, New York, 1968.

22. W. T. Cathy, *Optical Information Processing and Holography*, Wiley, New York, 1974.

23. R. J. Collier, C. B. Bunkhardt, and L. H. Lin, *Optical Holography*, Academic, New York, 1971.

24. M. Abramowitz and I. A. Stegun, *Handbook of Mathematical Functions*, National Bureau of Standards, Applied Mathematics Series No. 55, U.S. Government Printing Office, Washington, DC.

25. A. B. Porter, "On the Diffraction Theory of Optical Images," *Philos. Mag.*, Vol. 11, p. 154 (1906).

26. A. VanderLugt, "Operational Notation for the Analysis and Synthesis of Optical Data Processing Systems," *Proc. IEEE*, Vol. 54, p. 1055 (1966).

27. F. P. Carlson and R. E. Francois, Jr., "Generalized Linear Processors for Coherent Optical Computers," *Proc. IEEE*, Vol. 65, p. 10 (1977).

28. A. VanderLugt, "Design Relationships for Holographic Memories," *Appl. Opt*. Vol. 12, p. 1675 (1973).

29. A. VanderLugt, "Packing Density in Holographic Systems," *Appl. Opt.*, Vol. 14, p. 1081, (1975).

30. D. R. Scifres, C. Lindstrom, R. D. Burnham, W. Streifer, and T. Paoli, "Phase-Locked (GaAl)As Laser Diode Emitting 2.6W cw from a Single Mirror," *Electron. Lett.*, Vol. 19, p. 169 (1983).

31. R. S. Kardar, "A Study of the Information Capacities of a Variety of Emulsion Systems," *Photogr. Eng.*, Vol. 6, p. 190 (1955).

32. D. Casasent, "Spatial Light Modulators," *Proc. IEEE*, Vol. 65, p. 143 (1977).

33. A. R. Tanguay, Jr., "Materials Requirements for Optical Processing and Computing Devices," *Opt. Eng.*, Vol. 24, p. 2 (1985).

34. Special issue, "*Spatial Light Modulators for Optical Information Processing*," Applied Optics, Vol. 28, 1989.

35. N. George, J. Thomasson, and A. Spindel, "Photodetector for Real Time Pattern Recognition," U.S. Patent No. 3, 689,772 (1970).

36. J. T. Thomasson, T. J. Middleton, and N. Jensen, "Photogrammetric Uses of the Optical Power Spectrum," Proc. Soc. Photo-Opt. Instrum. Eng., Vol. 45, p. 257 (1974).
37. See, for example, A. Yariv, *Optical Electronics*, Holt, Rinehart and Winston, New York, 1985.
38. Carlton E. Thomas, "Optical Spectrum Analysis of Large Space Bandwidth Signals," *Appl. Opt.*, Vol. 5, p. 1782 (1966).
39. G. Lebreton, "Power Spectrum of Raster-Scanned Signals," *Opt. Acta*, Vol. 29, p. 413 (1982).
40. W. M. Brown, *Analysis of Linear Time-Invariant Systems*, McGraw-Hill, New York, 1963.
41. W. B. Davenport, Jr. and W. L. Root, *An Introduction to the Theory of Random Signals and Noise*, McGraw-Hill, New York, 1958.
42. A. Papoulis, *Probability, Random Variables and Stochastic Processes*, McGraw-Hill, New York, 1984.
43. H. L. VanTrees, *Detection, Estimation, and Modulation Theory*, Wiley, New York, 1968.
44. A. VanderLugt, "A Review of Optical Data-Processing Techniques," *Opt. Acta*, Vol. 15, p. 1 (1968).
45. A. VanderLugt, "Coherent Optical Processing," *Proc. IEEE*, Vol. 62, p. 1300 (1974).
46. Special Issue on Optical Computing, *Proc. IEEE*, Vol. 65, pp. 1-176 (1977).
47. Special Issue on Optical Computing, *Proc. IEEE*, Vol. 72, pp. 753–992 (1984).
48. Special Issue on Optical Pattern Recognition, *Opt. Eng.*, Vol. 23, pp. 687–838 (1984).
49. P. L. Jackson, "Analysis of Variable-Density Seismograms by Means of Optical Diffractions," *Geophys.*, Vol. 30, p. 5 (1965).
50. M. B. Dobrin, A. L. Ingalls, and J. A. Long, "Velocity and Frequency Filtering of Seismic Data Using Laser Light," *Geophys.*, Vol. 30, p. 1144 (1965).
51. D. G. Falconer, "Optical Processing of Bubble Chamber Photographs," *Appl. Opt.*, Vol. 5, p. 1365 (1966).
52. P. L. Jackson, "Diffractive Processing of Geophysical Data," *Appl. Opt.*, Vol. 4, p. 419 (1965).
53. H. J. Pincus and M. B. Dobrin, "Geological Application of Optical Data Processing," *J. Geophys. Res.*, Vol. 71, p. 4861 (1966).
54. A. Kozma and D. L. Kelly, "Spatial Filtering for Detection of Signals Submerged in Noise," *Appl. Opt.*, Vol. 4, p. 387 (1965).
55. B. R. Brown and A. W. Lohmann, "Complex Spatial Filtering with Binary Masks," *Appl. Opt.*, Vol. 5, p. 967 (1966).

56. A. W. Lohmann, D. P. Paris, and H. W. Werlich, "Binary Fraunhofer Holograms Generated by Computer," *Appl. Opt.*, Vol. 6, p. 1139 (1967).

57. F. Zernike, "Das Phasenkontrastverfahren bei der mikroskopischen Beobachtung," *Z. Tech. Phys.*, Vol. 16, p. 454 (1935).

58. J. Tsujiuchi, "Correction of Optimal Images by Compensation of Aberrations and by Spatial Frequency Filtering," in E. Wolf, ed., *Progress in Optics*, North-Holland, Amsterdam, 1963, Vol. II.

59. J. L. Horner, "Light Utilization in Optical Correlators," *Appl. Opt.*, Vol. 21, p. 4511 (1982).

60. P. D. Gianino and J. L. Horner, "Additional Properties of the Phase-Only Correlation Filter," *Opt. Eng.*, Vol. 23, p. 695 (1984).

61. A. Kozma, "Photographic Recording of Spatially Modulated Coherent Light," *J. Opt. Soc. Am.*, Vol. 56, p. 428 (1966).

62. C. S. Weaver and J. W. Goodman, "A Technique for Optically Convolving Two Functions," *Appl. Opt.*, Vol. 5, p. 1248 (1966).

63. W-H. Lee, "Sampled Fourier Transform Hologram Generated by Computer," *Appl. Opt.*, Vol. 9, p. 639 (1970).

64. A. L. Ingalls, "The Effects of Film Thickness Variations on Coherent Light," *Photogr. Sci. Eng.*, Vol. 4, p. 135 (1960).

65. E. N. Leith, "Photographic Film as an Element of a Coherent Optical System," *Photogr. Sci. Eng.* Vol. 6, p. 75 (1962).

66. R. E. Swing and M. C. H. Shin, "The Determination of Modulation Transfer Characteristics of Photographic Emulsions in a Coherent Optical System," *Photogr. Sci. Eng.* Vol. 7, p. 350 (1963).

67. J. H. Altman, "Microdensitometry of High Resolution Plates by Measurement of the Relief Image," *Photogr. Sci. Eng.* Vol. 10, p. 156 (1966).

68. A. VanderLugt, F. B. Rotz, and A. Klooster, Jr., "Character Reading by Optical Spatial Filtering," in Tippett et al., eds. *Optical and Electro-Optical Information Processing*, MIT Press, Cambridge, 1965.

69. G. L. Turin, "An Introduction to Matched Filters," *IRE Trans. Inf. Theory*, Vol. IT-10, p. 311 (1960).

70. A. VanderLugt and F. B. Rotz, "The Use of Film Nonlinearities in Optical Spatial Filtering," *Appl. Opt.*, Vol. 9, p. 215 (1970).

71. A. VanderLugt, *The Theory of Optical Techniques for Spatial Filtering and Signal Recognition*, Ph.D. thesis, University of Reading 1969.

72. F. B. Rotz and M. O. Greer, "Photogrammetry and Reconnaissance Applications of Coherent Optics," *Proc. Soc. Photo-Opt. Instrum. Eng.*, Vol. 45, p. 139 (1974).

73. A. VanderLugt, "The effects of Small Displacements of Spatial Filters," *Appl. Opt.* 6, 1221 (1967).

74. Y. W. Lee, *Statistical Theory of Communication*, Wiley, New York, 1960.

75. L. Brillouin, "Diffusion de la Luimiere et des Rayons X par un Corps Transparent Homogene," *Ann. Phys.* (*Paris*), Vol. 17, p. 88 (1922).

76. P. Debye and F. W. Sears, "Scattering of Light by Supersonic Waves," *Proc. Nat. Acad. Sci.*, Vol. 18, p. 409 (1932).

77. R. Lucas and P. Biquard, "Proprietes Milieux Solides et Liquides Soumis aux Vibrations Elastiques Ultra Sonores," *J. Phys. Radium*, Vol. 3, p. 464 (1932).

78. C. V. Raman and N. S. Nagendra Nath, "The Diffraction of Light by High Frequency Sound Waves: Parts I and II," *Proc. Indian Acad. Sci. Sect. A*, Vol. 2, p. 406 (1935); "Part III-Doppler Effect and Coherent Phenomena," *Proc. Indian Acad. Sci., Sect. A*, p. 75 (1936); "Part IV-Generalized Theory," *Proc. Indian Acad. Sci. Sect. A*, Vol. 3, p. 119 (1936); "Part V-General Considerations-Oblique Incidence and Amplitude Changes," *Proc. Indian Acad. Sci., Sect. A*, Vol. 3, p. 359 (1936).

79. A. Korpel, *Acousto-Optics*, Marcel Dekker, New York, 1988.

80. E. H. Young, Jr., and S-K. Yao, "Design Considerations for Acousto-Optic Devices," *Proc. IEEE*, Vol. 69, p. 54 (1981).

81. A. Korpel, "Acousto-Optics—A Review of Fundamentals," *Proc. IEEE*, Vol. 69, p. 48 (1981).

82. I. Fuss and D. Smart, "Cryogenic Large Bandwidth Acousto-Optic Deflector," *Appl. Opt.*, Vol. 26, p. 1222 (1983).

83. M. Amano, G. Elston, and J. Lucero, "Materials for Large Time Aperture Bragg cells," *Proc. SPIE*, Vol. 567, p. 142 (1985).

84. A. VanderLugt, "Bragg Cell Diffraction Patterns," *Appl. Opt.*, Vol. 21, p. 1092 (1982).

85. H. N. Roberts, J. W. Watkins, and R. H. Johnson, "High Speed Holographic Recorder," *Appl. Opt.*, Vol. 13, p. 841 (1974).

86. A. H. Rosenthal, "Application of Ultrasonic Light Modulation to Signal Recording, Display, Analyses and Communication," *IRE Trans. Ultrason. Eng.*, Vol. SU-8, p. 1 (1961).

87. R. M. Wilmotte, "Light Modulation System for Analysis of Information," U.S. Patent No. 3,509,453 (1970).

88. L. B. Lambert, "Wideband Instantaneous Spectrum Analyzers Employing Delay-Line Light Modulators," *IRE Int. Conv. Rec.*, Vol. 10, p. 69 (1962).

89. D. L. Hecht, "Multifrequency Acousto-Optic Spectrum Analysis," *IEEE Trans. Sonics Ultrason.*, Vol. SU-24, p. 7 (1977).

90. D. R. Pape, "Minimization of Nonlinear Acousto-Optic Bragg Cell Intermodulation Products," 1989 Ultrasonics Symposium Proceedings, p. 509.

91. G. A. Coquin, J. P. Griffin, and L. K. Anderson, "Wideband Acousto-Optic Deflectors Using Acoustic Beam Steering," *IEEE Trans. Sonics Ultrason.*, Vol. SU-17, p. 34 (1970).

92. R. B. Brown, A. E. Craig, and J. N. Lee, "Predictions of Stray Light Modeling on the Ultimate Performance of Acousto-Optic Processors," *Opt. Eng.*, Vol. 28, p. 1299 (1989).

93. E. N. Leith and J. Upatnieks, "Wavefront Construction with Diffused Illumination and Three-Dimensional Objects," *J. Opt. Soc. Am.*, Vol. 54, p. 1295 (1964).

94. M. King, W. R. Bennett, L. B. Lambert, and M. Arm, "Real-Time Electro-Optical Signal Processors with Coherent Detection," Appl. Opt., Vol. 6, p. 1367 (1967).

95. L. Slobodin, "Optical Correlation Technique," *Proc. IEEE*, Vol. 51, p. 1782 (1963).

96. R. Whitman, A. Korpel, and S. Lotsoff, "Application of Acoustic Bragg Diffraction to Optical Processing Techniques," *Symposium on Modern Optics*, Polytechnic Institute of Brooklyn, 1967, p. 243.

97. R. L. Whitman and A. Korpel, "Probing of Acoustic Surface Perturbations by Coherent Light," *Appl. Opt.*, Vol. 8, p. 1567 (1969).

98. A. Korpel and R. L. Whitman, "Visualization of a Coherent Light Field by Heterodyning with a Scanning Laser Beam," *Appl. Opt.*, Vol. 8, p. 1577 (1969).

99. A. VanderLugt, "Interferometric Spectrum Analyzer," *Appl. Opt.*, Vol. 20, p. 2770 (1981).

100. G. W. Anderson, B. D. Guenther, J. A. Hynecek, R. J. Keyes, and A. VanderLugt, "Role of Photodetectors in Optical Signal Processing," *Appl. Opt.*, Vol. 27, p. 2871 (1988).

101. G. M. Borsuk, "Photodetectors for Acousto-Optic Signal Processors," *Proc. IEEE*, Vol. 69, p. 100 (1981).

102. A. VanderLugt and A. M. Bardos, "Spatial and Temporal Spectra of Periodic Functions for Spectrum Analysis," *Appl. Opt.*, Vol. 23, p. 4269 (1984).

103. P. H. Wisemann and A. VanderLugt, "Temporal Frequencies of Short and Evolving Pulses in Interferometric Spectrum Analyzers," *Appl. Opt.*, Vol. 28, pp. 3800 (1989).

104. S. W. Golomb, *Shift Register Sequences*, Holden-Day, San Francisco, 1967, p. 86.

105. M. D. Koontz, "Miniature Interferometric Spectrum Analyzer," *Proc. SPIE*, Vol. 639, p. 126 (1986).

106. J. L. Erickson, "Linear Acousto-Optic Filtering with Heterodyne and Frequency-Plane Control," Ph.D. thesis, Stanford University (University Microfilms, 1981).

107. R. W. Brandstetter and P. G. Grieve, "Recursive Optical Notching Filter," *Proc. SPIE*, Vol. 1154, p. 206 (1989).

108. P. J. Roth, "Optical Excision in the Frequency Plane," *Proc. SPIE*, Vol. 352, p. 17 (1982).
109. T. P. Karnowski and A. VanderLugt, "Generalized Filtering in Acousto-Optic Systems Using Area Modulation," *Appl. Opt.*, Vol. 30, p. 2344 (1991).
110. C. S. Anderson and A. VanderLugt, "Hybrid Acousto-Optic and Digital Equalization for Microwave Digital Radio Channels," *Opt. Lett.*, Vol. 15, p. 1182 (1990).
111. M. Arm, L. Lambert, and I. Weissman, "Optical Correlation Technique for Radar Pulse Compression," *Proc. IEEE*, Vol. 52, p. 842 (1964).
112. E. B. Felstead, "A Simple Real-Time Incoherent Optical Correlator," *IEEE Trans. Aerosp. Electron. Syst.*, Vol. AES-3, p. 907 (1967).
113. E. B. Felstead, "A Simplified Coherent Optical Correlator," *Appl. Opt.*, Vol. 7, p. 105 (1968).
114. C. Atzeni and L. Pantani, "A Simplified Optical Correlator for Radar-Signal Processing," *Proc. IEEE*, Vol. 57, p. 344 (1969).
115. C. Atzeni and L. Pantani, "Optical Signal-Processing Through Dual-Channel Ultrasonic Light Modulators," *Proc. IEEE*, Vol. 58, p. 501 (1970).
116. R. M. Montgomery, "Acousto-Optic Signal Processing System," U.S. Patent No. 3,634,749 (1972).
117. R. A. Sprague and C. L. Koliopoulos, "Time Integrating Acousto-Optic Correlator," *Appl. Opt.*, Vol. 15, p. 89 (1976).
118. R. Manasse, R. Price, and R. M. Lerner, "Loss of Signal Detectability in Band-Pass Limiters," *IRE Trans. Inf. Theory*, Vol. IT-4, p. 34 (1958).
119. C. R. Cohn, "A Note on Signal-to-Noise Ratio in Band-Pass Limiters," *IRE Trans. Inf. Theory*, Vol. IT-7, p. 39 (1961).
120. F. B. Rotz, "Time-Integrating Optical Correlator," *Proc. SPIE*, Vol. 202, p. 163 (1979).
121. P. Kellman, "Time Integrating Optical Processing," Ph.D. thesis, Stanford University (1979).
122. D. Mergerian, E. C. Malarkey, P. P. Pautienus, J. C. Bradley, G. E. Marx, L. D. Hutcheson, and A. L. Kellner, "Operational Integrated Optical RF Spectrum Analyzer," *Appl. Opt.*, Vol. 19, p. 3033 (1980).
123. M. W. Casseday, N. J. Berg, and I. J. Abramovitz, "Space-Integrating Signal Processors Using Surface-Acoustic Wave Delay Lines," in N. J. Berg and J. N. Lee, eds., *Acousto-Optic Signal Processing*, Marcel Dekker, New York, 1983.
124. I. J. Abramovitz, N. J. Berg, and M. W. Casseday, "Coherent Time-Integration Processors," in N. J. Berg and J. N. Lee, eds., *Acousto-Optic Signal Processing*, Marcel Dekker, New York, 1983.
125. D. Mergerian and E. C. Malarkey, "Integrated Optics," in N. J. Berg and J. N. Lee, eds., *Acousto-Optic Signal Processing*, Marcel Dekker, New York, 1983.

126. J. M. Elson and J. M. Bennett, "Relation between the Angular Dependence of Scattering and the Statistical Properties of Optical Surfaces," *J. Opt. Soc. Am.*, Vol. 69, p. 31 (1979).
127. T. M. Turpin, "Time Integrating Optical Processing," *Proc. SPIE*, Vol. 154, p. 196 (1978).
128. T. M. Turpin, "Spectrum Analysis Using Optical Processing," *Proc. IEEE*, Vol. 69, p. 79 (1981).
129. W. T. Rhodes, "Acousto-Optic Signal Processing: Convolution and Correlation," *Proc. IEEE*, Vol. 69, p. 65 (1981).
130. J. D. Cohen, "Ambiguity Processor Architectures Using One-Dimensional Acousto-Optic Transducers," *Proc. SPIE*, Vol. 180, p. 134 (1979).
131. A. W. Lohmann and B. Wirnitzer, "Triple Correlations," *Proc. IEEE*, Vol. 72, p. 889 (1984).
132. R. A. K. Said and D. C. Cooper, "Crosspath Real-Time Optical Correlator and Ambiguity-Function Processor," *Proc. IEE*, Vol. 120, p. 423 (1973).
133. A. VanderLugt, "Crossed Bragg Cell Processors," *Appl. Opt.*, Vol. 23, p. 2275 (1984).
134. P. M. Woodward, *Probability and Information Theory, with Applications to Radar*, McGraw-Hill, New York, 1953.
135. D. Psaltis and K. Wagner, "Real-time Optical Synthetic Aperture Radar (SAR) Processor," *Opt. Eng.*, Vol. 21, p. 822 (1982).
136. Michael Haney and Demetri Psaltis, "Real-Time Programmable Acousto-Optic Synthetic Aperture Radar Processor," *Appl. Opt.*, Vol. 27, p. 1786 (1988).
137. A. VanderLugt, "Adaptive Optical Processor," *Appl. Opt.*, Vol. 21, p. 4005 (1982).
138. A. VanderLugt and A. M. Bardos, "Stability Considerations for Adaptive Optical Fitlering," *Appl. Opt.*, Vol. 25, p. 2314 (1986).

Bibliography

M. Amano and E. Roos, "32-Channel Acousto-Optic Bragg Cell for Optical Computing," *Proc. SPIE*, Vol. 753, p. 37 (1987).

Gordon Wood Anderson, Francis J. Kub, Rebecca L. Grant, Nicholas A. Papanicolaou, John A. Modolo, and Douglas E. Brown, "Programmable Frequency Excision and Adaptive Filtering with a GaAs/AlGaAs/GaAs Heterojunction Photoconductor Array," *Opt. Eng.*, Vol. 29, p. 1243 (1990).

B. Auld, *Acoustic Fields and Waves in Solids*, Wiley, New York, 1973.

T.R. Bader, "Acoustooptic Spectrum Analysis: A High Performance Hybrid Technique," *Appl. Opt.*, Vol. 10, p. 1668 (1979).

M. J. Bastiaans, "The Wigner Distribution Function Applied to Optical Signals and Systems," *Opt. Commun.*, Vol. 25, p. 26 (1978).

R. Barakat, "Application of the Sampling Theorem to Optical Diffraction Theory," *J. Opt. Soc. Am.*, Vol. 54, p. 920 (1964).

D. F. Barbe, "Imaging Devices Using the Charge-Coupled Concept," *Proc. IEEE*, Vol. 63, p. 38 (1975).

H. Bartelt, A. W. Lohmann and B. Wirnizter, "Phase and Amplitude Recovery from Bispectra," *Appl. Opt.*, Vol. 18, p. 3121 (1984).

W. R. Beaudet, M. L. Popek, and D. R. Pape, "Advances in Multi-Channel Bragg Cell Technology," *Proc. SPIE*, Vol. 639, p. 28 (1986).

N. J. Berg, J. N. Lee, M. W. Casseday, and E. Katzen, "Adaptive Fourier Transformation by Using Acousto-Optic Convolution," Proc. 1978 IEEE Ultrasonics Symp., IEEE No. 78CH1344-1, p. 91 (1978).

G. D. Boreman and E. R. Raudenbush, "Characterization of a Liquid Crystal Television Display as a Spatial Light Modulator for Optical Processing," *Proc. SPIE*, Vol. 639, p. 41 (1986).

Robert W. Brandstetter and Philip G. Grieve, "Excision of Interference from Radio Signals by Means of a Recursive Optical Notching Filter," *Opt. Eng.*, Vol. 29, p. 804 (1990).

H. J. Butterweck, "Principles of Optical Data-Processing," in E. Wolf, ed., *Progress in Optics*, North Holland, Amsterdam, 1981, Vol. XIX.

David Casasent, "Optical Information Processing Applications of Acousto-Optics," in N. J. Berg and J. N. Lee, eds., *Acousto-Optic Signal Processing*, Marcel Dekker, New York, 1983.

David Casasent and Giora Silbershatz, "Product Code Processing on a Triple-Product Processor," *Appl. Opt.*, Vol. 21, p. 2076 (1982).

David Casasent and James Lambert, "General I and Q Data Processing on a Multichannel AO System," *Appl. Opt.*, Vol. 25, p. 1886 (1986).

S. G. Chamberlain and J. P. Y. Lee, "A Novel Wide Dynamic Range Silicon Photodetector and Linear Imaging Array," *IEEE Trans. Electron. Devices*, Vol. ED-31, p. 175 (1984).

I. C. Chang and S. Lee, "Efficient Wideband Acousto-Optic Cells," Proc. 1983 IEEE Ultrasonics Symp., IEEE No. 83CH 1947-1, p. 427 (1983).

J. D. Cohen, "Incoherent Light Time Integrating Processors," in N. J. Berg and J. N. Lee, eds., *Acousto-Optic Signal Processing*, Marcel Dekker, New York, 1983.

M. G. Cohen, "Optical Study of Ultrasonic Diffraction and Focusing in Anisotropic Media," *J. Appl. Phys.*, Vol. 38, p. 3821 (1967).

L. J. Cutrona, E. N. Leith, L. J. Porcello, and W. E. Vivian, "On the Application of Coherent Optical Processing Techniques to Synthetic Aperture Radar," *Proc. IEEE*, Vol. 54, p. 1026 (1966).

R. W. Dixon, "Photoelastic Properties of Selected Materials and their Relevance for Applications to Acoustic Light Modulators and Scanners," *IEEE J. Quantum Electron.*, Vol. QE-3, p. 85 (1967).

Jerry Erickson, "Linear Acousto-Optic Filters for Programmable and Adaptive Filtering," *Proc. SPIE*, Vol. 341, p. 173 (1982).

Jerry Erickson, "Optical Excisor Performance Limits Versus Improved Signal Detection," *Proc. SPIE*, Vol. 639, p. 232 (1986).

Michael W. Farn and Joseph W. Goodman, "Optical Binary Phase-Only Matched Filters," *Appl. Opt.*, Vol. 27, p. 4431 (1988).

E. B. Felstead, "Optical Fourier Transformation of Area-Modulated Spatial Functions," *Appl. Opt.*, Vol. 10, p. 2468 (1971).

T. K. Gaylord and M. G. Mohoram, "Analysis and Applications of Optical Diffraction Gratings," *Proc. IEEE*, Vol. 73, p. 894 (1985).

J. W. Goodman, "Operation Achievable with Coherent Optical Information Processing Systems," *Proc. IEEE*, Vol. 65, p. 39 (1977).

A. P. Goutzoulis and M. S. Gottlieb, "Design and Performance of Optical Activity Based Hg_2Cl_2 Bragg Cells," *Proc. SPIE*, Vol. 936, p. 119 (1988).

B. D. Guenther, C. R. Christensen, and J. Upatnieks, "Coherent Optical Processing: Another Approach," *IEEE J. Quantum Electron.*, Vol. 15, p. 1348 (1979).

D. L. Hecht, "Spectrum Analysis Using Acousto-Optic Devices," *Proc. SPIE*, Vol. 90, p. 148 (1976).

D. L. Hecht and G. W. Petrie, "Acousto-Optic Diffraction from Acoustic Anisotropic Shear Modes in Gallium Phosphide," Proc. 1980 IEEE Ultrasonics Symp., IEEE No. 80CH1602-2, p. 474, 1980.

W. R. Klein and B. D. Cook, "Unified Approach to Ultrasonic Light Diffraction," *IEEE Trans. Sonics Ultrason.*, Vol. SU-14, p. 723 (1967).

P. Kellman, H. N. Shaver, and J. W. Murray, "Integrating Acousto-Optic Channelized Receivers," *Proc. IEEE*, Vol. 69, p. 93 (1981).

A. Korpel, S. N. Lotsoff, and R. L. Whitman, "The Interchange of Time and Frequency in Television Displays," *Proc. IEEE*, Vol. 57, p. 160 (1969).

M. A. Krainak and D. E. Brown, "Interferometric Triple Product Processor (Almost Common Path)," *Appl. Opt.*, Vol. 24, p. 1385 (1985).

John N. Lee, N. J. Berg, M. W. Casseday, and P. S. Brody, "High-Speed Adaptive Filtering and Reconstruction of Broad-Band Signals Using Acousto-Optic Techniques," *1980 Ultrasonics Symposium*, p. 488 (1980).

J. N. Lee and A. D. Fisher, "Device Developments for Optical Information Processing," *Adv. Electron. Electron Phys.*, Vol. 69, p. 115 (1987).

J. N. Lee, "Optical Architectures for Temporal Signal Processing," in J. Horner, ed., *Optical Signal Processing*, Academic, Orlando, 1988.

E. N. Leith and A. L. Ingalls, "Synthetic Antenna Data Processing by Wavefront Reconstruction," *Appl. Opt.*, Vol. 7, p. 539 (1968).

J. P. Lindley, "Applications of Acousto-Optic Techniques to RF Spectrum Analysis," in N. J. Berg and J. N. Lee, eds., *Acousto-Optic Signal Processing*, Marcel Dekker, New York, 1983.

H. K. Liu and T. H. Chao, "Liquid Crystal Television Spatial Light Modulators," *Appl. Opt.*, Vol. 28, p. 4772 (1989).

G. E. Lukes, "Cloud Screening from Aerial Photography Applying Coherent Optical Pattern Recognition Techniques," *Proc. Soc. Photo-Opt. Instrum. Eng.*, Vol. 45, p. 265 (1974).

R. J. Marks, J. F. Walkup, and M. O. Hagler, "A Sampling Theorem for Space-Variant Systems," *J. Opt. Soc. Am.*, Vol. 66, p. 918 (1976).

E. L. O'Neill and A. Walther, "The Question of Phase in Image Formation," *Opt. Acta*, Vol. 10, p. 33 (1963).

D. A. Pinnow, "Guide Lines for the Selection of Acousto-Optic Materials," *IEEE J. Quantum Electron.*, Vol. QE-6, p. 223 (1970).

Demetri Psaltis, "Acousto-Optic Processing of Two-Dimensional Signals," *J. Opt. Soc. Am.*, Vol. 71, p. 198 (1981).

J. E. Rau, "Real-Time Complex Spatial Modulation," *J. Opt. Soc. Am.*, Vol. 57, p. 798 (1967).

J. E. Rhodes, Jr., "Analysis and Synthesis of Optical Images," *Am. J. Phys.*, Vol. 21, p. 337, (1953).

J. F. Rhodes and D. E. Brown, "Adaptive Filtering with Correlation Cancellation Loops," *Proc. SPIE*, Vol. 341, p. 140 (1983).

J. F. Rhodes, "Adaptive Filter with a Time-Domain Implementation using Correlation Cancellation Loops," *Appl. Opt.*, Vol. 22, p. 282 (1983).

W. T. Rhodes and J. M. Florence, "Frequency Variant Optical Signal Analysis," *Appl. Opt.*, Vol. 15, p. 3073 (1976).

A. M. Tai, "Low-Cost Spatial Light Modulator with High Optical Quality," *Appl. Opt.*, Vol. 25, p. 1380 (1986).

David Slepian, "On Bandwidth," *Proc. IEEE*, Vol. 64, p. 292 (1976).

Robert Spann, "A Two-Dimensional Correlation Property of Pseudorandom Maximal Length Sequences," *IEEE Trans. Inform. Theory*, Vol. 11, p. 2137 (1965).

N. Uchida and N. Niizeki, "Acousto-Optic Deflection Materials and Techniques," *Proc. IEEE*, Vol. 61, p. 1073 (1973).

Jean-Charles Vienot, Jean-Pierre Goedgebuer, and Alain Lacourt, "Space and Time Variables in Optics and Holography: Recent Experimental Aspects," *Appl. Opt.*, Vol. 16, p. 454 (1977).

A. VanderLugt, G. S. Moore, and S. S. Mathe, "Multichannel Bragg Cell Compensation for Acoustic Spreading," *Appl. Opt.*, Vol. 22, p. 3906 (1983).

A. Walther, "The Question of Phase Retrieval in Optics," *Opt. Acta*, Vol. 10, p. 41 (1963).

Index